Science in
the Middle Ages

The Chicago History
of Science and Medicine
Allen G. Debus, editor

Science in the Middle Ages

Edited by
David C.Lindberg

The University of Chicago Press
Chicago and London

The University of Chicago Press, Chicago 60637
The University of Chicago Press, Ltd., London

DAVID C. LINDBERG is professor of the history
of science at the University of Wisconsin, Madison.
His numerous publications include *Theories of
Vision from Al-kindi to Kepler*, also published by
the University of Chicago Press.

Library of Congress Cataloging in Publication Data

Main entry under title:

Science in the Middle Ages.

(Chicago history of science and medicine)
Bibliography: p.
Includes index.
1. Science, Medieval. I. Lindberg, David C.
II. Series.
Q124.97.S35   509'.02   78–5367
ISBN 0–226–48232–4

Contents

Preface

The first serious large-scale exploration of the history of medieval science was that of the French physicist, philosopher, and historian Pierre Duhem (1861–1916). While searching for precursors of Leonardo da Vinci, Duhem discovered a wealth of manuscript and printed material which persuaded him that medieval scholasticism had by no means suffocated empirical scientific endeavor, as had theretofore been supposed. Quite the contrary: it now appeared to Duhem that the very foundations of modern science had been laid during the fourteenth century. He summarized his findings in the following words:

> A more exact knowledge of the doctrines taught in the . . . schools of the Middle Ages . . . teaches us that in the fourteenth century the masters of Paris, having rebelled against the authority of Aristotle, constructed a dynamics entirely different from that of the Stagirite; that the essential elements of the [mechanical] principles [commonly] thought to have received mathematical expression and experimental confirmation from Galileo and Descartes were already contained in this [fourteenth-century] dynamics; that at the beginning of the fifteenth century these Parisian doctrines were spread into Italy, where they encountered a vigorous resistance from the Averroists, jealous guardians of the Aristotelian tradition and the great Commentator [Averroes]; that they were adopted in the course of the sixteenth century by the majority of mathematicians; and finally that Galileo, in his youth, read several of the treatises where these theories . . . were presented.

(*Rendiconti della Reale Accademia dei Lincei*, Classe sci. fis. mat., vol. 22 [1913], p. 429)

Duhem's work not only established the history of medieval science as an autonomous field of scholarly endeavor; it also convinced a generation of followers that the medieval contribution to science had been to anticipate many of the basic conceptions and theoretical achievements of Galileo and his contemporaries. The result (as Edward Grant has put it) was to "infect the newly developing discipline of the history of medieval science with the pox of precursoritis, a disease destined to reach epidemic proportions." Duhem's successors approached their task more carefully and with more subtlety than had Duhem, but it remained a major goal of their work to demonstrate that modern science had its roots in the medieval period and that Galileo and the other supposed founders of modern science must be viewed as members of a continuous intellectual tradition emanating from the medieval university. Among those who made early contributions of monumental proportions to this endeavor were Lynn Thorndike, Anneliese Maier, and Marshall Clagett. In the first four volumes of his *History of Magic and Experimental Science* Thorndike devoted more than 3000 pages to a survey of medieval science, concentrating on its empirical aspects and the relationship between magic and science, and today (more than forty years after their publication) these volumes remain an indispensable source of information and interpretation. Moreover, his *Catalogue of Incipits of Mediaeval Scientific Writings in Latin* (with Pearl Kibre) revealed the great bulk and diversity of scientific materials available in manuscript. Of equal significance were Maier's five *Studien zur Naturphilosophie der Spätscholastik*, which offered important new interpretations of late-medieval natural philosophy and, in the process, taught historians of medieval science to read their sources against a broader and more accurate philosophical background. And Marshall Clagett made major contributions with his textual work and his influence on a flock of graduate students.

Duhem's continuity thesis was a strong polarizing force. If historians of medieval science could now self-confidently argue that early modern science was an outgrowth of the medieval scientific achievement, an inevitable counterattack was provoked among historians of early modern science, who were not disposed to yield easily on so important a point. The outcome was a vigorous polemic between medievalists, defending the originality and importance of their chosen period, and those who preferred to locate the origins of modern science in a revolution of the sixteenth and seventeenth centuries. The number of protagonists in the first six decades of the twentieth century may have been small, but the debate was spirited.

However, since the late 1950s several important changes have taken place. First, there has been an astonishing growth in the number of historians of medieval science. Whereas in 1950 the scholars engaged in serious research into the history of medieval science could be counted on the fingers of two hands, by 1975 they numbered in the scores. This has meant an enormous increase in the quantity of research and has brought us to the point where virtually every scientific discipline pursued during the Middle Ages has attracted a coterie of historians. Second, the polemics of the first half of the century have gradually subsided, and historians of medieval science no longer occupy themselves principally with the search for precursors but, rather, with the attempt to understand the medieval scientific achievement on its own terms. The value of comparing medieval and early modern ideas on a given question may still be acknowledged, but the medieval achievement is now recognized as interesting for its own sake, its worth no longer being associated with the anticipation of later developments. Third, the idea that the history of science (and, in particular, the history of medieval science) is an autonomous discipline pursued by scholars whose principal training is in modern science has been replaced by the recognition that medieval science, if it is to be properly understood, must be viewed within a broad social and intellectual context—that it influenced and was influenced by a myriad of social forces and other scholarly disciplines—and, therefore, that it must be pursued by scholars with a broad grasp of medieval social and intellectual history.

These recent investigations of medieval science have not yet borne fruit in the form of satisfactory general texts. An important early attempt at synthesis was A. C. Crombie's *Augustine to Galileo* (1952), reissued under the title *Medieval and Early Modern Science*. E. J. Dijksterhuis also treated medieval developments in his *The Mechanization of the World Picture*, first published in 1950 and translated into English in 1961. In 1971 Edward Grant published his *Physical Science in the Middle Ages*, a brief introduction to certain aspects of medieval science. And medieval achievements in physics and astronomy have been sketched in Olaf Pedersen and Mogens Pihl's *Early Physics and Astronomy: A Historical Introduction* (1974), a revised and updated translation of their earlier *Historisk Indledning til den Klassike Fysik* (1963). Critical editions and translations of original source materials also abound, as do specialized studies of particular problems; and these are, of course, the necessary preliminaries to broad interpretative efforts. But it appears that the time is ripe for a book that examines in some depth all major aspects of the medieval scientific enterprise.

Sixteen authors have collaborated in producing the present book. Virtually all have written on topics that have been central to their scholarly endeavors and have thus been able to write on their subjects with confidence and authority. The purpose has not been to produce an encyclopedic work on medieval science in which all important details are enumerated, but, rather, to set forth and defend generalizations about the structure and the principal aims and developments of medieval science—to interpret rather than to catalog. The aim has also been to write an introductory account, which will appeal to nonspecialists as well as to professional historians of science. As in any collaborative effort, the contributions vary in style and level of presentation, but it is the conviction of the editor that in this volume the stated aims have been achieved with a high degree of success and that readers will find themselves led with a sure hand through the various disciplines and the major developments of medieval science.

The medieval period can be conveniently divided into two parts: the early Middle Ages (roughly A.D. 400 to 1150 or 1200), a time of social, economic, and intellectual transition (and sometimes disorganization) in northern Europe, during which intellectual hegemony was held by Mediterranean civilizations, particularly Islam; and the High and later Middle Ages (1150 or 1200 to about 1500), when virtually the whole corpus of Greek and Arabic learning became available to Latin-speaking scholars, transforming the intellectual life of Western Christendom and shifting the center of intellectual leadership northward. The early period is treated in chapter 1, which describes the state of learning in the Latin-speaking world, traces the modification of Greek science by medieval Islam into the form in which it would eventually be communicated to the West, and analyzes the developments in technology and social and economic structure, and the changing cultural values, which contributed to the rise of a vigorous theoretical science in the High and later Middle Ages.

In chapter 2 the translating activity which brought the Greek and Islamic legacy to the West is discussed. Chapter 3 describes the origins and development of the institutional home of learning in the later Middle Ages—the medieval university, its curriculum, and the scholars who populated it. Chapter 4 provides an account of the major philosophical debates which form the immediate context of medieval science and the intellectual background against which it must be viewed.

Chapters 5 through 13 are devoted to the specific disciplines that
constituted the medieval scientific enterprise—mathematics, the science of weights, the science of motion, cosmology, astronomy, optics,
the science of matter, medicine, and natural history. These chapters
focus on developments in Western Christendom from the beginning
of the thirteenth century onward, for this was the period of most important progress in medieval science. Nevertheless, contributors have
been encouraged to adapt the chronological boundaries of their respective chapters to the subject matter. Thus Islamic, and even Greek,
materials figure quite importantly in certain chapters (chapter 10 on
optics, for example), whereas in others (chapter 7 on the science of
motion) they barely make an appearance. The aim has been to avoid·
imposing arbitrary chronological and geographical boundaries.

Chapter 14 contains an analysis of medieval debates over the meaning of "science" (*scientia*) and the proper classification of its subdivisions. And, finally, chapter 15 discusses the intellectual or high
magic of the Middle Ages and its relationship to natural philosophy.

It is inevitable that when sixteen scholars join forces to write on
fifteen aspects of a historical problem, they will not always agree.
However, because the chapters of this book are self-contained discussions of different aspects of medieval science, no great opportunity
for disagreement is presented; and even where overlap occurs, there is
a high degree of unanimity. Nevertheless, subtle differences of interpretation and emphasis must be expected, whether on small matters
of detail or on large perennial issues such as the boundaries and the
distinctive characteristics of medieval science, the relative importance
of reason and experiment in its development, and the degree to which
the course of medieval scientific thought was influenced by practical
concerns. No attempt has been made in this book to erase or conceal
these differences; none of them is major, and a healthy discipline must
learn to air, and to live with, alternative points of view. Each chapter
speaks for its own author or authors, and this preface speaks for the
editor of the volume.

The term *science* has thus far been left undefined, and from this
some might infer that its applicability to medieval intellectual endeavor is unproblematic. In fact, however, the term *science* has connotations in the twentieth century that are quite inapplicable to the
Middle Ages. It is necessary, therefore, to pause briefly and discuss
the conception of medieval science on which this book is premised.

First, medieval science is to be distinguished from medieval technology. It is true, as Brian Stock argues in the first chapter, that tech-

nology and science were not totally independent enterprises during the Middle Ages. Technological innovation made a prodigious contribution to the development of a prosperous medieval economy capable of supporting universities and theoretical scientific endeavor. Technology was also an important force in social change, contributing ultimately to a new social system and new cultural values compatible with, even supportive of, natural science. Moreover, there are rare instances during the Middle Ages of technology providing a device (the compass, for example) that cried out for empirical study or scientific explanation or both. And advances in technology surely contributed to the production of scientific instruments—the quadrant and astrolabe of the medieval astronomer, the razors and saws of the medieval surgeon, and the alembic and furnace of the alchemist.

But despite these important connections, science and technology lacked the close interplay so characteristic of them today. By and large, the theoretical scientist and the craftsman seem not to have perceived that they had any mutual concerns. There is no reason to believe that the craftsman or technologist (with the possible exception of the architect) had any knowledge of or any use for the theoretical understanding of the scientist; he could succeed quite well enough employing empirical and rule-of-thumb methods. On the other side of the aisle, it did not occur to the scientist (nor, in all probability, was it the case) that his theoretical knowledge could be used to produce new hardware or industrial processes that would contribute to the betterment of mankind; nor did he consider it his business to analyze practical contrivances that the craftsman might build—or to use these contrivances for gathering scientific data. His purposes and methods were primarily intellectual; he sought theoretical knowledge for its own sake, and his most characteristic activities were reading, reflecting, lecturing, and disputing. The disciplines that we now consider part of the scientific enterprise were largely "bookish" in the Middle Ages, and they were practiced principally by men whom we should regard as natural philosophers. We must not forget the exceptions— the development of a practical mathematics applicable to commerce and surveying, the astronomer who made observations, the naturalist who observed flora and fauna at close hand, and the physician who treated patients and attempted to adjust his medical theory to what he observed—but we must be aware of their exceptional status. This book, then, focuses not on applied science, but on the dominant current of theoretical science or natural philosophy. With the exception of chapter 1, where the contribution of technology to social and cultural change is considered, and chapter 11, which touches upon chemical technology, technological developments receive only incidental treatment.

Second, use of the expression "medieval science" must not be taken to imply that science was an autonomous discipline or an identifiable profession. A student in the arts faculty of the medieval university did not "major" in science. (Medicine, which was represented by an independent graduate faculty, was, of course, an exception.) The student might incline toward mathematical or physical subjects, but he was trained in all of the liberal arts and the whole of philosophy and emerged from his schooling equipped (at least in theory) to teach anything in the curriculum. It is true that various scholarly specialties now considered "scientific" were acknowledged—one finds frequent reference in medieval works to *mathematica, astronomia,* and *perspectiva*—but all of these were branches of philosophy (in its broader sense) and, as such, formed but a small part of the repertoire of a universal scholar.

It is instructive to examine the literary output of some of the major figures of medieval science and to observe their enormous range. Robert Grosseteste, an early and important figure in the English scientific movement, wrote a series of seminal tracts on scientific questions —the tides, the heat of the sun, colors, the rainbow, meteorology, comets, and the like. He also wrote commentaries on Aristotle's *Physics, Posterior Analytics,* and *On Sophistical Refutations*; biblical commentaries on the *Hexaemeron* (the six days of creation), the first hundred Psalms, the Gospels, and the Pauline Epistles; sermons; and a series of brief treatises on free will, potentiality and actuality, the liberal arts, phonetics, and a variety of other philosophical, theological, and pedagogical topics; on top of all this, Grosseteste translated a number of Greek treatises into Latin, including Aristotle's *Nicomachean Ethics* and the Neoplatonic works of the pseudo-Dionysian corpus. A generation after Grosseteste, another Englishman, Roger Bacon, included sections on mathematical and experimental science in his major work, the *Opus maius*; the same book also contains sections on the general causes of human error, the relationship of philosophy and theology, the study of language, and moral philosophy. In other works, Bacon treated grammar and logic and commented on a variety of Aristotelian books. John Buridan, a Parisian arts master of the fourteenth century whom Duhem considered the key figure in the development of medieval dynamics, wrote commentaries on Aristotle's *Physics, On the Soul, On Sense and Sensibles, Metaphysics, On the Heavens, Politics,* and *Nicomachean Ethics*; he also wrote extensively on logic. And finally, Nicole Oresme, perhaps the most distinguished French natural philosopher of the fourteenth century, wrote commentaries on Euclid's *Elements*, Sacrobosco's *Sphere*, and a half-dozen Aristotelian works; independent treatises on mathematics and cosmology; refutations of astrology and magic; a commentary on Peter Lombard's *Sentences*; sermons; and a variety of miscellaneous works on economics, preaching, and various theological and ecclesiastical problems. What these lists reveal is that Grosseteste, Bacon, Buridan, and Oresme were not specialists in science, but universal savants, and the same is true of most of their confreres. The point is important, because there is no separating Grosseteste the scientist from Grosseteste the biblical commentator, Bacon the scientist from Bacon the moral philosopher, Buridan the scientist from Buridan the logician, or Oresme the scientist from Oresme the theologian. Work in one realm inevitably influenced achievements in the others.

The interpenetration of disciplines could be very deep. Although medieval scholars knew perfectly well where the borders between, for example, physics and theology lay, they did not permit the borders to restrain their free movement. Thus, treatises written by a master of theology could draw freely on all of the disciplines that he had studied in the faculty of arts, and in a theological tract (especially a commentary on the *Hexaemeron* or the *Sentences* of Peter Lombard) we should not be surprised to find explicit and extended discussion of a wide range of physical problems. If, for example, we wish to locate medieval discussions of motion, we must turn not only to works entitled *De motu*, but also to commentaries on Aristotle's *Physics*, and *On the Heavens*, quodlibetal questions, *Sentence* commentaries, collections of sophisms, *Hexaemeral* literature, and a variety of works dealing with particular aspects of natural philosophy (such as Nicole Oresme's *Treatise on the Configurations of Qualities and Motions*) or with the whole of natural philosophy (such as John Dumbleton's *Summa of Logic and Natural Philosophy*). This freedom to roam the length and breadth of the intellectual world (with some caution required of a master of arts approaching theological territory) is a general characteristic of medieval learning, which we must keep in mind if we are to understand the nature and course of medieval science.

By the same token, there was no career as "scientist" for which one might train, except in medicine. There may have been a rare court astronomer/astrologer, but apart from such anomalies science was pursued by university professors (or former university professors)— men who regarded themselves as masters of the whole of human knowledge. It was in such an environment of broadly trained scholars, who had extraordinary freedom to cross present-day interdisciplinary boundaries in pursuit of whatever intellectual quarry captured their fancy, that the Greek scientific achievement, with its Islamic additions and modifications, was received, criticized, rearticulated, and reconstructed to produce the science of medieval Christendom. The details of this process are considered in the following chapters.

I wish to thank the following for permission to reproduce photographs of instruments or manuscripts in their keeping: the Bodleian Library, Oxford; the Warden and Fellows of Merton College, Oxford; the British Museum, London; the University Library, Cambridge; Gonville and Caius College, Cambridge; the University Library, Glasgow; the Bibliothèque Nationale, Paris; the Bibliothèque Interuniversitaire (Montpellier), Section Médecine; the Museum of Decorative Art, Copenhagen; and the Bayerische Staatsbibliothek, Munich. I thank also the Wellcome Institute of the History of Medicine, London, for making available photographic prints from the collection of Loren MacKinney, published in his book *Medical Illustrations from Medieval Manuscripts* (London, 1965).

Many friends have contributed to this volume. Michael Mahoney has been my advisor and confidant since the beginning of the project, and to him I express special appreciation. I am also grateful to many others who have read and advised me on various parts of the book: Marie-Thérèse d'Alverny, Marshall Clagett, Michael McVaugh, Victor Thoren, George Hourani, William Courtenay, Sabetai Unguru, and William Ashworth. I express deep gratitude to Brian Stock for supplying the photographs for figures 1, 14, 15, and 36 from his own collection of manuscript illustrations. My greatest debt, however, is to the scholars whose contributions have made the book a reality, some of whom have waited nearly five years for their contributions to appear and all of whom have uncomplainingly endured my editorial cavils.

David C. Lindberg

1

Brian Stock

Science, Technology, and Economic Progress in the Early Middle Ages

William Whewell, introducing the early Middle Ages in his *History of the Inductive Sciences* in 1857, stated: "We have now to consider . . . a long and barren period, which intervened between the scientific activity of ancient Greece and that of modern Europe; and which we may, therefore, call the Stationary period of Science."[1] In contrast, a recent authority has written: "The chief glory of the Middle Ages was not, as Henry Adams thought, its cathedrals, its epics, its vast structures of scholastic philosophy, or even its superb music . . . ; it was the building for the first time in history of a complex civilization which . . . envisaged . . . a labor-saving technology."[2]

These differing assessments are separated by a hundred years of scholarship. Yet putting them side by side takes one to the heart of the controversy surrounding early medieval science. For Whewell, science equaled scientific theory, the fully conceptualized heritage of Euclidean geometry and Ptolemaic astronomy. The early Middle Ages had almost no new theory; therefore, by definition, no new science. Lynn White, Jr., aware of this bias, presents the opposite view. Medieval science was really modern technology in disguise; as such, it is not to its discredit that it remained unconceptualized. Far from sliding into backwardness, in White's opinion, the period between Augustine and Abelard was one of the most dynamic and progressive in Western history. Neither of these views is acceptable in isolation from the other. The one focuses on purely intellectual achievement, while the other virtually excludes it from the discussion.

The separation raises a major theoretical issue in the contemporary history of science: the possible influence that technology and science may have on each other in the course of their mutual development. But it also poses a problem at a simpler historical level. Throughout the Middle Ages, theory and practice, in science as in law and theology, were closer to the realm of everyday reality than they are today. That fundamental Cartesian assumption of modern science— "the cleavage between the world as it presents itself in the perceptual experience of everyday life and the world as it is in scientific truth and 'in reality' "—was never framed in terms of rigid opposites.[3] The "scientific" character of reality was not conceived as something distinct and separate from the full range of experience in which man defined his relationship to the living universe of which he was a part. In other words, although, in reality, technology and science may not have been causally interrelated, neither were they as yet part of a civilization which made them opposite departments of thought and action.

The real question is, of course: What was their relation to each other? No one has a final answer. All that comparative history can do is propose a working hypothesis, which runs as follows. The story of technology *and* science between the decline of the Roman Empire and the rise of Western society in the twelfth century repeats, with some significant variations, a pattern that was first unfolded in the eastern Mediterranean area several centuries before. Technology preceded science, laying the foundations of material progress which later supported theoretical endeavor. The technical bases of Greek and Roman life—its agriculture, metallurgy, pottery, and textiles[4]— were inherited from the Neolithic and Bronze Ages. After the Homeric Age little fundamental innovation took place in technique, but there was a spectacular outburst of scientific theory. No one has offered a wholly satisfactory explanation of why Greek technology and science were so rigidly separated, but this pattern was passed on with little change to the Romans who, in science as in literature and art, were eager imitators of the people they had conquered. Under Rome, the Greek Hellenistic world made important contributions to earlier theory, and the Romans themselves adapted and improved inherited practice; but the two did not meet. When, after lengthy internal turmoil, the Arabs rose in the seventh century to rival Byzantium as heirs to Roman hegemony, theoretical science received a powerful stimulus. But technology was less radically affected, and the manner in which scientific activity was institutionalized maintained rather than broke with the traditions of the Hellenistic world.

Even when they were most clearly outdoing the Greeks, the Arabs looked back self-consciously to Athens. When Arabic natural philosophy began to lose its vigor in the eleventh century and the accumulated learning of the Greeks and Arabs passed to the West, there was no abrupt replacement or transformation of scientific theory—for the time-honored ideas of both peoples were precisely what the Latins wished to secure for themselves; rather, there was a shift in the geographical focus of science, from the Mediterranean basin to more northerly regions.

The early stages of technology and science in northwestern Europe seemed to indicate that there would be a repetition of the Mediterranean pattern. As in Greece and Rome, a long period of technical expansion preceded the creation of institutions capable of supporting continuous theoretical activity. But here the similarity ended. In the North, as early as the first century A.D., a new people and a new technology were adapting functionally to a new habitat. A tiny population, a large land mass, a cold climate, moist soils, and an abundance of basic metals all played their roles in the challenge offered to man, and he responded by putting into widespread use the stirrup, efficient harness, the heavy plow, three-field crop rotation, and the first stages of water and air power. The techniques were often old, but the extent of their utilization was new. Their gradual implementation allowed population to increase and provided the conditions through which nomadic peoples could settle down and evolve into a society. Tacitus described the Germanic *paterfamilias* as a restless warrior who willfully neglected his household and farm, a hybrid of bellicose energy and intensive idleness who was incapable of a rational use of the land. But when the *Völkerwanderung* was finished, a thin layer of people had been spread permanently over a formerly uninhabited continent, and the distinctive temper of the northern peasantry could already be discerned. Contrary to popular belief, the migrations of the ninth and tenth centuries did not interrupt the overall demographic growth. They provided an impetus for consolidating new types of social and political organization. When civilized culture, which had made a tentative reappearance under Charlemagne, was more securely reintroduced into this world, a larger economic unit, based in the North but incorporating the older Mediterranean trading area, had begun to take shape, importing the rich store of theoretical knowledge. Somewhat later, this northerly civilization also began to articulate the cultural values associated with economic progress.

Whitehead called this complex interaction of the old and the new, which reached maturity around 1150, the rebirth of "science in the modern world." Yet to those living through this troubled period, in which communications were poor, it was neither widely known nor appreciated that men were beginning to control and comprehend the universe in which they were living. Let us look at some of the factors that made it possible.

The Later Ancient World

Although neither the later ancient nor Islamic worlds radically altered the relationship between technology and science in the Mediterranean, both civilizations made significant contributions to the theoretical achievements of the Greeks, altering them permanently into the forms in which they eventually passed to the West.

Rome itself was a paradox, producing no universally acknowledged theorist and, excepting the apparently accidental discovery of cement, behaving with relative indifference toward labor-saving devices. Yet the extent of its practical success was unparalleled. The thinkers were to a man compilers of others' ideas: they substituted erudition for original investigation. Yet Roman dams, plumbing, irrigation, surgical tools, postal services, and engineering were unequaled down to the Renaissance. Natural philosophers were not wholly absent, of course: Lucretius's atomism was admired by the pioneers of modern physics, and Pliny still furnishes archeologists with their only comprehensive account of classical flora and fauna. But the most forceful expression of the Roman attitude toward nature is not found in imitations of Greek science. It is better reflected where it was thought about less: in the *mélange* of gossip, hardheadedness, and superstition of Cato's *De agri cultura*, or in art—the realistic touches in portraits, the accurate floral decoration of sarcophagi, and the frequent scenes of everyday life in later imperial mosaics and wall paintings. Here one senses that the ancient world is fading imperceptibly into the Middle Ages. For, despite an avowed admiration for Greek learning, at heart the Roman had little time for theory. He was interested in law and order, an efficient administration, a sanitary city, and his family. His daily experience led him to believe that nature's forces could be imitated, even placated: he was less sure they could be understood. In his most Italian poem, the *Georgics*, Vergil did not intellectualize farming; he merely sang of the plow, the harvest, the bee, the vine, and the flock. His ideal was not *scientia* but *otium*, the cultivated leisure of the country gentleman.

If Romans themselves added little, within the Greek-speaking regions of the Empire, and particularly at Pergamum and Alexandria, a
gradual modification was undertaken of Greece's greatest contribution
to science, the Aristotelian model of the universe. This took place
largely through the patient amassing of information and experience.[5]
For the Greeks, science had been mainly a theoretical pursuit; the
Hellenistic world brought it down to earth. With the exception of
optics, the major areas of improvement were practical: mechanics,
pneumatics, hydrostatics, medicine, geography, and mechanical devices. The leading figures were all Greeks: Euclid, Archimedes, and
Apollonius in mathematics; Herophilus and Erasistratus in anatomy
and physiology; Aristarchus and Hipparchus in astronomy; Ctesibius
and Hero in mechanical gadgetry; and the universal scholars, Eratos-
thenes, Ptolemy, and Galen. However, it was not only scientific interests that had changed. So had the conditions of research. In Alexandria, which, under the Ptolemies, emerged as the center of inquiry,
science was less disinterested than it had been in Athens and more
a subject of government patronage. Of course there were advantages.
Science lost its unhealthy connections with schools and sects; it benefited from continuity, institutionalization, and planning. But it was
directed more and more toward mundane ends. The areas in which
Alexandrian scientists excelled were measurement and calculation. The
measurement of space was chiefly useful to large landowners who were
determined to run their estates efficiently, the calculation of dates to
the official priesthood whose task it was to arrange the feasts of the
new gods, Isis and Serapis. Even the vast library, which was said to
incorporate Aristotle's books, functioned less as a research institution
than as a showpiece of Egyptian wealth. Technological improvements
were not encouraged. To the Ptolemies, labor-saving devices could
only disrupt the social order. As a result, "science tended more and
more to retreat from its function as man's weapon in the fight against
nature and confine itself to . . . being a mental discipline for the contemplative."[6]

Until the twelfth century the Latin West had almost no direct access to Alexandrian science. Yet, through the encyclopedists, an indirect contact was established with the doctrines of the two most
distinguished figures of the Hellenistic period, Ptolemy and Galen.
Ptolemy, who died about A.D. 170, took over the astronomical system
of Hipparchus and, with the aid of Apollonius's mathematics, produced the geocentric model of the heavens, which remained unchallenged until the publication of Copernicus's theories in 1543. The
thirteen books of the *Syntaxis mathematica*, or *Almagest*, include a

systematic summary of the achievements of astronomy down to Ptolemy's day, and the order in which they presented their contents remained unchanged for centuries—books 1 and 2 dealing with general theorems and the basic doctrine that the spherical earth rests at the center of a finite universe; book 3 with the sun; 4 and 5 with the moon; 6 and 7 the fixed stars; and 8 to 13 with the five planets. Ptolemy was a fervent critic of the unscientific elements in Aristotle. He also took over from his predecessors and considerably refined Plato's problem of "saving the phenomena," that is, of accounting for the apparently irregular motions of heavenly bodies by a system of mathematically regular circular motions. Ptolemy's major achievements were mathematical and calculatory: here he revealed himself to be a forerunner of modern science. Yet his system, like any great scientific synthesis, incorporated elements which remained unquestioned by lesser minds. In order to save appearances, he resorted to the theory of epicycles and eccentric circles, which never gave more than an approximation of the actual distances between planets. In his other major area of interest, geography, he often judged distances by the simple reports of travelers. It is not clear to what degree he really abandoned the older ways of thinking that he criticized: in the *Almagest*, he vacillated between observing the universe and admiring it as a Platonic *kosmos*, an ornament. The very authoritative bulk of the work discouraged men from taking up the problem again from the beginning. Yet Ptolemy remained the central authority in his field for well over a thousand years. His methods had to be learned painfully, first by the Arabs, later by the Europeans, before they could be bettered.

Galen was born in A.D. 129 or 130 at Pergamum, Alexandria's great rival. He died in 199 or 200. The son of an architect and surveyor, he divided his early studies between anatomy and the Neoplatonic schools of Albinus and Numisianus. At 28, having studied medicine for over a decade in various centers of Greek learning, he returned to Pergamum as physician to the gladiators, and in 161, near the beginning of the Antonine period, he arrived in Rome, where he quickly won court patronage, becoming, in turn, physician to the young Commodus, who became emperor in 180, and remaining on good terms with his successor, Septimius Severus. Galen was by far the most remarkable physician of later antiquity. His major method for introducing scientific elements into medicine was through philosophy. A platonizing Aristotelian, he rejected the Epicurean and Asclepiadean notions of an irrational nature and assumed that the system of bodily organs functioned teleologically. He refuted Erasistratus's idea that the *pneuma* flowed through the arteries and veins and established that the presence of blood in the heart and lungs was a normal condition. In diagnosis he went far beyond his chief mentor, Hippocrates. But Galen also had some important, if less well-known, limitations. His *Anatomical Procedures*, which describes the dissection of pigs, goats, and apes, could have easily been improved had his distaste for surgery not prevented him from taking apart cadavers. His diagnosis of fevers was remarkable for its period, but it placed too much emphasis on prognostication. In some areas he did not improve upon his predecessors; for instance, he did not follow Herophilus in the quantitative measurement of pulse. A true child of a great age of rhetoric—the Second Sophistic—he displayed at every turn a desire to unite rather than to separate science and philosophy. Yet Galen's scientific legacy to the West was considerable. He remained the foremost authority in many areas down to the Renaissance. He also left to the Latin West the fourfold scheme of the humors, the elemental qualities, and the seasons, whose actions paralleled Aristotle's four causes and whose direction was guided by the Stoic *demiurgos*.

During the second century A.D., scientific thought in the Latin and Greek worlds began to crystallize into the format in which it was known throughout the Middle Ages: the encyclopedia. The trend began as early as Theon of Smyrna, whose *Mathematical Knowledge Useful for the Learning of Plato* was absorbed without alteration by Calcidius. It was continued by Pliny, Ptolemy, and Galen. But its form and its characteristically obscure language did not really come into vogue until the fourth century A.D., when the Latin West began to be cut off permanently from the Greek-speaking East. At this point a vast program of codification took place. It was undertaken not by scientists but by *littérateurs*, and it provided the Latin world with virtually all it knew of Greek science until the intellectual revival of the twelfth century. The fourth-century *Matheseos* of Firmicus Maternus synthesized ancient astrology, and Calcidius, his contemporary, gave the early Middle Ages almost all it had of Plato, a translation of the *Timaeus* which terminated abruptly at 53b, followed by a lengthy essay on classical views of matter. *The Marriage of Philology and Mercury* of Martianus Capella, who died in 529, had greater pretensions. Under the veil of allegory it purported to review all the seven liberal arts as set forth by Varro. Book 8, *De astronomia*, remained a standard reference work on the heavens until the reintroduction of Ptolemy. Macrobius, another fifth-century pagan, used Cicero's *Dream of Scipio* as a point of departure for an allegorical commentary largely based on Porphyry, which is a virtual textbook of Neoplatonic science. The encyclopedic tradition was a mixed blessing. It summarized and expounded many ancient ideas which might otherwise have been lost, but it created a heavy dependence on authority. It also adopted a language of allegory which deliberately attempted to conceal scientific truth, just as phenomenal appearances were assumed to be covering an inner, formal reality. The best of the encyclopedic tradition is illustrated by the first-century herbalist, Dioscorides. Less allegorical but no less systematic than his later counterparts, he was translated in Italy as early as the sixth century, thus making available to the Middle Ages the single greatest compendium of Hellenistic remedies. Rearranged in alphabetical order in the eleventh century, the *Materia medica* remained useful and popular until the Renaissance. But if Dioscorides had assembled existing knowledge in order to advance it, the later encyclopedists merely compounded ancient error with their own naiveté.

With one exception: Anicius Manlius Torquatus Severinus Boethius (ca. 480–524), the last bilingual philosopher of the Empire, who, in life as well as thought, stands squarely at the dividing line between the ancient and medieval worlds. His father, a consul under Theodoric, died while he was a boy, and he was raised by a powerful aristocrat, Q. Aurelius Symmachus, whose daughter, Rusticiana, he eventually married. In 522, in circumstances that are still obscure, he mobilized his then considerable influence at court to defend another aristocrat, Albinus, and in the ensuing litigation fell from favor, was sentenced and put to death. His contributions to the Middle Ages were numerous. He provided translations of Aristotle's most fundamental logical works. He popularized a division of philosophy into *theorica* and *practica*, the one dealing with physics, mathematics, and metaphysics, the other with ethics, household management, and politics. He introduced into medieval logic the problem of universals, to which he gave a distinctive interpretation. Among the textbooks of the next seven centuries were his translations and commentaries on the *Isagoge* of Porphyry, and the *Categories, Interpretation*, and *Topics* of Aristotle. His own logical works included *On Categorical Syllogism, On Division, On Hypothetical Syllogism*, the *Prolegomena*, and *On Different Topics*. His *De arithmetica*, a rendering of Nicomachus of Gerasa, and his *De musica*, incorporating the harmonics of Ptolemy, were widely read, as were the several astronomical treatises and the *De geometria* ascribed to him. His *Consolation of Philosophy*, which was composed in prison, offered to subsequent thinkers an example of the selfless pursuit of wisdom in the face of earthly uncertainties, and, perhaps better than his logic or influential *Theological Tractates*, con-

Fig. 1   Boethius in his cell with Philosophia. Cambridge, University Library, MS Dd.VI.6, fol. 2v (12th c.).

veyed the essential spirit of the man. His fate at the hands of historians has alternated for a hundred years, the classicists assimilating him to an earlier period, the medievalists projecting him into a much later one, neither side entirely willing to accept the enigma of a sincere Christian who was also a serious Hellenist.

The case of Boethius raises a larger problem that has sustained much debate: the influence of Christianity on the science of the later Empire. In general, the force of the new religion has been exaggerated. The early Christians, who worked chiefly among the poorer urban masses, attended the pagan schools, whose essential curriculum did not change down to Justinian. They differed only in the moral training given to the young.[7] By the fourth century an educated class of converts had surfaced, and both Christians and pagans had begun to argue from clearly defined positions. Although it is difficult to generalize, a difference may be noted between the Greek East and the Latin West. In the East, the assimilation of the Hellenistic heritage was more thorough, and differences between the two sides assumed the character of a genuine philosophical dialogue on common problems. In the West, Christianity was less concerned with establishing a theology than it was in gaining spiritual control over men. Its primary questions were those of conduct, discipline, and ethics. The Latin theologians Irenaeus, Tertullian, and Lactantius actively polemicized against classical tradition. They considered pagan philosophy to be part of a morally dissolute world which could be redeemed by the second coming of Christ. Ambrose showed greater tolerance in works like his Genesis commentary, which embodied pagan cosmological views. But the most influential reassessment of the ancient heritage was made by Augustine. In *The City of God*, Greek learning, in particular Plato, was seen as a necessary but transient stage in Christian education, a psychological parallel, so to speak, to the progress of history itself from the obscurity of Greece and Rome to the clear light of the Christian commonwealth. Augustine gave a new meaning to history, but, from the standpoint of Aristotle, his views incorporated a number of rather unscientific elements. The world and its development, as in Plato, were conceived as a story, a series of events in time, not as in Ptolemy and Galen, as a perpetually moving system analyzable through laws. The phenomenal world was a set of appearances, beneath whose misleading contours a universe of stable forms was concealed, only to be fully revealed at the end of time. To study nature was to study a copy of reality, which, without the illumination of God, was a pointless enterprise. Plato had only wished to save the phenomena; Augustine appeared at times to pretend they did not exist.

However, it would be misleading to assess Christian influence, as many have, by Augustine alone. In practical areas Christians improved upon the aristocratic habits of the classical world. They opened Roman hospitals to the poor, thus extending the benefits of rational medicine and foreshadowing the hospitals of the later Middle Ages, which were among the first institutions of social welfare. In the realm of ideas, the most serious critic of Aristotle after Ptolemy was also a Christian, the monophysite John Philoponus. He inaugurated a new era in scientific thought dominated by three beliefs, which were shared alike by Jews, Christians, and Muslims: the universe is the creation of a single God; the heavens and earth have the same physical properties; and the stars are not divine. With these assumptions Philoponus led a sustained attack against his rival Simplicius, questioning Aristotle's dynamics and cosmology. Against Aristotle, Philoponus argued that there is no absurdity in the idea of motion in a void, that the velocity of descent is not proportional to weight, and that if a medium resists motion it cannot also cause motion (such as the motion of a projectile). Philoponus also argued that God, who transcends nature, created matter endowed with all its physical properties, possessing the capacity to develop according to natural laws from chaos to the present organized state of the universe. He thus anticipated the arguments with which William of Conches and Thierry of Chartres would attack the twelfth-century fortress of Augustinian Platonism. The meager collection of Philoponus's writings that have survived suggest that he employed the didactic methods of modern science and even controlled his observations in the manner of genuine experiments.

The Growth of Islamic Science

The first modifications of Greek science into the form in which it passed to the West were undertaken in the Hellenistic period. The later stages were enacted in Islam.[8]

In 622, a religious reformer called Muhammad, then in his forties, after attempting to propagate his ideas unsuccessfully for several years in Mecca, became "chief magistrate" in Medina, thereby gaining control of an urban community for the first time in his career. His popular acclaim coincided with the maturing of supra-tribal sentiment among the Bedouins. The result was a new, dynamic religious force, Islam. The spread of Islam began during the prophet's lifetime and gained enormous speed and momentum after his death, gradually extending Arab domination over a huge land mass stretching from con-

temporary Pakistan to Spain. By 650 it had overthrown the enfeebled
Persian Empire and taken Syria and Egypt from Byzantium. The conquest of North Africa was completed by 698. Within a few generations the scientific traditions of the Greek East began to pass slowly
into Arab hands, where they remained, as far as the West was concerned, until the retaking of Cordova by Ferdinand III of Castile in
1136. Around that date Islamic scientific texts began to find their
way northward, where they intermingled with translations directly
from the Greek.

"The tremendous effect the invasion of Islam had upon western
Europe has not, perhaps, been fully appreciated."[9] Pirenne's now-
famous words may be applied in particular to science. Throughout
the Middle Ages, the West's debt to Islam was too obvious to need
acknowledgement. But during the Renaissance, the humanist dislike
of Averroism together with the sudden rediscovery of antiquity tended
to obscure the theoretical continuities between East and West. However, the neglect of Islam was not only due to the accidents of history.
The early historians of science also minimized Arab achievement.
Carra de Vaux remarked that its mathematics "remind one of good
secondary textbooks," and Dreyer asserted that "they left astronomy
pretty much as they found it."[10] The view that Islamic science was an
ancilla to Western development received further support from the
medieval and modern habit of reducing its original philosophical vocabulary to Greco-Roman concepts. Behind these errors lay the refusal of Westerners to consider Islam a self-contained civilization
obeying its own laws of growth and change. Even Pirenne, who, despite the limitations of his thesis, drew the attention of his whole generation to Islam's importance, saw its development through essentially
Western eyes. The pivotal point of his argument was that cities had
declined in the fourth century and had been reborn in the eighth. But
viewed from the Mediterranean rather than the North, urban life
showed no such interruptions. The lacunae arose not in the life of
cities but in the Latin documents out of which the Western historian
constructed his picture.[11]

Where urban culture shifted, science was sure to follow. Alexandria
was taken during the Egyptian campaign of 642, but the tradition of
its great library was not extinguished. Similar research facilities were
created in Baghdad between 780 and 820. Even before the *hegira* the
persecutions of the Byzantine Empire, which affected pagan philosophers as well as minor Christian sects, forced a number of scientific
figures eastward. Far from disrupting the growth of interest in natural
philosophy, Islam brought into being an eastern Mediterranean focus

which had quietly been taking shape for some time. Situated at the intersection of the trade routes that linked East and West, Islam was the meeting point for Greek, Egyptian, Indian, and Persian traditions of thought, as well as the technology of China. It was to be expected that the strengths of Greek science would be affirmed; it would not be surprising if in some areas they were surpassed: this is precisely what happened.

Three interdependent factors stimulated the growth of an indigenous scientific tradition in Islam: translations, the cultural force of the new religion, and the successful assimilation of non-Arabs into various disciplines.

The translations were made possible by the availability of paper, a cheap, efficient medium suitable for a dry climate. Introduced into the Muslim world after the capture of Samarqand in 704, where only two years before the Chinese had set up a manufacture, it spread gradually westward, reaching Baghdad in 794. Soon a wide selection—including Egyptian wrapping tissue and a lightweight grade for pigeon post—was seen all over the Islamic world. Paper provided the means for transforming the largely oral Arab culture into a written form. It has been argued that Arabic is inherently a good language for expressing scientific constructions—a fact of considerable importance for mathematization. But what really made an Islamic scientific tradition possible was the shift from a tribal, nomadic way of life, with the clan as its center, to a more stable, geographically fixed and intensively urbanized culture, largely dependent on paper, writing, and administration. The great model for this transformation was the Koran itself. For the Muslim, Arabization was synonymous with the coming of the Book. The nomadic stage of development was looked back upon as *jahiliyya*, a time of ignorance and uncouthness. A certain amount of respect was reserved for the other "peoples of the Book," the Jews and Christians. Scribes, from the first, were a class apart. They not only transferred ideas from one tongue to another; they played a decisive role in "building up a complex and lucid Arabic philosophical vocabulary and laying the foundations of a philosophical Arabic style."[12] With paper cheap and plentiful, the translation of Greek, Indian, and Persian scientific and philosophical texts underwent a renaissance between 800 and 1000, after which the Muslim world possessed an incomparably finer library of the ancients than was available in Latin. Walzer cites the example of Galen, of whose works only one was known to Cassiodorus. By A.D. 900, the Arabs were actively studying some one hundred twenty-nine books by the same author.[13]

The intermingling of cultures, disciplines, and peoples contributed
to the internationality of Islamic science. Despite the anti-Aristotelian
element in Islam, which first appeared in al-Ashʿari and matured into
a theology in al-Ghazzali, scientific activity was not only permitted
but on occasion positively encouraged. Partly this was the result of a
distinction in Islamic education between "native" and "foreign" disciplines. In al-Nadim's *Fihrist*, which was written between 987 and 988,
knowledge was divided into ten parts. The first six were connected in
some way with Islam. They included the careful reading of the Koran
as well as grammar, history, poetry, dogmatics, and law. The other
four treated non-Islamic subjects: philosophy, science, literature, and
secular history. It is difficult to imagine a rigorous application of these
principles. Nor is it true, as some claim, that the most distinctive contributions to science in Islam were made by outsiders. Of course,
many of the leading scientists were not Arabs. Al-Biruni and Omar
Khayyam were Persians, al-Farabi a Turk, Avicenna from Bukhara,
Jabir ibn Hayyan a Sabaean, Mashaʾallah a Jew, and the Bakhtyishuʿ
family Nestorian Christians. Yet modern notions of nationality must
not be misapplied. Islam extended into a new environment and a
wider geographical framework a principle well-known in the Roman
Empire, namely, the conferring of citizenship onto men from different
backgrounds who shared a single cultural and political allegiance.
What they added was religious unity. In patronizing science, Baghdad,
Cairo, and Cordova did nothing that Rome, Alexandria, and Pergamum had not done. They just did it under a new banner, and on a
bigger scale.

The Muslims did much to expand the idea of science, to elaborate
and clarify its methods, and to affirm its preeminence, along with
philosophy, in the hierarchy of knowledge. The predominant division
of the sciences was Aristotelian and came to Islam either through
Aristotle himself or through the translation of John Philoponus's commentary on Porphyry's *Isagoge*. In general its scope was broadened
to include practical disciplines not mentioned by Aristotle or his Alexandrian students, and there was a greater interest in the application
of mathematics to other branches of the quadrivium. A seminal work
was the *Catalogue* of al-Farabi (d. 950), which came into vogue in
the West after the translation of Dominicus Gundissalinus between
1130 and 1150. Al-Farabi not only distinguished between substance
and accidents as in Aristotle, but insisted that the *scientia multiplicandi*
held the key to interrelating the two. Number governed the relations
of proportion as well as the manner of proceeding from non-being to
being. Next to arithmetic in importance was the art of measurement.

Thus, al-Farabi gave an account of the two basic sciences, arithmetic and geometry, by bringing together quantification and philosophy. From the two primary disciplines derived a third, the science of heavenly motions. And from astronomy one proceeded to music, the harmonious and proportionate division of sound, a subject to which he devoted an outstanding separate work. In this manner the four basic sciences of the *quadrivium* were related to number.[14] However, in assessing al-Farabi's contribution, a distinction must be made between his influence in the West and his role in tenth-century Islam. To the West he was the sole source of an advanced, that is, Aristotelian, classification. By his contemporaries he was more appropriately looked upon as one of the first who attempted to adapt Greek philosophy to current problems in Islam. His interests were not only scientific, but also moral and political. Those familiar with other Alexandrian commentaries on Aristotle may also note that his innovations stayed well within traditional lines of thought. His classification of the natural sciences repeated a number of Aristotelian formulas, including the activity and passivity of matter. He said nothing new on the relation between the four elements or about those aspects of the trivium and theology which touched upon science.[15]

Within Islam, and influencing the West only marginally, a more rigorous schema was attempted in the late eighth or early ninth century by the great alchemist Jabir ibn Hayyan. Paul Kraus, who first sorted out the various manuscripts attributed to Jabir, established that his vast corpus was the work of a school which held common assumptions about the laws of nature, and which, among its other activities, was heavily involved, as were the Brethren of Purity, in the Isma'ili propaganda that promoted the Fatimid caliphate in Egypt.[16] Although Berthelot succeeded in separating Jabir from the Latin Geber who is associated with a number of later texts, the study of the Chinese Taoist and Greek antecedents, Arabic development, and Latin survival is still the subject of intensive research. Jabir, it would appear, divided the sciences into two interdependent halves, religious knowledge ('ilm ad-din) and worldly knowledge ('ilm ad-dunya).[17] Religious knowledge was divided into forms of worship and of understanding. The forms of worship were subdivided into those within and those outside Islam, the principles of understanding into the numerical significance of the Arabic alphabet and the principles of interpretative criticism. The latter was separated once again into philosophy and metaphysics. Worldly knowledge was also subject to two major divisions, higher and lower disciplines. The higher was identified with alchemy, the lower with the other techniques. Alchemy was divided into dependent and self-sufficient principles, the latter being the elixir, the former comprising the production of drugs and the methods and procedures of science. The other techniques included those necessary for alchemy, and those which were independent but which functioned along analogous principles. Ideally, the art of alchemy was to lead the devotee to a higher form of knowledge which would not only redeem his soul but also further the transformation of the world into a political utopia. Despite its mysticism, the system put forward by Jabir and his school contained a number of inherently more scientific elements than the reinterpretation of Aristotle attempted by al-Farabi, with whose classification it is not often enough compared. The world of science was more clearly separated from philosophy; its truths were not the product of reason alone, but of experience and practice. This scientific manifesto was deliberately prefaced to the Latin translation of the early ninth-century *Secrets of Creation* associated with the "Geberian" Balinus.[18]

If one looks over the field of Muslim science as a whole, the most remarkable feature is not the frequently emphasized appearance of a handful of universal minds—al-Kindi, al-Khwarizmi, al-Razi, al-Farabi, al-Biruni, Avicenna, Alhazen, and Averroes—but, rather, that science in one or another form was the part-time or full-time occupation of so large a number of intellectuals. No comprehensive review can be attempted here, but the following may give a general idea of the place of individual sciences within Islamic culture and of their significance in influencing Western medieval development.

Mathematics

The intensity of mathematical research is clearly revealed from the following excerpt from the entry under Euclid in al-Nadim's *Fihrist*:

> [The *Elements*] was twice translated by al-Hajjaj ibn Yusuf ibn Matar: one translation, the first, is known under the name of Harunian, while the other carries the label Ma'munian and is the one to be relied and depended on. Furthermore, Ishaq ibn Hunayn also translated the work, a translation in turn revised by Thabit ibn Qurra al-Harrani. Moreover, Abu 'Uthman al-Dimashqi translated several books of this same work; I have seen the tenth in Mosul, in the library of 'Ali ibn Ahmad al-Imrani (one of whose pupils was Abu l-Saqr al-Qabisi, who in turn in our time lectures on the *Almagest*). Hero commented on this book [i.e., the *Elements*] and resolved its difficulties. Al-Nayrizi also commented upon it, as did al-Karabisi. . . . Further, al-Jawhari . . . wrote a

commentary on the whole work from beginning to end. Another commentary on book V was done by al-Mahani. . . . Furthermore, Abu Ja'far al-Khazin al-Khurasani . . . composed a commentary on Euclid's book, as did Abu'l-Wafa', although the latter did not finish his. Then a man by the name of Ibn Rahiwayh al-Arrajani commented on book X, while Abu'l-Qasim al-Antaqi commented on the whole work. . . . Further, a commentary was made by Sanad ibn 'Ali . . . and book X was commented upon by Abu Yusuf al-Razi. . . .[19]

The tradition outlined by al-Nadim fell into three stages: translation, exposition, and innovation. These are the categories still used by historians to describe the development of Muslim mathematics. The only aspect of the subject he omitted was the practical. He did not recall that the diffusion of Hindu-Arabic numerals and the decimal positional system was brought about by trade.[20] Nor did he mention that Muslim mathematicians, to a much greater degree than the Greeks, interested themselves in everyday problems. Masha'allah, the noted astrologer (d. ca. 815–20), was the author of a treatise on commodity prices. Abu'l-Wafa' combined original work on Euclid and Diophantus with books bearing such titles as *What is Necessary from the Science of Arithmetic for Scribes and Businessmen* and *What is Necessary from Geometrical Construction for the Artisan*. In these works the theory was old but the examples were new. One may doubt that the most refined theory penetrated commercial circles, but commerce stimulated the theorists and oriented them toward the concrete.

Astronomy

Early historians undervalued Arab astronomy because it failed to overthrow the geocentric system, but for the practical purposes of surveying and fixing the dates of feasts Ptolemy's model was adequate. Omar Khayyam, working within the Ptolemaic framework, even succeeded in making a solar calendar more accurate than the Gregorian. A large part of Muslim astronomy dealt with civil calculations and the construction of observatories and tables. It chiefly surpassed the Greeks in application, measurement, and instrumentation. Al-Fazari (d. ca. 777), who helped to introduce Hindu notation into Islam, was one of the first to construct astrolabes. Al-'Abbas and al-Nahawandi were primarily known for having taken part in the observations at Baghdad in 829–30, at Damascus in 832–33, and at Jundishapur. A distinctive achievement of early Islamic astronomy was to separate the observatory from the temple and from astrology and to make it part of a scientific tradition. One of the first was built by Sanad ibn ʿAli for al-Maʾmun, who died in 833. From his reign onward every caliph and many minor princes felt obliged to have an observatory, and there is a continuous tradition of excellent observational astronomy running from ʿAbd al-Rahman al-Sufi (d. 986) to Ibn Yunus in Cairo in the first half of the eleventh century and to the scientists of Ulugh Beg in the first half of the fifteenth. The observatory was one of the few institutions of Muslim science to survive the Mongol invasion and to pass directly to the modern world.

The tendency to improve the geocentric system without overthrowing it is well illustrated by the careers of al-Battani (d. 929) and al-Biruni (d. 1048). They also reveal the vicissitudes of Muslim astronomical achievement in the West. Al-Battani's father, Jabir ibn Sinan al-Harrani, was a famous instrument maker. His son was a pragmatic astronomer with a passion for measurement. His best-known book, the *Zij* (or Tables) is totally unlike the *Almagest*. In contrast to al-Farghani, whose *Elements*, when made available by John of Spain, was correctly greeted as a paraphrase of Ptolemy, al-Battani began with mathematical considerations: the division of the heavenly sphere into signs and degrees and the principles of astronomical calculation. His improvements on Ptolemy were also calculatory. He more accurately determined the magnitude of lunar eclipses, the moon's mean longitude, and the apparent diameters of the sun and moon throughout the year. Unlike Thabit ibn Qurra before him and Copernicus much later, he dismissed the trepidation of the equinoxes. Yet, if al-Battani's *De scientia stellarum* was accessible to the Latins, the works of al-Biruni were not. Thus, one of Islam's universal minds remained virtually unknown in the West until modern times. A Persian by birth, a rationalist in disposition, this contemporary of Avicenna and Alhazen not only studied history, philosophy, and geography in depth, but wrote one of the most comprehensive of Muslim astronomical treatises, the *Qanun al-Masʿudi*. In a spirited series of letters to Avicenna, his junior by seven years, al-Biruni also criticized at length Aristotle's conception of heavenly motion. Against the latter he denied the possibility of heavenly gravity, the notion that stars and planets move in circles, and the doctrine that each element has its own intrinsically suitable position. Observation and reason, he said, taught otherwise. Against Avicenna's defense of the Aristotelian tenet that matter may be divided infinitely, he cited al-Razi, a pragmatic clinician, who argued that such a division could not be carried out in actuality. Similar was his answer to the possibly elliptical shape of the heavens:

> Aristotle has mentioned in his second article that elliptical or lentil-shaped figures need a vacuum in order to have circular motion, while a sphere has no need of a vacuum. Such, however, is not the case, for the elliptical figure is formed by the rotation of that same ellipse about its minor axis. Therefore, if in the process of the revolution of the ellipses which form these figures, there be contradiction or infraction, what Aristotle has claimed does not hold true.[21]

Medicine

The oriental medical tradition began before Islam at Jundishapur, which was founded on the site of an army camp in the third century by the Sassanid king Shapur I. After 489, when the school of medicine at Edessa was closed by the Byzantine emperor, Nestorian physicians fled to the new town, bringing with them a number of Greek texts translated into Syriac. Jundishapur was further enriched after 529, when Justinian closed the pagan school at Athens, and in the seventh century, after the destruction of Alexandria and the opening of trade routes to India, along which passed new pharmacological information. Of all the sciences, medicine was held in the highest esteem by the new rulers. As a doctrine governing every aspect of life, Islam had to concern itself not only with worship but also with diet and personal hygiene. An anthology of the Prophet's writings on health is still popular in parts of the Arab world. The connection between morality and medicine was also preserved in the tradition of the hakim, who was at once a wise man and a physician. From the time of al-Mansur, who summoned Jirjis Bakhtyishuʿ, the Nestorian doctor, to Baghdad to cure his dyspepsia in 765, Muslim capitals became centers of clinical training. Unlike the Roman valetudinarium, an improvised hospital which was imported from the battlefield to the city, the Muslim bimaristan was a highly specialized institution where, as early as the tenth century, diagnosis was an established art and candidates were examined for degrees.[22] Also, drugs formed an important part of treatment. If Galen had brought pharmacology from an ars conjecturalis to an exact discipline, in Islam the pharmacy first became institutionalized as a dispensary for drugs from doctors' prescriptions.

More than other sciences, medicine was also overcrowded with outsiders. Ibn Khaldun, reviewing its development in the fourteenth century, remarked that one almost had to have a non-Muslim name to be taken seriously. But this was an exaggeration. Many famous physicians —the Bakhtyishuᶜ, Johannitius, Ishaq al-Israʾili—came from outside Islam. But the four greatest doctors were Muslims. The earliest, al-Razi (d. ca. 925), was the finest clinician and a pioneer in chemical medicine. He anticipated certain modern discoveries: his treatise on natural habit looks forward to the theory of the conditioned reflex, and his monograph on children's diseases to specialized pediatrics.[23] His famous study of measles and smallpox was printed regularly in Latin from 1489 to 1781, when it passed into several vernacular languages. Much of his enormous output was not translated directly into Latin but was absorbed into the encyclopedias of his successors. Chief among them was ᶜAli ibn al-ᶜAbbas (d. ca. 995), who was, after Avicenna, the most influential Arab physician in the West. His *Liber regius*, which was more systematic than al-Razi's *Continens* and more practical than Avicenna's *Canon*, was in common use down to the Renaissance. In surgery, the major contributor was Abu-l-Qasim (d. 1013), the Latin Albucasis, whose work had a greater vogue in the West than in the Orient, where there were more elaborate religious sanctions against the cutting of the body.

But earlier physicians were eclipsed in theory if not in practice by Ibn Sina or Avicenna (d. 1037), who was at once Islam's archetypical and most influential scientist-philosopher. Avicenna was the son of a tax collector in the eastern province of Balkh and was educated at Bukhara. Although known primarily as a philosopher and physician, he wrote on a wide variety of subjects, including natural history, physics, chemistry, astronomy, mathematics, and music. An important commentator on the Koran and Sufi doctrine (at least, so it is thought), he also composed excellent verse and, as a counselor to various princes, contributed to the solution of important economic and political problems. His two most influential works were the *Kitab al-Shifa*, the *Book of Healing* [*of the Soul*], which treated logic, physics, mathematics, and metaphysics; and the *Qanun fiʾl-Tibb*, the famous *Canon of Medicine*, which was translated by Gerard of Cremona between 1150 and 1170, and remained the standard textbook of Galenism down to Harvey. The *Canon* is divided into five books. Book 1 deals with generalities concerning the body, sickness, and health; book 2 with the pharmacology of herbs; book 3 with pathology; books 4 and 5 with fevers, signs, symptoms, diagnostics, and prognostics. Influenced by Aristotle, Plotinus, and al-Farabi, the *Canon* was highly philosophical in language and style; yet it remained close to the real problems of health and disease and, more than Galen, returned to the pragmatism of Hippocrates. Nor was Avicenna's influence limited to medicine: in his unique codification of previous tradition one can see the roots of virtually the whole of scholastic philosophy in the West. In contrast to al-Biruni's rationalism and al-Razi's clinical empiricism, both of which illustrated the separation of science and religion, Avicenna incorporated scientific knowledge into an Aristotelian framework that was ideally suited for acceptance in Christendom. Yet within Islam he represented a turning point in another sense. So universal was his achievement that a repetition of it was unthinkable. Already overconcerned with their relation to the ancients, Muslim authors now had to contend with a philosophical classic of their own.

The Method of Science

Islam made a wide variety of contributions to scientific method. Major figures like al-Battani were also expert makers of instruments that enhanced their powers of observation and calculation. The artisan-scientist, who was considered an aberration in the ancient world, was more of a norm in Islam. The dependence on well-made apparatus meant that theory and practice were brought closer together. On at least two occasions in the *Canon*, Avicenna indicated that they were different but interdependent parts of the same discipline.[24] Remarks of this kind were not limited to medicine, where an inevitable problem arose in relating Galenic theory to clinical practice. By introducing new disciplines like ophthalmology, or granting new status to old ones like geography, the Arabs made practice a more natural part of science itself.

  Within the *quadrivium* their contribution to method was threefold. They increased the range of mathematization, not only by developing algebra and trigonometry, but by attacking hitherto difficult questions through a quantitative approach. In physics al-Biruni made among the most accurate measurements of specific densities down to modern times; at a more practical level, the dispensaries of urban hospitals prescribed accurate amounts of drugs of controlled composition. Secondly, they restored Aristotle's philosophy of science to its original integrity, from which point it could stimulate the growth of new thinking in the West. This was the work of three philosophers of Muslim Spain, Avempace, Ibn Tufail, and Averroes (d. 1198). Averroes' influence on Western philosophers was so widely acknowledged that he was known simply as "the commentator" on Aristotle. His ideas turned up in Albert the Great, Thomas Aquinas, and a variety of Renaissance thinkers. A third contribution lay in the frequent resort to a prototype of the experimental method. Here perhaps the most serious single step forward after the Greeks was taken by Alhazen (d. 1039). By reorienting al-Kindi's concept of vision, Alhazen argued through a combination of geometry and physics that objects are seen by means of rays passing toward the eye, and not vice versa, as the Greeks had assumed. In the course of his observations, he used the term *i 'tabara* (literally, "to regard"), which was rendered *experimentare* in the Latin translation of his *Kitab al-Manazir*, available to the West from the early thirteenth century. The use of the word is not, of course, the same as the invention of the concept: but imperceptible progress toward it through Greek and Arabic had doubtless been made.[25]

The Decline of Scientific Thinking

Islamic science may be divided into a series of overlapping stages of development: translation, assimilation, original activity, and decline. The reasons for decline, which set in from the eleventh century, are numerous and complex. Throughout Islamic history science was institutionalized in such a way that its spectacular progress could be easily stopped by a number of factors. When its impetus finally waned, these external forces began to play a larger and larger role.

A persistent weakness resulted from the reluctance to innovate in basic technology or to apply inventions to activities performed by slaves or servants, whose numbers were always large. Instead, attention was focused on scientific instruments whose function was rather specialized. General political factors also weakened the continuity of Muslim science. The breakup of a unified empire into warring principalities was accompanied by the gradual decline of the rationalistic tradition after al-Ghazzali. But there were also limitations from within. With rare exceptions, the man of science was a universal savant, not a specialized professional. Science, for all its distinctiveness, was often not separated from philosophy but bound up with it in the encyclopedic framework. The system of teaching, when it was not autodidactic, was that of the master and disciple, not the classroom. The student had more respect for his guide, who personally embodied a subject, than for an abstract discipline. Institutions of learning in the cities depended on royal patronage; the House of Wisdom continued the tradition of the Alexandrian Museum, as did later the new universities, or Nizamiya. Under enlightened monarchs like 'Adud al-Dawla or al-Hakam II the system could flourish, but it had the innate fragility of scientific patronage under Roger of Sicily or Frederick II. The bulwark of Muslim society, the family, also played an ambiguous role. Practical disciplines by and large ran in families. Information was jealously guarded and passed from father to son. There are numerous examples: al-Naubakht, astrologer to al-Mansur, and his son al-Fadl; the Banu Musa, who devoted their wealth to translation and mathematics; the astronomers al-Marwarrudhi and ibn Amajur and their sons; Ibrahim ibn Sinan and Sinan ibn Thabit, grandson and son, respectively, of Thabit ibn Qurra; and, of course, the celebrated medical family, the Bakhtyishu͑. These cases are too widely separated in space and time to serve as a basis for general conclusions. But the strong family tradition suggests that science was often caught between vertically operating genealogies and horizontally designed institutions. It rarely separated itself from kinship and official patronage at once.

Technology and Cultural Values in the West

Early northern society in the Middle Ages surpassed the Mediterranean not in the theoretical interpretation of nature, but in the success with which periodic waves of new techniques and inventions improved the human condition.

The decline of ancient science has traditionally been attributed to rivalry between Christians and pagans. But its real weakness may have been its incapacity to overcome the impediments to human progress through labor-saving devices. Vitruvius hit the problem exactly when he wrote that good architecture depends on a careful combination of *fabrica* and *ratiocinatio*, of practice and theory.[26] But his contemporaries paid little attention. The technical treatises of antiquity—the *Mechanical Problems* falsely attributed to Aristotle, and the works of Ctesibius, Archimedes, and Hero—openly avoided discussing applications for their inventions. "Hero describes some useful instruments— a fire pump and a water organ—but all the rest are playthings, puppet shows, or apparatuses for magic."[27] It was not the world of Benjamin Franklin, but of Apuleius: sophist, magician, and priest of Isis. Even in agriculture, where the availability of superior tools can be proven, their adoption was at best reluctant. The water mill, the wheeled plow, and the Gallic harvesting machine were all known before the first century A.D., but their use was limited to outlying areas, short in manpower, while within the Empire a high level of efficiency was attained without technical improvement.[28] The failure of Greece and Rome to increase productivity through innovation is as notorious as the inability of historians from Gibbon to the present to account for it. Slavery, the low status of craftsmen, the lack of professional training

apart from legal studies, and the dearth of investment capital outside
the complacent landowning classes have all been cited as factors. But
whatever the causes, the result was the same: "neither technique nor
productivity nor economic rationalism made an advance in those final
centuries of antiquity."[29]

Too little is known of the working relations between technology
and economic change in Islam to speak authoritatively about attitudes
toward innovation. More than the Greeks or Romans, or their heirs,
the Byzantines, but perhaps less than the contemporary Chinese, the
Arabs recognized technology as a legitimate branch of science. But
the agricultural revolution that followed swiftly upon the religious,
linguistic, and political conquest was not founded on technological
innovation. Rather, it was at once capital- and labor-intensive.[30] At
its core was the displacement of a whole series of food and fiber crops
—rice, hard wheat, sugarcane, watermelon, spinach, lemons, and
many other plants—from India and Persia to the Near East and North
Africa.[31] Under Islam, the new crops, when combined with effective
irrigation and a summer growing season, permitted a more intensive
use of the land, allowed the rural population to rise, and eventually
transformed all the economies between Transoxania and Spain. The
new agriculture was stimulated by a number of factors: a legal corpus
on irrigation protecting individual rights and granting a lower rate of
taxation for land watered by the noria—a water-raising device using
chains and buckets—than by hand; and the prevalence of small holdings, of sharecroppers and free farmers, as opposed to the *latifundia*
of antiquity with their scores of slaves. Royal gardens were also used
as experimental farms, especially in Muslim Spain.[32] But the very
factors which made Arab agriculture a success militated against a
concentration on machinery alone. Although water-raising devices
were very widespread, they were limited to two ancient types: the
Archimedean screw and the water-wheel, usually drawn by an animal.
Despite state financing, no systematic attempt was made under the
Abbasids or Fatimids to use what water there was as a power source.
The plow was invariably a light type lacking colter or moldboard;
normally it had no wheels, and animals were employed without efficient harness. Even though private capital was available, as it had not
been under Rome, for investment in technology, the successful adaptation of older techniques and the abundance of the rural labor force
made the effort unnecessary.

Conditions in Europe were different after the first century A.D.,
when there arose in the north a new center of gravity for trade, industry, and agriculture. The pacification of the Empire and the de-

velopment of roads allowed capital and production to shift from Italy proper to remoter areas where gold, lead, tin, and iron were plentiful. The newly exploited regions of Gaul and the Rhineland gradually became the workshops of Rome for pottery, glass-blowing, bronze ware, and metallurgy. Galen noted that the best steel for surgical instruments came from Noricum, the same place that later gave birth to the famous pattern-welded steel of the Merovingians. If Rome remained the administrative head of the Empire, Lyon and Cologne emerged as more rapidly expanding areas of manufacture, and the Rhine became for the first time the connecting link between the Atlantic and Baltic lands and the Mediterranean. Until the seventh century the essential unity of the Mediterranean economy was still intact, but afterward a combination of demographic and agrarian change succeeded in creating a new focus for civilization in the north. Allowing for minor respites during the Scandinavian and Saracen migrations of the ninth and early tenth centuries, the global expansion continued unabated until the first half of the twelfth. During the same period, the agrarian system of the Mediterranean, which had been imported to the northern extremities of the Empire, was definitively abandoned, and a new agriculture based on the heavy plow, the open fields, the use of the horse, and efficient harness, as well as triennial rotation, imperceptibly became the norm. From the tenth century, when villages in the long-settled areas began spilling over, large amounts of forested and swampy land were gradually brought under cultivation. This movement reached its apogee in the Cistercian expansion of the twelfth century. From the eleventh century as well, urban clusters began to dot the landscape, first as trading grounds for the remonetizing economy but later as focal points for market concentration. Land reclamation, the rebirth of towns, and the freeing of the rural population, especially artisans, from servility compelled people of all social classes —those who labored, prayed, or fought—to seek new solidarities. And during these profound changes, which were frequently all taking place at once, technology acted as a persistent, subterranean force. If political instability had done much to unsettle long-established habits, the adaptation to a hardy environment and the recurrent shortage of manpower made labor-saving devices a necessity. As Europe slowly emerged from prehistory into history, as men were freed from an absolute bondage to the soil and gradually found a voice for their hitherto inarticulate desires, mere grains of sand began to give evidence of a perceptible inner design. There emerged a new set of values which— for better or worse—regarded the transformation of this world as a sufficient condition for salvation in the next.

The major areas of innovation were weaponry, agriculture, and machinery for utilizing natural sources of power. Although the developmental stages of each are enveloped in obscurity, it is possible to see the tenth century as a rough dividing line. Before that time invention itself may have been rapid, but diffusion was slow and irregular. Afterwards, new ways of doing things became more widespread, and the devices were applied to an ever-increasing variety of tasks. References to the water mill were infrequent before 1000, but by 1086 the *Domesday Book* recorded 5624 mills for some 3000 English communities alone.

In warfare, the single most important innovation was the stirrup, which allowed a mounted rider to deliver a blow from a spear supported under his shoulder, thus combining his own forward thrust with the momentum of a galloping steed.[33] The union of the stirrup, which probably came from China, and the mounted *clibanarius*, who was known in the later Empire, may have taken place as early as the eighth century, but its use was not generalized until later. From archeological evidence it has been suggested that the spurred Carolingian wing-spear, suitable for cavalry, also appeared about this time.[34] An even more formidable invention was the Merovingian long sword. Its essential feature was a type of steel built up by welding together thin strips of metal, which were then cut, bent, and forged into the desired pattern. The resulting pattern-welded steel was stronger than the oriental, crystal-welded variety.[35] But the new military techniques took centuries to put effectively into practice. Their superiority was not decisively proven until the battle of Hastings in 1066. And their actual origins remain obscure. Edgar Salin has spoken of the Merovingian sword as a *"chef d'oeuvre de technique"* whose roots lay in the unfathomable depths of Germanic culture,[36] the world of *Beowulf* and *Nibelungenlied*. Employing prehistoric methods of smelting and forging, the northern tribesmen had revived the Roman mines of Styria, Carinthia, Franconia, Swabia, Lorraine, and Burgundy, breathing life, so to speak, into the myth of Wieland. In the Burgundian and Salic Laws, the *faber*, the specialized smith, was given a high *wergeld* and social status.[37] The cost of weapons and, as time went on, of the heavy armor that accompanied them, could only be born by the nobility, who hardened into a warlike class and legitimized their activities through the ideals of chivalry. But the real importance of Frankish and later Norman militarism arose from its connection with "feudalism." From the seventh century, monasteries and great lay estates were regarded as capital resources which at any moment could be converted into military expenditures for defense.[38] Directly or indirectly, an interdependence was built up between property and "defense-spending." Technical innovation was stimulated through warfare.

The transformation of agriculture, as noted above, arose from a response to a fundamentally different environment. "Classical antiquity was a civilization of dry-farming, stone, bronze and textiles of vegetable origin, while the medieval West had drainage farming and utilized chiefly wood, iron, leather and textiles of animal origin."[39] The most significant innovations were the heavy plow and the efficiently harnessed horse. In general, the plows of the ancient and medieval world may be divided into two families, the *aratrum* and the *carruca*.[40] The *aratrum* was a light plow, easily assembled and worked by a single man, suitable for light farming in areas lacking heavy draft animals and specialized artisans for making metal implements. All early plows, those used in the Near East, the classical world, and the prehistoric north of Europe, belonged to this type, which conserved moisture by moving the soil without turning it over. The *carruca*, which appeared in the north from the sixth century, gradually acquired a coulter, an iron blade fixed in front, a moldboard for guiding the furrow slices to one side, and wheels which, inserted under the beam by a series of peg holes, allowed for adjustments of depth. It achieved an economy of manpower and eliminated the need for cross-plowing, but it was a much more costly and elaborate machine. When hitched to a horse, or later a team, it constituted the first stage in the mechanization of agriculture. But the horse had to be properly harnessed. In antiquity horses had been strapped around the throat and abdomen. If the animal brought its whole weight to bear on a load, it ran the risk of strangling itself. In replacing the traditional oxen with a more efficient equine force, the medieval peasant introduced a type of breast-strap harness which allowed the horse to realize its greater speed and endurance over other draft animals. These two basic changes ushered in others: new crops like rye and barley and, as a result of diversification, improved nutrition; the first attempts at three-field rotation, better suited to the northern climate than the Mediterranean two-field system; larger, faster, and more reliable transport vehicles, which shortened the distance between the producer and consumer of cash crops; and the freeing of the farmer from the tyranny of having to live next door to his fields. With improved communications he could now live in a large village or town. The result of these combined innovations was Europe's first "green revolution." The lowering of the man/land ratio and improved productivity had by the eleventh century increased some yields by up to four times what they had been under Charlemagne.[41]

The third phase of early medieval technology was the systematic harnessing of water and air power for human ends by means of mechanical devices. To the six basic *machinae* of antiquity, the lever, wedge, screw, pulley, winch, and inclined plane, the Middle Ages made no addition. The change came partly in distinguishing more clearly between implements and machines, between devices which used muscular effort and those with more complicated parts that employed natural forces. But it was also an altered social psychology. Medieval inventions invariably originated elsewhere, but only in the West were they extensively exploited for saving labor, increasing productivity, and eventually rationalizing economic activity. This was not only true of the water mill, which was known to the Romans, and the windmill, which was imported from China via Persia. It was also characteristic of those inventions of the twelfth and thirteenth centuries to which the fourteenth and fifteenth were so indebted: hydraulic hammers, screw-jacks, screw-presses, toothed wheels, transmission shafts, gears, springs, and possibly even the crank and connecting rod. The best documented example of specialization from a known technique is the water mill. As early as the tenth century it was used for irrigation, for grinding pigments, for producing oil and malt, for wood-turning, and for grinding cutlery. With a variety of cams and gear-wheels it was adapted for fulling and tanning, for producing hemp and iron, for sawing, and even for paper-making.[42] In coastal areas tidal mills also appeared as early as the eleventh century, and from the 1180s, the air-powered device which swept across Europe in the thirteenth and fourteenth centuries, the windmill. The application of simple machines to the progressive mechanization of economic tasks was paralleled much later in the Industrial Revolution. That is why Roger Bacon's prediction of motor-powered ships, vehicles, and airplanes is as apt a reflection on his own age as on ours. But if mechanical devices saved labor and increased productivity, they also made money. The actual sources of the large investments necessitated by mills and plowing apparatus are not well known, but there is no doubt who drew the profits. Lay and ecclesiastical lords attempted to monopolize revenues through a system of bans and tithes.[43] Poor peasants, reluctant to pay these dues, often resorted to techniques which they could operate themselves, like cottage hand mills. The clandestine use of such devices may well have slowed down the construction of bigger mills. Yet on the whole, the tug and pull of class tensions, which increased throughout the twelfth century, seems to have had a beneficial effect. Before 1200, the major foundations were laid for the triumphs of Western power technology down to the eighteenth century.

This revolution could not have been brought about without changes in the status of artisans, and in wider attitudes toward work, time, discipline, and the active life. The "internal" and "external" components have also become clearer as Westerners have deepened their understanding of Eastern technology. Of the three inventions that Francis Bacon claimed changed civilization—printing, gunpowder, and the magnetic compass—all were Chinese and two were known to medieval men. China also anticipated by centuries Western use of astronomical instruments, the stirrup, efficient harness, the wheelbarrow, the water-powered chain-drive, belt and crank, the metallurgy of iron and steel, the stern-post rudder, and chemical apparatus.[44] What was new in the West was not inventiveness but the social context of techniques.

Since Max Weber it has been fashionable to associate the economic rationality characteristic of Western society with religious attitudes that made their appearance in the Middle Ages and grew to maturity during the Reformation, when capitalism, in an international sense, was born. Even Weber's most enthusiastic disciples would no longer accept his modernization theory without qualification, but the supposed connection between religious and economic behavior helps to explain many otherwise unaccountable features of medieval technology and social change. Manual labor, the mechanical arts, and Western monasticism were united as early as Benedict of Nursia, who founded the monastery at Montecassino in 529. In opposition to the aristocratic cultural ideal of antiquity he bluntly stated, "*Otiositas inimica est animae*" (Inactivity is an enemy of the soul).[45] Christians were fond of recalling the opening chapters of Genesis, in which, as penance for his sin in Eden, man was commanded by a stern, unyielding God to earn his daily bread by the sweat of his brow. In obedience, the monastic day was divided into *lectio divina* and *labor manuum*. The work of the hands was released from servility and given a new dignity.

Yet the admonitions of Benedict did not become a widespread force for activism until the late eleventh century when, under the weight of six hundred years of slow agrarian change, they provided a point of departure for new values, religious groupings, and social solidarities. Gioacchino Volpe caught the essential flavor of the renewal in a single phrase: "a new people, a new religious sentiment."[46] But it would be equally apt to summarize the changes as a new technology, social mobility, and altered cultural values. For none of the religious movements, orthodox or heretical, that sprang to life were entirely free from some association with the economic changes that were motivating them or inspiring their revolt. The growth of the Lyon textile industry was coterminous with the appearance of the Waldensians, that of the Low Countries with the Beghards and Beguines. The Umiliati of Milan began a highly disciplined workers' alliance in the factories for processing wool and within a few decades controlled the capital financing of their industry. The Templars, later famous bankers, owned the first iron mill recorded in Champagne and actively contracted Sicilian wheat to Genoese merchants. One may even regard the guilds as a subtle adaptation of economic necessity to chivalresque initiation rites and religious discipline. However, one movement stands apart in the degree to which it captured the imaginations of men and epitomized to later generations the contradiction of simultaneous withdrawal from the world and involvement in it. So swift was its growth, so complete its triumph, that it has been justly said that it was defeated by its own success.

The Cistercians were founded in 1110. Their principles of government were defined in a simple revolutionary document, the *Carta caritatis*, which asserted, in contrast to the hierarchical interdependence of the Benedictine houses, that each monastery was economically independent within a wider framework of spiritual control. Formerly isolated in a vast countryside, monastic communities now became a family of individuals with a collective direction. Both in ideology and in the allocation of resources they were future-oriented. Capital-consuming ornament, time-consuming liturgy, and emotion-consuming gesture were all reduced to a minimum. Economic dependence on vassals, the local market, and the pilgrimage route was superseded by a risk-capital venture, the location of monasteries in hitherto unsettled areas. Thus Christianity moved in one leap from the crossroads of civilization to the hinterland. Having long colonized, the monks became colonials. The new estates were models of planning and efficiency. The system of granges located within easy reach of the mother house permitted the development of a rational system of production, storage, and distribution. New areas, like the famous Burgundian vineyards Chablis and Clos de Vougeot were brought under cultivation, and others, like the Yorkshire uplands, under pasturage. Capital investment in church construction was economically justified; building time was greatly reduced by the use of local artisans, accessible materials, and a standardized plan. In addition to functional architecture, the monks were pioneers in standardization and industrial autarky. Liturgy, prayerbooks, and scripts were unified. Regular production lines with forges, wheat and fulling mills, and tanneries grew up at larger houses like Royaumont and Fontenay. The rapid diffusion of the order, which had over three hundred houses at the time of St. Bernard's death in 1153, helped to spread triennial rotation, the use of machinery, and literacy. In a sense, the entire movement was made possible by the exploitation of a single natural power source, water. An anonymous and idealistic description of Clairvaux stated that every abbey should be located near a river, which, entering one side, should in turn run the corn mill, the beer-boiler, the fulling machines, the tannery, and other departments, whether for cooking, rotating, crushing, watering, washing, or grinding, before carrying away the refuse.[47] "Out of thirty French documents of the twelfth century concerned with hammer-forges and iron metallurgy twenty-five were drawn up by Cistercian monks."[48] That singular engineer of the thirteenth-century, Villard d'Honnecourt, was also a Cistercian. Thus, three centuries before the mechanical clock, these early "Puritans" had virtually perfected a time-disciplined microsociety. Their success was so spectacular that within two generations they became rich in wine, wool, grain, and gold, vying with moneylenders as purveyors of credit. But so obvious a contradiction could not long rest unnoticed. The Cistercian became the target of the reforming Joachite, the Provençal heretic, and the mendicant friar. In the shadow of his achievement arose the conflicts over poverty, property, and usury. Europe's first debate on economic change and cultural values had begun.

But new religious attitudes were not only promoters of change; they were also products of it. In the nascent towns, where the artisans and the bourgeoisie made their appearance, the measure of popular values was the craftsman. On the great country estates even as late as the twelfth century the artisan was primarily a servile domestic who was allowed to sell his wares only after fulfilling his services to the local lord. In the towns he acquired civil rights, corporate status, and access to a concentrated market where the laws of supply and demand could freely operate. The monasteries may have provided indirect support; Bernard of Tiron, a wandering preacher who died in 1117, founded a house specifically as a haven for craftsmen. But the real model for change came once again from Islam, in which the status of the artisan had changed from that of the slave to that of the free laborer. Documents from the medieval Jewish quarter of Cairo, the earliest that are known, point out that artisans were instrumental in transferring techniques throughout the highly mobile Muslim world. A pilgrimage was the normal obligation of every Muslim, rich or poor. To the cultural stimulus was added the commercial. "During the High Middle Ages men, goods, money, and books used to travel far and almost without restrictions throughout the Mediterranean area. In many respects, the area resembled a free-trade community."[49] The loose machinery of the state, the personal rather than territorial conception of law, and the liberal treatment of foreigners all promoted cultural interchange. If large-scale industry was rare, the skilled laborer was normally financially independent and earned about twice as much as an unskilled worker.

As commercial activity quickened in the north, and the possibilities of new techniques were gradually realized, a Western equivalent of the Mediterranean artisan appeared, as yet less refined in ornamentation and products of luxury, but characterized by that specialization of function within the division of labor which lies at the root of all modern industrial development. As early as 1121 a charter of Louis VI gave the water merchants of Paris, who defined themselves as a Roman *collegium*, the right to impose a tax.[50] Throughout the twelfth and thirteenth centuries, one industrial trade after another became incorporated. Etienne Boileau, who was made provost of Paris in 1254, was able to weave an unparalleled tapestry of workers' guilds, whose charters in some cases underwent no alteration until the eighteenth century. The emergence of a strong free force of artisans at a critical point in commercial evolution aided the bourgeoisie in creating a counterweight to the landed aristocracy. Together the two groups had a profound effect on inherited patterns of thought and action. Nothing could have been better calculated to bring men back to reality from the otherworlds of chivalry and mysticism. If earlier monastic writers tried to connect the active and reflective sides of life in a society whose institutions opposed them, the twelfth-century artisan and merchant, as inhabitants of a secular world, made the union seem natural and legitimate. From the master-builders of the cathedrals to the commercially inspired genius Leonard of Pisa, they did so without unnecessary theoretical fuss. In this atmosphere of hardy doers rather than pure thinkers, who inherited centuries of inarticulate but successful handling of natural forces, the corpus of Greek and Arabic natural philosophy began to find its way into the West.

The Rebirth of Theoretical Science

Before the Twelfth Century

The period between the decline of Rome and the twelfth-century "renaissance" has traditionally been looked upon as the "dark age" of theoretical science. There is some accuracy in this view. Compared to the earlier achievements of the Greeks and the contemporary activities of the Arabs, Latin natural philosophy remained at a rudimentary level. Yet recent paleographical studies have demonstrated the existence of numerous students of scientific texts. Given the instability of political conditions, the persistent copying and interpreting of the ancients bears witness to a determined community of learning.

A rigid identification of "science" with the perpetuation of the classical tradition should also be resisted. During this period, Europe was largely an oral society. As in contemporary "primitive" societies, there was little or no literate tradition of natural philosophy. But primitive Western science was not all magic and superstition. It also contained a good deal of rational and empirical knowledge, differing from the period's science not so much in observational accuracy and results as in its method of organization and presentation. One useful lens through which this "science" may be viewed is the rise of Christianity, for Christianity held out the possibility of an alternative means of controlling the natural universe and subtly absorbed many of the oral traditions into its fabric. Christian interdependence with the world of magical science can be seen in many places: in miracles, the standard evidence to the unconverted of God's limitless power; in the ameliorative powers of baptism, the eucharist, and the other sacraments, which, because of their inner mysteries, generated a subset of parasitic beliefs; in physical objects like amulets, talismans, charms, and relics, which, like the legendary magic touch of the king, made up for the inadequacies of medical services, herbals, and bloodletting; in semilegitimate forms of divination, to which the Church closed its eyes and which functioned in areas where normal decision-making processes failed; in oaths, as a guarantee of testimony, and ordeals, as a guarantee of justice, both of which posited a moral authority above human justice to whom all forms of good and evil were immediately revealed; in forms of prognostication such as dreams, visions, omens, and judgments, which offered certain knowledge of the future in an uncertain society; in prayer itself, which was often a source of help for material and temporal difficulties and not merely a subjective experience of God; and, in general, in Christianity's disposition to see the world as a meaningful order revealing God's purpose and physically sensitive to his will, an idea into which animistic and particularistic demonology was often incorporated. Christianity, in order to convert, imitated the practices of oral tradition while laying the foundations of literate culture. It absorbed and transformed an already existing world of oral science.

Christianity was also the major vehicle for preserving and spreading ancient science. This was normally done through the study of the quadrivium, that is, arithmetic, geometry, astronomy, and music, the four liberal arts that were often allied with medicine. Modest revivals of the quadrivium took place in monastic and cathedral schools between the fifth and twelfth centuries. The earliest was the work of Isidore of Seville (d. 636). Born of a noble Hispano-Roman family, Isidore received what was for his times an outstanding classical education. He incorporated his scientific material into two encyclopedias that also treated nonscientific subjects. In both format and content they served as models to many later medieval authors. The shorter, entitled *De rerum natura*, dealt with the interrelations between the elements, the humors, the planets, and man. The longer, called the *Etymologiae* or *Origines*, brought under one cover the teachings of the ancients on grammar, rhetoric, mathematics, medicine, and history. Isidore is sometimes portrayed as an illogical symbolist, but he clearly distinguished between the worlds of empirical and nonempirical reality. In the seventh century an efflorescence of scholarly activity took place in the British Isles, and Isidore's encyclopedic efforts were complemented by those of Bede (d. 735). Although Bede spent almost his entire life in the monasteries of Wearmouth and Jarrow, he was considered to be the most learned European of his time. His major interest was in church history. But he also made an important contribution to chronology, and, within his historical works, made scientific observations of often refreshing originality. For instance, he noted for the first time that tides had to be measured from individual ports. His *De natura rerum*, a revision of Isidore, was avidly studied at the School of Chartres in the twelfth century. His *De temporibus* and *De temporum ratione* remained standard studies of dating and time reckoning for centuries. Largely owing to Bede's influence, events began to be dated from the birth of Christ.

The scientific activities of Bede's age formed part of a widespread missionary program. The spearheads of the movement were the monasteries. Here many texts of Roman science that might otherwise have been lost were assiduously copied and protected. Monks also furthered the goals of literacy in an oral-aural world. Their missionary activity spread in different directions. Under Columbanus (d. ca. 615), Irish monks, using an indigenous *Rule,* began to establish centers in Scotland and France. The clear Irish uncial script offered a model to be imitated on the Continent, and Irish monks were responsible for two of the seventh century's most influential abbeys, Bobbio and St. Gall. At the suggestion of Gregory the Great, Augustine, then prior of St. Andrews in Rome, set out for England in 597. By the time Theodore of Tarsus, first archbishop of Canterbury, died in 690, the rudiments of the English Church, and therefore of an educational system based on writing, were organized. Benedict Biscop, who also died in 690, founded Wearmouth and Jarrow, and the Irish, who had been active in Cornwall and Wales, founded an abbey at Malmesbury. In 635 Lindisfarne became a missionary outpost, and in 664 the Benedictine *Rule* was laid down at York, where the famous cathedral school began to flourish in 735. Continental centers also emerged. Reichenau, situated on Lake Constance, was founded in 724, and the extensive travels of Boniface bore fruit in the abbey of Fulda, headed by his favorite pupil, Sturmi, from 744. However, the character of monastic activity should not be mistaken. Isolated incidents like the accusation of Vergil of Salzburg in the eighth century of possessing heretical scientific books do not indicate a rebirth of science itself. Despite the fusion of pagan and Christian ideas that took place at every interface of the oral and written cultures, the monastic tradition remained an essentially passive one, transmitting rather than seriously studying scientific texts.

A true revival of the liberal arts in the West did not become a possibility until the Carolingian Renaissance, which patronized both sacred and profane subjects anew in an attempt to raise the standards of education throughout the newly organized empire. Under Charlemagne, a leader of genius, and Alcuin, the gifted head of his palace school from 782, a brilliant generation of poets, philosophers, and prelates appeared. Charlemagne's capital, Aix-la-Chapelle, laid down norms that were universally emulated. But Alcuin was a great teacher rather than an original mind, and, as far as science was concerned, the Renaissance turned out to be less a revival of classical antiquity than an outgrowth of late antique Christianity. Although the quadrivium was taught, the purpose was not to innovate but to simplify and explain what was known. The attitude toward pagan knowledge merely repeated the combination of admiration and distrust so characteristic of the Fathers. The *Hortulus* of Theodulph of Orleans took a sincere interest in botany, but it was less typical of the period than the unoriginal encyclopedia of Rabanus Maurus. Even John Scotus Eriugena, the one truly great mind of the age, was largely indifferent to science. What Duhem thought so original, his statement that Jupiter, Mars, Venus, and Mercury circle directly around the sun, was probably based on an ignorance of Ptolemy, not on observation or calculation.[51] His idea of nature arose not out of physics but out of theology; it was an extension of similar notions in Gregory of Nyssa, Maximus the Confessor, and the pseudo-Dionysius. The most permanent contributions to scientific learning were indirect: the habit of copying and exchanging texts to improve their accuracy, a major concern of Lupus of Ferrières, and the invention of the Caroline minuscule hand, which provided a standard of precise communication that was imitated, not always successfully, down to the Renaissance.

The Carolingian period was followed by a century or more of political and administrative confusion. Yet it is a short step from the textual refinement of its most typical document, the *polyptique*, to the renewed interest in geometry and measurement in the tenth and eleventh centuries. The revival is signaled by the *Computus* of Helperic of St. Gall, an elementary set of tables for astronomical and calendrical calculations which was much admired by subsequent writers. About the same time there was a series of isolated attempts at quantification, culminating in Walcher's measurement of the longitudinal differences separating Italy and England by means of the lunar eclipse of 19 October 1091. In the eleventh century, partly, as Gerbert himself demonstrated, through the rediscovery of the *Corpus agrimensorum*, a new beginning was made in the application of geometry to field surveying. This science seems to have been taught at Liège or Cologne. Around 1050 a certain Franco of Liège composed an elementary *De quadratura circuli*. Ignorant of *Pythagoras's Theorem*, he solved his problem by piecing together bits of parchment.[52] He had apparently read a series of letters between two students at Chartres, Raimbold and Raoul, debating whether the sum of the internal angles of a triangle equaled two right angles. This unsophisticated correspondence, which was copied elsewhere, provided the first example for many centuries in the West of the simultaneous investigation of a scientific question by different parties in contact with one another. Other treatises began to appear, like the *Second Geometry* attributed throughout the Middle Ages to Boethius, but actually composed by an eleventh-century student somewhere in Lorraine. A compilation of Euclid, lessons on the abacus, and the *Corpus agrimensorum*, it provided the earliest Latin occurrence of Arabic numerals.[53] But interest in mathematization was not limited to the North. From the tenth century Muslim Spain provided southern centers with treatises on the quadrant with cursor derived from Arabic originals.[54] In a more general sense, the problem-solving spirit also began to make a more frequent appearance in man's imaginative life. In the eleventh-century romance *Ruodlieb*, one of the hero's tests was an energetic game of chess with the king in whose palace he was a guest. On winning, he had a symbolic foretaste of success in an otherwise uncertain future. About the same time a new game, *Rithmomachia*, came into vogue, played on an oversized chessboard and based on the number theory of Nicomachus as transmitted by Boethius.

The new ways of thinking were epitomized by the intellectual development of a single figure, Gerbert of Aurillac (ca. 930–1003). A practicing mathematician who taught at Rheims from 972, and the first French pope, Silvester II, he exemplified the growth of scientific culture in the North and the larger role the hitherto amorphous Gallic world was to play in church government. In addition to solving a number of mathematical problems, Gerbert made contributions to three areas of great interest in the succeeding two centuries: the abacus, the algorism, and the astrolabe. Like Abbo of Fleury and Heriger of Laubach, his rough contemporaries, and Hermann Contractus, a half-century later, he wrote a treatise on the calculating instrument with which all Europe was then fascinated, and which did so much to transform commercial transactions up to the introduction of double-column accounting by Leonard of Pisa. The abacus was a polished board divided into thirty equal columns. Through a system of semicircular arcs groups of three columns were united, and thus, even using Roman numerals, the operations of addition, subtraction, and multiplication could be performed. Gerbert also realized the importance of Hindu-Arabic numerals (excepting the zero). Until the eighteenth century the common term for this new system of notation was algorism, which is derived from a corruption of al-Khwarizmi (that is, *algoritmi*) in Latin translation. In Gerbert and in Hermann Contractus, who knew of the astrolabe, the chilinder (a portable sundial), and the quadrant with cursor, one feels the pragmatic bent of the Western intellectual taking its first genuine steps toward science.

The Twelfth Century

The tenth and eleventh centuries not only terminated a phase of northern agrarian expansion; they also inaugurated its first industrial, commercial, and artistic revolution. In the wake of these changes, the great bulk of Greek and Arabic science, hitherto unknown, finally reached the West.

The twelfth-century renaissance, because it was both renaissance and reformation at once, was a many-sided affair. Its achievements were both religious and secular. The settling, however impermanently, of the investiture controversy, and the death of the reformist pope Gregory VII, had left his successor, Urban II, free to attempt to unify Christianity in the pursuit of new ideals. These were transformed into action in the First Crusade in 1096, which managed at once to promote the *nova religio* and to find an outlet for the aggressive tendencies of the North's far too numerous nobility. In modern economic terms the crusade could be described as Europe's first successful attempt to export her wars and unemployment. Under a protective, apparently progressive, and as yet unchallenged ideology of expansion, the militarism that conquered England in 1066, Sicily in 1091, and Muslim Spain by 1118 created Europe's first colonial dependencies in North Africa and the Near East. The new cause appealed not only to the nobility. Robert Guiscard and Raimond Bérenger le Vieux merely translated into *Realpolitik* the sentiments, passions, and unfulfilled dreams that rose from below. What prepared the way and politicized the countryside was a new force, the wandering preacher. Peter the Hermit, evangelist and prophet, inspired by omens, signs, and prognostications, had gone from village to overcrowded village with a single message: the heavenly and the earthly Jerusalem were not incompatible. It was essentially the same message to which Bernard of Clairvaux, preaching a half-century later, gave rhetoric, canonical authority, and ecclesiastical organization.

Where the crusaders went, the merchants followed. Furnishers of armaments, traders in raw materials and textiles, they crowded in after the troops, attracted by the pleasant prospect of a market that had been closed to them for centuries. While the nobility bullied each other, or grew impatient at the Arabs' refusal to yield before their God's inexorable will, the economic motivation of the crusades was gradually unfolded in the merchants' pragmatic and permanent implantation along the whole Syrian coast from Antioch to Jaffa. The Asian trade routes still eluded them, but the ports of call were within their grasp. Their success created immense pools of capital in the Italian cities and accelerated the growth of commerce throughout Europe. Long dependent on Byzantium to furnish via the Arabs the necessities of its economic life, the West could now depend on its own suppliers. Speculators, encouraged by customs privileges, made fortunes, and, after a lapse of several hundred years, European enterprises were once again able to finance their own growth from internal sources. With the towns as their nerve centers, Europe also awoke to a new intellectual universe in which ideas, like goods, began to flow easily across international boundaries. The same commerce that remonetized the economy established, for the first time since antiquity, a self-conscious community of intellectuals whose uninhibited communication with each other was the necessary condition for the advancement of learning. The essential feature of the new intellectual was that, like the urban craftsman, he was a specialist. In Rome the only real professionals had been lawyers; in Islam, lawyers, physicians, and theologians. Although the potential for greater specialization was present in Islam, in practice most scientific figures preferred some version of the Hellenistic universal education. With the rebirth of commercial life in the North, first in the cathedral schools and later in the universities, the sheer weight of traditional material began to force a disinterested progress in ideas through increasing abstraction, classification, and specialization of language. Tolerance helped. The towns, which had achieved autonomy after long struggles with local lords, and which within a century would be forced to submit to the centralizing authority of the monarch, provided, at this fortuitous moment, a haven of liberalism.

"At no point," wrote Haskins, "is the intellectual revival of the twelfth century more marked than in the domain of science."[55] But this distinguished historian saw the growth of science in terms of transfer, translation, and assimilation alone. He displaced the center of gravity from Europe itself to various points of contact with the Arab world: Syria, Sicily, Constantinople, and Toledo. It was a renaissance from outside. This view was partially corroborated by a large body of manuscript evidence. Of the new material, quite aside from the attitudes of the men who looked for it, the overwhelming amount came from translations. The Latin world was in some respects like seventh-century Islam, an empty goblet waiting to be filled with the ambrosia of Greek rationalism. But there is no further similarity. The Islamic assimilation of Greek science was so complete that Muslim thinkers are best viewed as an extension of Hellenistic education, which is how they viewed themselves. Conditions in northern Europe were different. The translations were invariably poorer than the Arabic versions, and their appearance followed no pattern. Often lesser authors like Abu Ma'shar were received with uncritical enthusiasm, while truly important men like al-Razi or al-Biruni remained unknown. Inevitably, the translations affected different disciplines in different ways. The clearest case for the Haskins thesis lies in the allied fields of mathematics and astronomy. Early in the twelfth century Euclid's *Elements, Data,* and *Optics* were made available. In 1126 Adelard of Bath translated al-Khwarizmi's trigonometry and astronomical tables, and in 1145 Robert of Chester latinized his *Algebra.* Together with the abacus, the algorism, and the renewed interest in uniform standards demanded by Roman law, these texts paved the way for a wide variety of scientific advances in the thirteenth century. By the 1160s, the astronomical tables of al-Battani and al-Zarqali were known, and the *Almagest* itself appeared in translation from the Greek in 1160 and from the Arabic in 1175. In these areas light clearly came from the East.

In the other sciences—music, geometry, geography, and medicine— the debt to the Arabs was intermingled with the changing attitudes of the Latin world itself. A good example is physical theory with its evident links with music and medicine. Aristotle's *Physics* was not available *in toto* until around 1200, but by this time problems were already orienting themselves around the issues that were to dominate the following two centuries. Fragments of Aristotle's *Libri naturales* had been made available through medical writers, and Calcidius, the standard gloss on Plato, offered a lengthy review of classical theories of matter. With these rough guides, twelfth-century minds began to rethink physical theory along new lines. William of Conches, who in- herited from Galen the Stoic conception of matter, used it to implant in the four elements the notions of activity and passivity, mixing, ten- sion, and rational causality. He also limited his treatment to the empirically definable world. His words were echoed by Thierry of Chartres and Bernard Silvester, each of whom incorporated his physi- cal notions into literary works. Thierry wrote a commentary on Gen- esis, in the opening paragraph of which he declared his scientific in- terests: "I propose to elucidate, according to the letter and to physics, the first part of Genesis, which treats the seven days and the distinct activities of the six kinds of works. . . . I shall wholly pass over both the moral and the allegorical reading, which the holy commentators have clearly dealt with exhaustively. . . . [For] the utility of this book is a knowledge of God by means of the things he has made."[56] Thierry's physics, like William's, revealed a strong Stoic and Hermetic influence, and suggested that the universe, as created by God, was composed essentially of matter and number. In his *Cosmographia*, Bernard Silvester, commenting on the same question, transformed the Calcidean doctrine of *hyle/silva* into a theory of material change which similarly accounted for the inner formal and the outer tangible shape of things. William, Thierry, and Bernard each returned in his way to Plato's problem of saving the phenomena, but with a new twist. It was no longer the model that accounted for appearances but the living world that saved the model.

The twelfth-century admixture of tradition and innovation can easily be discerned in the career of the earliest representative of the "new science," Adelard of Bath. Born in England, a student at Tours and an instructor at Laon, Adelard mastered the learning of the cathedral schools and then set off in search of Arabic texts, certainly as far as Sicily and perhaps to the Near East itself. Haskins credited him with an original spirit of inquiry, a secular, rationalistic habit of mind, and a capacity for experiment and observation.[57] In reality Adelard transferred to the West mathematical and astronomical knowledge that he did not fully assimilate. Although he often spoke of relying on his senses, in practice he remained a humanist who derived his information from traditional authorities. To his translations of Euclid, al-Khwarizmi, and the Arab astrologers one must inevitably compare his treatise on falconry, which, by Haskins' own observation "ignores eastern experience and concerns itself chiefly with the old English recipes for the diseases of hawks."[58] Nor was there a radical break with tradition in either of his own writings, the *De eodem et diverso* or the *Questiones naturales*. The first concerned itself with the Platonic question of identity and difference; the second was an attempt to introduce a heterogeneous range of natural-philosophic subjects into the older format of Seneca. Neither is written as a new type of discourse. The *De eodem* is a *prosimetrum*, a medieval form that goes back to Boethius. The *Questiones* is a dialogue between an inquisitive uncle and a conservative nephew. The explanation of how to determine the height of a tower with which *De eodem* concludes hardly conceals the fact that its major theme is the common twelfth-century debate on whether reality is what the senses reveal or the hidden formal qualities perceptible only to the mind. Philocosmia, who takes the former position, eventually yields to Philosophia—not, from a scientific viewpoint, much of an advance. In the *Questiones naturales* allegory is abandoned for a more systematic approach, and on some issues, like the *aplanos*, the outer rim of the firmament, Adelard moves out of the older Platonism with genuine audacity. But he is less original on a typical question like number 66 on the nature of thunder and lightening. Unaware of Aristotle's bookish treatment, which had elicited the contempt of al-Kindi, he merely sharpens the answer of Agobard (d. 840), who dismissed superstition in favor of harmonizing the biblical accounts to eliminate their contradictions.[59]

The most renowned center of indigenous scientific activity from the mid-eleventh to the mid-twelfth century was the school at Chartres. The qualities for which Chartres is best known—philosophical naturalism, the accurate reading of ancient texts, and the refusal to compromise the humanistic ideal of a complete education for the onesidedness of logic—are also found at Tours, Orleans, and Paris. But none of these centers were able to rival the fame of Chartres. A continuous tradition of distinguished minds began with Fulbert (d. 1028), who was both bishop and founder. It was carried on by Bernard of Chartres, his putative brother Thierry, and William of Conches. In the twelfth century it included such well-known fellow travelers as Adelard of Bath, Bernard Silvester, Hermann of Carinthia, and John of Salisbury. The most influential master, if influence is measured by the number of extant copies of his works, was Gilbert of Poitiers.[60] The aura of Platonism in which historians have enveloped the school has perhaps obscured its real contribution to science, which lay as much in method as in doctrine. Certainly Plato played an important role. In the *Heptateuchon*, the school's charter, Thierry wrote that an attempt was made to unite *scientia* and *eloquentia, ratio* and *verbum*. But that was not all Chartres achieved. It really adapted the cathedral school to a new set of cultural ideals. As formulated by Bernard they were essentially three in number: the defense of secular studies as an integral part of Christian education, the delimitation of the knowable to disciplines understandable through reason alone, and the reintroduction of the idea of nature as a consistent system obeying a set of verifiable laws. These principles had been laid down by Cassiodorus, Boethius, and Bede. In Bernard, whose unfailing standards and personal warmth radiated through an entire generation of students, they received new meaning and commitment. They also stood for freedom of thought at a time when more authoritarian voices were taking the opposite view. Although vindicating the best traditions of humanism, Chartres acquired a reputation for permitting the study of controversial subjects. Bernard Silvester and Hermann of Carinthia dedicated scientific treatises to Thierry, and William of Conches's youthful rationalism became a target of the Cistercians.

During the first half of the twelfth century, partly owing to the activity at Chartres but common to many other centers, especially Paris, a number of ideas destined to play a large role in the subsequent history of medieval science made their appearance. One was the eternity of the world, which entered twelfth-century thought from different directions. For the translators of Ptolemy it was an assumption, as it was for the numerous *medici* and *physici* descended indirectly from Galen. It was widely popularized by the Cathars and promoted by the pantheists David of Dinant and Amaury of Bène. It received an immense impetus when the first translations of Aristotle began to filter into learned circles, and was condemned along with other natural-philosophic doctrines of the Stagirite in 1210. But it also made a subtle accommodation with Christianity. After all, it had been a tenet of Boethius, as a lengthy twelfth-century commentary on the *Consolation* pointed out;[61] and it was compatible with God's creation of the universe, provided that afterward the world was assumed to continue running according to natural law. This was essentially the position of the Chartrain cosmologists and of the audacious Honorius of Autun. Through their use of euphemisms like *perpetuus* and *sempiternus* for the absolute *eternus*, which was reserved for God, they helped to make the uncompromising orthodox conception a minority view.

But the world's eternity was indirectly related to a deeper issue: whether the universe could be interpreted, following Plato, as a story, or whether a scientific understanding would not require that it be analysed as a dynamic state, independent of the past. In the end the latter won. After the mid-century there was a parting of the ways between religion, committed to a diachronic view of reality, and natural philosophy, which gradually adopted a synchronic one. Typical was the fate of Macrobius. Battered from the one side by truly original astronomers like Raymond of Marseilles, who had access to the Arabs, he was also subjected to a series of devastating critiques from within the Latin World.[62] In rejecting a long-established, traditional mode of thought, much beneficent humanism was lost, but science finally achieved independence from the literary disciplines and took a decisive step toward mathematization. Historians have viewed this transition as the decline of Plato's influence and the rise of Aristotle's. But much more was at stake. Between the Platonism of the twelfth century and the Aristotelianism of the thirteenth there is no abrupt transition.[63] The specific problems remain the same. But the individual *scientiae* begin to find their own languages and internal relations; they break away from the encyclopedic tradition and become autonomous disciplines. Bernard Silvester, in purely scientific terms a minor thinker, mirrors this change dramatically in his *Cosmographia*, in which the *dramatis personae* have ceased to be mere abstractions and have begun to represent university subjects.

Many new attitudes crystallized around a single conception, *natura*, which made its first serious reappearance since antiquity. But it was much changed. Most remarkable was its sudden proliferation into a number of areas at once: lyric poetry, biblical exegesis, political theory, the *roman courtois*, cathedral sculpture, and allegory. In cosmology Nature had become God's vicegerent who transformed his divine plan into reality and presided over the creation of the macrocosm and microcosm. The doctrine of the absolute transcendence of God in Augustine was thus modified by an intermediary that symbolized the laws of the sublunary world. But the notion would not have made so widespread an appearance if the natural, in both theory and practice, had not invaded so many areas of life. The most important starting point was law. The reintroduction of Roman law by Irnerius of

Bologna and the compilation of the *Decretum* by Gratian placed in the hands of lawyers, theologians, and administrators a powerful system of analysis with the idea of nature at its core. The relation between natural philosophy and law was direct: a standard gloss on *natura* in the *Decretum* cited the definition popularized at Chartres, namely, that nature consisted of a power inherent in things possessing the capacity to create like from like.[64] From law and cosmology it was a short step to aesthetics. "For just as a craftsman wishing to fashion something first arranges it in his mind, then, after seeking out the materials, creates it according to his mental image, so the creator, before making anything, framed it in his mind and afterwards filled it out as a work."[65] In making man, God was an artisan and an engineer; in imitating nature, man did the same. For nature, by the twelfth century, was no longer a primitive force to be shunned as the Germanic tribesmen had once feared the great forests. The settling of the countryside and the evident success of new techniques had suggested that nature, if still at times a redoubtable enemy, could on occasion be subdued and put to work in man's service. The growth of institutions depending on literacy had also given men a new faith in *ordo* and *ratio*. The universe seemed to be an organized entity: perhaps it was.

The idea of nature, as both a formal and phenomenal reality, is also the key to the twelfth-century criticism of the ancients. The period's authors are often pictured in Bernard of Chartres's metaphor as dwarfs standing on the shoulders of giants.[66] It is less often pointed out where they saw further. One area was the reassessment of the value of the evidence of the senses. The Greeks, by and large, had relegated the senses to second place. The Arabs, more practical in bent, trusted what they saw and could measure, even when philosophy taught them differently. Twelfth-century writers, who, like the Arabs, inherited a platonized Aristotelian universe, tried to introduce an earthly, tangible notion of empirical reality. Paradoxically, although William of Conches, Thierry of Chartres, and later Daniel of Morley are all, in some sense, commenting on Plato, they affirm that the starting point for understanding the real world is its concrete manifestation to all five senses. Their position also accorded well with the nominalism of Abelard. As a criticism of the ancients, this rough blend of empiricism and pragmatism bore no immediate fruit. In the Aristotelian revival of the thirteenth century the reality of the world revealed by the senses once again retreated before the abstract question of universals. But the twelfth-century position reappeared in Roger Bacon's distrust of Aristotle, in Grosseteste's methods and, along with

Alhazen's theory of vision, contributed to the cognitive empiricism of
the fourteenth century. If the fourteenth, not the thirteenth, century is
seen as the high point of medieval philosophy, then its natural predecessor is perhaps the twelfth.

The reassessment of phenomenal reality went hand in hand with
the growth of interest in the mechanical arts. The twelfth century
stands midway between the contempt for the *artes mechanicae* in antiquity and their full acceptance in the Renaissance. The author who
most clearly enunciated the new doctrine was Hugh of St. Victor. In
his *Didascalicon*, written in the 1120s, he divided philosophy into the
theoretical, the practical, the mechanical, and the logical.[67] His radical simplification of the Boethian classification elevated technology to
equal status among the sciences. Hugh said little that was new about
the mechanical arts, which he enumerated as weaving, weaponry, commerce, agriculture, hunting, medicine, and theatricals. The significance
arose from the space he devoted to them in a treatise on theoretical
education. Like Theophilus and monastic theologians of work, he
accounted for the rise of the mechanical arts through man's desire to
redeem his fallen nature by active labor. The mechanical arts, even if
adulterine, could work as a potent force in the pursuit of virtue, and
could help to reveal that aspect of divinity which man reflected on
earth. "Indeed, man's reasoning shines forth much more brilliantly in
inventing these very things than ever it would have had man naturally
possessed them."[68] In contrast to Augustine, who regarded curiosity
as a kind of pride, Hugh suggested that the inquiry into nature was
an important stage in man's intellectual development. Hugh's classification was reproduced in an early *Ysagoge in theologiam*, in an exposition of Martianus Capella, and in the *Aeneid* commentary attributed to Bernard Silvester.[69] His theory also accorded well with monastic practice. In a manuscript of Reun, Austria, a Cistercian house, an
illustration celebrated in one place hunting, painting, cooking, and
praying.[70]

In reevaluating the mechanical arts, Hugh, like other twelfth-century authors, was only bringing inherited theory into line with ac-
cumulating practice. The novelty did not arise from any one discipline
but from a changed habit of mind. The era of pure theory had come
to an end; the normative, the experiential, the rule of thumb had
emerged as legitimate guides to action. The term *ingeniator*, engineer,
first appeared in 1086,[71] and within a century the cathedral builders
had left all over Europe indelible traces of man's new capacity to control nature. But the growth of practice was also noticeable elsewhere.
The bookish geography of the scholar and churchman, drawn largely

from the stylized *mappaemundi* of the handbooks, gradually yielded to the more informed travel experience of the merchant, the pilgrim, and the soldier.[72] In music, attempts were made from the eleventh century to adapt the monochordic system of Boethius to the needs of liturgical plainsong. By the mid-twelfth century an obscure monk called Theophilus had provided the West with a full-fledged technical treatise including patient descriptions of how to build apparatus that were later admired by Cornelius Agrippa. But the two most influential areas were law and pastoral theology. In England the common law provided Glanvil with one model based on custom, precedent, and experience; on the continent, Roman and canon law created a whole new body of exact information that had to be readapted continually to changing circumstances. The twelfth century also saw the first efforts at standardizing confession, penance, and pastoral care, and the century's most influential religious leader, Bernard of Clairvaux, never tired of repeating the value of interrelating the active and contemplative life. If Benedict had rejected the aristocratic ideal of *otium*, the leisured study so dear to classical antiquity, Bernard gave the idea a new twist in stating that "the leisure of wisdom (*sapientiae otia*) is a time of activity; and the more leisured wisdom, the more preoccupied. . . ."[73] The practical spirit also invaded medicine. From 1137, when Adalbert, archbishhop of Mainz, went to Montpellier to study at Europe's first university clinic, new life was breathed into Avicenna's dictum on the value of an intimate relation between doctor and patient.

But science, if it is to advance, cannot live by attitudes alone; it must also have a method. Ideas must be sorted out, the useful separated from the otiose. They must be classified according to their logical properties. The twelfth-century sorting system was the scholastic method, which was essentially a means of comparing different approaches to a question so as to elicit a logically correct choice. Elements of it appeared as early as Eriugena, and refinements of its technical vocabulary did not take place until the thirteenth century, but its fundamental principles were all laid down between the schools of Anselm of Laon and Peter Lombard. In logic the most important contribution was made by Peter Abelard (1079–1142). He attempted to answer three questions from Porphyry's *Isagoge*: Do universals exist? Are they corporeal or incorporeal? And are they part of the sensible world or not? His answer in each case was both yes and no. Genera and species, he argued, are only products of the mind, but they also signify particular things in the real world. Universals, inasmuch as they are nouns, are uttered words and therefore corporeal, but their capacity to signify groups of particulars is not. Incorporeal things are of two kinds, those within the sensible world, like the forms of bodies, and those outside it, like God and the soul.[74] Abelard had not really solved the problem of reconciling Plato and Aristotle, but he started philosophy, and with it science, along a labyrinth that would take at least two centuries to unravel. In Abelard's linguistic philosophy the Western intellectual tradition suddenly grew to full maturity. Abelard was someone who could be set beside the two other great twelfth-century minds, Averroes and Moses Maimonides. Perhaps for this very reason the men of his own age and of ours have found it difficult to remain dispassionate in their estimation of his work. At once arrogantly brilliant and optimistic, but unsure of his way, he was too large and unconventional a figure for his age. What spread its shadow across the following centuries was not the arid achievement of his logic, but the simple anguish of his conscience, which reappeared, little changed, in Galileo and Giordano Bruno.

Conclusion

In the West, the Middle Ages was a period of transition, but not only, as the medievals themselves would have us believe, from ancient to modern, from Greek science to a Greek renaissance. It was a more subtle metamorphosis from a Mediterranean to a northern civilization in which, inevitably, the institutions of science underwent a profound change. Like the early Greek world, the medieval West passed from technology to science, and, like Islam, from translating a heritage to indigenous creation. Yet unlike the Mediterranean, in which technology preceded science and for social reasons remained relatively isolated from it, the West experienced both transformations, if not at once, at least closely enough together for the one to have influenced the other. If the documentary evidence is to be trusted, the two rarely if ever met before Theophilus, and even afterward relations were the exception rather than the rule. But if Latin science is assessed at the third quarter of the twelfth century, it is clear that an indirect cross-fertilization has taken place. There is a new native empiricism, a widespread interest in theory and practice, and a tacit acceptance of the link between technical innovation and economic progress. In Alexander Neckam, whose *De rerum natura* mentions the magnetic compass, and who was pragmatic enough to write a treatise on *The Names of Utensils*, we have a foreshadowing of the practical bent that is a common feature in Jordanus, Leonard of Pisa, and Roger Bacon. Ahead lie the triumphs of Aristotelian natural philosophy and, in the

Renaissance, the discovery of the scientific method. But even here the connection with the earlier, less well articulated goals of medieval science can be demonstrated. The medievals would have regarded the trinity of religion, economics, and science on which the modern world is founded as an unholy one, but the forces they unleashed from below contributed decisively to its success.

Notes

1. 3d ed. (New York, 1890), 1:185. Originally published in 1837.
2. Lynn White Jr., *Dynamo and Virgin Reconsidered* (Cambridge, Mass., 1968), pp. 70–71.
3. A. Gurswitch, *Phenomenology and the Theory of Science*, ed. L. Embree (Evanston, 1974), p. 34.
4. Gordon Childe, *What Happened in History*, with a new foreword by G. Clark (Harmondsworth, 1964), p. 31.
5. S. Sambursky, *The Physical World of Late Antiquity* (London, 1956), p. xi.
6. Benjamin Farrington, *Greek Science: Its Meaning for Us* (Harmondsworth, 1966), p. 197.
7. H. I. Marrou, *A History of Education in Antiquity*, trans. G. Lamb (New York, 1964), pp. 419–21.
8. The following section is intended chiefly to be an introduction to the historical context of Islamic science. For a more detailed treatment, see the other chapters of this book. The author is indebted to Prof. M. E. Wickens for a number of valuable comments.
9. *Medieval Cities*, trans. F. D. Halsey (New York, n.d.), p. 25.
10. *Legacy of Islam*, ed. T. Arnold and A. Guillaume (Oxford, 1931), p. 378; and *A History of Astronomy from Thales to Kepler* (Cambridge, 1906), p. 249. This view is not repeated in the second edition of the *Legacy of Islam*, ed. Joseph Schacht and C. E. Bosworth (Oxford, 1974).
11. Maurice Lombard, *Espaces et réseaux du haut moyen âge* (Paris and The Hague, 1972), p. 63.
12. Richard Walzer, *Greek into Arabic: Essays on Islamic Philosophy* (Oxford, 1962), p. 8.
13. Ibid., pp. 236–37.
14. Respectively, *De Ortu Scientiarum* 1.1, 1.2 (*scientia mensurandi*), 1.3 (*astronomia*), 1.4 (*musica*), ed. C. Baeumker, *Beiträge zur Geschichte der Philosophie des Mittelalters*, vol. 19, pt. 3 (Münster, 1916), pp. 18–20.
15. Ibid., 1.6 (theology), 2.2 (the trivium) and 4 (the four elements): respectively, pp. 21–22, p. 22, and pp. 23–24.
16. Paul Kraus, "Studien zu Jabir ibn Hayyan," *Isis* 15 (1931):11–14.
17. Ibid., pp. 12–20.
18. Paris, Bibl. Nationale, MS Lat. 13,951, fol. 13v.
19. Quoted in John Murdoch, "Euclid: Transmission of the Elements," *Dictionary of Scientific Biography*, 4:438.
20. David Eugene Smith and L. C. Karpinski, *The Hindu-Arabic Numerals* (Boston, 1911), p. 99.

21. Quoted in Seyyed Hossein Nasr, *Science and Civilization in Islam* (Cambridge, Mass., 1968), p. 136.

22. A. I. Bey, *Histoire des bimaristans (Hôpitaux) à l'époque islamique* (Cairo, 1928), pp. 9, 12, 13–15.

23. Cyril Elgood, *A Medical History of Persia and the Eastern Caliphate* (Cambridge, 1951), p. 202.

24. *Liber Canonis* (Venice, 1555), book 1, preface, fol. 3r; book 2, fen 3, chap. 1, fol. 56r.

25. A. I. Sabra, "Ibn al-Haytham," *Dictionary of Scientific Biography*, 6:190–91. On the relation of Alhazen's theory to ancient ideas, see chap. 10, below.

26. *De Architectura* 1.1.1.

27. A. G. Drachmann, *The Mechanical Technology of Greek and Roman Antiquity* (Copenhagen, 1963), p. 311.

28. K. D. White, *Roman Farming* (London, 1970), pp. 450–54.

29. M. I. Finley, "Technical Innovation and Economic Progress in the Ancient World," *Economic History Review* 18 (1965):44.

30. A. M. Watson, "The Arab Agricultural Revolution and Its Diffusion," *The Journal of Economic History* 34 (1974):16.

31. Ibid., p. 9.

32. Ibid., pp. 27–31.

33. Lynn White, Jr., *Medieval Technology and Social Change* (Oxford, 1962), pp. 1–2.

34. Ibid., p. 27.

35. R. J. Forbes, "Metallurgy," in *A History of Technology*, ed. Charles Singer et al., vol. 2 (Oxford, 1957), p. 63.

36. *La civilisation mérovingienne*, pt. 3 (Paris, 1957), p. 107.

37. *Lex Burgundiorum* 21.2; *MGH Leges* 2.1, p. 60. *Lex Salica* 11.2; ibid., 4.2, p. 49.

38. A. J. Raftis, "Western Monasticism and Economic Organization," *Comparative Studies in Society and History* 3 (1961):456.

39. Bertrand Gille, "The Medieval Age of the West (Fifth Century to 1350)," in *A History of Technology and Invention*, ed. M. Daumas, vol. 1 (New York, 1969), p. 426.

40. K. D. White, *Agricultural Implements of the Roman World* (Cambridge, 1967), p. 126.

41. Georges Duby, *Rural Economy and Country Life in the Medieval West*, trans. C. Postan (Columbia, S.C., 1968), pp. 99–103.

42. R. J. Forbes, "Power," in *History of Technology*, ed. Singer, 2: 610–11.

43. Marc Bloch, "The Advent and Triumph of the Watermill," in *Land and Work in Mediaeval Europe*, trans. J. E. Anderson (London, 1967), pp. 152–58.

44. Joseph Needham, *The Grand Titration* (Toronto, 1969), p. 76.

45. *Regula* 48.

46. *Movimenti religiosi e sette ereticali nella società medievale italiana, secoli XI–XIV* (Florence, 1922), p. 37.

47. *Descriptio . . . monasterii clarae-vallensis*, in J.-P. Migne, ed., *Patrologiae cursus completus, series Latina*, vol. 185 (Paris, 1855), cols. 570–72.

48. Forbes, "Power," p. 610.
49. S. D. Goitein, *A Mediterranean Society*, vol. 1 (Berkeley, 1967), p. 66.
50. E. Boileau, *Livre des métiers*, ed. R. de Lespinasse and F. Bonnardot (Paris, 1879), p. iv; see Jacques Le Goff, *Pour un autre Moyen Age* (Paris, 1977), pp. 46–79, 108–30.
51. J. L. E. Dreyer, "Medieval Astronomy," in *Studies in the History and Method of Science*, ed. Charles Singer, vol. 2 (Oxford, 1921), p. 105.
52. Paul Tannery, "La géometrie au XIe siècle," *Mémoires scientifiques*, ed. J. L. Heiberg, vol. 5 (Paris, 1922), p. 87.
53. Menso Folkerts, *"Boethius" Geometrie II* (Wiesbaden, 1970), pp. ix–x.
54. J. Millás Vallicrosa, "La introducción del cuadrante con cursor en Europa," *Isis* 17 (1932):228. Cf. Emannuel Poulle, "Les instruments astronomiques de l'Occident latin aux XIe et XIIe siècles," *Cahiers de civilisation médiévale* 15 (1972):27–28.
55. *The Renaissance of the Twelfth Century* (Cambridge, Mass., 1927), p. 303.
56. *De septem diebus*, chap. 1; trans. Brian Stock, *Myth and Science in the Twelfth Century* (Princeton, 1972), pp. 240–41.
57. *Studies in the History of Mediaeval Science* (Cambridge, Mass., 1927), p. 42.
58. Ibid., p. 347.
59. *De grandine et tonitruis*, in Migne, *Patrologia Latina*, vol. 104 (Paris, 1951), cols. 147ff; *Questiones naturales* 46, ed. M. Müller, *Beiträge zur Geschichte der Philosophie und Theologie des Mittelalters*, vol. 31, pt. 2 (Münster, 1934), p. 60.
60. N. Häring, "Paris and Chartres Revisited," in *Essays in Honour of A. C. Pegis*, ed. J. R. O'Donnell (Toronto, 1974), pp. 299–313, 329.
61. *Saeculi Noni Auctoris in Boetii Consolationem Philosophiae Commentarius*, ed. Edmund T. Silk (Rome, 1935), p. 159.
62. Haskins, *Studies*, pp. 96–103.
63. Tullio Gregory, *Anima mundi: La filosofia di Guglielmo di Conches e la scuola di Chartres* (Florence, 1955), p. 263.
64. *Glossa ordinaria decreti*, ad dist. 1, chap. 7, s.v. *natura*.
65. William of Conches, *Glosae super Platonem*, ed. Edouard Jeauneau (Paris, 1965), chap. 32, p. 99.
66. Edouard Jeauneau, " 'Nani gigantum humeris insidentes.' Essai d'interprétation de Bernard de Chartres," *Vivarium* 5 (1967): 79–99.
67. *Didascalicon* 2.1 and 3.1, ed. C. H. Buttimer (Washington, D.C., 1939), pp. 24, 48.
68. 1.9, trans. Jerome Taylor (New York, 1961), p. 56.
69. Stock, *Myth and Science*, pp. 36–37, n. 42.
70. Vienna, Nationalbibliothek, MS lat. 507, fol. 2v.
71. R. Latham, *Revised Medieval Latin Word-List* (London, 1965), p. 249, s.v. *ingenium*.
72. J. K. Wright, *The Geographical Lore of the Time of the Crusades* (New York, 1925), p. 3.

73. *Sermones in Cantica Canticorum* 85.3.8, ed. J. Leclercq, C. H. Talbot, and H. M. Rochais (Rome, 1958), p. 312.

74. Etienne Gilson, *History of Christian Philosophy in the Middle Ages* (London, 1955), pp. 155, 159.

David C.
Lindberg

**The Transmission of
Greek and Arabic
Learning to the West**

**The Separation
between East
and West**

In the Roman world there was a certain appreciation of Greek philosophy and science, but little
need for translations from Greek to Latin, since the
Roman scholar was generally able to read Greek.[1]
Nevertheless, a few translations were executed: for
example, Cicero (106–43 B.C.), a well-trained Hellenist, translated at least the first third of Plato's
Timaeus and the *Phaenomena* (a poem dealing with
astronomical phenomena and the weather) of
Aratus of Soli. But of greater significance than the
translation of Greek books was the adaptation of
Greek learning. There developed in Latin a thin
body of knowledge based on Greek sources—evident in such works as the encyclopedias of Varro,
Celsus, and Pliny and the slightly more specialized
writings of Lucretius, Vitruvius, and Seneca.[2] However, with the gradual collapse of Roman civilization, knowledge of Greek became ever scarcer, and
the Latin-speaking world was increasingly limited to
the meager sources already in its possession. Those
few who were still able to read or translate Greek
philosophical and scientific classics generally had no
interest in doing so. The learning of East and West
had been severed.

The problem facing us in this chapter is thus
posed. The philosophical and scientific culture of
western Europe in the twelfth to fifteenth centuries
was not created out of its own limited resources, but
resulted from the reintroduction of Greek learning
(with Islamic additions and modifications) into a
Christian theological tradition that had flourished
for a thousand years with only a minimum of outside interference. If, therefore, we hope to gain a

full appreciation of the medieval scientific achievement, we must examine closely the process by which Greek and Islamic learning were transmitted to the West.[3] Many scholars have applied their talents to the investigation of specific personalities and events in this process, but thus far there have been few attempts to write its history in general terms.[4] Perhaps the time is not yet ripe for such a history, but I hope, by surveying the specific findings of a host of scholars, to shed some light on the broader features of the transmission movement.

Before the knowledge of Greek disappeared from the western half of the disintegrating Roman Empire, a few books of considerable significance passed from Greek to Latin. Among the most crucial for determining the philosophical outlook of the early Middle Ages was the translation by Calcidius (fourth century?) of the first half of Plato's *Timaeus*; this translation, along with Calcidius's commentary on it, gave the Middle Ages its most systematic account of natural philosophy until the further translations of the twelfth century.[5]

Of equal importance were the translations of Anicius Manlius Severinus Boethius (ca. 480–524), Roman consul and advisor of Theodoric, the Ostrogothic King of Italy.[6] Boethius hoped to make available in Latin as many of the works of Aristotle and Plato as he could obtain and to write a handbook for each of the quadrivial disciplines —arithmetic, music, geometry, and astronomy. Boethius offended Theodoric and was put to death before he could complete his plan, but he did succeed in translating a very substantial portion of the Aristotelian logical corpus. His well-authenticated logical translations include Aristotle's *Categories, On Interpretation, Prior Analytics, Topics*, and *Sophistical Refutations*, and Porphyry's *Isagoge*. He also wrote handbooks on arithmetic and music, which still survive, and seems to have translated at least part of Euclid's *Elements*.[7] But Boethius's principal contribution was in the realm of logic; he gave the Middle Ages a substantial logical corpus and just about all the Aristotle they would have until the twelfth century.

Boethius's contemporary and friend Cassiodorus (ca. 480–ca. 575) was also devoted to the preservation of ancient secular learning, and his attitude toward it was more typical of the early Christian Middle Ages than was that of Boethius: whereas Boethius seems to have valued secular learning for its own sake, Cassiodorus valued it as the handmaiden of biblical studies.[8] Cassiodorus, like Boethius, was a civil servant in the regime of Theodoric; with the reconquest of Italy by the Byzantine emperor Justinian, he retired from public life and founded a monastery, Vivarium, near Squillace in Calabria. There he established a *scriptorium* and made the copying of books an essential part of monastic life; the Bible and the writings of the church fathers received primary attention, but secular literature was also valued and copied. In his handbook written for the monks at Vivarium, *Institutiones divinarum et humanarum litterarum*, Cassiodorus devoted most of his space to sacred literature, but he did offer a brief discussion of each of the seven liberal arts and an account of the available sources, some of which were to be found at Vivarium.[9] Cassiodorus should perhaps be viewed principally as a compiler, but the tradition of monastic scholarship that he established at Vivarium was to have a lasting influence on medieval learning.[10]

With the death of Boethius in 524, the translation of philosophical and scientific works from Greek to Latin virtually ceased for a period of six hundred years. The exception to this generalization is the field of medicine, where a group of important translations was executed in Italy, especially Ravenna, between the fifth and seventh centuries.[11] It appears that a corpus of Galenic and Hippocratic writings was rendered into Latin by a group of physician-translators—such works as the *Aphorisms, Prognostica*, and *Regimen* of the Hippocratic corpus; Galen's *De methodo medendi*; and commentaries on a number of Galenic works. During the same period there were also translations of medical works by Oribasius, Alexander of Tralles, and Rufus of Ephesus.

Italy always had enclaves of Greeks, among whom there was the possibility of translating activity, but in northern Europe knowledge of Greek became rare indeed.[12] Knowledge of the Greek alphabet and the ability to employ a few Greek phrases may have been rather common among scholars and churchmen, but competent Hellenists were very scarce. One such person was John Scotus Eriugena (fl. 850), an Irish scholar who taught at Laon. At the request of the Carolingian king Charles the Bald, he produced a competent Latin translation of the Neoplatonic works circulating under the name of Dionysius the Areopagite; he also translated a few other books.[13] The scholar who wished to acquire a mastery of Greek comparable to that of John Scotus faced formidable obstacles. There were no Greek grammars for Latins, and although one might acquire an elementary knowledge of Greek through contact with the tradition of Greek studies kept alive by Irish scholars, for a mastery of the language the help of a Greek-speaking tutor would be required; John probably had such a tutor, as did Robert Grosseteste and William of Moerbeke in the thirteenth century.[14] It is true that an elementary Greek vocabulary might be obtained from a Greco-Latin glossary, such as the *Hermeneumata* containing medical terms.[15] There were also phrase books for travelers, and a few bilingual texts (especially of the Psalms or parts of the New Testament);[16] but the knowledge of Greek gained from such sources was a totally inadequate preparation for translation. It must be pointed out, finally, that such knowledge of Greek as existed was generally put to religious, rather than philosophical or scientific, use—for biblical exegesis, hagiography, ecclestiastical politics, and the study (and possibly the translation) of the Greek fathers of the church.[17] Further translation of scientific materials would have to wait for the renewed activity at the beginning of the twelfth century.

The Transmission of Greek Learning to Islam

Beginning as early as the fourth century B.C. and continuing for a thousand years, Greek culture and learning were slowly diffused eastward into Asia.[18] Military campaigns and trading missions contributed to this diffusion. Of much greater importance, however, were religious factors: missionary activity of the eastern Christian church, which created communities of Syriac-speaking Christians as far east as the Persian highlands; and persecutions, which drove heretical Christian groups, especially the Nestorians, eastward beyond the reach of their Byzantine persecutors. In 457, Nestorians from Edessa in Syria fled across the Persian border and established a theological school at Nisibis (where, because of the semantic nature of many of the disputes, Aristotelian logic took firm root);[19] at Jundishapur, another Nestorian center, a school was established offering instruction in the full range of Greek philosophical and scientific learning, including medicine and astronomy.[20] Within Persia, Nestorian communities did not merely survive as enclaves of Hellenism within a hostile environment, but succeeded in imparting a taste for Greek culture and learning to influential Persians.[21]

With the rapid expansion of Islam after Muhammad's death in 632, regions where Greek learning had previously been deposited came under Islamic domination. Under the Abbasid caliphs, after 749, Muslims established fruitful intellectual contact with both Christians and Hellenized Persians. In 762, Caliph al-Mansur moved his capital from Damascus to Baghdad, close enough to Jundishapur to permit physicians from Jundishapur to treat royal patients in Baghdad. Indeed, in 765 al-Mansur brought the Nestorian head of the hospital at Jundishapur, Jurjis ibn Jibril ibn Bakhtishuʿ, to Baghdad as court physician, a position subsequently held by other members of the Bakhtishuʿ family.

Translation into Arabic began during the reign of al-Mansur (754–775) and continued under Harun al-Rashid (786–809). It reached new heights during the reign of Harun's son, al-Ma'mun (813–33), who developed Harun's library into a formal research center, the "House of Wisdom," staffed principally by Nestorians.[22] The Abbasid caliphs and other patrons of this translating activity were, of course, interested in works with an immediate practical utility—technical treatises on medicine, astrology, logic, and the mathematical sciences. But, partly through the growth of scholastic theology (*kalam*), Islamic interests quickly expanded to encompass the whole of Platonic and Aristotelian philosophy.[23] Texts already available in Syriac (still the literary and liturgical language of the Nestorians) were translated into Arabic. Works not available in Syriac were rendered directly from Greek to Arabic, or first into Syriac and subsequently into Arabic. On several occasions, expeditions were sent to Byzantium to secure copies of Greek works not otherwise available or to secure better copies of texts available only in corrupt versions.

One of the most skillful and productive of the translators in Baghdad was Hunain ibn Ishaq (d. 873/877), a Nestorian physician who for a time directed the House of Wisdom—though much of his support seems to have come from the Banu Musa (sons of Musa), wealthy patrons of learning.[24] Hunain was in all probability bilingual (in Syriac and Arabic) from childhood. As a young man he studied medicine under Yuhanna ibn Masawaih at Jundishapur; he left Jundishapur, perhaps for Alexandria, and returned with an excellent knowledge of Greek and the techniques of Greek textual criticism; he went to Basra to perfect his knowledge of Arabic grammar, after which he moved to Baghdad and began a career as physician and translator. Hunain's concern for establishing an accurate text before undertaking its translation is revealed in the following remark about Galen's *De methodo medendi*, which he translated from Greek to Syriac: "For the first six books only a single manuscript, and besides a very faulty one, was at my disposal at the time. I was therefore unable to produce these books in the manner required. Later I came across another manuscript and collated the text with it and corrected it as much as possible. It would be better if I could collate a third manuscript with it if only I were fortunate enough to find one."[25] A fourteenth-century biographer, al-Safadi, contrasted the skillful translations of Hunain with the crudely literal translations of some of his predecessors:

The translators use two methods of translation. One of them is that of Yuhanna b. al-Bitriq, Ibn an-Na'imah al-Himsi and others.

According to this method, the translator studies each individual Greek word and its meaning, chooses an Arabic word of corresponding meaning and uses it. Then he turns to the next word and proceeds in the same manner until in the end he has rendered into Arabic the text he wishes to translate. This method is bad for two reasons. First, it is impossible to find Arabic expressions corresponding to all Greek words, and therefore, through this method many Greek words remain untranslated. Second, certain syntactical combinations in the one language do not always necessarily correspond to similar combinations in the other; besides, the use of metaphors, which are frequent in every language, causes additional mistakes.

The second method is that of Hunain b. Ishaq, al-Jauhari and others. Here the translator considers a whole sentence, ascertains its full meaning and then expresses it in Arabic with a sentence identical in meaning, without concern for the correspondence of individual words. This method is superior, and hence there is no need to improve the works of Hunain b. Ishaq. The exception is those dealing with the mathematical sciences, which he had not mastered, in contrast with works on medicine, logic, natural science and metaphysics whose Arabic translations require no corrections at all.[26]

Hunain is credited with an enormous list of translations—into both Syriac and Arabic. He was most interested in medicine, but also translated books on philosophy, mathematics, and astronomy; an Arabic version of the Old Testament is also attributed to him. He translated ninety-five Galenic works from Greek to Syriac and another thirty-nine into Arabic—if we take at face value a letter to ʿAli ibn Yahya in which Hunain listed all the Galenic works known to him (and their translators).[27] In addition, Hunain trained a circle of followers, including his nephew Hubaish and his son Ishaq, whose output approached his own. By the end of the tenth century, through the combined efforts of Hunain, Hubaish, Ishaq, their confreres in Baghdad, and their counterparts in other times and places, most of the Greek philosophical and scientific works available to us today had been put into usable Arabic versions, and the process of assimilation and criticism had begun.

The dissemination, to other parts of the Islamic empire, of the newly translated works and the philosophical and scientific discussions they spawned is poorly understood. Most important for our purposes is the spread of secular learning to Spain, where much of the translation from Arabic to Latin would later occur.[28] Muslim armies entered Spain from North Africa in 711 and subdued it within the next few years. Under the Umayyad Amirs, especially Abd al-Rahman III (912–61), a brilliant court was established at Cordova and a pluralistic society tolerated. Under al-Hakam II (961–76) learning was generously patronized and educational institutions founded. Al-Hakam's agents scoured the Islamic world for books, and the royal library is reputed to have contained 400,000 volumes (probably an exaggerated figure). Spanish Muslims were sent to Baghdad and Cairo for medical study and returned with books and broad philosophical interests. Physicians were also brought from Baghdad to Cordova. With the collapse of the Umayyad dynasty early in the eleventh century, al-Hakam's library was destroyed and the scholars under his patronage dispersed, but some of the books and many of the scholars turned up in the royal courts of the smaller states that replaced the Umayyad—some of which continued to patronize learning on a significant scale. Conditions were ripe for the beginning of transmission to Latin Christendom, in the north of Spain and across the Pyrenees.

The Beginnings of Translation from Arabic to Latin

The principal regions of translation during the Middle Ages were the linguistic borderlands—regions where different languages coexisted and bilingual (or trilingual) people could be found. The bilingual native of such a region could serve the translation enterprise in any of several ways: as translator, as collaborator in a cooperative translation using the local vernacular as an intermediary, or as language instructor to a would-be translator who had emigrated from elsewhere and required linguistic instruction before he could begin his work. For translation from Arabic to Latin, there were three important borderlands—Spain, Sicily, and the Latin kingdoms established by crusaders in the Near East.

The least significant of these as a source of translations was the last mentioned. Indeed, only two Near Eastern translators of works relevant to the history of science are known. In the first half of the twelfth century Stephen of Antioch translated the *Dispositio regalis* of the physician ʿAli ibn al-ʿAbbas (Haly Abbas) and a glossary of *materia medica*; and in the first half of the thirteenth century Philip of Tripoli translated the pseudo-Aristotelian *Secretum secretorum*.[29] Of considerably greater significance than the Near East was Sicily, which had been successively under Byzantine, Muslim, and Norman control and had a population (and, more importantly, a civil service) in which Greek, Arabic, and Latin were all represented. Latin and Greek were also used in some parts of the Italian mainland, and perhaps (on rare occasions) Arabic as well, for Stephen of Antioch points out in the prologue to the aforementioned glossary that "in Sicily and at Salerno . . . there are both Greeks and men familiar with Arabic."[30] Although Sicily and southern Italy became an important locus of Arabo-Latin translating activity, it never equaled Spain in that respect—probably because it had never had an Islamic culture as vigorous as that of Spain and because learned Sicilians and Italians preferred to drink from the original sources and made their greatest contribution in translation from Greek to Latin.

In many respects, Spain was ideal as a center of Arabo-Latin translation. Because Muslim Spain (al-Andalus) possessed a brilliant Arabic culture, it contained an ample supply of Arabic texts for translation—Arabic versions of Greek originals, the works of the great mathematicians, astronomers, and physicians of the Arabic east, and scientific works produced in al-Andalus. Moreover, al-Andalus contained communities of Christians, known as Mozarabs, whose culture and literary language were Arabic and who could serve as intermediaries between Muslims and Latin Christendom. Spain also had a population bilingual in Arabic and the romance vernacular or occasionally (among the clergy, both within al-Andalus and at the borders between al-Andalus and the small Christian states in the north) Arabic and Latin. Finally, al-Andalus was not a closed realm, for in times of peace commercial and cultural relations flourished between it and the Christian regions to the north.

Contact between al-Andalus and the north is illustrated by the diplomatic exchange between ʿAbd al-Rahman III, caliph of Cordova, and the German emperor Otto the Great.[31] In 953 Otto sent an ambassador, the monk John of Gorze (later Abbot of Gorze and a monastic reformer), to negotiate with ʿAbd al-Rahman regarding Muslim piracy in the Mediterranean and certain Alpine passes. In the course of the negotiations John remained in Cordova for almost three years and there became acquainted with a Spanish Jew named Hasdeu, who knew both Arabic and Latin, and also with a Mozarab official in the caliph's chancery, Recemundus, who was later sent to Otto's court in Frankfurt. What is important is that the two envoys, John and Recemundus, the one a German monk (from Lorraine) who spent almost three years in Cordova, the other a Cordovan Mozarab who spent nearly a year in Germany, were both scholars. During a previous mission to southern Italy, John of Gorze had returned with Greek manuscripts of Aristotle's *Categories* and Porphyry's *Isagoge*; for his part, Recemundus was an accomplished mathematical astronomer, co-author of the *Calendar of Cordova*.[32] Whether any scientific treatises were actually carried from the one region to the other in this ambassadorial exchange cannot be determined, but it is inconceivable that knowledge of Arabic learning in Lorraine was not in some way enhanced.[33] This conclusion is strengthened when we recognize that in the eleventh century Lorraine became an important locus of mathematical and astronomical activity based on Arabic sources.[34]

Another instance of contact between al-Andalus and the Christian north is the celebrated and much debated case of Gerbert of Aurillac. Gerbert (ca. 945–1003), later Pope Sylvester II, was educated at a Benedictine convent in Aurillac.[35] In 967 he went to Catalonia, in northern Spain at the foot of the Pyrenees, to study mathematics and music under Atto, Bishop of Vich.[36] There can be no doubt that Gerbert encountered Arabic learning during his three years with Atto, possibly through the monastery of Santa Maria de Ripoll located some twenty-five miles from Vich. From the fact that he went to Catalonia to study mathematics, one may surmise that it was his intent to make at least indirect contact with Arabic learning. And the remainder of his career provides constant testimony to his interest in, and knowledge of, Arabic mathematics and astronomy: for example, his correspondence for the year 984 contains a letter to Lupito of Barcelona, asking Lupito to send the book on astrology that he had translated (obviously from Arabic), and letters to Bishop Bonfil of Gerona and Abbot Giraldus of Aurillac, asking for copies of a book *De multiplicatione et divisione numerorum* written by Joseph the Spaniard.[37]

More problematic is the question of Gerbert's exact relationship to a manuscript now preserved in Barcelona, originally MS 225 of the monastery of Santa Maria de Ripoll.[38] The manuscript itself is of considerable importance for the history of Arabo-Latin translating activity, for it contains the earliest extant copies of scientific treatises translated from Arabic to Latin. The manuscript, which dates probably from the second half of the tenth century or possibly from early in the eleventh century (and is apparently a copy of an earlier manuscript), contains a set of Latin treatises on the astrolabe and the quadrant, some of which are translations and others adaptations of Arabic treatises. In Vernet's judgment, it represents "the manual of studies of the monks of Santa Maria de Ripoll."[39] Whatever the precise origins of this manuscript, it is clear that in the second half of the tenth-century Spain began to yield up her scientific treasures to Latin Christendom. As for Gerbert, his name has sometimes been associated with one of the works of the Ripoll manuscript, *De utilitatibus astrolabii*, and although it is now generally agreed that it cannot have been authored by him,[40] one can speculate that Gerbert, through contact with the monastery of Santa Maria de Ripoll, carried this and other treatises across the Pyrenees at the end of his period of study. Millás has argued, "He thus seems to have been the first ambassador who carried this new Arab science across the Pyrenees; to it he owed his great scientific reputation; he taught it to his numerous disciples, especially at Rheims, and they introduced it to learned circles in Lorraine, which soon showed signs of having been initiated into the new techniques of mathematics and astronomy."[41] The picture is admittedly speculative, and it contains unresolved issues, such as the relative importance to be given to Gerbert and the earlier diplomatic exchanges between the courts of ʿAbd al-Rahman and Otto in the transmission of Arabic mathematics and astronomy to Lorraine. What is clear is that Arabic science had established a beachhead in the north. Fifty years later we see a direct reflection of the Ripoll tradition in Germany, for Hermann Contractus (1013–54) produced an adaptation of the treatise *De utilitatibus astrolabii* in his monastery in Reichenau.[42]

A century after Gerbert made his way to Catalonia to study Arabic mathematics, Constantine the African (fl. 1065–85), a North African turned Benedictine monk, began a prolific career as translator of medical works from Arabic to Latin. We know little of Constantine's origins. According to Peter the Deacon, historian of the monastery of Monte Cassino, Constantine left his native city of Carthage for thirty-nine years of study in the East, where he mastered the grammar, dialectics, geometry, arithmetic, astronomy, necromancy, and music of the Chaldeans, Arabs, Persians, Saracens, Egyptians, and Indians. Returning to Carthage with such knowledge, he provoked hostility and had to flee for his life to Salerno, where he lived in poverty until noticed by Duke Robert Guiscard. He later entered the monastery of Monte Cassino, where most of his translations were executed.[43] According to another account of his life by a Salernitan physician Matthaeus (twelfth or thirteenth century?), Constantine was a Saracen merchant who visited Salerno in southern Italy. There he learned of the dearth of medical literature in Latin, returned to North Africa, and several years later again appeared in Salerno bearing Arabic medical texts, which he proceeded to translate into Latin. Shortly after his return to Italy he became a Christian and entered Monte Cassino and there undertook his translating endeavors.[44]

Constantine's medical output was large and significant. His authentic translations and compilations include three works of Isaac Israeli, Hunain ibn Ishaq's *Isagoge to Galen's Tegni* and *Liber de oculis*, three works of Hippocrates (*Aphorisms, Prognostica*, and *Regimen*) and Galen's commentaries thereon, the *Pantegni* of ʿAli ibn al-ʿAbbas (Haly Abbas), Galen's *Megategni* in an abbreviated version, the *Viaticum* of Ibn al-Jazzar, and a few other short treatises.[45] The impact of these translations was very large. In the short run, Constantine supplied the sources on which Salerno would found its reputation as one of the foremost medical schools in Europe.[46] In the long run, he supplied a basic foundation of medical literature on which the West would build for several centuries.

Translation from Arabic to Latin in the Twelfth and Thirteenth Centuries

The high-water mark of translation from Arabic to Latin came during the twelfth century. The movement broadened its topical scope beyond medicine and astronomy to embrace the full range of philosophical and scientific learning; at the same time it broadened its geographical basis to become a European-wide movement.

One of the pioneers of the Arabo-Latin translating movement was the Englishman Adelard of Bath (fl. 1116–42). As a young man Adelard went to France and studied at Tours and later taught at Laon.[47] By his own account, he traveled for seven years after leaving Laon, his itinerary including Salerno and Sicily, Cilicia, Syria, and possibly Palestine. It is obvious that he found an opportunity to learn Arabic during his travels, but we have no details. He claimed to have encountered Arabic learning in the course of these travels, and it is possible that he returned from them bearing Arabic books. Nothing else is known of his personal life except that he was for a time tutor to the future Henry II of England. Before he became a translator, Adelard was a scholar of encyclopedic interests, which he revealed in several early works that exhibit little familiarity with Arabic learning: *De eodem et diverso, Natural Questions*, and a work on the abacus, *Regule abaci*. Adelard's translations include two astrological works, one of them being Abu Maʿshar's *Shorter Introduction to Astronomy*. In the realm of mathematical astronomy, Adelard translated the *Astronomical Tables* of al-Khwarizmi in the revision of the Cordovan astronomer Abu ʾl-Qasim Maslama al-Majriti.[48] The fact that al-Majriti's version was used has suggested to some that Adelard visited Spain. It is just as likely, however, that the Arabic text was brought to England by Petrus Alphonsi, a converted Jew who served as physician to King Henry I and may himself have begun to translate the same work.[49] In any case, this translation was of crucial importance, for through it Western astronomers first learned how to use astronomical tables.[50] Finally, Adelard gave the West its first complete translation of Euclid's *Elements*,[51] thus supplying a foundation for Western mathematical studies comparable in importance to that which he supplied for mathematical astronomy.

Adelard was exceptional among twelfth-century translators from Arabic to Latin in that his career seems not to have included a sojourn in Spain—for Spain remained the focus of Arabo-Latin translating activity. Besides its traditional advantages of bilingual inhabitants and a supply of Arabic books, Spain in the twelfth century provided increasingly easy access for Christian scholars to Arabic sources. This was the result of the *reconquista*, the reconquest of Muslim Spain by the Christian states in the north, especially Leon, Castile, and Aragon, which gradually opened up centers of Muslim learning. Toledo was an early Christian conquest in 1085, fitful southerly progress was made throughout the twelfth century, and Cordova and Seville finally fell into Christian hands in the first half of the thirteenth century.

Some of the translators were native Spaniards. John of Seville (fl. 1133–42) was probably a Mozarab, born in the Muslim south, who spent much of his career in the Christian north—in Limia and Toledo. His principal interest was astrology, and he produced Latin translations of a large collection of astrological texts: al-Farghani's *De scientia astrorum*, Masha'allah's *De receptione planetarum sive de interrogationibus*, Abu Ma'shar's *Great Introduction to the Science of Astrology*, and others.[52] He also translated two short tracts, *De regimine sanitatis* (an extract from the pseudo-Aristotelian *Secret of Secrets*) dedicated to a queen of Spain and Qusta ibn Luqa's *De differentia spiritus et anime* dedicated to Raymond, Archbishop of Toledo. Another Spaniard with interests in astrology was Hugh of Santalla (fl. 1145), whose work was apparently patronized by Michael, Bishop of Tarazona (in the Christian state of Aragon), to whom most of his translations were dedicated. Hugh translated the pseudo-Ptolemaic astrological work known as the *Centiloquium*, Masha'allah's *De nativitatibus*, a work on spatulomancy (divination from the shoulder blades of animals), and the Hermetic work known (in Hugh's translation) as *Liber Apolonii de secretis nature et occultis rerum causis*.[53]

If John and Hugh represent the native Spanish contribution to the twelfth-century movement of translation in Spain, there were also foreigners who journeyed to Spain (without, so far as we know, any prior knowledge of Arabic) and remained to participate in the translation effort. Robert of Chester (fl. 1141–50) came from England; his friend Hermann the Dalmatian (fl. 1138–43) was a Slav from Carinthia; and Plato of Tivoli (fl. 1132–46) was presumably an Italian. As Haskins has pointed out, Robert and Hermann constituted a "literary partnership."[54] They were discovered by Peter of Cluny in the Ebro Valley and enlisted by him to produce a translation of the Koran (completed by Robert in 1143) and of several theological tracts (one rendered by Robert and two by Hermann), which were attached to Robert's version of the Koran. In addition, Robert translated al-Khwarizmi's *Algebra* and al-Kindi's *De iudiciis astrorum*; he was eventually appointed archdeacon of Pamplona and seems to have made trips to England as well.[55] Hermann probably studied under Thierry of Chartres; by 1138 he was in Spain engaged in translation. His translations include Ptolemy's *Planisphere* and Abu Maʿshar's *Great Introduction to the Science of Astrology* (translated a little earlier by John of Seville), and probably Euclid's *Elements*.[56] Plato of Tivoli seems to have passed his career as a translator principally in Barcelona, where he collaborated with the Jew Abraham bar Hiyya (Savasorda). His translations were principally astrological, including Ptolemy's *Quadripartitum* and ʿAli ibn Ahmad al-ʿImrani's *De electionibus horarum*; he also translated al-Battani's *De motu stellarum*, the *Book of Areas* of Savasorda, and perhaps Archimedes' *De mensura circuli*.[57] In a number of these translations Plato appears to have been assisted by Savasorda.

In the second half of the twelfth century the focus of translating activity appears to have shifted to Toledo—though one should avoid exaggerating this activity into a veritable "school of translation" surrounding Archbishop Raymond.[58] There is only one translation unmistakably dedicated to Raymond (John of Seville's translation of Qusta ibn Luqa's *De differentia spiritus et anime*), and the activity in Toledo extended long after Raymond's death in 1151. To see the Toledo translations in their proper light, one must understand why Toledo provided such a favorable environment for the translator. Millás Vallicrosa has written:

> The strong Mozarabic atmosphere prevailing at the time of the Reconquest had been succeeded by the powerful influence of the Romanizing, Cluniac, Francophile element. This is evidenced by the attitude of the first Bishops of Toledo and the substitution of the Roman for the Mozarab liturgy. The Mozarab element remained strong, however, supported as it was by relations with southern Spain. Even well into the thirteenth century the Christians and Jews of Toledo were using Arabic for a good deal of their private writing, and it was probably due to this combination of interests

between the Mozarabs and Jews of the city, who were bilingual and familiar with the secrets of Arab science, and the prelates who, for all their foreign influence, were eager to master that new science, that Toledo became such a flourishing center of translations. The cultural movement thus established had the support of the bishops of Toledo—and not merely of the celebrated Don Raimundo.[59]

Given its advantages, Toledo could hardly be ignored by a prospective translator choosing the site of his activity from general considerations in the middle of the twelfth century. But there is no evidence to suggest that there was a school (or even a society of translators) in Toledo for him to join. An examination of the work of several translators who settled there will serve to illustrate the level of activity and the achievements of the Toledo translators.

Dominicus Gundissalinus or Gundisalvo, archdeacon of Cuellar, was immersed in Jewish and Arabic learning. He lived for a time in Toledo, where he collaborated with a learned Jew named Avendauth to translate Avicenna's *De anima*, dedicated to Archbishop John of Toledo (1151–66).[60] Both Gundissalinus and Avendauth are credited with other translations—Gundissalinus with versions of Avicebron's *Fons vitae* and al-Ghazali's *Aims of the Philosophers*, both in collaboration with a certain Johannes (probably Johannes Hispanus, another archdeacon of Cuellar), and Avendauth with the preface of Avicenna's philosophical encyclopedia, *Kitab al-Shifa*, in collaboration with an unnamed assistant. Gundissalinus also wrote several philosophical treatises of his own, based on Arabic, Jewish, and Christian learning.

The greatest of the Toledo translators was undoubtedly the Italian Gerard of Cremona (ca. 1114–87). A eulogy written by his associates and disciples shortly after his death and attached to his translation of Galen's *Tegni* maintains that Gerard had been

trained from childhood at centers of philosophical study and had come to a knowledge of all of this that was known to the Latins; but for the love of the *Almagest* [Ptolemy's great astronomical text], which he could not find at all among the Latins, he went to Toledo; there, seeing the abundance of books in Arabic on every subject, and regretting the poverty of the Latins in these things, he learned the Arabic language, in order to be able to translate. In this way, combining both languages and science, . . . he passed on the Arabic literature in the manner of the wise man who, wandering through a green field, links up a crown of flowers, made from not just any, but from the prettiest; to the end of his life, he continued to transmit to the Latin world (as if to his own beloved heir) whatsoever books he thought finest, in many subjects, as accurately and as plainly as he could.[61]

Gerard discovered Ptolemy's *Almagest* in Toledo and produced a Latin translation of this enormously long and technical work. But that was only a tiny fraction of his output. Gerard's students attached to the above-quoted eulogy a list of Gerard's translations containing seventy-one entries—and this list is known to be incomplete.[62] Some of these translations may have been executed by Gerard's assistants under his general direction. Others were cooperative ventures: Daniel of Morley, who studied astronomy under Gerard, reported that Gerard was assisted by a Mozarab named Galippus.[63] Whatever the precise details, Haskins is undoubtedly correct in asserting that "more of Arabic science . . . passed into western Europe at the hands of Gerard of Cremona than in any other way."[64]

It is neither desirable nor possible to give a complete enumeration of Gerard's translations, but we must at least sample the list of works attributed to him by his students. The list includes twelve works on astronomy, including Ptolemy's *Almagest*, al-Farghani's *De scientia astrorum* (previously translated by John of Seville), and two works by Thabit ibn Qurra. Seventeen works on mathematics are included in the list: Euclid's *Elements*, the *Sphere* of Theodosius, al-Khwarizmi's *Algebra*, Thabit ibn Qurra's *On the Kariston* (or Roman balance), al-Kindi's optical work *De aspectibus*, and others. In the philosophical realm, Gerard's students credit him with fourteen translations, including Aristotle's *Posterior Analytics, Physics, On the Heavens, On Generation and Corruption*, and *Meteorology*; two works by al-Kindi; and three by al-Farabi. In medicine, Gerard is credited with twenty-one translations, including nine works by Galen, two by Isaac Israeli, three by Rhazes (including the *Liber ad Almansorem*), and Avicenna's great medical encyclopedia, the *Canon of Medicine*. Finally, Gerard translated several works on alchemy and geomancy. These works cover a wide spectrum of Greek and Arabic philosophy and science; many of them are long and technical, requiring detailed knowledge of the subject matter, and some would individually be a noteworthy achievement for a lifetime of scholarship.

But Gerard's brilliance must not be permitted to obscure the achievements of many others who labored in the same fields. Marc, canon of Toledo (fl. 1191–1216), was supported in his work by the archbishop of Toledo, Rodrigo Jimenez, who recognized the utility of translations in his crusade against Muslim infidels. Marc translated the Koran, a theological work by Ibn Tumart, and several medical opuscules.[65] Late in the twelfth century or early in the thirteenth, Alfred of Sareshel, an Englishman, traveled to Spain (possibly, but not necessarily, to Toledo) in search of Arabic learning. In his commentary on Aristotle's *Meteorology* Alfred refers to his master, Salomon Avenraza, obviously a Jew, who may have tutored him in Arabic or assisted him in his translations. Alfred translated the pseudo-Aristotelian *De plantis* and a portion of Avicenna's *Kitab al-Shifa* that was appended to Aristotle's *Meteorology* and referred to as *De mineralibus*.[66] Michael Scot (fl. 1217–35) held benefices in England, Scotland, and Ireland, and it is not clear from which of these lands he originated.[67] He divided his career as translator between Spain and Italy: about 1217 he was in Toledo, where (with the help of a Jew, Abuteus Levita) he produced a translation of al-Bitruji's *De motibus celorum*; in Toledo he also translated (with or without help, we do not know) Aristotle's *De animalibus*. It seems that Michael enjoyed the patronage of Archbishop Rodrigo Jimenez, for in 1215 he was sent to the Fourth Lateran Council in Rome as an acolyte of the archbishop.[68] Contemporary with Michael Scot in Toledo was Salio, canon of Padua, who translated three astrological works, two from Arabic and one from Hebrew.[69]

In the thirteenth century the pace of translation from the Arabic decreased in Spain while increasing in Italy and Sicily. The court of Frederick II (1198–1250) was the center of activity, and its best known translator was Michael Scot.[70] Although Michael's early translating took place in Spain, we find him in Italy in the 1220s, perhaps as Frederick's court astrologer, certainly as translator of Avicenna's *De animalibus* dedicated to Frederick.[71] Frederick's court was inhabited by other translators as well—a certain Theodore of Antioch, who translated an Arabic work on falconry, and Jacob Anatoli, a relative of the Ibn Tibbon family (discussed below), who produced Hebrew versions of several astronomical treatises and of Averroes' commentaries on Aristotle's *Organon*.[72]

Hebrew and Vernacular Languages

The Jewish role in the transmission of Islamic learning is frequently conceived merely as one of mediation: Spanish Jews, many of whom knew Arabic, were in a position to serve Western Christendom by translating Arabic books into Hebrew or the romance vernacular (Castilian or Catalan)—an intermediary, from which somebody else could immediately or subsequently put them into Latin. This view captures part of the truth, for, indeed, Jewish scholars occupied themselves as hired translators, supplying Christian patrons with desired translations. But Jews also had an intellectual life of their own, and Arabic treatises were most typically translated into Hebrew, not so that they could be subsequently rendered into Latin, but to serve the needs of a Jewish community that had as much interest in their contents as neighboring Christians. These two aspects of Jewish involvement in translation, intertwined but distinguishable, must be kept in mind if we are to understand the Jewish contribution to the transmission of learning.

We can see both aspects in the career of one of the early Jewish translators, Abraham bar Hiyya, surnamed Savasorda (fl. 1133–45). Savasorda spent most of his life in Barcelona, though he was also in touch with Jewish communities in southern France and expressed concern over the ignorance of Arabic language and learning in Provence.[73] In an effort to relieve some of that ignorance, both in Provence and in his native Catalonia, Savasorda wrote books in Hebrew on cosmography, astronomy, and geometry. In 1145 the latter was translated into Latin in abbreviated form by Plato of Tivoli, under the title *Liber embadorum (Book of Areas)*, probably with the assistance of the author; this is the first known instance of a scientific work translated from Hebrew to Latin.[74] Savasorda also collaborated with Plato of Tivoli on the translation of scientific works from Arabic to Latin—chiefly astrological tracts, but also al-Battani's *De motu stellarum*, one of the finest achievements of Arabic mathematical astronomy.

Savasorda's contemporary Abraham ibn Ezra, born in Tudela (in northern Spain), trod a similar path, though with a somewhat stronger bent toward astrology.[75] He wrote many books in Hebrew, based on Arabic science, including several on numbers, two on the calendar, one on the astrolabe, and more than fifty on astrology. A century later the astrological works were translated into French by a Jew, Hagin, employed by Henry Bate of Malines, then into Latin by Henry Bate and others. Ibn Ezra was widely traveled, visiting (between about 1140 and 1167) Rome, Salerno, Lucca, Pisa, Mantua, Verona, Béziers, Narbonne, Bordeaux, Angers, Dreux, London, and Winchester. One of his occupations during these travels was to construct astronomical tables for the meridians of various cities.[76] Finally, as a translator Ibn Ezra rendered into Hebrew the commentary of Ibn al-Muthanna on the astronomical tables of al-Khwarizmi and Masha'-allah's *Book on Eclipses*.

The need for Hebrew translations of Arabic scientific works was felt with particular intensity in southern France and Italy, where large

Jewish communities flourished but knowledge of Arabic was scarce. Many of the required translations were provided by members of the Ibn Tibbon family, which came from Spain to southern France (Languedoc and Provence) in 1150.[77] Judah (1120–ca. 1190), who moved the family from Granada to escape unrest, translated a few philosophical and theological works. His son Samuel (1150–1232) rendered into Hebrew the *Guide of the Perplexed* of Moses Maimonides, written originally in Arabic; Galen's *Microtegni* with the commentary of ʿAli ibn Ridwan; and Yahya ibn Bitriq's Arabic version of Aristotle's *Meteorology*. Samuel's son Moses (fl. 1240–83) translated an enormous collection of Arabic books into Hebrew, including Averroes' commentaries on Aristotle's *Physics, On the Heavens, On Generation and Corruption, Meteorology, On the Soul, Parva naturalia,* and *Metaphysics*; half-a-dozen medical works of Maimonides (later turned into Latin by John of Capua, a converted Jew, and possibly by the Montpellier physician Armengaud Blasi);[78] Euclid's *Elements*; Theodosius's *Sphere*; and medical works by Hunain ibn Ishaq, Rhazes, and Avicenna.

The most remarkable of the Tibbonides was Moses' son, Jacob ben Machir ibn Tibbon, known as Don Profeit or Profatius Judaeus (ca. 1236–1305).[79] Profatius studied medicine at Montpellier and seems to have lived in Spain for a year. He translated from Arabic to Hebrew Euclid's *Elements* and *Data*, Alhazen's *On the Configuration of the World*, Autolycus of Pitane's *On the Moving Sphere*, and works of al-Ghazali, Averroes, Qusta ibn Luqa, Ibn al-Saffar, al-Zarqali, and Jabir ibn Aflah. Following the example of Abraham bar Hiyya, he collaborated with Christian scholars to produce Latin versions of Arabic treatises on the exact sciences, and some of his own works as well. In at least some of these cooperative translations the vernacular was employed as an intermediate language. This is clear from the explicit of the translation of al-Zarqali's *Saphea*, where we read: "This work was translated . . . from Arabic to Latin in the year of our Lord Jesus Christ 1263, Profatius of the Hebrew people vulgarizing [that is, rendering into the vulgar or vernacular language] and John of Brescia rendering into Latin."[80] Profatius's own compositions were translated into Latin in his lifetime: his treatise on the quadrant was translated into Latin by Armengaud Blasi in 1290; and his *Almanach*, inspired by al-Zarqali's almanac, with emendations and improvements, presenting tables calculated for the meridian of Montpellier, was translated several times into Latin and widely used through the sixteenth century.[81]

We have seen the use of the vernacular language as intermediary in the translating efforts of Profatius Judaeus and John of Brescia. This method was frequently employed in cooperative translations from Arabic to Latin: a Jew or Mozarab who knew Arabic but not enough Latin to complete the translation alone would communicate orally with a Latin-speaking cleric through the vernacular.[82] The process is described in some detail in a letter of dedication attached to the translation of Avicenna's *De anima*. Here the Jew Avendauth writes that he "put the text into the vulgar tongue one word at a time, while the archdeacon Dominicus [Gundissalinus] converted the individual words into Latin."[83] Gundissalinus either wrote down the Latin text himself or dictated it to an amanuensis.

Sometimes the vernacular was the intended language of the final version, rather than an intermediate language. In the thirteenth century, Arabic works were translated into Castilian at the court of King Alfonso X of Castile (1252–84).[84] The translators were mostly Jews, and they translated the Arabic into Castilian, often with the help of one of the king's clerks, who was presumably responsible for the final written version. Sometimes the vernacular version was later translated into Latin by Alfonso's notaries. In the fourteenth century Arabic and Hebrew texts were translated into Catalan. In 1313 King James II of Aragon paid a fee to a Jewish physician of Barcelona, Yehuda or Jafuda Bonsenyor, for the translation of a medical work, probably the *Surgery* of Abulcasis, from Arabic to Catalan.[85] Later in the century Peter IV of Aragon also patronized Catalan translations of scientific treatises. The same process was repeated with Latin as the original language and other vernaculars as the final language of the translated work. The books translated into the vernacular were generally on subjects of wide popular interest, such as arithmetic, practical geometry, medicine (chiefly surgery), and astrology. To give but a single example, Guy de Chauliac's *Great Surgery*, written in Latin in the fourteenth century, was rendered into French, Provençal, Catalan, Italian, Dutch, English, Irish, and Hebrew.[86]

Translation from Greek to Latin

With the exception of a few medical translations executed in Ravenna, the translation of scientific treatises from Greek to Latin virtually ceased from the sixth century to the twelfth. The conditions that gave rise to the revival of translating activity must be briefly described. The most important seat of activity was Italy. In Sicily and southern Italy the Normans had established a kingdom as a result of conquests

between 1060 and 1091, and in Palermo they established a cosmo-
politan court, where Greek, Arabic, and Latin were all in use. Haskins has pointed out that "King Roger . . . drew to his court men
of talent from every land, regardless of speech or faith: an Englishman, Robert of Selby, stood at the head of his chancery; . . . a Greek
monk, Nilus Doxopatres, wrote at his command the history of the five
patriarchates which was directed at the supremacy of the Roman see;
a Saracen, Edrisi, prepared under his direction the comprehensive
treatises on geography which became celebrated as 'King Roger's
Book.' "[87] There were substantial communities of Greek-speaking people, and Greek books could be found in the royal capital at Palermo
as well as in monastic libraries.[88] Southern Italy and Sicily were also
in contact with Byzantium: travelers passed from the one to the
other, and commercial relations were maintained. Finally, as we shall
see, some of the Norman kings became vigorous patrons of learning.
There was also contact between northern Italy and Byzantium, chiefly
theological, diplomatic, and commercial, and Constantinople had both
a Venetian and a Pisan quarter.[89]

That there was a relationship between the Byzantine connections of
northern Italy and translating activity can be seen from an account of
a theological disputation held in Constantinople in 1136. Among
those present were "not a few Latins, among them three wise men
skilled in the two languages and most learned in letters, namely James
a Venetian, Burgundio a Pisan, and the third, most famous among
Greeks and Latins above all others for his knowledge of both literatures, Moses by name, an Italian from the city of Bergamo, and he
was chosen by all to be a faithful interpreter for both sides."[90] James
of Venice, Burgundio of Pisa, and Moses of Bergamo are all known
as translators. Moses was probably the oldest of the three, his scholarly labors going back to the beginning of the twelfth century.[91] His
known translations are limited to a theological tract. James (fl. 1136–
48), a canon lawyer, was sometimes referred to as "Grecus," signifying either that he was of Greek descent or had spent considerable
time in Greek-speaking lands.[92] James was in touch with Byzantine
philosophers, among whom Aristotelianism had undergone a recent
revival, and he was one of the major figures in introducing Aristotelian philosophy to Latin Christendom. He gave the West its first translation of Aristotle's *Physics, On the Soul, Metaphysics* (at least the
first three-and-a-half books), major portions of the *Parva naturalia*,
and perhaps *Posterior Analytics*.[93] He retranslated the *Sophistical
Refutations* and probably the *Topics* and *Prior Analytics*. His translations, sometimes in revised form, became the standard versions of

Aristotle for centuries.[94] The last of our trio, Burgundio of Pisa (fl. 1136–93), translated a number of important theological works.[95] He also retranslated *De natura hominis* of Nemesius of Emesa, earlier put into Latin by Archbishop Alphano of Salerno (1058–85), a work destined to exercise a strong influence over medieval psychological theory. But Burgundio's principal contribution to the scientific movement was his translation of medical books—the *Aphorisms* of Hippocrates, previously translated from the Greek by a member of the Ravenna group and from the Arabic by Constantine the African, and ten Galenic works, including the *Therapeutica, De temperamentis, De crisibus,* and *De differentiis pulsuum.*[96]

In southern Italy and Sicily, royal patronage played an important role in supporting translation, or at least translators. Emir (or Admiral) Eugene, an important official in the Norman administration, translated Ptolemy's *Optica* from Arabic to Latin.[97] However, Eugene's native language was Greek, and he was active in Greco-Latin translation as well. He translated two pieces of oriental literature, which had previously been translated into Greek (one from Chaldean and the other from Sanskrit).[98] Far more significant for the history of science was the assistance he lent to a Greco-Latin translation of Ptolemy's *Almagest.*

In the preface to this translation, the anonymous translator recounts the history of his efforts.[99] While a student of medicine at Salerno, he learned that the Byzantine emperor (probably Manuel Comnenus) had presented a manuscript containing the *Almagest* to the Sicilian king, carried by the hand of the minister Aristippus. After locating Aristippus near Mount Etna (investigating its volcanic eruptions), our translator discovered that he was not yet equipped to undertake a translation of the *Almagest* and set out to prepare himself by studying (or possibly by endeavoring to translate) Euclid's *Data, Optics,* and *Catoptrics* and Proclus's *Elements of Physics.*[100] When he was ready to return to Ptolemy's astronomical text, he encountered "by divine providence, a gracious expositor, Eugene, a man extraordinarily skilled in both Greek and Arabic and not ignorant of Latin"—undoubtedly Emir Eugene.[101] It is not clear whether Eugene's assistance was required for linguistic or astronomical purposes, or both, but with his help the translation into Latin was completed. This was the first translation of the *Almagest,* dated by Haskins to about 1160, fifteen years before Gerard of Cremona's translation from the Arabic.[102]

The envoy who carried the Greek *Almagest* to Sicily, Henry Aristippus, was himself a translator of significance. Aristippus became archdeacon of Catania in 1156 and from 1160 to 1162 served as the king's chief minister.[103] His known translations are Plato's *Meno* and *Phaedo*, the only medieval Latin versions of these two dialogues, and the fourth book of Aristotle's *Meteorology*. In the prologue to the *Meno*, Aristippus reports that he has also been asked by the king to translate Gregory of Nazianzus and by the king's minister Maio and the archbishop of Palermo to translate Diogenes Laertius's *Lives of the Philosophers*.[104] Clearer testimony to royal interest in learning and royal patronage of translation could hardly be found.

A final group of twelfth-century translations must be mentioned— Greco-Latin versions of Euclid's *Elements, Data, Optics*, and *Catoptrics*. Since the latter three were among the works used propaedeutically by the translator of the *Almagest* in Sicily, it is reasonable to suppose that the translations are of Sicilian or southern Italian provenance. An examination of the Latin words used to render Greek particles in these three translations has led to the conclusion that all were probably executed by the same person (possibly Emir Eugene).[105] A similar analysis suggests that Euclid's *Elements* was put into Latin by the same anonymous translator who rendered the *Almagest*.[106]

In the thirteenth century, Greek influence at the Sicilian court was diminished,[107] and such Greco-Latin translating activity as occurred generally occurred elsewhere. An exception is the work of Bartholomew of Messina, official translator to King Manfred (d. 1266), who translated a number of pseudo-Aristotelian works and also a treatise on veterinary medicine.[108] In England, early in the century, Robert Grosseteste managed to acquire a sufficient mastery of Greek (probably with the assistance of a tutor, Nicholas of Sicily, himself the translator of the pseudo-Aristotelian *De mundo*) to translate Aristotle's *Nicomachean Ethics* (accompanied by several Greek commentaries) and the first two books of *On the Heavens* (with the commentary of Simplicius); *De fide orthodoxa* of John of Damascus (previously translated by Burgundio of Pisa); the pseudo-Dionysian corpus; and several other minor works.[109]

Undoubtedly the outstanding translator from Greek to Latin in the thirteenth century was William of Moerbeke, a Flemish Dominican.[110] Nothing is known about Moerbeke's education or the nature of his linguistic training. We first encounter him at Nicea in Asia Minor in 1260, and again at Thebes later in the same year. In 1267 and following he was in Viterbo, the seat of the papal curia, and from 1272 (or earlier) to 1278 he was chaplain and confessor to the pope. In 1278 he was appointed archbishop of Corinth, a position held until his death. His translations span this entire period, the earliest datable one having been completed in Nicea in 1260 and the last shortly before his death in 1286. Moerbeke set out to provide Latin Christendom with a complete and adequate collection of Aristotelian treatises. He translated five that had never before been rendered into Latin, including the *Politics, Poetics,* book 11 of the *Metaphysics,* and two works on animals; five more that had been previously translated from the Arabic, including the first three books of the *Meteorology, On the Heavens* (books 3 and 4), and three works on animals; and five works that already existed in Greco-Latin translations, including the *Categories, On Interpretation,* two versions of the *Rhetoric,* book 4 of the *Meteorology,* and the first two books of *On the Heavens.* He also revised many earlier translations of Aristotle, such as Boethius's version of the *Sophistical Refutations,* Grosseteste's translation of the *Nicomachean Ethics,* and James of Venice's versions of the *Posterior Analytics, Physics,* and *On the Soul.* Added to this output were Aristotelian commentaries by Alexander of Aphrodisias, Themistius, Ammonius, Philoponus, and Simplicius; four works by Proclus (thus introducing the West to a major body of Neoplatonic literature); seven tracts by Archimedes; the *Catoptrics* of Hero of Alexandria; and a handful of other works. Altogether almost fifty distinct translations or revisions issued from his pen. As for the quality of Moerbeke's translations, Minio-Paluello has written that he

> was meticulous in his quest for exactitude; he would search the Latin vocabulary with a sound critical sense and great knowledge, in order to find words which would convey to the intelligent reader the meaning of the Greek terms. If his search failed to produce the necessary results, he would form new Latin words by compounding two terms, or adding prefixes and suffixes on the Greek pattern, or even combining Greek and Latin elements. . . . In extreme cases he would resort to that great source of enrichment of a language, the transliteration, with slight adaptations, of foreign—in this case Greek—words.[111]

He was equally meticulous when revising rather than translating: "in most cases where he introduced a change, a misinterpretation was put right, a serious mistake corrected, or a more appropriate shade of meaning introduced."[112]

A steady stream of medical translations continued to issue from Italy in the thirteenth and fourteenth centuries, especially from the Angevin court in Naples. Under Charles I (1268–85) and Charles II (1285–1309) Greek medical books were assiduously collected and translations sponsored; it is reported that "a house in Naples was actually taken over to lodge the scribes engaged on the transcription of medical books for the King."[113] It was under Robert (1309–43), however, that translation reached its peak. Robert employed a number of translators, some of whose names are known, but the only one whose translations have come down to us is Niccolò da Reggio. Niccolò was from Calabria, undoubtedly (since he is referred to as "Grecus") from the Greek-speaking part of the population. He was trained as a physician, perhaps at Salerno, and joined the court of Charles II and then of Robert as royal physician and translator. Over a period of approximately thirty-five years he translated some fifty medical treatises, most of them Galenic. Included among his translations are *De utilitate particularum* (or *De usu partium*), previously translated only in part; *Commentary on the Aphorisms of Hippocrates; De passionibus uniuscuiusque particule*; and a large number of short tracts consisting of a few leaves each.[114]

Niccolò's last dated translation is from 1345. It did not, of course, mark the end of the translation of scientific works from Greek to Latin—one need only recall the burst of translating activity in the fifteenth and sixteenth centuries, which enlarged the corpus of Greek mathematical and medical works available to the West. But Niccolò was one of the last translators from Greek to Latin whose work would substantially influence medieval scientific thought and practice. The translations of the fifteenth and sixteenth centuries would not shape the medieval world view, but help to bring its downfall.

Conclusion

Our data on the transmission of learning are often little more than lists of translators and translated works, with an occasional indication of place or date; such data do not lend themselves to easy generalization about the nature of the movement they represent. Indeed, what must impress one is the individuality and specificity of the transmission process: scholars pursued personal interests, often on their own initiative, with whatever assistance and support they could discover or devise. In short, there was no organized movement to lend coherence to the process, and the story of the transmission of learning to the West must remain largely a tale of individual scholars responding in personal ways to unique historical circumstances. However, we need not abandon all efforts to generalize.

First, that order and coherence were lacking is itself an important generalization. There was no bureau of translation to coordinate efforts and insure the orderly and systematic acquisition of Greek and Arabic learning. Scholars translated what interested them or their patrons, and what was ready at hand. Thus Haskins has written: "In this process of translation and transmission accident and convenience played a large part. No general survey of the material was made, and the early translators groped somewhat blindly in the mass of new matter suddenly disclosed to them. Brief works were often taken first because they were brief and the fundamental treatises were long and difficult; commentators were often preferred to the subject of the commentary. . . . Much was translated to which the modern world is indifferent, something was lost which we should willingly recover, yet the sum total is highly significant."[115] Duplicate translations are one measure of this lack of system. The Italian Gerard of Cremona traveled to Spain in search of Ptolemy's *Almagest* (which he ultimately translated from the Arabic), although a Latin version (from the Greek) already existed close to home, in Sicily. Many other works also exist in multiple translations—for example, Euclid's *Elements* and *Optics*, Abu Maʿshar's *Great Introduction to the Science of Astrology*, Hippocrates' *Aphorisms, Prognostica*, and *Regimen*, and many of the works of Aristotle. Sometimes retranslation was undertaken in an effort to improve the text, but frequently it was done in ignorance of previous translation; and, as Haskins notes,[116] the more faithful version was not always the more popular, for historical accident played as large a part in the dissemination of translated literature as in the original selection of materials for translation.

Because of a dearth of biographical data on the translators, we can usually only speculate on the origins of their linguistic abilities. It is clear that some knew from childhood the language out of which they translated; this is perhaps most apparent in the Greco-Latin translations by southern Italians or Sicilians who came from Greek-speaking communities, but it is undoubtedly true of many translations from Arabic (into Latin, Hebrew, or the vernacular) as well. It is equally clear that others had to learn as adults the language from which they translated; Gerard of Cremona and William of Moerbeke are good examples. Those who translated into Hebrew or the vernacular were translating into a native tongue; those who translated into Latin were translating into a scholarly language acquired at school.

But where and how did those who had not known Arabic or Greek from childhood acquire sufficient skill to translate from these languages? In most cases we simply do not know. It is clear, however, that a teacher would usually be required, and the best place to find one would be a bilingual or trilingual region such as Spain or Sicily, although occasionally teachers were brought to northern Europe.

Robert Grosseteste is known to have had at least two men of Greek extraction around him—Robertus Grecus and Nicolaus Grecus, the latter as a member of his household—and to have collaborated with John of Basingstoke, who had himself studied in Athens.[117] Roger Bacon claimed that in his day (the third quarter of the thirteenth century) teachers of Greek and Hebrew were easy to find "in Paris and in France and in all other regions."[118] Although they do not appear to have had a major impact on translation, the *studia linguarum* (language institutes) of the friars provided a place for some to learn Arabic in the second half of the thirteenth century.[119] In 1236 the General Chapter of the Dominican Order at Paris prescribed the study of languages for missionary purposes, and we know that several *studia* were organized.[120] In 1250 the Provincial Chapter of Toledo sent eight friars to the *studium arabicum* in Tunis, including Ramon Marti, who was to become a great Arabist. One example of science benefitting as a by-product of such linguistic programs is the translation (in 1270 or 1271) of an ophthalmological work, the *Liber oculorum* of ʿAli ibn ʿIsa (Jesu Haly), by Friar Dominicus Marrochini in the *studium* of Murcia in southern Spain.[121]

Virtually all of the translators were scholars, many of them with a prior interest in the subject of the translated work, for one could not successfully translate what was not first thoroughly understood. Many were also clerics: in northern Europe university students were generally members of the clergy, so that most scholars would have clerical status, and some translators are known to have held ecclesiastical benefices or to have been ordained; many Mozarabic translators were probably also clerics, especially those who knew some Latin; among Italian and Jewish translators members of the clergy or rabbinate were numerous, but here the practice of medicine was a significant alternative.

This brings us to the question of livelihood. Since there can scarcely have been a secure and predictable market for their product, many translators would have required support or subsidy from other sources. We have caught glimpses throughout this chapter of various forms of support, but a brief summary may be useful. Those who had another profession that provided sufficient leisure for translating activity would, of course, require no further support: physicians, civil servants, and the holders of ecclesiastical posts would fit this description. One need only recall Boethius and Emir Eugene, who held high governmental office, and Robert Grosseteste and William of Moerbeke, who managed to fit translation into busy ecclesiastical careers. But patronage by high officials, royal or ecclesiastical, was also a factor of vital importance. An impressive list of royal courts patronized translating activity: the Carolingian court of Charles the Bald; the courts of Roger, Frederick II, and Manfred in Sicily; the Angevin court in Naples; and the courts of Alfonso X in Castile and James II in Aragon. On the ecclesiastical side, Michael, Bishop of Tarazona, supported the work of Hugh of Santalla; John of Seville and Michael Scot secured patronage from two archbishops of Toledo, Raymond and Rodrigo Jimenez, respectively; and monastic institutions supported the translations of Constantine the African and the collaborative efforts of Robert of Chester and Hermann the Dalmatian.

Finally, we must inquire briefly into the quality of the translations that resulted from all of this effort. Perhaps the most striking characteristic of the translations is their tendency toward literalism. Boethius had argued at the beginning of his second commentary on Porphyry's *Isagoge* that the translator must render his text word for word and sacrifice elegance for fidelity, and this became the standard for medieval translators.[122] Gerard of Cremona and William of Moerbeke are examples of important translators who followed Boethius's advice. Robert Grosseteste is another; Callus has commented on the extreme literalism of his translations, "holding fast to the exact order of the words and representing the Greek in every detail, to the extent of disregarding Latin idiom and syntax."[123] To these can be added the anonymous translator of Euclid's *Elements* from the Greek. Regarding this translation Murdoch notes: "Rare is it . . . that a particle, no matter how insignificant, finds no mate in the Latin; . . . and at times the ordering of the words is dictated by the Greek at hand and not the custom (even a medieval one) of Latin syntax. Nor is this predilection for precision and fidelity a mere enthusiastic beginning that wears thin and becomes forgotten as the propositions push on; it is retained, literally, to the bitter end."[124] There was little variety in the vocabulary employed by the literalists; it was safer to render a given Greek or Arabic phrase in a fixed and determined manner. Transliteration was frequent when the translator could find no exact Latin equivalent, and this had the benefit of introducing Arabic and Greek technical terms into the Western vocabulary.

However, not all medieval translators shared these ideals. There were a few whose command of the languages was such as to permit them to translate *ad sensum* instead of *ad verbum*. Hugh of Santalla was an exceptional stylist, with a large vocabulary at his disposal, and he made no attempt to adhere strictly to the original. Others aimed only to communicate the substance of the translated work and willingly omitted a phrase or even an entire section if it appeared to contribute nothing of substance. The same motives could justify condensation, expansion, or correction. This pattern is seen vividly in the work of Constantine the African, whose method has been described by McVaugh:

> He was by no means intent upon exactly reproducing whatever text was in question; rather, as his prefaces reveal, he saw himself as *coadunator*, with the responsibility of sumarizing or expanding the substance of the original, perhaps adding material from other sources, in whatever way was best suited to the needs of an essentially ignorant Western audience. It would certainly be wrong to look for any consistent plan, any impulse to systematization or comprehensiveness, in his writings; he was composing primarily to satisfy requests or to fill whatever practical and pedagogical needs arose; and it was this that produced so many explanatory additions.[125]

And, of course, a translator sometimes omitted words and phrases simply because he did not understand their meaning or import.[126]

As for fidelity to meaning (as opposed to word-order and syntax), it is not possible to say much more than that translations varied from good to bad. Some translators, skilled in the subject matter as well as the relevant languages, were able to capture exactly the sense of the original; others, having an imperfect command of the languages or the subject matter, produced unintelligible results. Viewed as a whole, however, the translations provided Western Christendom with an adequate knowledge of the Greek and Arabic intellectual achievement— and thus with the basic materials out of which its own system of philosophy and natural science would be constructed.

Notes

1. This chapter could not have been written without the generous assistance of Marie-Thérèse d'Alverny, who has both guided and criticized my efforts. She is not, however, to be held accountable for its deficiencies.

2. On Roman science, see William H. Stahl, *Roman Science* (Madison, Wis., 1962).

3. Transmission was of two sorts: geographical transmission within the boundaries of a single linguistic region, as from eastern to western Islam or from southern to northern Europe; and translation across linguistic boundaries. This chapter, though touching upon both aspects of the transmission process, will concentrate on the latter. On the distinction between these two aspects of the transmission process, see George F. Hourani, "The Early Growth of the Secular Sciences in Andalusia," *Studia Islamica* 32 (1970):143.

4. I have found the following works, of a more or less general nature, to be useful. For transmission from Greek to Arabic: F. E. Peters, *Allah's Commonwealth: A History of Islam in the Near East, 600–1100* A.D. (New York, 1973), pp. 266–396; Peters, *Aristotle and the Arabs: The Aristotelian Tradition in Islam* (New York, 1968); Majid Fakhry, *A History of Islamic Philosophy* (New York, 1970), pp. 12–31; ʿAbdurrahman Badawi, *La Transmission de la philosophie grecque au monde arabe* (Paris, 1968); Richard Walzer, "Arabic Transmission of Greek Thought to Medieval Europe," *Bulletin of the John Rylands Library* 29 (1945–46): 160–83; Max Meyerhof, "Von Alexandrien nach Baghdad. Ein Beitrag zur Geschichte des philosophischen und medizinischen Unterrichts bei den Arabern," *Sitzungsberichte der preussischen Akademie der Wissenschaften, Berlin*, Phil.-hist. Klasse, 1930, pp. 389–429; and De Lacy O'Leary, *How Greek Science Passed to the Arabs* (London, 1949). For transmission from Greek and Arabic to Latin: José M.ª Millás Vallicrosa, "Translations of Oriental Scientific Works (to the End of the Thirteenth Century)," in *The Evolution of Science*, ed. Guy S. Métraux and François Crouzet (New York, 1963), pp. 128–67 (translated from "La Corriente de las traducciones cientificas de origen oriental hasta fines del siglo XIII," *Journal of World History* 2 [1954]:395–428, and nearly identical to chap. 5 of Millás's *Nuevos estudios sobre historia de la ciencia española* [Barcelona, 1960]); Charles H. Haskins, *The Renaissance of the Twelfth Century* (Cambridge, Mass., 1927), chap. 9; Amable Jourdain, *Recherches critiques sur l'âge et l'origine des traductions latines d'Aristote*, new ed. (Paris, 1843); George F. Hourani, "The Medieval Translations from Arabic to Latin Made in Spain," *The Muslim World* 62 (1972):97–114; and F. Gabrieli, "The Transmission of Learning and Literary Influences to Western Europe," in *Cambridge History of Islam*, ed. P. M. Holt, Ann K. S. Lambton, and Bernard Lewis, vol. 2 (Cambridge, 1970), pp. 851–89.

5. Virtually nothing is known of Calcidius except that he translated the *Timaeus*, even the period of his activity being open to dispute. See William H. Stahl, "Calcidius," *Dictionary of Scientific Biography*, 3:14–15; Stahl, *Roman Science*, pp. 142–50.

6. On Boethius, see Lorenzo Minio-Paluello, "Boethius, Anicius Manlius Severinus," *Dictionary of Scientific Biography*, 2:228–36; Stahl, *Roman Science*, pp. 195–202.

7. Minio-Paluello, "Boethius," p. 230. For a convincing defense of a Boethian translation of Euclid's *Elements*, see Menso Folkerts, *"Boethius" Geometrie II: Ein mathematisches Lehrbuch des Mittelalters* (Wiesbaden, 1970), pp. 67 ff.; see also John E. Murdoch, "Euclid: Transmission of the Elements," *Dictionary of Scientific Biography*, 4:443–44. A letter from Cassiodorus (or King Theodoric) to Boethius has led some scholars to attribute to Boethius translations of Ptolemy's *Almagest* and certain works of Archimedes (see *The Letters of Cassiodorus*, trans. Thomas Hodgkin [London, 1886], p. 169); however, these probably represent intentions, rather than actual achievements. See Stahl, *Roman Science*, pp. 196–97; Minio-Paluello, "Boethius," pp. 228–29.

8. On Cassiodorus, see Stahl, *Roman Science*, pp. 202–11; Cassiodorus, *An Introduction to Divine and Human Readings*, trans. Leslie W. Jones (New York, 1946), a translation of the *Institutiones*, with an excellent

introduction; and Phillip Drennon Thomas, "Cassiodorus Senator, Flavius Magnus Aurelius," *Dictionary of Scientific Biography*, 3:109–10.

9. For a useful survey of the *Institutiones*, see Stahl, *Roman Science*, pp. 205–11. On the works of Greek authorship known to Cassiodorus, see Pierre Courcelle, *Les lettres grecques en occident de Macrobe à Cassiodore*, 2d ed. (Paris, 1948), pp. 328–36, 382–88.

10. For a cautious assessment of this influence, see Jones's introduction to Cassiodorus, *Introduction to Divine and Human Learning*, pp. 47–50.

11. The basic research is by Augusto Beccaria, *I Codici di medicina del periodo presalernitano (secoli IX, X e XI)* (Rome, 1956); Beccaria, "Sulle tracce di un antico canone latino di Ippocrate e di Galeno," *Italia medioevale e umanistica* 2 (1959):1–56; 4 (1961):1–75; and 14 (1971): 1–23. See also Henry Sigerist, "The Latin Medical Literature of the Early Middle Ages," *Journal of the History of Medicine and Allied Sciences* 13 (1958):133–35.

12. On medieval knowledge of Greek, see Bernhard Bischoff, "The Study of Foreign Languages in the Middle Ages," *Speculum* 36 (1961): 209–23; Bischoff, "Das griechische Element in der abendländischen Bildung des Mittelalters," *Byzantinische Zeitschrift* 44 (1951):27–55. Both articles by Bischoff are reprinted in his *Mittelalterliche Studien*, vol. 2 (Stuttgart, 1967), pp. 227–45 and 246–74, respectively. See also M. L. W. Laistner, *Thought and Letters in Western Europe*, A.D. *500–900*, 2d ed. (London, 1957), pp. 238–50; and R. R. Bolgar, *The Classical Heritage and Its Beneficiaries from the Carolingian Age to the End of the Renaissance* (Cambridge, 1954), pp. 93–94, 99–100, 122–23.

13. Maïeul Cappuyns, O.S.B., *Jean Scot Érigène: Sa vie, son oeuvre, sa pensée* (Paris, 1933), esp. pp. 59–66, 128–79; Laistner, *Thought and Letters*, pp. 245–47.

14. Bischoff, "Study of Foreign Languages," *Mittelalterliche Studien*, 2: 234. On Grosseteste and Moerbeke, see below.

15. Beccaria, *I Codici*, p. 459.

16. Bischoff, "Study of Foreign Languages," *Mittelalterliche Studien*, 2:236–40; Laistner, *Thought and Letters*, pp. 241–43.

17. J. T. Muckle, C.S.B., "Greek Works Translated Directly into Latin Before 1350," *Mediaeval Studies* 4 (1942):34–42; A. van de Vyver, "Les plus anciennes traductions latines médiévales (Xe–XIe siècles) de traités d'astronomie et d'astrologie," *Osiris* 1 (1936):662–64.

18. This section summarizes a story that has been often recounted. See the works cited under Greco-Arabic transmission in n.4 above. Still a basic source of data on the Greco-Arabic translating movement, though it has been superseded in many details, is Moritz Steinschneider, *Die arabischen Übersetzungen aus dem griechischen*, originally published in four parts, 1889–96, and reissued in a single volume (Graz, 1960).

19. Peters, *Aristotle and the Arabs*, p. 38.

20. On the relationship of the school at Jundishapur to the Greek academy at Alexandria, see Meyerhof, "Von Alexandrien nach Baghdad"; and Meyerhof, "La fin de l'Ecole d'Alexandrie d'après quelques auteurs arabes," *Archeion* 15 (1933):1–15.

21. This point has been forcefully made by Peters, *Aristotle and the Arabs*, pp. 41–48.

22. See D. Sourdel, "Bayt al-hikma," *The Encyclopaedia of Islam*, new ed., 1:1141.

23. On the Aristotelian tradition, see Peters, *Aristotle and the Arabs.* On the Platonic tradition, see Peters, *Allah's Commonwealth*, pp. 286–307; Richard Walzer, "Platonism in Islamic Philosophy," in Walzer's *Greek into Arabic: Essays on Islamic Philosophy* (Oxford, 1962), pp. 236–52. Muslim philosophers generally had access only to paraphrases of Plato's dialogues; by contrast, they knew Aristotle's works firsthand.

24. On Hunain, see Gotthelf Bergsträsser, *Hunain ibn Ishak und seine Schule* (Leiden, 1913); Bergsträsser, *Hunain ibn Ishaq über die syrischen und arabischen Galen-Übersetzungen* (Leipzig, 1925); Max Meyerhof, *The Book of the Ten Treatises on the Eye ascribed to Hunain ibn Is-hâq (809–877 A.D.)* (Cairo, 1928), pp. xvi–xxiii; Meyerhof, "New Light on Hunain ibn Ishaq and His Period," *Isis* 8 (1926):685–724 (which summarizes Bergsträsser); Lufti M. Saʾdi, "A Bio-Bibliographical Study of Hunayn ibn Is-haq al-Ibadi (Johannitius) (809–877 A.D.)," *Bulletin of the Institute of the History of Medicine* 2 (1934):409–46; and G. Strohmaier, "Hunayn b. Ishak al-ʿIbadi," *The Encyclopaedia of Islam*, new ed., 3:578–81.

25. Quoted by Franz Rosenthal, *The Classical Heritage in Islam*, trans. Emile and Jenny Marmorstein (London, 1975), p. 21.

26. Ibid., pp. 17–18.

27. Meyerhof, "New Light on Hunain."

28. The most important works on this subject are Hourani, "Early Growth of the Secular Sciences," pp. 143–56; Hourani, "Medieval Translations from Arabic to Latin," pp. 97–114; and Juan Vernet, "Les traductions scientifiques dans l'Espagne du Xe siècle," *Cahiers de Tunisie* 18, nos. 1–2 (1970):47–50.

29. Charles H. Haskins, *Studies in the History of Mediaeval Science*, 2d ed. (Cambridge, Mass., 1927), pp. 131–40.

30. Ibid., pp. 132–33.

31. My account of this exchange follows James Westfall Thompson, "The Introduction of Arabic Science into Lorraine in the Tenth Century," *Isis* 12 (1929):187–91; D. M. Dunlop, *Arabic Science in the West (Karachi*, n.d. [1959?]), pp. 23–26; and Vernet, "Les traductions scientifiques," p. 55.

32. On the *Calendar of Cordova*, see Vernet, "Les traductions scientifiques," pp. 50–51.

33. One indication of Recemundus's influence at the German court is that he there met Liutprand, whom he persuaded to write a historical work, entitled *Antapodosis*, which Liutprand dedicated to Recemundus. About this same time a German nun, Hrosvitha of Gandersheim, wrote a poem in Latin on the life and death of the Mozarab Pelagius (d. 925), in which she described Cordova and referred to its caliph, ʿAbd al-Rahman; clearly, she had been in touch with somebody who had been to al-Andalus. See Dunlop, *Arabic Science in the West*, pp. 25–26.

34. Mary Catherine Welborn, "Lotharingia as a Center of Arabic and Scientific Influence in the Eleventh Century," *Isis* 16 (1931):188–99; Haskins, *Studies in Mediaeval Science*, pp. 334–35; Thompson, "Introduction of Arabic Science into Lorraine," p. 191. Illustrative of the scien-

tific tradition in eleventh-century Lorraine is the school of Liège, which counted among its members Radulf (known for his mathematical correspondence with Ragimbold of Cologne, on which see chap. 5, below).

35. On Gerbert's career, see F. Picavet, *Gerbert, un pape philosophe* (Paris, 1917); the introduction to *The Letters of Gerbert, with his Papal Privileges as Sylvester II*, trans. Harriet Pratt Lattin (New York, 1961), pp. 3–20; Uta Lindgren, *Gerbert von Aurillac und das Quadrivium: Untersuchungen zur Bildung im Zeitalter der Ottonen (Sudhoffs Archiv: Zeitschrift für Wissenschaftsgeschichte*, Beiheft 18) (Wiesbaden, 1976); and Dirk J. Struik, "Gerbert," *Dictionary of Scientific Biography*, 5:364–66.

36. He was accompanied by Borrell II, Count of Barcelona, who had visited the monastery in Aurillac and had been so impressed with Gerbert that he secured permission from the abbot to take Gerbert to Catalonia for further education. Borrell placed Gerbert's education in the hands of Atto. See Lattin's introduction to *Letters of Gerbert*, p. 3.

37. *Letters of Gerbert*, trans. Lattin, pp. 64, 69, 70. According to Lattin, Lupito was a wealthy archdeacon of the cathedral of Barcelona. The work by Joseph the Spaniard was probably concerned with the use of the abacus.

38. The manuscript is now in the Archivio de la Corona de Aragon. For a description of it and a discussion of its contents and significance, see José M.ª Millás Vallicrosa, *Assaig d'història de les idees físiques i matemàtiques a la Catulunya medieval* (Barcelona, 1931); Millás, *Nuevos estudios*, pp. 93–101; and Millás, "Translations of Oriental Scientific Works," pp. 138–44.

39. Vernet, "Les traductions scientifiques," p. 56.

40. Millás and Vernet deny Gerbert's authorship of this treatise, as does van de Vyver, "Les plus anciens traductions," p. 665.

41. Millás, "Translations of Oriental Scientific Works," p. 143; cf. *Nuevos estudios*, pp. 98–99.

42. Millás, "Translations of Oriental Scientific Works," pp. 145–46; *Nuevos estudios*, pp. 103–4; cf. A. van de Vyver, "Les premières traductions latines (Xe–XIe s.) de traités arabes sur l'astrolabe," in *Premier Congrès international de géographie historique*, vol. 2, *Mémoires* (Brussels, 1931), pp. 266–290. On Hermann, see also Claudia Kren, Hermann the Lame," *Dictionary of Scientific Biography*, 6:301–3.

43. See Petrus Diaconis, *Chronica Monasterii Casinensis* and *De viris illustribus Casinensis Coenobii*, in *Patrologiae cursus completus, series latina*, ed. J.-P. Migne, vol. 173 (Paris, 1854), cols. 766–68, 1034–35. On Constantine, see also Michael McVaugh, "Constantine the African," *Dictionary of Scientific Biography*, 3:393–95; Rudolf Creutz, "Die Ehrenrettung Konstantins von Afrika," *Studien und Mitteilungen zur geschichte des Benediktiner-Ordens* 49 (1931):25–44; Creutz, "Der Arzt Constantinus von Montekassino. Sein Leben, sein Werk und seine Bedeutung für die mittelalterliche medizinische Wissenschaft," *Studien und Mitteilungen zur Geschichte des Benediktiner-Ordens* 47 (1929):1–44.

44. McVaugh, "Constantine the African," p. 393. On the identity of Matthaeus, see Paul Oskar Kristeller, "The School of Salerno: Its Development and Its Contribution to the History of Learning," *Bulletin of the History of Medicine* 17 (1945):151, 157.

45. Much work remains to be done on the corpus of Constantine's medical translations. I here follow Heinrich Schipperges, *Die Assimilation der arabischen Medizin durch das lateinische Mittelalter (Sudhoffs Archiv für Geschichte der Medizin und der Naturwissenschaften,* Beiheft 3) (Wiesbaden, 1964), pp. 17–54. On Constantine's works as adaptations rather than literal translations, see below.

46. Kristeller, "School of Salerno," p. 155, argues that "after the middle of the twelfth century the translations of Constantine became the common property of the Salerno school and even the center of its medical teaching."

47. The standard work on Adelard's life is Franz Bliemetzrieder, *Adelhard von Bath* (Munich, 1935). See also Haskins, *Studies in Mediaeval Science,* pp. 20–42; and Marshall Clagett, "Adelard of Bath," *Dictionary of Scientific Biography,* 1:61–64.

48. On Adelard's translations and his own compositions, see Haskins, *Studies in Mediaeval Science,* pp. 20–33. On al-Majriti, see Juan Vernet, "Al-Majriti," *Dictionary of Scientific Biography,* 9:39–40.

49. José M.ª Millás Vallicrosa, *Estudios sobre historia de la ciencia española* (Barcelona, 1949), pp. 213–17; Millás, *Nuevos estudios,* p. 107; Millás, "Translations of Oriental Scientific Works," pp. 148–49; Otto Neugebauer, *The Astronomical Tables of al-Khwarizmi (Historisk-filosofiske Skrifter udgivet af Det Kongelige Danske Videnskabernes,* vol. 4, pt. 2) (Copenhagen, 1962), *passim* but especially pp. 230–32. On the astronomical work of Petrus Alphonsi, see also Haskins, *Studies in Mediaeval Science,* pp. 115–19.

50. See below, chap. 9.

51. Indeed, three different versions of Euclid's *Elements* are associated with Adelard's name: see Marshall Clagett, "The Medieval Latin Translations from the Arabic of the *Elements* of Euclid, with Special Emphasis on the Versions of Adelard of Bath," *Isis* 44 (1953):16–42; John E. Murdoch, "The Medieval Euclid: Salient Aspects of the Translations of the *Elements* by Adelard of Bath and Campanus of Novara," *Revue de synthèse,* ser. 3, nos. 49–52 (1968), pp. 69–74.

52. Lynn Thorndike, "John of Seville," *Speculum* 34 (1959):20–38; Manuel Alonso Alonso, S.J., "Juan Sevillano, sus obras propias y sus traducciones," *Al-Andalus* 18 (1953):17–49 (which must be used with caution); and Haskins, *Studies in Mediaeval Science, passim.*

53. Haskins, *Studies in Mediaeval Science,* pp. 67–81; Thorndike, *History of Magic,* 2:85–87.

54. *Studies in Mediaeval Science,* pp. 11–12, 55–56.

55. Ibid., pp. 120–23; Thorndike, *History of Magic,* 2:83; Louis C. Karpinski, *Robert of Chester's Latin Translation of the Algebra of al-Khowarizmi,* in University of Michigan Studies, Humanistic Series, vol. 11 (Ann Arbor 1930), pp. 25–32; Angel J. Martín Duque, "El Ingles Roberto, traductor del Coran," *Hispania* 88 (1962): 483–506.

56. Haskins, *Studies in Mediaeval Science,* pp. 43–66; Clagett, "Medieval Latin Translations from the Arabic of the *Elements* of Euclid," pp. 26–27; Murdoch, "Euclid: Transmission of the Elements," p. 447.

57. Haskins, *Studies in Mediaeval Science,* p. 11; Thorndike, *History of Magic,* 2:82–83; Lorenzo Minio-Paluello, "Plato of Tivoli," *Dictionary*

of Scientific Biography, 9:31–33; Millás, *Estudios sobre historia de la ciencia*, pp. 259–62.

58. On the translating activity in Toledo, see Valentin Rose, "Ptolemaeus und die Schule von Toledo," *Hermes* 8 (1874):327–49. Rose maintains (p. 329) that "Toledo . . . was in fact a sort of university, even if it was not a *studium generale*." Jourdain, *Recherches critiques*, p. 108, refers to "ce collège de traducteurs." Their view has recently been repeated by D. M. Dunlop, "The Work of Translation at Toledo," *Babel* 6 (1960):55–59. For a protest against this opinion, see Lorenzo Minio-Paluello, "Aristotele del mondo arabo a quello latino," in *L'occidente e l'Islam nell'alto medioevo* (Settimane di studio del Centro italiano di studi sull'alto medioevo, XII, Spoleto, 2–8 April 1964) (Spoleto, 1965), 2:606–8. For a sense of the philosophical and scientific riches to be found in twelfth-century Toledo, see José M.ª Millás Vallicrosa, *Las Traducciones orientales en los manuscriptos de la Biblioteca Catedral de Toledo* (Madrid, 1942).

59. "Translations of Oriental Scientific Works," p. 154.

60. On Gundissalinus, see Juan F. Rivera, "Nuevos datos sobre los traductores Gundisalvo y Juan Hispano," *Al-Andalus* 31 (1966):267–80; Manuel Alonso Alonso, S.J., "Traducciones del arcediano Domingo Gundisalvo," *Al-Andalus* 12 (1947):295–338; Millás, "Translations of Oriental Scientific Works," pp. 154–56; Claudia Kren, "Gundissalinus, Dominicus," *Dictionary of Scientific Biography*, 5:591–93. On Avendauth, see M.-T. d'Alverny, "Avendauth?" in *Homenaje a Millás-Vallicrosa*, vol. 1 (Barcelona, 1954), pp. 19–43. On the collaboration between Gundissalinus and Avendauth, see below.

61. Translated by Michael McVaugh, in *A Source Book in Medieval Science*, ed. Edward Grant (Cambridge, Mass., 1974), p. 35.

62. See ibid., pp. 36–38, for an annotated translation of the list of Gerard's translations.

63. Haskins, *Studies in Mediaeval Science*, p. 15. For the text of Daniel of Morley in which he reveals Gerard's use of an assistant, see Rose, "Ptolemaeus und die Schule von Toledo," pp. 347–49.

64. Haskins, *Studies in Mediaeval Science*, p. 15.

65. M.-T. d'Alverny and Georges Vajda, "Marc de Tolède, traducteur d'Ibn Tumart," *Al-Andalus* 16 (1951):99–140, 259–307; 17 (1952):1–56. See also Millás, "Translations of Oriental Scientific Works," p. 157.

66. On Alfred's life and translations, see James K. Otte, "The Life and Writings of Alfredus Anglicus," *Viator* 3 (1972):275–91.

67. For differing views on his origins, see Lorenzo Minio-Paluello, "Michael Scot," *Dictionary of Scientific Biography*, 9:361; Haskins, *Studies in Mediaeval Science*, pp. 272–73; Lynn Thorndike, *Michael Scot* (London, 1965); p. 11. On the use of the epithet "Scot" to mean "Irish," see the entry in the *Oxford English Dictionary*. See the works of Minio-Paluello, Haskins, and Thorndike for Michael's translations.

68. Juan F. Rivera, "Personajes hispanos assistentes en 1215 al IV Concilio de Letrán," *Hispania sacra* 4 (1951):349, 354–55.

69. On Salio's translations, see Lynn Thorndike, "A Third Translation by Salio," *Speculum* 32 (1957):116–17. Carmody (*Arabic Astronomical and Astrological Sciences*, p. 136) places the translation of *De nativitati-*

bus of Albubather or Alcabitius in Padua, but the explicit of the treatise (quoted by Carmody) reveals that the translation was completed "in barrio Iudeorum" ("barrio" being the term typically employed to designate the Jewish quarter); moreover, another of Salio's translations is explicitly located in Toledo (see Thorndike, p. 116). *De nativitatibus* was executed with the help of a certain David.

70. On science at Frederick's court, see Haskins, *Studies in Mediaeval Science*, pp. 242–71.

71. At some point in his career, Michael also translated several of Averroes' Aristotelian commentaries; see Minio-Paluello, "Michael Scot," pp. 362–63, for details.

72. Haskins, *Studies in Mediaeval Science*, pp. 246–51; Romeo Campani, "Il 'Kitab al-Farghani' nel testo arabo e nelle versioni," *Rivista degli studi orientali* 3 (1910):214–18.

73. Martin Levey, "Abraham bar Hiyya Ha-Nasi," *Dictionary of Scientific Biography*, 1:23. On Savasorda, see also Millás, *Estudios sobre historia de la ciencia*, pp. 219–62; Moritz Steinschneider, *Gesammelte Schriften*, ed. Heinrich Malter and Alexander Marx, vol. 1 (Berlin, 1925), pp. 327–87. On Jewish translators and Hebrew translations, the basic source is Moritz Steinschneider, *Die hebraeischen Uebersetzungen des Mittelalters und die Juden als Dolmetscher* (Berlin, 1893).

74. The *Liber embadorum*, according to Levey, "Abraham bar Hiyya," p. 22, was "the earliest exposition of Arab algebra written in Europe" and "among the earliest works to introduce Arab trigonometry into Europe."

75. On Ibn Ezra, see Steinschneider, *Gesammelte Schriften*, 1:407–506; Martin Levey, "Ibn Ezra, Abraham ben Meir," *Dictionary of Scientific Biography*, 4:502–3.

76. Millás, *Estudios sobre historia de la ciencia*, pp. 289–347.

77. On the Tibbonides, see David Romano, "La transmission des sciences arabes par les Juifs en Languedoc," in *Juifs et judaïsme de Languedoc, XIIIe siècle–début XIVe siècle*, ed. Marie-Humbert Vicaire and Bernhard Blumenkranz (*Cahiers de Fanjeaux*, vol. 12) (Toulouse, 1977), pp. 363–86; Juan Vernet, "Ibn Tibbon, Jacob ben Machir" and "Ibn Tibbon, Moses ben Samuel," *Dictionary of Scientific Biography*, 13:400–402.

78. Ernest Wickersheimer, *Dictionnaire biographique des médecins en France au moyen âge* (Paris, 1936), 1:40–41; Steinschneider, *Hebraeischen Uebersetzungen*, p. 772. John of Capua certainly translated from Moses ibn Tibbon's Hebrew version. Armegaud Blasi knew some Arabic, and it is not clear whether he made use of the Hebrew version or translated directly from the Arabic text; see the incipits and explicits of several of Armengaud's translations, quoted by M.-T. d'Alverny, "Avicenna Latinus," *Archives d'histoire doctrinale et littéraire du moyen âge* 30 (1963):223.

79. On Profatius, see Steinschneider, *Hebraeischen Uebersetzungen*, pp. 607–14, 976; Romano, "Transmission des sciences arabes," pp. 374–77; and Vernet, "Ibn Tibbon, Jacob ben Machir."

80. Don Profeit Tibbon, *Tractat de l'assafea d'Azarquiel*, ed. José M.ª Millás Vallicrosa (Universitat de Barcelona, Facultat de Filosofia i Lletres,

Biblioteca hebraico-catalana, 4) (Barcelona, 1933), p. 152; also Millás, *Estudios sobre historia de la ciencia*, p. 104.

81. Lynn Thorndike, "Date of the Translation by Ermangaud Blasius of the Work on the Quadrant by Profatius Judaeus," *Isis* 26 (1936): 306–9. On Profatius's *Almanach*, see José M.ª Millás Vallicrosa, *Estudios sobre Azarquiel* (Madrid, 1943–50), pp. 356–62; G. J. Toomer, "Prophatius Judaeus and the Toledo Tables," *Isis* 64 (1973):351–55.

82. For a protest against the view that cooperative translations through the medium of the vernacular were common, see Minio-Paluello, "Aristotele dal mondo arabo a quello latino," pp. 608–9.

83. *Avicenna Latinus: Liber de anima seu sextus de naturalibus, I–II–III*, ed. S. van Riet (Louvain-Leiden, 1972), p. 4. An illustration of the limited knowledge of Latin that a Mozarab translator might possess is the case of Peter of Toledo, who in 1142 translated the work of an oriental Christian author for Peter the Venerable, Abbot of Cluny. Peter's Latin proved to be inadequate, and, therefore, the Abbot assigned his secretary, Peter of Poitiers, the task of producing an adequate final version; see M.-T. d'Alverny, "Deux traductions latines du Coran au moyen âge," *Archives d'histoire doctrinale et littéraire du moyen âge* 16 (1947–48):72.

84. For scientific activity at the court of Alfonso X, see David Romano, "Le opere scientifiche di Alfonso X e l'intervento degli Ebrei," in *Oriente e occidente nel medioevo: filosofia e scienze* (Accademia Nazionale dei Lincei, Fondazione Alessandro Volta, Atti dei Convegni, 13: Convegno internazionale, 9–15 April 1969) (Rome, 1971), pp. 677–711; Evelyn S. Procter, "The Scientific Works of the Court of Alfonso X of Castille: The King and his Collaborators," *The Modern Language Review* 40 (1945): 12–29; and Procter, *Alfonso X of Castile: Patron of Literature and Learning* (Oxford, 1951).

85. José Cardoner, "Nuevos datos acerca de Jafuda Bonsenyor," *Sefarad* 4 (1944):287–93.

86. See David C. Lindberg, *Theories of Vision from al-Kindi to Kepler* (Chicago, 1976), p. 271, for vernacular translations of this and other surgeries.

87. *Studies in Mediaeval Science*, p. 156. On the intellectual and linguistic characteristics of Sicily and southern Italy, see ibid., pp. 155–90; also Haskins, *The Normans in European History* (Boston, 1915), pp. 218–49; Haskins, *Renaissance*, pp. 291–96; and Philip K. Hitti, *History of the Arabs*, 7th ed. (London, 1960), pp. 606–14.

88. On libraries of Greek books, see Robert Weiss, "The Translators from the Greek of the Angevin Court of Naples," *Rinascimento* 1 (1950): 200–202, 205–6, 210–12.

89. Haskins, *Studies in Mediaeval Science*, pp. 194–97.

90. Quoted by Haskins, *Renaissance*, p. 294; for the Latin text, see Haskins, *Studies in Mediaeval Science*, p. 197.

91. The available evidence regarding Moses is collected by Haskins, *Studies in Mediaeval Science*, pp. 197–206.

92. On James, see Lorenzo Minio-Paluello, "Iacobus Veneticus Grecus: Canonist and Translator of Aristotle," *Traditio* 8 (1952):265–304; Minio-Paluello, "James of Venice," *Dictionary of Scientific Biography*, 7:65–67;

Gudrun Vuillemin-Diem, "Jacob von Venedig und der Übersetzer der *Physica Vaticana* und *Metaphysica media*," *Archives d'histoire doctrinale et littéraire du moyen âge* 41 (1974):7–25; and Ezio Franceschini, "Il Contributo dell'Italia alla transmissione del pensiero greco in occidente nei secoli XII–XIII e la questione di Giacomo chierico di Venezia," in Franceschini, *Scritti di filologia latina medievale*, vol. 2 (Padua, 1976), pp. 560–88.

93. A translation of the *Posterior Analytics* is sometimes attributed to Boethius, but it did not, in any case, survive; it is thus certainly through James's translation that the West first effectively came into possession of this work. See Minio-Paluello, "James of Venice," p. 66.

94. Minio-Paluello, "Iacobus Veneticus Grecus," p. 264: "Ten generations of Latin-speaking scholars and philosophers read the *Posterior Analytics* almost exclusively in his translation; his versions of the *Physics, De anima*, three books of the *Metaphysics*, and some of the *Parva Naturalia* held the ground almost unchallenged for more than a century, and, not substantially revised, for two more centuries."

95. On Burgundio and his translations, see Haskins, *Studies in Mediaeval Science*, pp. 206–9.

96. Several of the Galenic works were also retranslations.

97. On Eugene, see Evelyn Jamison, *Admiral Eugenius of Sicily: His Life and Work* (London, 1957). The term "emir," Jamison points out (p. 33 ff.), was applied to high administrative officers, always men of Greek Christian descent, who were commanders in time of war and also the chief fiscal officers of the Norman administration. Eugene came from a line of emirs.

98. Haskins, *Studies in Mediaeval Science*, pp. 173–78.

99. Ibid., pp. 157–65.

100. On the translator's purpose in using the four works, see John E. Murdoch, "Euclides Graeco-Latinus: A Hitherto Unknown Medieval Latin Translation of the *Elements* Made Directly from the Greek," *Harvard Studies in Classical Philology* 71 (1966):268–69; Haskins, *Studies in Mediaeval Science*, pp. 178–79. Since the translator refers to the *Data, Optics,* and *Catoptrics* by their Greek titles (*Dedomena, Optica, Catoptrica*), rather than their customary Latin titles (*Data, De visu, De speculis*), it is probable that he read them in the Greek.

101. Haskins, *Studies in Mediaeval Science*, p. 191.

102. Ibid., p. 161. Despite the fact that the Greco-Latin translation was earlier, the Arabo-Latin translation was much more widely known and used.

103. On Aristippus, see ibid., pp. 160–68; Lorenzo Minio-Paluello, "Henri Aristippe, Guillaume de Moerbeke et les traductions latines médiévales de 'Météorologiques' et du 'De generatione et corruptione' d'Aristote," *Revue philosophique de Louvain* 45 (1947):206–35; Jamison, *Admiral Eugenius*, pp. xvii–xxi.

104. Haskins, *Studies in Mediaeval Science*, p. 166. Whether or not these translations were completed is unknown.

105. Murdoch, "Euclides Graeco-Latinus," pp. 263–67.

106. Ibid.

107. Haskins, *Studies in Mediaeval Science*, p. 244.

108. On Bartolomeo, see Gerardo Marenghi, "Un Capitolo dell'Aristotele medievale: Bartolomeo da Messina traduttore dei *Problemata physica*," *Aevum* 36 (1962):268–83; Haskins, *Studies in Mediaeval Science*, p. 269; S. Impellizzeri, "Bartolomeo da Messina," *Dizionario biografico degli Italiani*, 6:729–30.

109. Daniel A. Callus, "Robert Grosseteste as a Scholar," in *Robert Grosseteste, Scholar and Bishop*, ed. Callus (Oxford, 1955), pp. 33–34, 36–66; D. J. Allan, "Medieval Versions of Aristotle, *De caelo*, and of the Commentary of Simplicius," *Mediaeval and Renaissance Studies* 2 (1950):82–120. On Nicholas of Sicily, see Lorenzo Minio-Paluello, "I Due traduttori medievali del *De mundo*: Nicola siculo (greco), collaboratore di Roberto Grossatesta, e Bartolomei da Messina," *Rivista di filosofia neo-scolastica* 42 (1950):232–37.

110. On William, see Martin Grabmann, *Guglielmo di Moerbeke O.P., il traduttore delle opere di Aristotele* (Rome, 1946); Lorenzo Minio-Paluello, "Moerbeke, William of," *Dictionary of Scientific Biography*, 9: 434–40; Marshall Clagett, *Archimedes in the Middle Ages*, vol. 2 (Philadelphia, 1976), pp. 3–13; 28–53. I have followed principally Minio-Paluello in the account that follows.

111. Minio-Paluello, "Moerbeke," p. 436.

112. Ibid.

113. Weiss, "Translators from the Greek of the Angevin Court," p. 205; see Weiss for a good discussion of translating activity at the Angevin court.

114. On Niccolò's Galenic translations, see Lynn Thorndike, "Translations of Works of Galen from the Greek by Niccolò da Reggio (c. 1308–1345)," *Byzantina Metabyzantina* 1 (1946):213–35.

115. Haskins, *Renaissance*, p. 289.

116. Ibid.

117. Callus, "Grosseteste as a Scholar," pp. 39–40.

118. Roger Bacon, *Opera quaedam hactenus inedita*, ed. J. S. Brewer (London, 1859), p. 434. Bacon knows that in Italy there are many places where "the clergy and the people are purely Greek," but the presence of Greek-speaking tutors elsewhere in Europe makes it unnecessary to go there.

119. On the *studia linguarum*, see Ugo Monneret de Villard, *Lo Studio dell'Islam in Europa nel XII e nel XIII secolo* (Vatican City, 1944), pp. 35–49; José M.ª Coll, O.P., "Escuelas de lenguas orientales en los siglos XIII y XIV," *Analecta sacra Tarraconensia* 17 (1944):115–38; 18 (1945):58–89.

120. The Dominicans were imitated by others, including the Franciscans and Alfonso I of Castille.

121. Lynn Thorndike, "Berne 216, a Manuscript not used in Pansier's *Collectio ophtalmologica*," *Medievalia et Humanistica* 16 (1964):56. A little later, Dominicus's student Rufinus of Alexandria translated Hunain's *Isagoge in medicinam* with Dominicus's help, also at Murcia; see the ex-

plicit of the Erfurt manuscript, quoted in Wilhelm Schum, *Beschreibendes Verzeichnis der amplonianischen Handschriften-Sammlung zu Erfurt* (Berlin, 1887), p. 180.

122. Boethius, In *Isagogen Porphyrii commenta*, ed. Georgia Schepss and Samuel Brandt (Corpus scriptorum ecclestiasticorum latinorum, vol. 48) (Vienna-Leipzig, 1906), p. 135. It has been argued by Haskins (*Studies in Mediaeval Science*, pp. 150–52), on the basis of Burgundio of Pisa's preface to his translation of St. John Chrysostom's *Homilies on the Gospel of John*, that the painful literalism of medieval translations does not reflect an inability of medieval scholars to translate otherwise, but a reverence for the ancient sources. This is no doubt true in some cases, but it strains credibility to suppose that the average medieval translator self-consciously selected a method of translation on the basis of philosophical or moral principle. To Haskins's explanation of the literalism of medieval translations I should therefore like to suggest three additions. First, people learn how to translate (as they learn so many other skills) by imitation— by observing other practitioners and attempting to follow identical procedures, without much reflection on the underlying philosophy. In short, medieval translators strove for a literal rendering because their models had done so. Second, a word-for-word translation was safer, and medieval translators often lacked the self-confidence to free themselves from the syntax of the original. Third, when two translators cooperated through the oral use of an intermediate language, the only reasonable and efficient method was to proceed sequentially a word (or perhaps a phrase) at a time.

123. Callus, "Grosseteste as a Scholar," p. 46.

124. "Euclides Graeco-Latinus," pp. 252–53.

125. McVaugh, "Constantine the African," pp. 393–94.

126. For example, it is clear that the anonymous translator of Alhazen's *De aspectibus*, while capturing the substance of the original with considerable skill, failed to translate certain phrases, either because they were considered unnecessary or because their precise meaning was unclear; I owe this understanding to Professor A. I. Sabra, who has made extensive comparisons of the Arabic and Latin texts.

3

William A. Wallace, O.P.

The Philosophical Setting of Medieval Science

Unlike the science of the present day, which is frequently set in opposition to philosophy, the science of the Middle Ages was an integral part of a philosophical outlook. The field of vision for this outlook was that originally defined by Aristotle, but enlarged in some cases, restricted in others, by insights deriving from religious beliefs—Jewish, Islamic, and Christian. The factors that framed this outlook came to be operative at different times and places, and they influenced individual thinkers in a variety of ways. As a consequence, there was never, at any period of the Middle Ages, a uniform philosophical setting from which scientific thought, as we now know it, emerged. Rather, medieval philosophy, itself neither monolithic, authoritarian, nor benighted —its common characterizations until several decades ago—underwent an extensive development that can be articulated into many movements and schools spanning recognizable chronological periods. Not all of this development, it turns out, is of equal interest to the historian of science. The movement that invites his special attention is a variety of Aristotelianism known as High Scholasticism, which flourished in the century roughly between 1250 and 1350, and which provides the proximate philosophical setting for an understanding of high and late medieval science. Explaining such a setting will be the burden of this chapter: how it came into being, why it stimulated the activity that interests historians of science, and how it ultimately dissolved, giving way in the process to the rise of the modern era.[1]

The scope of medieval philosophy was quite broad, encompassing everything that can be known speculatively about the universe by reason alone, unaided by any special revelation. Theology, or sacred doctrine, thus fell outside the scope of philosophy, and so did practical arts and disciplines such as grammar, mechanics, and medicine. Ethics, then as now, pertained to philosophical discourse, as did logic, natural philosophy, and metaphysics. Epistemology as we know it did not yet exist, although the problem of human knowledge, its objects and its limits, interested many thinkers during this period. Psychology, the study of the soul, was regarded as a branch of natural philosophy, as were all the disciplines we now view as sciences, that is, astronomy, physics, chemistry, and biology. Even mathematics was seen as a part of philosophy, broadly speaking, but there was no general agreement on the way in which mathematical reasoning was related to natural philosophy, and disputes over this led ultimately to the mathematical physics of Galileo and Newton, which has become paradigmatic for much of modern science.[2]

Medieval Philosophy before the Reception of Aristotle

By the year 1250 the works of Aristotle were well diffused and understood in the Latin West, and the materials from which medieval science would take its distinctive form were then ready at hand. It would be a mistake, however, to think that there was no philosophy in the Middle Ages before the thirteenth century. Much of this, it is true, had developed in a theological context, when the teachings of Greek philosophers and the teachings of the Scriptures were juxtaposed. Some of the early Church Fathers were openly hostile to Greek philosophy; Tertullian saw only error and delusion in secular learning and asked, on this account, "What has Athens to do with Jerusalem?" Others, such as Clement of Alexandria and Gregory of Nyssa, having been trained in rhetoric and philosophy, saw in Christianity the answers to questions raised in those disciplines and so proposed to use them, at least as preparatory studies, in the service of revealed truth.[3] Generally the thought of Plato was seen as best approximating Christian wisdom, mainly because of the creation account in the *Timaeus* and the teachings on the soul and its immortality in the *Phaedo*.[4] The most complete blending of Neoplatonic and Christian doctrines that emerged from such syncretism appeared in the early sixth century in a series of works ascribed to Dionysius the Areopagite, the philosopher recorded as being baptized by St. Paul in Athens.[5] Because of this ascription, false though it was, these works (*On the*

Divine Names, On the Celestial and Ecclesiastical Hierarchy, and *On Mystical Theology*) were accorded unprecedented authority in the later Middle Ages.[6]

Among the Latin Fathers, the writer who gave most systematic expression to Neoplatonic thought was St. Augustine, bishop of Hippo (354–430). Trained in rhetoric and Latin letters, Augustine was first attracted to the sect of the Manichees because of the solution they offered to the problem of evil in the world. His early philosophical leanings were toward skepticism, but he turned from this to Neoplatonism after reading some "Platonic treatises" (probably the *Enneads* of Plotinus [204–70]) that had been translated into Latin by Marius Victorinus. This experience prepared for his conversion to Christianity by convincing him of the existence of a spiritual reality, by disclosing the nature of evil as a privation and not something positive, and by showing how evil did not rule out the creation of the world by a God who is all good. Convinced that philosophy is essentially the search for wisdom, itself to be found only in the knowledge and love of God, Augustine wrote his *Confessions,* his *City of God,* and many smaller treatises wherein reason and faith are closely intertwined but which, nonetheless, have a recognizable philosophical content.[7]

Augustine's early encounter with skepticism led him to anticipate Descartes' *cogito* with the similar dictum *si fallor sum*—that is, if he could be deceived, he must exist. He was aware of the limitations of sense knowledge but did not think that man's senses err; rather error arises from the way in which his soul judges about the appearances that are presented to it. A world of eternal truths exists, and the soul can grasp these because it is illumined by God to see them and to judge all things in their light. Augustine's theory of knowledge thus utilized the themes of light and illumination and was readily adaptable to Plotinus's view of creation as an emanation analogous to the diffusion of light from a unitary source. For Augustine eternal ideas are in the mind of God, where they serve as exemplars for creation and also, through the process of illumination, exercise a regulative action on the human mind; they enable it to judge correctly and according to changeless standards without themselves being seen. With regard to the created universe Augustine held that God had placed *rationes seminales* or germinal forms in matter at the beginning and that these actualize their latent potentialities in the course of time. One of Augustine's best examples of philosophizing is, indeed, his treatment of the nature of time and the paradoxes that are presented by discourse about its existence prior to the creation of the world.[8]

Plato was thus the first of the Greek philosophers to be baptized and to enter the mainstream of medieval thought under the patronage of Augustine, the pseudo-Dionysius, and other Neoplatonic writers. Aristotle, by contrast, exerted little influence in the early Middle Ages. Some of his logical writings (the "old logic," the *Categories* and *On Interpretation*) were made available to the Latin West through the translations of Boethius (ca. 480–524), however, and an interest was thereby whetted in his thought. In many ways Boethius was a mediator between ancient culture and Scholasticism, introducing once again the liberal arts (the *trivium* and the *quadrivium*), promoting the use of logic in rational inquiry, and setting the stage for the controversy over universals that was to recur in later centuries. He commented on the *Isagoge* of Porphyry and wrote a number of theological tractates that were influential for their views on the division of philosophy. He also proposed to translate into Latin all the writings of Aristotle and Plato, aiming to show the basic agreement of these thinkers in matters philosophical. The lines of his reconciliation are seen in his discussion of general ideas or universals as these were treated by Porphyry; pondering the question whether genus and species are real or simply conceptions of the mind, Boethius leaned toward the Platonic view that they not only are conceived separately from bodies but actually exist apart from them.[9]

The problem of universals posed by Boethius assumed considerable importance for Peter Abelard (1079–1142), who insisted that universality must be attributed to names, not things, thus anticipating Nominalism, though not in the precise form in which this movement achieved prominence in the fourteenth century. As Abelard saw it, there are no universal entities or things, for all existents are singular. So, if man has exact and vivid representations of objects, these apply to individuals alone; only his weak and confused impressions can fit the members of an entire class. Man's confused grasp of natures, moreover, can never match God's universal concepts of the substances that he alone creates. Such knowledge, therefore, must be associated with a word that has at best a significative or pragmatic import, seeing that it does not strictly denominate anything in the order of existents. An expert dialectician, Abelard also produced a work entitled *Sic et non*, variously translatable as "Yes and No," "For and Against," "This Way and That." Based on the conviction that controversies can often be solved by showing how various authors used the same words in different senses, Abelard's work gave powerful stimulus to the development of scholastic method.[10] The stylized proposal of questions, with arguments for and against, that resulted was used extensively in the interpretation of biblical, canonical, and philosophical texts, and is best illustrated in the *Decretum* of Gratian and the *Sentences* of Peter Lombard, which became highly successful textbooks in canon law and theology, respectively, during the later Middle Ages.[11]

Aristotle and High Scholasticism

The early Scholasticism of Abelard and others, such as St. Anselm of Canterbury (ca. 1033–1109), was developed in the Parisian School of Saint-Victor and in the cathedral school of Chartres and prepared the way for the full flowering of Scholasticism at Paris and Oxford in the thirteenth century. Around 1200 such schools were organized as the guild or "university of the masters and scholars of Paris," and that city quickly became the prestigious center of medieval European learning. Hitherto unknown works of Aristotle, together with commentaries and treatises composed by Arab and Jewish thinkers, became available there in Latin translation, and the Schoolmen were suddenly confronted with vast branches of new learning that had to be assimilated into their existing syntheses. A like situation developed at Oxford, where a university was legally constituted in 1214, at which time Robert Grosseteste (ca. 1168–1253), who was among the first to set himself to making Aristotle Catholic, assumed its leadership.[12]

Grosseteste is a convenient figure with which to begin discussion of High Scholasticism, for his Aristotelianism was more allied to Augustinian ways of thought than the Aristotelianism that developed at Paris, and at the same time it had fruitful implications for the growth of medieval science. Unlike many Scholastics Grosseteste knew Greek, translated many works from that language into Latin, including the *Nicomachean Ethics* and the *De caelo* of Aristotle and the major treatises of the pseudo-Dionysius, and commented on parts of Aristotle's "new logic" (the *Posterior Analytics* and *On Sophistical Refutations*) and the *Physics*. One could say that he pioneered in introducing Aristotelian learning to Oxford, even though his interpretations were strongly influenced by Neoplatonism. Augustine's theme of illumination pervades Grosseteste's writings; and, indeed, the resulting philosophy has been aptly described as a "metaphysics of light." In Grosseteste's view *lux* is the first form to come to primary or primordial matter, and it multiplies its own likeness or species in all directions, thereby constituting corporeal dimensionality and the entire universe according to determined laws of mathematical proportionality. Even the human soul is a special manifestation of light, although to possess knowledge it must also be illumined by God, the source of all light. Moreover, since the multiplication of force or species follows geometrical patterns, the world of nature has a mathematical substructure, and so the key to natural science will be found in mathematics. The science of geometrical optics provides an ideal illustration of the required methodology: observation and experience (*experimentum*) can provide the facts (*quia*), but mathematics is necessary to see the reason for the facts (*propter quid*). Grosseteste's ideas were taken up by the Oxford Franciscans, especially Roger Bacon and John Pecham, who did important work in optics; his emphasis on mathematics also stimulated the fourteenth-century development of physics at Merton College in Oxford, to be discussed later.[13]

Grosseteste was a secular master, but with close connections to the Franciscans, a mendicant order that had come into being along with the Dominicans early in the thirteenth century. Both orders of mendicants were welcomed at Paris by another secular master, William of Auvergne (1180s–1249), a somewhat eclectic theologian who had been made bishop of that "city of books and learning." The Paris Franciscans, whose luminaries included Alexander of Hales (ca. 1185–1245) and St. Bonaventure (ca. 1217–74), had some knowledge of Aristotle, but they preferred to work within the older Augustinian tradition and were not particularly receptive to the newly available scientific learning. The same cannot be said of the Dominican masters who assumed chairs at the University of Paris, especially St. Albert the Great and St. Thomas Aquinas. These men, as it turned out, were the architects of a new Aristotelianism at Paris that put secular learning on an almost equal footing with revealed truth and laid firm foundations for the growth of medieval science.[14]

Albert the Great (ca. 1200–1280) was apparently the first to realize how Greco-Arabic science could best serve Christian faith by granting it proper autonomy in its own sphere. He was quite willing to accord Augustine primacy in matters of faith and morals, but in medicine, as he said, he would much rather follow Galen or Hippocrates, and in physics Aristotle or some other expert on nature.[15] Remarkable for his range of interests and for his prodigious scholarly activity, Albert was called "the Great" in his own lifetime and was commonly given the title of Universal Doctor. Much of his fame derived from his encyclopedic literary activity; he made available in Latin, for example, a paraphrase of the entire Aristotelian *corpus* ranging from metaphysics through all of the specialized sciences. Himself an indefatigable observer and cataloger of nature, he added to these accounts and generated interest and enthusiasm in his students for their further development.[16]

Among Albert's students was the young Italian Dominican, Thomas Aquinas (ca. 1225–74), destined to become the greatest theologian in the High Middle Ages. From Albert, Aquinas derived his inspiration to christianize Aristotle, and subsequently became so proficient in Aristotelian methodology that his ultimate theological synthesis can almost be seen as an Aristotelianization of Christianity. Aquinas's basic metaphysical insight consisted in a thorough grasp of the Aristotelian principles of potentiality and actuality, which he first used to refine the distinction between essence and existence, and then applied the resulting doctrine in a novel way to a whole range of problems, from those relating to God and creation to that of the human soul and its activities. While treating Augustine respectfully, Aquinas preferred to speak of the potency of matter rather than of *rationes seminales*; he saw the human soul as the unique entelechy, or active principle, in man; and he substituted a theory of abstraction, effected through each man's *intellectus agens*, for Augustine's theory of divine illumination. While best known in the present day for his metaphysical and theological innovations, Aquinas was regarded in his own day as a competent logician and natural philosopher also, and his commentaries on Aristotle's *Posterior Analytics, Physics, De caelo, De generatione et corruptione*, and *Meteorology* rank among the best produced within the medieval period.[17]

For both Albert and Aquinas the natural philosophy contained in Aristotle's *Physics* was important for laying the foundations of metaphysics and theology, but it was even more important for the general theory it provided for the scientific study of the entire world of nature. In their view physics was prior to metaphysics, and they were concerned to preserve the autonomy of physics from the more abstract disciplines of mathematics and metaphysics. The difference between these disciplines is set out by Aquinas in his commentary on Boethius's *De trinitate* in terms of his theory of intellectual abstraction. Physics is the least abstract of the speculative or theoretical sciences, in that it always considers material objects that have sensible matter as part of their definition; mathematics is more abstract, in that it leaves aside such sensible matter to construct numbers and figures in the imagination out of a matter that is pure extension—called "intelligible matter" because bereft of sensible qualities; and metaphysics is most abstract, in that it separates its objects from matter entirely and considers them purely under the aspect of being. Apart from these three disciplines

Aquinas also allowed for sciences intermediate between them, which he referred to as *scientiae mediae*. Astronomy and optics would be examples of these, being situated between mathematics and physics and so using mathematical principles to attain an understanding of physical objects. Apart from such mathematical principles, of course, there would also be physical principles that are proper to natural philosophy, and these would guarantee the autonomy of natural philosophy from mathematics and metaphysics. In insisting on this autonomy, Albert and Aquinas were consciously at variance with Grosseteste and the Oxford school, not sharing in their "light metaphysics" and the mathematicism this implied.[18]

For the medievals, as has been mentioned, physics was a speculative science, but this did not mean that it consisted in haphazard and groundless speculation, as some have caricatured it. For Aquinas it took its roots from experience, which means that it had to be based on sense knowledge. His epistemology was realist in this regard, for when Aquinas spoke of sensible matter, he meant matter as possessed of qualities that are directly apprehended by the senses. Such qualities were the attributes or accidents of the individual substances in which they were perceived. Another type of accident, for him, and this in the case of corporeal substances, would be their quantitative extension, and since a quality such as heat would require bodily extension to be present in a substance, such extension could also indirectly quantify the quality and, thus, be the basis for the latter's quantification or measurement.[19] Thus, Aquinas did not rule out the possibility of a quantitative physics, even though the sense experience to which he referred was primarily qualitative, and the natural philosophy he elaborated, following Aristotle, shared in this characteristic. But the methodology he advocated was still basically empirical, since he had rejected Augustine's theory of divine illumination and held that no natural knowledge could come to man's mind without first originating in his senses.[20]

For Aquinas, again, natural philosophy was a science, a *scientia*, and as such could yield true and certain knowledge of the material universe. It would have to do so, of course, through principles that could be discovered in experience by rational inference and that would have a self-evident character, thus being able to serve as premises in a strict proof or demonstration.[21] The experience on which he relied was that of ordinary sense observation, wherein man was a passive spectator and not an active interrogator in an experimental way. Although sometimes Aquinas used the Latin word *experimentum* (as did other medievals) to refer to such experience, and even spoke of the resulting knowledge as *scientia experimentalis*, these expressions should not be taken to connote systematic and controlled experimentation in the modern sense. It should be noted also that Aquinas was acquainted with hypothetico-deductive methodology, and pointed out the conjectural character of arguments that had not been so recognized by Aristotle. Yet he identified such examples mainly in the *scientiae mediae*, such as Ptolemaic astronomy, which he recognized as merely "saving the appearances" and not as demonstrating, for example, the reality of eccentrics and epicycles.[22]

The key problem for the natural philosopher, in Aquinas's view, was that of understanding motion in terms of its causes, taking motion in the widest possible sense to mean any change perceptible in sense experience.[23] Since Aristotle had classified causes into four kinds—material, formal, efficient, and final—the investigation of the causes of motion amounted to identifying the material subject that underlies it, its formal definition, the agents that produce it, and the purposes or ends that it serves. Because of Aristotle's inclusion of final causality in this classification, his physics is generally labeled as teleological. For Aquinas, however, natural processes exhibit only an immanent teleology, in the sense that they terminate, for the most part, in forms that are perfective of the subject that undergoes the change. This he would differentiate from an extrinsic teleology, which would be some further goal or end to which the process or its product could be put. Thus, the fully grown olive tree (not the use to which olives may be put by man) is the final cause of the plant's growth from a seed, and a body's attainment of its natural place is the final cause of its movement under the influence of gravity or levity. Since natural processes are radically contingent and can always be impeded by defects in either matter or agent, Aquinas held that the natural philosopher is usually restricted to demonstrating *ex suppositione finis*, that is, on the supposition of an effect's attainment. Thus, for example, supposing that a perfect olive tree is to be generated, he can reason to all the causes that would be required to bring the generative process to completion, and this counts as scientific knowledge through causes, even though such causes might not be actually effective in the individual case.[24]

When considering the material cause of motion, we come immediately to Aristotle's hylomorphism, or matter-form theory, which pervaded all of Scholastic science.[25] In explaining the coming to be and passing away of substances by the natural processes of generation and corruption, themselves examples of substantial change, Aristotle was led to maintain that in all such changes something perdures or is conserved, whereas something else changes. The enduring substrate he called *hulē* or matter, and the changing but determining factor he called *morphē* or form; the word *hylomorphic* merely transliterates these terms and so refers to the matter-form composition of corporeal substances. It is important to note here that neither matter nor form as so conceived is itself an existing thing or substance; rather, they are substantial principles, that is, factors that enter into the composition of a substance and so are parts of its nature or essence. For this reason, and to differentiate *hulē* from the matter of ordinary experience, the Scholastics referred to it as primary matter (*materia prima*). Similarly, they called the form that gave primary matter its substantial identity and made it a substance of a particular kind the substantial form (*forma substantialis*). Primary matter, for them, is a material principle; it is undetermined, passive, and, being the same in all bodies, can serve to explain such common features as extension and mobility. Substantial form, by contrast, is a formal principle; as determining and actualizing, it can account for specific properties that serve to differentiate one kind of body from another.

For Aquinas, whose analysis of existence (*esse*) was considerably more refined than Aristotle's, *materia prima* had to be a purely potential principle, bereft of all actuality, including existence, and incapable of existing by itself even through God's creative action. Since God himself is Pure Actuality, he is at the opposite pole of being from primary matter, which is pure potentiality. Unlike Bonaventure, who followed in this the Jewish thinker Ibn Gebirol, Aquinas did not countenance the existence of a spiritual matter of which angels and human souls would be composed. Matter for him was the basic substrate of changeable or corporeal being, the proper subject of physical investigation. The presence of this qualityless protomatter that is conserved in all natural change, moreover, did not rule out a composition of integral parts: corporeal substances could also have elemental components, and to the extent that these might be separated out and themselves made to undergo substantial change, even such elements were essentially composed of primary matter and substantial form.

In Aquinas's view, therefore, as in that of other Scholastics, the four elements of Greek natural philosophy were also considered as a type of material cause.[26] These correspond roughly to the states of matter that fall under sense observation: fire (flame), air (gas), water (liquid), and earth (solid). To each of these could be assigned pairs of qualities that seemed most obvious and pervasive throughout the world of nature, that is, hot and cold, wet and dry. So fire was thought to be hot and dry; air, hot and wet; water, cold and wet; and earth, cold and dry. The elements could be readily transmuted into one another, moreover, by conserving one quality and varying the other: so fire could be changed into air by conserving its hotness and converting its dryness to wetness; air could be changed into water by conserving its wetness and converting its hotness to coldness; and so on. Such elements, for Aquinas and for Aristotelians generally, did not exist in a pure state; all of the substances that come under sense experience are composites (*mixta*) of elements in varying proportions. The combination of the elements' primary qualities thus gave rise to the entire range of secondary qualities that are observed in nature. Because Aquinas held to the unicity of the substantial form in each natural substance, this raised for him the problem of how the forms of elements could be present in compounds, and he was led to propose that they are present not actually but only virtually—a position that was widely accepted in the later Middle Ages. Like most medieval philosophers, Aquinas believed that material substances are continuous, and thus rejected Democritus's theory of matter being composed of atoms with interstitial voids. He did believe in *minima naturalia*, however, maintaining that minimum quantitative dimensions are required for the existence of most natural substances, and it is noteworthy that Albert the Great identified such *minima* with the atoms of Democritus.[27]

With regard to the formal definition of motion, Aquinas followed Aristotle in defining motion as an imperfect actuality or act, that is, the actuality of a being whose potentiality is actualized while still remaining in potency to further actualization.[28] This definition is not easy to comprehend, and, indeed, it raises many questions about the reality and existence of motion (as formulated, for example, in Zeno's paradoxes): Is motion anything more than the actuality or terminus that is momentarily attained by it at any instant, and should it be conceived as a *forma fluens* or as a *fluxus formae*? The latter question was adumbrated by Albert the Great and answers to it divided the nominalists and the realists of the fourteenth century. Ockham, for example, thought of motion as nothing more than the forms successively acquired by a subject, and so he defined motion from the nominalist viewpoint as merely a *forma fluens*. By contrast, Walter Burley, while admitting that motion could be viewed in the Ockhamist way, made the realist claim that motion is also a flux, a *fluxus formae*, that is, an actual transformation by which these new termini or forms are being successively acquired.[29] Burley saw motion also as a successive quantity which is continuous in the same way as corporeal substances appear to be; whereas bodies were static continua, however, motions (and likewise time) came to be regarded as "flowing" continua. The ways in which these various entities could be said to be constituted of quantitative parts quickly gave rise to all the problems of the continuum, many of which could not be solved before the invention of the infinitesimal calculus and modern theories of infinity.[30]

The definition of motion in terms of actuality and potentiality also had profound implications for investigating the agent causes that produce change. Since a thing could not be in both actuality and potentiality to the same terminus at the same time, it seemed obvious that no object undergoing change could be the active source of its own motion: rather, it would have to be moved by an agent that already possessed the actuality it itself lacked. Water, for example, could not be the active cause of its own heating, whereas fire could be, since fire was actually hot and could reduce the water's potentiality for heat from potency to act. This insight led medieval thinkers into an imaginative search for the movers behind all motions observed in nature, especially when such movers were not patently observable. Typical queries would be: What causes the continued motion of a projectile after it has left the hand of the thrower? What causes the fall of heavy bodies? and What causes hot water to cool when left standing by itself?[31]

Many of these problems, of course, had already been broached by Aristotle, but they acquired new interest for medievals in view of the arguments Aristotle had presented in the last part of his *Physics* to prove the existence of a First Unmoved Mover. For Aquinas, not only did Aristotle's principles open up the possibility of rendering intelligible all of nature's operations, but they could even lead one to a knowledge of the Author of nature by purely rational means. In a word, they made available reasonable proofs for the existence of a mover that was incorporeal, immaterial, of infinite power, and eternal in duration, who could be identified with the God of revelation. Small wonder, then, that for him Aristotelian physics held the greatest of promise. It allowed one to reassert the autonomy of reasoning based on sense experience, it explained the magnificent hierarchy of beings from the pure potentiality of primary matter through all the higher degrees of actuality, and it even provided access to the Pure Actuality, God himself, *Ipsum Esse Subsistens*, who had revealed the details of his inner being to all who accepted on faith his divine revelation.[32]

This brief sketch of the natural philosophy of the thirteenth century may help to explain the enthusiasm with which Aristotle's *libri naturales* came to be accepted at Paris and at similar centers in Christendom. Topics other than those already indicated occupied the attention of philosophers and theologians alike, and these cumulatively constituted the subject matter that would later become modern science. Among such topics were the nature of space and time, the existence of a vacuum and the possibility of motion through it, the kinematical and dynamical aspects of local motion, the various forces and resistances that determine a body's movement, the intensification of qualities, and a variety of problems associated with the structure of the cosmos and the relationships between celestial and terrestrial motions.[33] Such topics were approached anew with great confidence, for Greek learning now appeared to be buttressed by Catholic faith, and an all-knowing God seemed to be beckoning men, as it were, to uncover the rationality and intelligibility hitherto concealed in his material creation. Historians have seen in this situation the basic charter that underlies the whole scientific enterprise.[34] Whether this be true or not, there seems little doubt that between Oxford and Paris some fundamental contributions had already been made, such as highlighting the problem of the role of mathematics in physical science, asserting the primacy of empirical investigation in studying the world of nature, and granting physics its autonomy from metaphysics and theology as a source of valid knowledge concerning the cosmos.

The Crisis of Averroism

Mention has been made of Islamic influences on the development of natural philosophy, and these now need to be examined in some detail. The problem of the relationship between reason and belief came to a head earlier in Islam than it did in Latin Christendom, with results that very often asserted the primacy of reason over faith, rather than the other way around, as accepted without question by the Latins during the long period from Augustine to Aquinas. The main inspiration behind the Arab position was Averroes (1126–98), whose thought has been characterized as a twelfth-century rationalism similar in some respects to the later movement in modern Europe. Apart from Averroes' polemical writings against Arab divines such as Algazel, however, what is most significant is that he commented in detail on all the works of Aristotle, with such skill that when his writings were made available in Latin they earned for him the undisputed title of "Commentator," the interpreter of Aristotle *par excellence*. The Aristotle he interpreted, moreover, was somewhat Neoplatonized, but he had not been baptized and, thus, not contaminated with elements surreptitiously derived from Christian doctrine. Where Aquinas, therefore, had made room for a rational understanding of the universe in the light of Aristotle but in a general thought context provided by faith, Averroes pushed the claims of unaided reason even further. And where Aquinas felt compelled to question Aristotle's authority and even to modify his teachings as the occasion demanded, Averroes was under no such constraint; indeed, he regarded Aristotle as a god, the summit of all rational understanding, an infallible guide to knowledge of the world of nature.[35]

Averroes' contributions, of course, were made in an Islamic milieu and were critical of the thought of other Arabs, such as Avicenna, rather than that of the early Scholastics. Among his distinctive theses was the teaching that there is only one intellect for all men. Avicenna (980–1037) had taught that there was a single *intellectus agens,* or active intellect, but that each individual possessed his own *intellectus possibilis,* or passive intellect, wherein he would have his own ideas. Averroes disagreed with this teaching, holding that both the passive and the agent intellects were a separated substance, one and the same for all mankind, and so denying the accepted basis for man's personal immortality. Another of Averroes' theses was the eternity of the universe, for he followed Aristotle literally in maintaining that the heavens, motion, time, and primary matter had no temporal beginning or end; so they were not created, nor would they ever cease to exist. He disagreed too with Avicenna's teaching on essence and existence, and this affected his understanding of the relationships between God and the universe, effectively necessitating God's action in ways that would turn out to be contrary to Christian teaching.[36]

Averroes' commentaries on Aristotle were known to William of Auvergne, Albert the Great, and others who first advanced the cause of Aristotelianism at the University of Paris. The fact that Averroes' teachings could be inimical to Catholic belief was not immediately recognized, but as Averroes' influence increased along with the reception of Aristotle, his distinctive interpretations gradually got a hearing. Not only that, but they soon attracted adherents within the faculty of arts. Thus, a movement got underway that has been characterized as Latin Averroism or heterodox Aristotelianism, whose chief proponents were Siger of Brabant and Boethius of Sweden. The Latin Averroists taught the oneness of the passive intellect for all men, the eternity of the universe and of all its species (with the result that there would be no first man), and the necessity of God's causality in the world, although they admitted contingency in the sublunary regions because of matter's presence there. These teachings had ramifications that were opposed to Christian faith, and it is noteworthy that the

Latin Averroists never denied this faith, although they held that their philosophical conclusions were probable, or even necessary.[37]

The theologians at the University of Paris reacted, predictably, to such teachings, and Bonaventure, Albert the Great, Aquinas, Giles of Rome, and others all wrote polemical treatises against Siger and his followers. Alarmed at the growth of naturalistic rationalism, ecclesiastical authorities also attempted to halt heterodox Aristotelianism with a series of condemnations. In 1270 the bishop of Paris, Etienne Tempier, reinforcing some earlier prohibitions, condemned thirteen propositions that contained Averroist Aristotelian teachings, namely, the oneness of the intellect, the eternity of the world, the mortality of the human soul, the denial of God's freedom and providence, and the necessitating influence of the heavenly bodies in the sublunary world. Tempier followed this on March 7, 1277, with the condemnation of 219 propositions, all linked with philosophical naturalism, including some theses upheld by Aquinas. In the prologue to the condemnation, Tempier accused the Averroists of saying that what was true according to philosophy was not true according to the Catholic faith, "as if there could be two contrary truths," thus giving rise to what has been called a theory of double truth, although such a theory was not explicitly advocated by Averroes, Siger, or anybody else in the movement.[38]

It is undeniable that the condemnation of 1277 had an effect on the development of medieval science, although not as profound as was maintained by Pierre Duhem, who actually proposed 1277 as the birthdate of modern science. Duhem's argument was that among the condemned propositions were two that bore on subject matters later of interest to scientists: one, denying that God had the power to move the universe with a rectilinear motion for the reason that a vacuum would result; the other, denying God's power to create more than one world. By thus opening up the possibility that theses rejected by Aristotle were true, and asserting God's omnipotence as a factor that would henceforth have to be taken into account when deciding cases of possibility or impossibility in the cosmological order, Tempier stimulated the scientific imagination, as Duhem saw it, and so opened the way to a full consideration of various alternatives to an Aristotelian cosmology.[39]

It is generally agreed that Duhem's thesis is extreme, for there is no indication of a spurt in scientific thought or activity following 1277, and it is doubtful whether any authoritarian restriction on cosmological teachings could have stimulated the free spirit of inquiry that is generally seen as characteristic of modern science. There is no doubt, however, that the condemnation of 1277 did have profound consequences on the development of natural philosophy in the decades that followed, and particularly on the relationships between philosophical and theological thought, as will be discussed presently.[40]

While on the subject of Averroism, it should be noted that Averroes had a number of distinctive teachings relating to natural philosophy that were incorporated into commentaries and questionaries on Aristotle's *Physics*, thus becoming part of the Aristotelian tradition that would be taught in the universities to the end of the sixteenth century. Among these were the theses that substantial form is prior to and more knowable than the substance of which it is a part, which takes its entire essence or quiddity from the form; that motion through a vacuum would be instantaneous (which has the effect of ruling out motion in a vacuum, since no motion can take place in an instant); that the substantial form is the principal mover as an active principle in the natural motion of heavy and light bodies, and that the gravity and levity of such bodies are secondary movers as instruments of the substantial form; that a projectile is moved by the surrounding medium, and so there can be no projectile motion in a vacuum; that mover and moved must always be in contact, with the result that action at a distance is impossible; and that curvilinear motion cannot be compared to rectilinear motion because the two are incommensurable. Many of these teachings were propounded by John of Jandun (ca. 1275–1328) at Paris in the early fourteenth century, and they were commonly accepted in northern Italy, especially at Padua, from the fourteenth through the sixteenth century. Not all commentators, of course, subscribed to them. Thomists, for example, basing their interpretations of Aristotle on the many writings of Aquinas, taught that the essence or quiddity of a natural substance must include matter as well as form; that motion through a vacuum would not be instantaneous; that the principal mover in natural motion is the generator of the moved object (that which brought it into being) and that internal forms such as gravity serve only as passive principles of such motion; that a projectile can be moved by an internal form, or impetus, analogous to the form of gravity for falling motion; and that actual physical contact between mover and moved is not essential, but that in some cases a virtual contact (*secundum virtutem*) suffices. These few examples may serve to show that even within Aristotelianism there was no uniform body of doctrine, and that different schools, such as Averroists and Thomists, had their own distinctive teachings in natural philosophy, despite the fact that they took their basic inspiration from Aristotle.[41]

The Critical and Skeptical Reaction

To come now to other schools that figured in late medieval science, we must return again to the Franciscans, who took the ascendancy in the late thirteenth and early fourteenth centuries through the writings of John Duns Scotus and William of Ockham, themselves Franciscan friars. Both the Scotistic and the Ockhamist movements were part of a critical and skeptical reaction in philosophy, primarily motivated by theological interests, and following on the condemnation of 1277. The High Scholasticism of the thirteenth century had seen Aristotle welcomed enthusiastically as "the master of all who know." Now the theologians had pointed out what disastrous consequences could attend the uncritical acceptance of Aristotle's teachings in matters that touched on their discipline. There was no doubt, then, that Aristotle had erred in matters theological; might it not be the case that he likewise erred in matters philosophical? Both Scotus and Ockham implicitly answered this question in the affirmative, and in so doing set Scholasticism on a different course from that which it had been following under the inspiration of Albert the Great and Thomas Aquinas.[42]

Duns Scotus (ca. 1265–1308) is known as the Subtle Doctor, and his writings abound in fine distinctions and closely reasoned arguments that put him very much in the Scholastic mold. He was a critical thinker, moreover, and quite concerned with elaborating a systematic metaphysics, which, in turn, had important consequences for his natural philosophy. He understood Arabic thought quite well, consistently favoring Avicenna over Averroes. Again, though he denied the theory of divine illumination, he was much indebted to Augustine, Bonaventure, and the Oxford tradition within Franciscan thought. Unlike Aquinas, he was not so much interested in assimilating philosophy into theology as he was in preserving the autonomy, and, indeed, the very possibility, of theology against the encroachments of a naturalistic rationalism.[43]

Scotus attempted to do this by developing his own theory of knowledge, which focused on being as a univocal concept; by stressing the primacy of the will, to assure God's absolute freedom and also the preeminence of freedom in man; and by viewing God under the aspect of infinity as his essential characteristic. He held to primary

matter but did not conceive this as pure potentiality, as had traditional Aristotelians; rather, he saw it as a positive reality and actuality
capable of receiving further perfection. Moreover, apart from matter
and form, Scotus held that every concrete reality also has metaphysical components of universality and particularity, thereby reopening the
debate over universals. Every being, in his view, contains a common
nature (*natura communis*) that is itself indifferent to such universality and particularity but that is rendered particular by an individuating principle, which he referred to as "thisness" (*haecceitas*). Scotus
developed also a complex system of distinctions, including a novel
one, the formal distinction, which he proposed as intermediate between the real distinction and the distinction of reason generally invoked by the Scholastics.[44] He denied the necessary validity of the
principle, "Whatever is moved is moved by another," saying that this
is true of violent but not of natural motions, and, thus, he did not use
this principle, as Aquinas had, to prove the existence of a First Unmoved Mover. His own proof for the existence of God was distinctive,
being based on an analysis of the notion of possibility, and showing
how this entails the existence of a necessary uncaused being.[45]

Even from this brief summary it can be seen that Scotus was more
the metaphysician than the natural philosopher, and that in some
ways his desire to guarantee the possibility of theology as a science
led him to negate the gains made by Aquinas and Albert the Great in
their attempts to maintain the autonomy of reason against the Augustinians. Moreover, though Scotus and his followers did not neglect natural philosophy entirely, their importance for medieval science derives
less from their positive contributions to that discipline than from the
skeptical reaction they provoked from William of Ockham. Ockham
(ca. 1285–ca. 1349) had been exposed to Scotistic teaching during
his years of training in the Franciscan Order, and he developed his
own thought in conscious opposition to that of the Subtle Doctor. Like
that of Scotus, however, Ockham's intent was theological from the
beginning, and although his critique of Aristotle was indeed philosophical, its motivation is directly traceable to the condemnation of
1277. Again, like Scotus, Ockham stressed the traditional Franciscan
themes of divine omnipotence and divine freedom and was concerned
to eliminate any element of necessity from God's action, as this had
been found in Neoplatonic emanationism and in Arab thought generally.[46]

In working out his own philosophical position, Ockham consistently
invoked two main theses. The first was that God has the power to
do anything whose accomplishment does not involve a contradiction.

The net effect of this teaching was to admit that, in the order of nature, whatever is not self-contradictory is possible; thus, there is no *a priori* necessity in nature's operation, and whatever is the case must be ascertained from experience alone. Ockham's second thesis was a principle of parsimony, referred to as "Ockham's razor" and commonly expressed in the maxim "Beings are not to be multiplied without necessity." The application of this principle led Ockham to formulate a new logic, similar to the nominalism of Abelard, wherein he no longer sought to find real counterparts in the universe for all the categories, as most Aristotelians after Boethius had done.[47] Concepts or universals for Ockham became simply words, and the only real existents were "absolute things" (*res absolutae*), which he conceived of as individual substances and their qualities. All other categories were to be regarded as abstract nouns, used for the sake of brevity in discourse but having no real referent other than substance and quality. Much of Ockham's polemic was, in fact, directed against Scotus's "common nature" and his formal distinction between such a nature and its individuating principle. Ockham also denied, however, Aquinas's real distinction between essence and existence, and most other metaphysical distinctions that had come to play a dominant role in thirteenth-century Scholasticism.[48]

Fourteenth-century nominalism, as fathered by Ockham and quickly taken up by others, thus incorporated a view of the universe that was radically contingent in its being, where the effect of any secondary cause could be dispensed with and immediately replaced by God's direct causality. The theory of knowledge on which it was based was empiricist, and the problems it addressed were mainly those of the philosophy of language. While Scholastic in setting, Ockham's philosophy was thoroughly modern in orientation. Referred to as the *via moderna*, in opposition to the *via antiqua* of the earlier scholastics, it has been seen as a forerunner of the modern age of analysis—indeed, as a fourteenth-century attempt to unite logic and ontology in ways that had to await the twentieth century for their more rigorous formulation.[49]

In natural philosophy, following Scotus, Ockham accorded actual existence to primary matter and saw form as providing geometrical extension, more as *figura* than as *forma substantialis*, which in the Thomistic understanding confers actual existence on the composite. Motion, for Ockham, could not be an absolute entity, and, thus, it was not a reality distinct from the body that is in motion. Most of the difficulties in prior attempts to define motion, as Ockham saw it, arose from the inaccurate use of language, from speaking of motion as if it

were something different from the body moved and the terminus it
attains. In effectively rejecting Aristotle's definition of motion as the
actualization of the potential *qua* potential, Ockham also had to dis-
pense with the need for a mover to produce local motion, whether
this be located in the medium through which the body moves or within
the moving body itself. Some have seen in this rejection of motor
causality an adumbration of the concept of inertia, but this seems un-
warranted, as Ockham dispensed with a moving cause not merely in
the case of uniform motion in a straight line, but in all instances of
local movement.[50] And, in the long run, Ockham's analysis of motion
was not very profound—its logic led to the rejection of the very
reality it was devised to explain—although it did have some consequences for the development of medieval science at Merton College
in Oxford and at Paris, as will be explained later.[51]

Ockham's philosophy may be viewed as the first consistent attempt
to renovate Aristotelian Scholasticism, but it was neither the most
critical nor the most radical. Other thinkers who reacted yet more
skeptically include John of Mirecourt and Nicholas of Autrecourt.
John of Mirecourt (fl. 1345) insisted on the merely probable character of most human knowledge—this because he had extreme views as
to what might guarantee certitude and because he, like Ockham,
wished to make due allowance for the unlimited freedom of God. In
philosophy, for John, there is little hope of reaching anything better
than probability, since sense knowledge is deceptive, truths can rarely
be reduced to the principle of noncontradiction, and God can always
intervene miraculously to produce a different result. A similar strain
runs through the writings of Nicholas of Autrecourt (d. after 1350),
who likewise held that all knowledge arises from sensation and that
there can be only one valid criterion of certitude, again the principle
of noncontradiction. Since the senses deceive and since it is difficult to
resolve any arguments to noncontradictory assertions, human knowledge must be essentially limited to probabilities. Applying these principles to Aristotelian physics, and particularly to its teaching on causality, Nicholas argued that the atomism that was rejected by Aristotle
is just as likely as his theory of matter-form composition, and that
even if causality exists in nature it can never be demonstrated. Because of such skeptical views Nicholas has been seen as the "medieval
Hume," the forerunner of modern empiricism.[52]

By the middle of the fourteenth century, then, under the critical
and skeptical attacks of philosophers as diverse as Scotus, Ockham,
and Nicholas of Autrecourt, the Scholastic program that had been

initiated with such enthusiasm by Grosseteste, Albert, and Aquinas effectively came to an end. The newer, more critical movements did not negate the basic Aristotelian insights that had enlivened the theology of the High Middle Ages, and, indeed, a new philosophy of nature was about to emerge that would be eclectic in many particulars, but still would contain the seeds from which modern science could arise in the early seventeenth century. By this time, however, High Scholasticism had already peaked and had begun to disintegrate, fragmented into many schools and opposing factions, and given over to endless subtleties of disputation that were to be caricatured by Renaissance humanists and moderns alike. The final impression would be that Scholastic Aristotelianism had failed, initially because reason had claimed too little, ultimately because it had claimed too much, as it competed with the Christian faith in its efforts to seek an understanding of the world of nature.

The New Natural Philosophy

At this stage the elements were at hand for the forging of new and innovative patterns of scientific thought, which would broaden natural philosophy beyond the bounds set for it by Aristotle. The critical and skeptical reactions to Aristotle were the immediate forerunners, but they had little direct influence on the new natural philosophy that was making its way through the tangle of traditional ideas. They did reinforce, however, the general impression already produced by the condemnation of 1277, namely, that Aristotle's views had to be examined critically, corrected, reformulated, and sometimes rejected entirely—and this when they came into conflict not only with divine revelation but also with the manifest data of experience. The principal innovations that resulted in the natural philosophy of the mid-fourteenth century, when this finally assumed recognizable contours, were the emergence of new mathematical methods for use in physical investigations and the introduction of kinematical and dynamical concepts that would finally raise the mechanics of moving bodies to the status of a science. Neither of these innovations was especially fostered by Scotus or Ockham, whose philosophies had little need of mechanical concepts and who assigned to mathematics a role in physics no larger than that given it by Albert and Aquinas and far smaller than that given it by Grosseteste. As it turned out, however, the latter's mathematical tradition was still alive at Oxford, and it was there, at Merton College, that the first innovations were made. These quickly passed to the University of Paris, where they merged with new mechanical concepts, and thence were transmitted to other centers of learning in western Europe.[53]

A fuller elaboration of the resulting conceptual development is taken up in chapter 7 of this volume. Here only a few general observations need be made about this development as it relates to matters already discussed. At Oxford the principal contributors to the new natural philosophy were Walter Burley, Thomas Bradwardine, William of Heytesbury, and Richard Swineshead.[54] Burley (ca. 1275–ca. 1345), as we have seen, is noteworthy for his "realist" reaction to Ockham's nominalism and for reopening most of the problems relating to the causal agents involved in local motion that Ockham had sought to bypass. Whereas Burley thus initiated a traditionalist revival at Oxford, Bradwardine (ca. 1295–1349), on the other hand, was more the innovator. His own interests were heavily mathematical, and he set himself the problem of resolving some of the internal contradictions that were detectable in Aristotle's so-called dynamical laws. In Bradwardine's day many Arabic and Latin commentators were interpreting Aristotle's statements, especially those in the fourth and seventh books of the *Physics*, to imply precise quantitative relationships between motive force, resistance, and velocity of movement. In such a context Bradwardine proposed an ingenious interpretation of Aristotle that used a relatively complex mathematical function to render consistent the various ratios mentioned in his writings. This formulation, while incorrect from the viewpoint of Newtonian mechanics, introduced the concept of instantaneous velocity and adumbrated some of the computational apparatus of the infinitesimal calculus. Bradwardine was followed at Merton College by Heytesbury and Swineshead, who presupposed the validity of his dynamic analysis and extended it to a fuller examination of the comparability of all types of motions or changes. In so doing they discussed the intension and remission of forms, and spoke of the "latitude of forms," regarding even qualitative changes as traversing a distance (*latitudo*) and, thus, as quantifiable. They also employed a letter calculus that lent itself to the discussion of logical subtleties—the *sophismata calculatoria* soon to be decried by humanists. Such "calculations" were of unequal value from the viewpoint of natural philosophy, but they did suggest new techniques for dealing with the problems of infinity; they also led to a sophisticated terminology for describing rates of change that would have important applications in mechanics.[55]

Mertonians such as Bradwardine, Heytesbury, and Swineshead were sympathetic to nominalism, and, as a consequence, they did not have the realist concerns that were to become influential at Paris in the mid-fourteenth century. They were highly imaginative in their treatment of kinematical problems, but did so in an abstract mathematical way, generally without reference to the motions actually found in nature. By contrast, a group of thinkers at the University of Paris devoted themselves rather consistently to investigating the physical causes of motion, introducing the concept of impetus and quantifying, in ways suggested by the Mertonians, the forces and resistance involved in the natural movements of bodies. Foremost among these were the Scotist, Franciscus de Marchia, and Jean Buridan, Albert of Saxony, and Nicole Oresme, the last three generally referred to as "Paris terminists."[56] Terminism is sometimes equated with nominalism, and it is true that these thinkers were all nominalist in their logic, making extensive use of Ockham's *logica moderna*, but in natural philosophy they rejected the nominalist analysis of motion and developed realist views of their own.

Jean Buridan is regarded as the leader of the Paris group, playing a role there similar to that of Bradwardine at Oxford, and being best known for his development of the concept of impetus as a cause of projectile motion and of the acceleration of falling bodies. He also defended, against Nicholas of Autrecourt, the character of natural philosophy as a science *secundum quid* (that is, in a qualified sense) because based on evidence *ex suppositione*, thus using Aquinas's methodological expression; his concern was not merely with the contingency found in nature, however, but rather with the possibility of nature's order being set aside miraculously through divine intervention.[57] Buridan's pupils, Albert of Saxony and Nicole Oresme, showed greater competence than he in mathematics and applied Mertonian techniques to the discussion of both terrestrial and celestial motions. Oresme (ca. 1320–82) pioneered, in fact, in the development of geometrical methods of summing series and integrating linear functions, and adumbrated some of the concepts of analytical geometry. The writings of the Paris terminists were widely diffused throughout western Europe and were much discussed in commentaries and questionaries on the *Physics* because of their obvious relevance to the problems of natural philosophy. It is for this reason that pioneer historians of medieval science, such as Duhem, spoke of them as the "Parisian precursors of Galileo."[58]

By the end of the fourteenth century, therefore, a considerable body of new knowledge had become available that was basically Aristotelian and yet had been enriched by mathematical and dynamical concepts showing considerable affinity with those of modern science.

The history of the diffusion of this new natural philosophy throughout the fifteenth and sixteenth centuries is quite complex, and, indeed, has yet to be written in detail. Suffice it to say that the diversity of schools and movements continued, although with a noticeable relaxation of the fierce partisan loyalties that had characterized debates in the late thirteenth and early fourteenth centuries. A tendency toward eclecticism began to manifest itself, with most commentators picking and choosing theses that suited their purposes and seemed most consistent with their own experiences. In such an atmosphere, full-length commentaries on the *Physics* gave way to shorter tracts on various subjects, and treatises entitled *De motu* appeared in increasing numbers. Some of these were nominalist, others Thomist or Scotist, yet others Averroist in inspiration, but all covered essentially the same subject matter—motion, its definition, its causes, its quantitative aspects. Such treatises were taught in the universities when Galileo and the other founders of modern science pursued their formal studies, and they provided the proximate background for the emergence of the "new science" of the seventeenth century.[59]

This, then, completes our account of the philosophical setting of medieval science. The ideas discussed were developed over a span of a thousand years, from Augustine to Oresme, though they had received their initial formulation in Greek antiquity, in the writings of Plato and Aristotle. What we now call medieval science, as it was understood over most of this period, was actually identical with natural philosophy, or *scientia naturalis*, except for the ancillary role played by the *scientiae mediae* in the development of mathematical methodology.[60] It is perhaps noteworthy that most of the problems of natural philosophy, and particularly those formulated by Aristotle, still resist definitive solution in the present day, and in the main they have passed into a related discipline known as the philosophy of science, where realists and nominalists (now called positivists) continue to be divided over the basic issues.[61] How seventeenth-century science succeeded in disentangling itself from philosophy and in defining its own limits in apparent independence of philosophical thought still awaits adequate treatment. As far as the science of the Middle Ages is concerned, however, this problem did not present itself. It was part and parcel of man's attempt to comprehend the world of nature with the light of unaided reason. Once this is understood, the Middle Ages can no longer be regarded by historians of science as the Dark Ages, but, rather, must be seen as a period of gradual enlightenment, culminating in the thirteenth and fourteenth centuries, when recognizable foundations were laid for the modern scientific era.

Notes

1. A good general introduction to philosophy in the Middle Ages is Frederick Copleston's *A History of Medieval Philosophy* (London, 1972), which contains an extensive and up-to-date bibliography relating to all philosophers and schools mentioned in this chapter. Also basic as a reference work, although it accents metaphysics to the neglect of natural philosophy, is Etienne Gilson's *History of Christian Philosophy in the Middle Ages* (New York, 1955). General surveys of intellectual history of the Middle Ages will be found in David Knowles, *The Evolution of Medieval Thought* (New York, 1962), and in Gordon Leff, *Medieval Thought: St. Augustine to Ockham* (Baltimore, 1958). Shorter treatments that are practically classics in the field are Etienne Gilson's *Reason and Revelation in the Middle Ages* (New York, 1938), which focuses on Augustine, Anselm, Averroes, and Aquinas; Paul Vignaux, *Philosophy in the Middle Ages: An Introduction*, trans. by E. C. Hall (New York, 1959), which explains well the period from Anselm to Ockham; and Fernand Van Steenberghen, *Aristotle in the West: The Origins of Latin Aristotelianism*, trans. by Leonard Johnston (Louvain, 1955), which covers in detail the period from 1200 to 1277. For individuals, concepts, and movements, brief but informative summaries are to be found in specialized encyclopedias, especially the *New Catholic Encyclopedia*, ed. W. McDonald, 15 vols. (New York, 1967) (hereafter *NCE*), which gives extensive coverage to medieval philosophy and its relation to science; see also the supplement, vol. 16 (Washington, 1974). The *Encyclopedia of Philosophy*, ed. Paul Edwards (New York, 1967), and the *Dictionary of Scientific Biography*, ed. Charles C. Gillispie (New York, 1970–) (hereafter *DSB*) likewise include up-to-date articles on the more important personages; in the latter see, for example, G. E. L. Owen et al., "Aristotle," 1:250–81.

2. See chapter 14. For a comprehensive survey of natural philosophy in the Middle Ages and its relation to medieval science, see James A. Weisheipl, *The Development of Physical Theory in the Middle Ages* (New York, 1959), which contains a guide to further reading. Also helpful is the article by Olaf Pedersen, "The Development of Natural Philosophy, 1250–1350," *Classica et Mediaevalia* 14 (1953):86–155. The collected papers of Ernest A. Moody, published under the title *Studies in Medieval Philosophy, Science, and Logic* (Berkeley, 1975), contain considerable material relating to the themes of this chapter, and are particularly good on William of Ockham. William A. Wallace, *Causality and Scientific Explanation*, vol. 1. *Medieval and Early Classical Science* (Ann Arbor, 1972), stresses elements of methodological and epistemological continuity from the thirteenth to the seventeenth century. More detailed studies are to be found in Anneliese Maier, *Studien zur Naturphilosophie der Spätscholastik*, 5 vols. (Rome, 1949–58).

3. Gilson, *Reason and Revelation*, pp. 3–33; Copleston, *History of Medieval Philosophy*, pp. 17–26.

4. *Timaeus* 52D–57C; *Phaedo* in its entirety; see also D. J. Allan, "Plato," *DSB* 11:22–31.

5. Acts 17:34; see F. X. Murphy, "Pseudo-Dionysius," *NCE*, 11:943–44.

6. Copleston, *History of Medieval Philosophy*, pp. 50–56.
7. See E. McMullin, "St. Augustine of Hippo," *DSB*, 1:333–38, and A. C. Lloyd, "Plotinus," ibid., 11:41–42.
8. *Confessions*, bk. 11, par. 17–20; Copleston, *History of Medieval Philosophy*, pp. 27–49.
9. L. Minio-Paluello, "Boethius," *DSB*, 2:228–36; Knowles, *Evolution of Medieval Thought*, pp. 107–15.
10. L. Minio-Paluello, "Pierre Abailard," *DSB*, 1:1–4; E. A. Synan, "Universals," *NCE*, 14:452–54.
11. I. C. Brady, "Scholasticism," *NCE*, 12:1153–58; J. A. Weisheipl, "Scholastic Method," ibid., pp. 1145–46.
12. Gordon Leff, *Paris and Oxford Universities in the Thirteenth and Fourteenth Centuries* (New York, 1968).
13. A. C. Crombie, "Robert Grosseteste," *DSB*, 5:548–54; A. C. Crombie and J. D. North, "Roger Bacon," ibid., 1:377–85; David C. Lindberg, "John Pecham," ibid., 10:473–76; Lindberg, *Theories of Vision from al-Kindi to Kepler* (Chicago, 1976), pp. 94–102.
14. Gilson, *Reason and Revelation*, pp. 69–99.
15. *Librum II sententiarum*, dist. 13, art. 2, in *Omnia opera*, ed. A. Borgnet, vol. 27 (Paris, 1894), p. 247a.
16. James A. Weisheipl, "Albert the Great," *NCE*, 1:254–58; William A. Wallace, "Albertus Magnus," *DSB*, 1:99–103.
17. William A. Wallace, "Thomas Aquinas," *DSB*, 1:196–200; James A. Weisheipl, *Friar Thomas D'Aquino: His Life, Thought and Works* (New York, 1974).
18. Weisheipl, *Development of Physical Theory*, pp. 48–62.
19. There are two resulting types of measurement of a quality such as heat, one based more directly on the quantitative extension of the body, called quantity of heat, and the other based on the qualitative intensity itself, called degree of heat (the modern notion of temperature). See R. F. O'Neill, "Quality," *NCE*, 12:2–5; C. F. Weiher, "Extension," ibid., 5:766–67; and William A. Wallace, "Measurement," ibid., 9:528–29.
20. Pedersen, "Development of Natural Philosophy," pp. 98–100.
21. William A. Wallace, "Science (*Scientia*)," *NCE*, 12:1190–93; M. A. Glutz, "Demonstration," ibid., 4:757–60; E. Trépanier, "First Principles," ibid., 5:937–40.
22. Pedersen, "Development of Natural Philosophy," pp. 90–98.
23. M. A. Glutz, "Change," *NCE*, 3:448–49.
24. William A. Wallace, "Aquinas on the Temporal Relation Between Cause and Effect," *Review of Metaphysics* 27 (1974):569–84.
25. William A. Wallace, "Hylomorphism," *NCE*, 7:284–85; V. E. Smith, "Matter and Form," ibid., 9:484–90; A. Robinson, "Substantial Change," ibid., 13:771–72.
26. Pedersen, "Development of Natural Philosophy," pp. 102–3; for Aquinas's understanding of how elements persist in compounds, see *A Source Book in Medieval Science*, ed. Edward Grant (Cambridge, 1974), pp. 603–5.
27. *Librum I de generatione*, tract. 1, cap. 12, in *Omnia opera*, ed. Borgnet, vol. 4 (Paris, 1890), p. 354b; for the relation of *minima* to atomism, see William A. Wallace, "Atomism," *NCE*, 1:1020–24.

28. M. A. Glutz, "Motion," *NCE*, 10:24–27.
29. Note that on p. 100b of my article on Albert cited in n. 16, these positions are incorrectly reversed. On *forma fluens* and *fluxus formae*, see chapter 7, below.
30. For a survey, see the articles by John Murdoch and Edith Sylla on Walter Burley and Richard Swineshead, *DSB*, 2:608–12, and 13:184–213.
31. James A. Weisheipl, "The Principle *Omne quod movetur ab alio movetur* in Medieval Physics," *Isis* 56 (1965):26–45.
32. Weisheipl, *Development of Physical Theory*, pp. 31–48.
33. See chapters 6–8, below; also Pedersen, "Development of Natural Philosophy," pp. 107–14.
34. Alfred North Whitehead, *Science and the Modern World* (New York, 1967), pp. 13–18.
35. Gilson, *Reason and Revelation*, pp. 35–66; Copleston, *History of Medieval Philosophy*, pp. 104–24.
36. B. H. Zedler, "Arabian Philosophy," *NCE*, 1:722–26, and "Averroes," ibid., pp. 1125–27; L. Gardet, "Avicenna," ibid., pp. 1131–32. See also R. Arnaldez and A. Z. Iskandar, "Ibn Rushd," *DSB*, 12:1–9.
37. A. Maurer, "Latin Averroism," *NCE*, 1:1127–29.
38. "The Condemnation of 1277," in Grant, *Source Book*, pp. 45–50; B. H. Zedler, "Theory of Double Truth," *NCE*, 4:1022–23.
39. Pierre Duhem, *Etudes sur Léonard de Vinci*, vol. 2 (Paris, 1909), p. 412.
40. Edward Grant, *Physical Science in the Middle Ages* (New York, 1971), pp. 27–35, 84–90.
41. These observations are based on my studies of sixteenth-century commentaries and questionaries on Aristotle's *Physics*, which I have not yet had the chance to publish; some background information is provided by James A. Weisheipl in "Motion in a Void: Aquinas and Averroes," *St. Thomas Aquinas Commemorative Studies 1274–1974* (Toronto, 1974), 1:469–88, and in the article cited in note 31.
42. Vignaux, *Philosophy in the Middle Ages*, pp. 146–213; Copleston, *History of Medieval Philosophy*, pp. 213–56.
43. Gordon Leff, "John Duns Scotus," *DSB*, 4:254–56; C. Balic, "John Duns Scotus," *NCE*, 4:1102–06, and "Scotism," ibid., 12:1226–29.
44. A real distinction exists between things when they are nonidentical in their own right, apart from any insight of human reason; thus, there is a real distinction between a dog and a man, and between Peter and John. A modal distinction is also considered a real distinction, though weaker than that between one thing and another; it is the difference between a thing and its mode, for example, between a line segment and the point that terminates the segment, or between a stick and its ends. A distinction of reason, by contrast, originates in the mind that understands or reasons about things, and formulates a proposition such as "Man is man" or attributes to Peter predicates such as "body" and "living." The distinction between "man" as it is the subject of the proposition and "man" as it is the predicate is called a distinction of reason reasoning (*rationis ratiocinantis*) because it arises *only* in the mind formulating the proposition. The distinction between "body" and "living" as said of Peter, on the

other hand, is said to be a distinction of reason reasoned about (*rationis ratiocinatae*) because, though "body" and "living" are both really the same as Peter, there is an objective foundation in Peter, that is, in the thing reasoned about, that gives rise to the diverse predicates. Now Scotus's formal distinction is said to be midway between the modal real distinction and the distinction of reason reasoned about; on this account it is called the "intermediate distinction," that is, intermediate between the lesser of the real distinctions and the greater of the distinctions of reason. According to Scotus, the formal distinction is what differentiates the individuating principle, "thisness," from the common nature. The fineness of the distinction perhaps gives some indication as to why Scotus is referred to as the Subtle Doctor. See J. J. Glanville, "Kinds of Distinction," *NCE*, 4:908–11.

45. Roy R. Effler, *John Duns Scotus and the Principle "Omne quod movetur ab alio movetur"* (St. Bonaventure, N.Y., 1962); Johns Duns Scotus, *Philosophical Writings: A Selection*, trans. Allan Wolter (New York, 1962).

46. Ernest A. Moody, "William of Ockham," *DSB*, 10:171–75; Gordon Leff, *William of Ockham: The Metamorphosis of Scholastic Discourse* (Manchester, 1975).

47. According to Aristotle, the categories are the ten different classes of predicates that represent the ultimate ways of speaking about things, that is, as substance or as the nine different types of accident, for example, quantity, quality, relation, action, and so on. In addition to these being modes of predicating (hence, likewise, called predicaments), they were also commonly regarded as modes of being, and, thus, every existent entity would have to be located in one way or another within the categories; again, to each of the categories there would have to correspond some type of entity in the real order. See R. M. McInerny, "Categories of Being," *NCE*, 3:241–44.

48. Weisheipl, *Development of Physical Theory*, pp. 63–69; Pedersen, "Development of Natural Philosophy," pp. 120–21.

49. Moody, *Studies in Medieval Philosophy, Science, and Logic*, pp. 300–302, 316–19.

50. E. J. Dijksterhuis, *The Mechanization of the World Picture*, trans. C. Dikshoorn (Oxford, 1961), pp. 175–76; Edmund Whittaker, *Space and Spirit* (Edinburgh, 1946), pp. 45–47, 139–43.

51. Pedersen, "Development of Natural Philosophy," pp. 121–28.

52. Copleston, *History of Medieval Philosophy*, pp. 260–66; Pedersen, "Development of Natural Philosophy," pp. 128–34.

53. Copleston, *History of Medieval Philosophy*, pp. 270–75.

54. John E. Murdoch, "Thomas Bradwardine," *DSB*, 2:390–97; Curtis A. Wilson, "William Heytesbury," ibid., 6:376–80; John E. Murdoch and Edith D. Sylla, "Walter Burley," ibid., 2:608–12, and "Richard Swineshead," ibid., 13:184–213. Murdoch's "Mathesis in philosophiam scholasticam introducta: The Rise and Development of the Application of Mathematics in Fourteenth-Century Philosophy and Theology," in *Arts libéraux et philosophie au moyen âge* (Paris, 1969) is particularly good for explaining the new role played by mathematics in the fourteenth-century development.

55. Pedersen, "Development of Natural Philosophy," pp. 134–42.

56. G. F. Vescovini, "Francis of Marchia," *DSB*, 5:113–15; Ernest A. Moody, "Jean Buridan," ibid., 2:603–8, and "Albert of Saxony," ibid., 1:93–95; Marshall Clagett, "Nicole Oresme," ibid., 10:223–30.

57. William A. Wallace, "Buridan, Ockham, Aquinas: Science in the Middle Ages," *The Thomist* 40 (1976):475–83.

58. *Etudes sur Léonard de Vinci*, vol. 3 (Paris, 1913), p. 583.

59. William A. Wallace, "Mechanics from Bradwardine to Galileo," *Journal of the History of Ideas* 32 (1971):15–28; "The 'Calculatores' in Early Sixteenth-Century Physics," *British Journal for the History of Science* 4 (1968–69):221–32; "The Concept of Motion in the Sixteenth Century," *Proceedings of the American Catholic Philosophical Association* 41 (1967):184–95; "The Enigma of Domingo de Soto: *Uniformiter difformis* and Falling Bodies in Late Medieval Physics," *Isis* 59 (1968): 384–401; and "Galileo Galilei and the Doctores Parisienses," forthcoming.

60. John E. Murdoch, "Philosophy and the Enterprise of Science in the Later Middle Ages," in *The Interaction between Science and Philosophy*, ed. Yehuda Elkana (Atlantic Highlands, N.J., 1974), pp. 51–74.

61. William A. Wallace, "Philosophy of Science," *NCE*, 12:1215–19; "Toward a Definition of the Philosophy of Science," in *Mélanges à la mémoire de Charles De Koninck* (Quebec, 1968), pp. 465–85.

4

Pearl Kibre and Nancy G. Siraisi

The Institutional Setting: The Universities

Origins of the Universities

From the close of the twelfth century onward, science found its chief institutional home in the universities or corporate associations of scholars—students or masters (from the Latin term *magister* for teacher).[1] These universities deserve our attention not only as the transmitters of the knowledge of the past but also and more specifically as the nuclei and breeding ground for the creative and dynamic forces that were to coalesce and produce the subsequent scientific achievements of Western culture. Although their exact beginnings are obscure, the universities of scholars were both reminiscent of earlier precepts and characteristic products of the age in which they appeared. Their emergence as autonomous corporate associations in Paris, Oxford, Bologna, Padua, and elsewhere coincided with the formation of guilds in industry and commerce, and of self-governing communes in the civic sphere. The term *university* (*universitas*) itself originally applied to the totality of any group of persons with a common aim or function, such as a craft or merchant guild or a self-governing association of citizens organized as a legal entity with a right to sue or be sued, under precepts derived from Roman law. These self-governing associations, in conformity with contemporary views regarding a guild or sworn brotherhood of men performing a like function or following a common occupation—in this instance, teaching and studying a prescribed curriculum—set their own requirements and steps for the achievement of their specified objectives. In some respects, these progressive stages in the universities of scholars, namely, matriculation, bachelor, and master (one who re-

received the degree or license to teach upon the successful completion of such set intellectual exercises as the disputations, determination or defense of the thesis, and formal inception into the guild of teaching masters), paralleled the steps in a craft guild of apprenticeship, journeyman, and finally master workman, following the completion of a perfect piece of work (a shoe, a chest, or the like).

However, despite their outward resemblance to the craft or merchant guilds, the universities of scholars differed from them in the status of their members. The latter, unlike the members of craft or merchant guilds, were specially privileged, since they were concerned with the acquisition and transmission of knowledge. In recognition of this distinction, scholars had since antiquity, and particularly under precepts of Roman law, been accorded special exemptions and privileges; and these were augmented in the High Middle Ages by Frederick Barbarossa in the *Authentica habita* granted in 1158, and by successive grants of privilege by pontiffs, monarchs, and municipalities. Scholars, indeed, constituted a new privileged class in society, the intellectuals; they were the new nobility, as suggested by contemporary references to the "new chivalry," distinguished from the feudal aristocracy since they boasted no military prowess nor material wealth, although they borrowed feudal terminology and ceremony. The scholar on the first rung of the ladder of the academic hierarchy was named "bachelor," a term originally signifying a squire; and the ceremony attending the bestowal of the degree or license to teach was thought to confer "a sort of intellectual knighthood."[2]

The recipients of this intellectual knighthood, despite their contemporary trappings, were in their preparatory training heirs to a long tradition embodied in the Greco-Roman ensemble of the seven liberal arts, divided into the trivium—grammar, rhetoric, and logic—and the quadrivium—arithmetic, geometry, music, and astronomy. These arts, related to "productive reason," and "ordained to knowledge," were believed to be the indispensable foundation for higher and more specialized learning, and they had been so taught in the imperial and municipal schools of Rome. However, with the collapse of the Roman political structure in the West and the virtual disappearance of the imperial and municipal schools, leaving only a thin stream of lay teaching, the task of transmitting the arts fell to the agencies of the Roman church, under the expressed belief that they and other elements of Greco-Roman thought were essential for a rational understanding of scripture. St. Basil (d. 379), for example, in his *Hexaemeron* (a discussion of the six days of creation) utilized Greek scientific theories on cosmology, light, the four elements, and the four qualities, as well as the deviant conception of matter as uncreated and eternal, to bridge the gap between pagan thought and Christian doctrine and to elucidate the account of the creation of the physical universe in Genesis. Moreover, as in the words of St. Augustine (d. 430), the liberal arts were held to be a concomitant means for drawing men to truth and to God.[3]

However, under the new auspices of monastic, abbey, and cathedral schools, the basic thrust of the arts was shifted from their purely speculative to their religious and functional significance. This was probably due not only to the religious milieu but also to the exigencies of a less sophisticated and more largely rural society than that envisaged by those who had originally formulated the canon of the seven liberal arts—Plato in the *Republic* (book 7), Martianus Capella in the fifth century of our era, and Boethius in the fading light of Roman splendor in the early sixth century. The shift in emphasis was most notable in the mathematical arts of the quadrivium. Thus, arithmetic, which was earlier thought essential for anyone pursuing philosophy, was taught primarily in relation to calculation, principally for determining the date of the movable feast of Easter, also for use in weights and measures. Similarly, geometry was pursued not, according to the Boethian formula, as an example of continuous magnitude without motion, but as a useful instrument for land surveying or measurement, and for geography, surgery, and architecture. And astronomy, the study of the courses of the stars, was not studied primarily as a phenomenon of continuous magnitude in motion (as Boethius had put it), but, rather, for the practical purpose of calculating the times of movable church feasts and festivals according to the phases of the moon, as well as for specific religious offices. And finally, music was considered as an art to be performed, rather than as the philosophy of harmony or acoustics, with emphasis on the rules for both choral and instrumental performance.[4]

By the twelfth century, especially in the cathedral schools of such burgeoning urban centers as Chartres, Laon, Canterbury, Rheims, and Paris, the spectrum of studies had broadened. Masters and students were studying, in greater and ever widening perspective, mathematics, astronomy, and the more speculative sciences of nature on the basis not only of Plato's *Timaeus* and the writings of Boethius, Bede, Gerbert, and other traditional texts, but also, increasingly, of the works of Aristotle and the Arabs in the translations by John of Seville, Gundissalinus, Hermann the Lame, and others. Also, in these same cathedral schools we can see the beginnings of the characteristic method of scholastic teaching, namely the disputation based on logic or dialectics, popularized by such a master as Abelard.[5]

Despite the hospitality of the cathedral and other ecclesiastical schools to the new developments in learning and scholarship, they were limited in the task of expanding the boundaries of that learning because their primary function was the preservation, propagation, and elucidation of the faith. It remained, therefore, for the corporate associations of scholars or universities to accomplish the task of developing, adapting, and broadening the aims and content of the body of knowledge. And, interestingly enough, they were greatly aided in this task by the newly established mendicant orders, the Dominicans and Franciscans, whose members, in order to enhance their evangelical and missionary efforts, encouraged scholarship in a variety of fields, including the natural sciences. The religious orders established houses of study in university centers and provided leisure and facilities for such prominent scholars as Roger Bacon, Albertus Magnus, Thomas Aquinas, John Pecham, and a host of others. Furthermore, the achievements of these and other members of the orders greatly enriched the intellectual life of the universities, despite occasional friction between the friars and secular university masters. Moreover, the friars supplemented papal efforts to establish and provide for studies in Arabic, Greek, Hebrew, and Chaldean (Aramaic or Syriac) at Paris, since Dominican and Franciscan evangelicalism fostered interest in and study of non-Western languages as well as a continuation of the contacts with the Muslim and Byzantine world that had been established earlier by the reconquest of much of the Spanish peninsula and the efforts of the crusaders in the Near East.[6] The intellectual expansion engendered by these continued contacts with non-Western peoples was early reflected in the curriculum of the schools, especially in the fields of natural philosophy and the mathematical sciences.

Before proceeding to these areas of the teaching program, we should draw attention to the fact that although the universities shared many features, they differed in several aspects which affected their teaching. For example, of the four archetypal universities—so called because they served as models for others—Paris, Oxford, Bologna, and Padua, the two northern universities were made up of masters or teachers primarily in the liberal arts, with student participation in university activities only through a master, while the earliest universities in Bologna and Padua were self-governing associations of students in law, with faculty prerogatives limited, at least in the statutes. The universities of arts and medicine at Bologna and Padua, also made up of students, were not organized until later in the thirteenth century. Moreover, in both Paris and Oxford, university members were largely clerics, supported, for the most part, by the returns from their ecclesiastical benefices, from which they had permission to absent themselves for purposes of study for a period of five to seven years. They were, thus, not dependent upon student fees or municipal salaries, as masters in Bologna and Padua were. In both the latter cities the masters or doctors (another term for teacher from the Latin *docere*, to teach) formed *collegia*, or colleges, of their own.

These doctors' *collegia* should not be confused with the colleges in all university centers founded by philanthropists to provide food and lodging and a small stipend for poor scholars or for those coming from a specified locality. As examples of those for poor scholars at Paris may be noted the College of Eighteen (Collège des Dix-huit); the College of the Good Children of S. Honoré, founded by Étienne Belot and his wife; and Ave Maria College, which took in small boys as well as those attending the university schools. Illustrative of foundations for scholars from particular localities were those of Beauvais, of Upsala, and of Skara—all at Paris. On the other hand, the Collège de la Sorbonne was founded by Robert de Sorbon to accommodate poor scholars pursuing advanced studies in the faculty of theology. In time, several of the colleges also became places of instruction as well as of lodging, as was true of the Collège de la Sorbonne. And at Oxford, Merton College, founded by Walter de Merton, became the principal center of mathematical and astronomical studies.[7]

Paris and Oxford

Since the differences between the northern and southern universities, noted above, were to have some effect upon the nature of the teaching, we shall draw attention first to the two northern university centers, and second to those south of the Alps. The consideration of the former first is not intended to imply that they were established earlier, a moot point at best, but, rather, to draw attention to the fact that the universities of Paris and Oxford appeared to have reached the zenith of their importance in mathematical, natural, and physical science, and medicine, in the thirteenth and fourteenth centuries and thereafter gave evidences of decline. On the other hand, many of the developments in science and medicine that had given great promise earlier in Paris and Oxford were carried forward in the later fourteenth, fifteenth, and sixteenth centuries in the university centers of Bologna and Padua. One might suggest that the failure of the two northern universities to fulfill their earlier promise may be attributed to the cumulative and deleterious effects of the Hundred Years War, political upheaval, social unrest, the great Western schism, and the recurrent outbreaks of the Black Death.[8] For although the same or similar catastrophes afflicted the Italian cities, their effects were not augmented to the same extent by the devastation, exhaustion, and impoverishment resulting from the pursuit of the war on French soil, and the accompanying economic depression and political and social unrest that were rampant in England. Moreover, in the fourteenth and fifteenth centuries a number of new universities were established in the Germanies, in Bohemia, and elsewhere, for natives of those countries who formerly had gone to Paris, thus greatly reducing the internationalism of the University of Paris. Since the latter, nonetheless, provided the model for the newly founded universities, a brief overview of the principal features of that university, with reference also to Oxford, may be in order.

The University of Paris had emerged on the left bank of the Seine River at the close of the twelfth century as an autonomous association of masters or teachers primarily of liberal arts. Originally attached to the cathedral school of Notre Dame, these masters, as a result of the influx of students attracted from all parts of Europe by the fame of such teachers as Anselm of Laon, William of Champeaux, Peter Abelard, William of Conches, Gilbert de la Porrée, Peter Lombard, and others, had found it necessary to seek teaching quarters outside the cathedral walls in the open streets, under the shadow of the Abbey of Mont Ste. Geneviève, the Petit Pont, and in the vicinity of the Abbey of St. Germain des Prés. Thus removed from the immediate confines of the cathedral, the masters formed an association in the fashion of the contemporary guilds and sought and obtained recognition as a corporate body, independent of the jurisdiction of the cathedral chancellor, who was charged with the supervision of the cathedral school. Continuity with tradition was preserved, insofar as the cathedral chapter was concerned, through the retention by the chancellor of the right to confer the license to teach (*licentia docendi*) on candidates whose fitness had been determined by the masters. The university masters were encouraged and aided by the papacy, particularly popes Innocent III, Honorius III, and Gregory IX. The university also received royal recognition in the Great Charter of Privileges granted by King Philip Augustus in 1200, which extended royal protection and exemptions from local civic obligations and levies as well as from the jurisdiction of the local magistrates, to all members and associates or clients of the university. The provost or mayor of the city of Paris was named as protector of these royal privileges. Moreover, one of the bishops other than the bishop of Paris, chosen in turn from among the bishoprics around Paris, served as conservator of the university's apostolic privileges. Hence, by the mid-thirteenth century the members of the university, under guarantees of royal and papal protection, were subject only to the jurisdiction of their own elected head, the rector. However, as clerics, they were still subject to the ecclesiastical guidance of the bishop of Paris and the chancellor of the Cathedral of Notre Dame.[9]

Many of the features described above were shared by the University of Oxford, although the latter differed from Paris in certain respects. For example, in the position of the chancellor, who, although relieved at Paris from the supervision of the teaching by university masters, nevertheless continued, by virtue of the fact that the masters were clerics, to exercise with the bishop of Paris some ecclesiastical guidance over them. At Oxford, by contrast, since both a cathedral and a cathedral school were lacking, the chancellor, appointed by the bishop of Lincoln, became the nominal head of the university, taking the place of the rector, the elected head at Paris. Also, of the two universities, Paris was more international in makeup, as evidenced by its division into the four nations: French, English-German, Picard, and Norman, representing the localities from which the masters came; while at Oxford, which remained largely insular in makeup, there was only the division of the masters into northern and southern parts of the British Isles. Paris, therefore, had the unique advantage, at least through the mid-fourteenth century, of sharing in the teaching services of distinguished scholars from all of continental Europe as well as from the British Isles.[10]

Both universities were organized for teaching purposes into the four faculties of arts, medicine, law, and theology. The arts faculty provided the basis and preparation for the three higher faculties of medicine, law, and theology; it also provided opportunities for preparatory studies in natural or physical science. A student without previous university work who arrived in Paris from England or one of the continental countries would first be assigned, by the proctor or head of the nation representing the locality from which he came, to a master of arts in that nation. After payment of a small fee (unless he had declared himself a pauper and was exempted from payment), the student's name would be entered on the matriculation roll maintained by the proctor. It is probable that the student would normally follow lectures in the preparatory arts of the trivium and quadrivium before going on to the newer elements in the arts curriculum, including *physica* or natural science. At Paris the preparatory studies in grammar, or latin syntax, and literature, were gradually being reduced in scope in the later thirteenth and fourteenth centuries, and they were frequently relegated to the colleges or to special grammar schools maintained and supervised by the university.[11] Students preparing for the examinations for the baccalaureate and for the master of arts degree were in the early thirteenth century still required to have studied the time-honored Donatus and Priscian. But by the close of the thirteenth century, the classical grammarians were replaced at Paris by the newer *Graecismus* of Eberhard of Bethune and the *Doctrinale* of Alexander of Villa Dei. The Oxford curriculum continued to include Donatus and Priscian through the fifteenth century. On the other hand, at both Paris and Oxford, rhetoric, the art of persuasive discourse, although still required according to the early thirteenth-century statutes, appears to have disappeared from the statute requirements in the later thirteenth century, to reappear in the fifteenth century.[12]

Of the arts of the trivium, included in Paris under the new rubric of rational philosophy, only logic appears to have gained in scope and prestige. It was victor in both the allegorical and the actual battle of the seven arts. The chief explanation for this lay in the fact that logic provided the methodological basis for both philosophy and science, the two subjects that had fired the imagination and captured the enthusiastic attention of students and masters at Paris from the twelfth century onward. Hugh of St. Victor had suggested that logic should come first among the seven liberal arts, since, in his words, "it provides ways of distinguishing between modes of argument and the trains of reasoning themselves . . . it teaches the nature of words and concepts, without both of which no treatise of philosophy can be explained rationally."[13] And to this view was added the authority of such renowned thirteenth-century scholars and scientists as Robert Grosseteste, Bishop of Lincoln, prominent both at Oxford and Paris, and the two distinguished Dominican scholars, Albertus Magnus and Thomas Aquinas. All three held that since the study of logic provided the method for all sciences it should be placed first. Robert Grosseteste had apparently developed his own methodological approach to science from his prior study of Aristotle's *Posterior Analytics*, the work that was to provide the logical basis for the general physical theory of nature during the High and later Middle Ages. And Thomas Aquinas specifically asserted that the appropriate pedagogical sequence places logical topics first, since logic teaches the method for all philosophy and scientific inquiry.[14]

A differing view was that of Roger Bacon, who held to the priority of the study of mathematics. Bacon argued on the basis of the authority of Ptolemy and Boethius that mathematics was necessary to every discipline and science, since, according to Aristotle, the essential parts of philosphy, both natural and divine, are mathematical. Furthermore, he held that logic required mathematics, since its central core is based on the *Posterior Analytics*, which teaches the mathematical art of demonstration. Similarly, he continued, since logic begins with the *Book of Categories*, it is clear that the category of quantity cannot be known without mathematics. Indeed, there is no question in Bacon's mind but that mathematics is needed in every science. This, he affirmed, is further attested in the sixth book, twenty-second proposition of Euclid's *Elements*. And in Bacon's view, this necessity extends to all four parts of mathematics, namely, geometry, arithmetic, music, and astronomy. These, he held, were all founded when the human race began. Hence, according to Bacon, they should be studied first because through them one may advance to other sciences.[15]

The university documents are mute as to the outcome of this discussion on the order of study of the sciences. What does emerge is the preponderance of the texts required for logic, the subject that provided methodological rules for disputations and techniques for the pursuit of science. In 1215 works in the "old" logic, that is, the works in use in the West before about 1128, comprising Aristotle's *Categories* and *On Interpretation*, Porphyry's *Isagoge*, and the *Book of Six Principles* attributed to the logician Gilbert de la Porrée, were required, together with the "new logic," the works introduced in the later twelfth and early thirteenth centuries, namely, Aristotle's *Prior* and *Posterior Analytics, Topics,* and *Sophistical Refutations*—works which covered the subjects of syllogisms and the analysis of demonstration and proof. The number of works required in logic increased in succeeding years according to the statutes, with the addition principally of a number of systematic manuals and contemporary writings on logic, of which the most significant was the *Summulae logicales* of Petrus Hispanus, later Pope John XXI (d. 1277). This work, which dealt with the rules of syllogism and the doctrine of suppositions, was fruitful in developing a dialectical method of interpreting Aristotelian science, and perhaps also, as one author has noted, of explaining the results of direct observation.[16]

Although the statutes of Paris and Oxford leave little doubt as to the preponderance of logic, they are less explicit on the subject of the mathematical arts, except to note that at Paris these might be lectured on during feast days. Both Robert Grosseteste and Roger Bacon maintained that mathematics provides the gateway and key to all other sciences, and, as indicated earlier, Bacon urged that mathematics be studied before logic. However, despite the comparative lack of information in the documents regarding the mathematical or quadrivial arts, other evidence, such as the biographies of prominent scholars who lectured on these subjects at both Paris and Oxford, together with an extant list of required texts for the master of arts examination noted by an anonymous Parisian master, and other sources, testify to the teaching of these arts. For arithmetic, the required texts were the Latin translation or paraphrase of the Greek *Arithmetic* of Nicomachus of Gerasa by Boethius, on the theory of numbers, and the more elementary and practical abacus arithmetics and books on calculations, algebra, and Hindu numerals by the thirteenth-century author Jordanus de Nemore. In addition, the *Algorismus* of John of Sacrobosco, and the versified *Algorismus* of Alexander of Villa Dei, appear to have been utilized.[17]

For the study of geometry, the principal text was Euclid's *Elements*, according to the anonymous Parisian master. This work was also included in the Oxford curriculum with the specific admonition in the statutes that the candidate for inception as a teaching master affirm under oath that he had studied the first six books. The Euclidean *Elements* in the version of Boethius had long been utilized in the pre-university schools. However, the work had been newly translated from the Greek in the course of the twelfth century, and was possibly also translated in the thirteenth century by Campanus of Novara, who wrote a commentary on it, as did Albertus Magnus and Roger Bacon. Thirteenth-century manuscripts of Campanus's commentary at Paris suggest its use there, especially since Campanus was a canon of the Cathedral of Notre Dame. In addition, at Paris and Oxford the twelfth-century *Practica geometriae* ascribed to Hugh of St. Victor may have been utilized, as well as the work of the same title by the thirteenth-century Italian Leonardo of Pisa. It is not clear whether two other works, namely, the tracts on optics of Alhazen and Witelo, either of which could be substituted for Euclid according to the stipulations of the Oxford curriculum before 1350, were also included at Paris. The importance of geometry, in addition to its practical uses carried over from the earlier teaching in the monastic and cathedral schools, was asserted by Roger Bacon and other ecclesiastics on the grounds that it was a useful instrument for the demonstration of theological truth. Moreover, both Robert Grosseteste and Roger Bacon insisted upon the necessity of geometry for a knowledge of natural philosophy, and agreed that of the mathematical sciences, geometry in particular could explain the factual knowledge acquired through the physical sciences.

Of the other mathematical sciences, music continued to be taught in the thirteenth century from Boethius's *De musica*, which was still specified as the text in the fifteenth-century curriculum requirements for inception as master of arts at Oxford. Among the reasons set forth by Robert Grosseteste for including music in the curriculum was the importance of music in medicine and in promoting health. The question, too, of the effects of music on mind and body and the specific role of harmony in arousing the passions or elevating moral virtues provided a topic for discussion or disputation in the faculty of theology.

Finally, for astronomy, the fourth of the mathematical arts, the principal text, according to the anonymous Parisian master, was Ptolemy's *Almagest*, one of the most important sources of astronomical knowledge transmitted from antiquity. It had beome available for use by Western scholars in Latin translations from both the Greek and the Arabic in the course of the twelfth century. At Oxford, the curriculum also included a *Theorica planetarum*, or *Theory of the Planets*, which may refer to the work of that title by Campanus of Novara or to a work variously ascribed to the twelfth-century translators Gerard of Cremona and John of Seville, and to thirteenth-century authors such as John of Sacrobosco and Robert Grosseteste. The *Theorica planetarum* may also have been in use at Paris, where not only Ptolemy, but also the works of the Muslim astronomers Alfraganus, Albategni, and Alpetragius, and the Greek astronomer and geometer Theodosius of Bithynia, had become available in Latin translation in the course of the twelfth and thirteenth centuries. Hence, although Ptolemy was in the forefront, other works, in all likelihood, also provided the substance for disputations and lectures and the basis for compositions by contemporary authors. This was true, for example, of Sacrobosco's *Sphere*, which utilized Ptolemy and the Muslim authors Alfraganus and Albategni, and became the best known and most widely disseminated textbook in astronomy.

Mathematical studies, or the quadrivial arts, appear to have been followed by studies in physical or natural science, although the matter must remain somewhat conjectural. In any case, under the heading of natural philosophy, the arts curriculum in 1255 incorporated practically the entire Aristotelian corpus of natural philosophy, in the partly Arabo-Latin and partly Greco-Latin versions that had earlier been condemned at Paris largely because, upon first interpretation, they were deemed contrary to the faith by the Parisian ecclesiastical authorities.[18] The works on natural philosophy were also incorporated into the curriculum at Oxford, where no prior condemnation had been issued. Among the works included in 1255 at Paris were Aristotle's *Physics, On the Heavens, Meteorology, On the Soul, On Generation, On Animals, On Sense and Sensibles, On Sleep and Waking, On Memory and Remembering, On Life and Death*, and the tracts of doubtful authenticity, *On Causes* and *On Plants*. Aristotle's *Metaphysics*, the subject of which constituted the third division in natural philosophy according to Thomas Aquinas, was also included among the prescribed texts in 1255. In addition, there was included with the Aristotelian writings the Arabic treatise in Latin translation *On the Difference between Soul and Spirit*, by Qusta ibn Luqa, a work relating to physiological psychology.[19]

Following natural philosophy in the arts curriculum at Paris was moral philosophy, which at both Paris and Oxford comprised Aristotle's *Ethics, Politics*, and *Economics*.[20]

In the foregoing account, attention has been drawn principally to the subjects and texts utilized in the arts faculty in the thirteenth and early fourteenth centuries at Paris and Oxford, the archetypal universities north of the Alps. Little change appears to have been made in offerings in the later fourteenth and early fifteenth centuries. The practice of Paris in this regard was followed by other universities, such as Toulouse, Montpellier, and the newly founded universities north of the Alps in the fourteenth and fifteenth centuries. Similarly, the methods and techniques of instruction that had been developed in the twelfth and thirteenth centuries remained much the same. These comprised the use of the lecture and commentary on a specific text, followed by the repetition or review of the matter covered and the *collatio*, or discussion, and conference. The master's function was chiefly to explain the text and to resolve difficult points. The lectures were usually divided into ordinary lectures, those given in the morning as part of the regular curriculum by members of the faculty, and extraordinary or cursory lectures, given usually in the late afternoon or on feast days by bachelors rather than by masters.[21] In addition, the lectures were supplemented by disputations, which applied rational methods of inquiry in the presentation, explanation, and proving of specific assertions or propositions, and the answering of objections raised against them. Frequent references were made in these disputations to the standard authorities—the Bible, the Fathers, Aristotle, and others. In addition, there were two other types of disputations held periodically during the school year. In the "Quodlibetal" disputation, at a public session, questions chosen at random from the leading topics of the day were put to the master in charge and were first tentatively answered by a student closely associated with him. Then, at a later session, either twenty-four or forty-eight hours later, the master or professor made his formal presentation of the question in the form outlined above. In the other form of disputation, the "Disputed Questions," of which the *Questions on Truth* of Thomas Aquinas are an example, the master or professor set his own topic and then proceeded to give a formal disputation, again in the form noted above.[22]

The preceding survey of the subjects, textbooks, and techniques of teaching in the faculty of arts should make abundantly clear the generally profane or worldly nature of the curriculum of that faculty, despite the fact that at both Paris and Oxford, the masters were predominantly members of the clergy. The survey should also dispose of the claim often made that university scholars concentrated on theological studies to the exclusion of all else. At the same time, one should be mindful of the fact that the medieval concept of the ultimate aim of all learning was the discovery of truth, which in St. Augustine's formulation, influential throughout the Middle Ages, was synonymous with the love of God. Nevertheless, the masters of the individual faculties were zealous in their desire to maintain the institutional separation of the various fields of endeavor. They had a strong sense of the hierarchy of subject matter. At the pinnacle was the faculty of theology, the queen of the sciences; and to make this explicit, there were statutory prohibitions at Paris against the use of theological matter for lectures by the masters of the faculty of arts.[23] Theology and theological studies were reserved for the seasoned members of the theological faculties or for the schools of the monastic and mendicant orders. At Paris the faculty of theology was the smallest of the faculties, and in some of the other universities it was not included in the academic circle until the second half of the fourteenth century. The number of students who wished to undertake the long and arduous study required of candidates for the degree in theology was always small.

While there were prohibitions against members of the other faculties delving into and teaching matters pertaining to theological doctrine, there was no prohibition against theologians concerning themselves with profane subjects. Indeed, much of what they learned in their preparatory studies in the arts faculty was utilized by theologians in the exposition of theological texts. For example, theologians brought their knowledge of natural or physical science into lectures or commentaries on books of the Bible, especially on Genesis, dealing with the six days of creation, and into their commentaries on the *Sentences* of Peter Lombard (mid-twelfth century), based on the

Bible. Four distinctions or divisions (12–15) of book 2 of the *Sentences* relate to the work of corporeal creation. Hence, the lectures or commentaries on those sections by such distinguished theologians as Albertus Magnus, Thomas Aquinas, Bonaventura, and Duns Scotus (like the earlier Hexaemeral treatises) covered such matters of scientific interest as astronomical theories, physics (now incorporating Aristotelian material), optics, and biology.[24] Albertus Magnus, in particular, was convinced of the importance of profane science. In commenting on the *Sentences*, he had asserted that in matters of faith he would follow Augustine, but in matters of science, he preferred a scientific master: for medicine, Hippocrates or Galen; and for natural philosophy, Aristotle. Albertus Magnus also departed somewhat from the view that all investigation of the natural world should be for the service of God, in asserting that he was undertaking his investigations to satisfy his students' curiosity and, we may assume, his own.[25]

Besides the concern with natural science or philosophy by members of the arts faculties of Paris and Oxford, and by members of the faculty of theology, there was also a manifest interest in this subject by members of the faculty of medicine. Although at Paris and Oxford the institutional separation of arts and medicine into two separate faculties was maintained, there was, nevertheless, a close association between them. Students matriculating in medicine were required to have had preliminary training in arts, and this same requirement was applied to candidates for the license to teach medicine as well as to those wishing to practice it. Certain physicians of the late twelfth and thirteenth centuries appear to have taught, probably sequentially, in both the arts and medical faculties at Oxford and perhaps also at Paris. For example, Alexander Neckham, Alfred of Sareschal, and Raoul de Longchamps reportedly did so. Although they were already physicians, these men were also among the earliest university scholars to utilize the newly translated Aristotelian works in natural philosophy.[26]

This is not the place to assess the actual achievements of those who lectured, disputed, and observed natural phenomena in the course of their sojourn in the university centers of Oxford and Paris. But there can be little doubt, if the regulations were obeyed, of the breadth and depth of the curriculum in logic, mathematics, and natural philosophy.

Bologna and Padua

In medieval Italy, too, from the thirteenth century onward, the university, as an institutionalized association of scholars, was the chief focus of learning in the sciences. Of the numerous and sometimes short-lived universities of medieval Italy, those of Bologna and Padua were among the oldest and most prestigious; they provided not only an institutional model but also a continuing source of intellectual stimulation for similar associations in other cities.

At Bologna and Padua the first formal associations of scholars were of foreign (that is, non-Bolognese and non-Paduan, respectively) students of civil and canon law. Subsequently, at Bologna, probably at some time during the last forty years of the thirteenth century, they were followed by another university made up of those studying liberal arts and medicine. It was chiefly by the members of these universities of arts and medicine and their instructors that study and teaching of the sciences was pursued, as the following discussion will make plain.

The student universities of law and of arts and medicine were, like the universities of masters in northern Europe, self-governing corporations electing their own officials—the rectors, proctors, and consilarii —and making their own rules and regulations pertaining to teaching practices and curriculum. The members of the student universities also claimed the right to elect professors to a small number of salaried chairs. In the universities of arts and medicine of Bologna and Padua, only students who had studied medicine might vote for professors of medicine, only students who had studied logic for professors of logic, and so on.[27] Nevertheless, student control of faculty and curriculum, however specifically and impressively asserted in the statutes drawn up by the students themselves, was, in practice, greatly weakened by municipal payment of professors' salaries and by the fact that the conduct of examinations remained in the hands of the faculty.[28] Probably at about the same time as a student university of arts and medicine appeared at Bologna, some of the teaching masters of arts and medicine, who were citizens of Bologna and, hence, not normally members of the foreign students' university, organized themselves into a doctoral college. A similar pattern was followed at Padua, where some form of association of students of arts and medicine had been established by 1262, although the University of Arts and Medicine did not achieve full legal sanction until 1399. At Padua, too, a doctoral college of arts and medicine was formed; it was already in existence by 1307, when its members included professors of medicine (at least one of whom also taught philosophy and astrology), logic, and grammar. Padua, however, differed from Bologna in that its citizens were officially debarred from the professorial chairs by a frequently violated municipal statute.[29] It must be reemphasized that the doctoral colleges were quite separate and distinct from colleges of another type, namely, those founded by specific benefactors to provide stipends and lodgings for students. Colleges of the latter type came into existence in Bologna and Padua as well as at Oxford and Paris.

The professional and intellectual opportunities provided for teaching and study of arts and medicine in both Bologna and Padua attracted men of learning from far and near. Not all of those who came were formally members either of a doctoral college of arts and medicine or of the corresponding student university. The university locale in which those institutions functioned also embraced such a man as the surgical writer Theodoric of Lucca (d. 1295), a bishop residing outside his see, who flourished and probably taught in the city of Bologna for many years without apparently having official affiliation with any academic association.[30]

The curricula of the universities of arts and medicine included the study of mathematical and physical as well as medical science. By the close of the thirteenth century, in Italy as elsewhere, the normal arts curriculum comprised not only the verbal and mathematical disciplines of the trivium and quadrivium, but also the study of philosophy and natural science in the Aristotelian works on metaphysics, physics, the soul, the heavens, animals, and so forth. Moreover, the scientific portion of this curriculum was studied not only by those pursuing a degree in arts as their final goal but also by all those intending to proceed to the study of medicine. Logic, the fundamental tool of scholastic methodology in all branches of learning, astrology (essential to medicine in an age in which the reality of planetary influences upon human physiology and psychology was almost universally accepted), and natural philosophy were regarded as particularly important preparatory studies for physicians. As Peter of Abano (d. ca. 1316), professor of philosophy, astrology, and medicine at Padua, remarked, while all the arts and sciences are necessary to medicine, logic, astrology, and natural science (*scientia naturalis*) are "most necessary."[31] Accordingly, the 1405 statutes of the Bolognese student university of arts and medicine listed texts for medicine (theoretical and practical), for philosophy, and for astrology in a single section. (Set books for logic, studied from the works of Aristotle, and surgery are prescribed in other, separate, sections.) For the compilers of these statutes, "philosophy" meant the study of Aristotle's works on natural science and selections from his *Metaphysics*. "Astrology," a basic course in arithmetic, geometry, and astronomy (including the use of astronomical instruments), was presumably designed to equip the student with the necessary knowledge to make his own astrological determinations for medical or other purposes. The works assigned him for the study or mastery of this subject include an algorism, or arithmetic, the first three books of Euclid's *Elements*, the *Sphere* (presumably of Sacrobosco), the *Theory of the Planets* (perhaps that of Campanus of Novara), the *Centiloquium* ascribed to Ptolemy and part of his *Almagest*, Messahala's treatise on the astrolabe, and the Alfonsine Tables. Among the textbooks for theoretical medicine were the *Aphorisms* and *Prognostics* of Hippocrates, the *Tegni* and other works of Galen, and extensive selections from the *Canon* of Avicenna. The latter work was also the sole text prescribed for the study of practical medicine. Surgery, too, required the student to be familiar with portions of the *Canon* as well as with the seventh book of the *Liber ad Almansorem* of Rasis and the more recent work of the thirteenth-century writer Bruno Longoburgo of Calabria.[32]

The close union of natural philosophy or science with arts and medicine is further exemplified in the descriptive terminology applied to the many individuals with degrees in "arts and medicine" or "medicine and philosophy." Moreover, it was quite usual for a master to give lectures on the Aristotelian works on natural science or on the liberal arts (usually logic or astronomy) as well as on medicine during the course of his career. For instance, of fourteen doctors involved in formulating the 1378 statutes of the medical branch of the Bolognese College of Doctors of Arts and Medicine, seven are recorded as having degrees in philosophy as well as medicine, and five are known to have taught logic and Aristotelian natural philosophy along with medicine.[33]

A number of works by masters associated with the Italian faculties of arts and medicine who are primarily identified as physicians testify both to the dialectical and philosophical training of their authors and to the breadth of their interest in natural science. To give only three examples, an abbreviated Italian translation of the *Nicomachean Ethics* was prepared by Taddeo Alderotti (d. 1295), professor of medicine at Bologna; an extensive and learned commentary on the natural problems attributed to Aristotle was written by Peter of Abano; and Jacopo de'Dondi, professor of medicine at Padua, produced treatises on tides and hot springs. It seems that an individual master might teach more or less any combination of logic, mathematical and physical science, speculative natural philosophy, and medicine. In some instances the teaching of arts and natural philosophy occurred at an early stage of a master's career before he was qualified to teach medicine, as was the case with William of Brescia, who taught logic and philosophy at Padua in the 1270s and 1280s before proceeding to study medicine at Bologna,[34] but this sequence does not appear to have been invariable. Of course, not every professor of physical or mathematical science or natural philosophy in the Italian schools also taught or wrote on medicine, as the surviving output of some of these men makes clear. Thus, in the fourteenth century a number of discussions of questions of natural philosophy were produced by Bolognese scholars who apparently left no medical works, while, for example, among professors at Padua, Blasius of Parma (d. 1416) seems to have written only on mathematics and natural philosophy—and Prosdocimo de'Beldomandi (d. 1428) only on mathematics, astronomy, and music.[35] Nonetheless, the only inviolable division within the faculties of arts and medicine in terms of personnel was apparently between the teachers of grammar and rhetoric on the one hand and those of logic, mathematical and physical science, and medicine on the other. Moreover, since both natural philosophy and theoretical medicine were approached as branches of speculative science, the organization of teaching in these two fields had many features in common. Scholars in both disciplines lectured and produced commentaries upon authoritative works and disputed questions of particular interest. The distinctive feature of medical instruction was the division into theory and practice, which, as noted above, was embodied in the curriculum at Bologna by 1405.

Despite the institutional, intellectual, and personal union of mathematical and physical science with medicine in the universities of arts and medicine, the separate identity of each branch of study was carefully preserved by statute and, apparently, in practice. The Bolognese and Paduan universities of arts and medicine reflected in their statutes a desire on the part of the compilers to prevent the dilution of the curriculum through the blurring of the lines between disciplines and the admission of unqualified students to the more advanced studies. The subjects of the curriculum were arranged in a hierarchy in which medicine held the highest place. Only scholars who had studied (and at Bologna only masters who had taught) a particular subject might take part in public disputations pertaining to that discipline.[36] And although astrology and natural science were regarded as peculiarly appropriate preparation for the would-be medical student, the content of study in those areas was certainly not narrowed to serve purely medical ends. On the contrary, significant independent development took place. At Bologna an important group of radical Aristotelian or Averroist masters of arts and teachers of philosophy and natural science emerged during the early decades of the fourteenth century. The members of this group appear to have been closely in touch with contemporary developments in philosophy and natural science in the university centers north of the Alps.[37] At Padua, Jacopo and Giovanni de'Dondi, professors of medicine, who, like Peter of Abano before them, also taught astrology, made contributions to astronomy and related sciences that went far beyond medical astrology. Both Jacopo, the father, and Giovanni, the son, were renowned for their achievements as clockmakers; the younger Dondi designed and constructed an elaborate mechanical device to illustrate the movements of the planets.[38] In addition, mathematics at Padua had emerged as a separate discipline, independent of music and astrology, by 1389. In that year a professorial chair was apparently established in the subject. However, according to the university statutes of 1495, extraordinary—that is, junior—lecturers could still be appointed at random to teach any one of the subjects of philosophy, astrology, mathematics, or medicine.[39]

Arts, natural philosophy, and medicine could be studied either simultaneously or consecutively. A degree in arts seems to have called for four years study and a degree in medicine for four or five. At Bologna, students were permitted to specialize by concentrating on particular disciplines within the general category of arts and natural philosophy. For example, a student could choose to be examined either in all the arts and philosophy, in logic and philosophy alone, in grammar and rhetoric, in philosophy and astrology, or in medicine and one or more of the arts. For a general arts degree, a two-day examination was required, a similar examination for a degree in logic and philosophy. Grammar with rhetoric, or philosophy with astrology, also required only a single examination of one day's duration. If any three subjects were combined, two examinations were required, while a degree in medicine and all the other arts was granted only after three examinations. As for medicine itself, "if anyone wants to pass with distinction, then he is to take two examinations"—perhaps one in theory and the other in practice.[40]

The main features of the curriculum just described endured with relatively minor additions and modifications for several centuries. From the thirteenth century onward, the university centers acted as magnets to draw together all those concerned with Aristotelian natural philosophy, astronomy or astrology, and medicine. The grouping of disciplines in the faculties of arts and medicine thus had a lasting influence on Italian scientific life, and one to which a variety of developments, both intellectual and social, can be traced. For instance, scientists educated in the Italian university centers very commonly had medical training, even though their principal achievements were sometimes in quite unrelated fields: Copernicus is an excellent example. Moreover, the association of liberal arts and natural philosophy with medicine in university teaching and organization must surely have reinforced the belief of many academic physicians that medicine itself was an intellectual science, related to natural philosophy in its principles and methodology. Taddeo Alderotti, for example, explained that theoretical medicine derives its principles from, and, hence, is a subdivision of, natural science (*scientia naturalis*). Natural science, he maintained, deals with bodies in general, including the human body, while medicine deals with the human body alone.[41] Taddeo's statement is indicative of the importance attached to physical or natural science as well as logic as a foundation for medical study. It may be added that in the view of some historians the dialectical training of learned physicians in the universities of arts and medicine contributed significantly to the development of scientific methodology.[42]

At Bologna and Padua the study of mathematics and physical science was more likely to be undertaken with a vocational purpose than was true for Oxford or Paris. In the two Italian universities, as in the university centers north of the Alps, these disciplines were classified as branches of speculative philosophy, whose function it was to yield knowledge for its own sake. But this aim was modified in the medically oriented faculties of arts and medicine in Italy. Scholars there had, in addition, a very practical incentive to study astrology (including astronomy) and Aristotelian natural philosophy, since these subjects served as preparation for the prestigious and lucrative medical profession. No doubt, many future physicians treated the study of nonmedical sciences as a preliminary stage of their education, to be passed through as quickly as possible. But some, such as Taddeo Alderotti and Peter of Abano, seriously concerned themselves with "reconciling the differences of the philosophers and physicians."[43] This endeavor can perhaps be seen as an attempt to create a unified science of man, which would draw impartially upon the scientific works of Aristotle and of Galen and other medical authorities.

The link between arts, natural philosophy, and medicine also meant that in some cases the economic and social arrangements supporting the teaching of mathematical and physical science differed in the Italian university centers from those in their northern counterparts. In the north, as has been noted above, almost all the thirteenth- and fourteenth-century scientists associated with the university centers were clerics, many of them members of religious orders; their scientific activities and teaching were thus supported by ecclesiastical resources. This was not necessarily the case in Italy. There, from the time the academic universities and colleges of arts and medicine were established, their senior members were in many cases married laymen. That this became the norm is demonstrated by the statutes of the Bolognese College of Doctors of Arts and Medicine (1378), which give preference to the sons of members,[44] and by the rise of veritable professional and learned dynasties such as, in medicine, the del Garbo and Santa Sofia families, and, in astrology and astronomy, the Dondi family. Municipal governments in the Italian university centers normally funded a few salaried chairs in logic, philosophy, astrology, and, at Padua in the later fourteenth and fifteenth centuries, mathematics. In addition, the Bolognese student university statutes of 1405 make provision for the collection by faculty members of fees from students on an individual basis.[45] It is plain, however, that medical practice was the chief means of support of many teachers of medicine[46] and, therefore, conceivably of some teachers of mathematical and physical science, since certain of these men, as noted above, also taught medicine. In Italy as in the north, those learned in astrology could, of course, also contribute to their support by preparing prognostications for private clients or by entering the service of a prince. The latter course, for example, is said to have been chosen by the famous astrologer Guido Bonatti (d. after 1282), who was probably a professor at Bologna.[47] In 1405 professors of astrology at Bologna were obliged by university statute to provide individual prognostications or "judgments of the year" without charge for scholars of the University of Arts and Medicine, a requirement that may indicate that it was normal for professors to prepare such judgments for other clients for a fee.[48]

On the whole, the institutionalization of the sciences in association with medical education in the Italian faculties of arts and medicine probably benefited the development of science. Because the institutional and intellectual links between logic, astrology, natural science, and medicine drew to the study of the sciences many who were attracted by a medical career, authors and teachers who wished to specialize in astrology (and astronomy), mathematics, and natural philosophy were assured of a constant flow of students, and, hence, of a secure institutional position. This, in turn, made it possible to extend the scientific part of the curriculum. For example, the provision of a separate chair of mathematics at Padua—a chair held at different times by such noted mathematical authors as Blasius of Parma, Prosdocimo de'Beldomandi, and Regiomontanus (d. 1476)—developed as an offshoot of the chair of astrology. And the demand for the teaching of astrology, as already demonstrated, was directly linked to the needs of medical education as then conceived.

Notable scientists flourished in the environment of the medieval universities—such men as Buridan, Oresme, and the members of the fourteenth-century Merton school come to mind. Yet in the long run, the contribution of the medieval university to the development of science was perhaps less in the achievements of its great men than in the fact that it was the first educational institution in the history of Europe to impose some elements of systematic and organized training in scientific subjects upon large numbers of people. Upon this foundation, the legacy of the Middle Ages, later generations could build.

Notes

1. Pearl Kibre has prepared the introductory portions of this essay as well as the section relating to the universities north of the Alps; Nancy Siraisi has written the sections on the universities south of the Alps. However, both authors have attempted to integrate their efforts throughout.

2. Hastings Rashdall, *The Universities of Europe in the Middle Ages*, ed. Frederick M. Powicke and A. B. Emden, 3 vols. (London, 1936), 1:4–8, 151–52, 220–31, 283–87; Pearl Kibre, "Scholarly Privileges: Their Roman Origins and Medieval Expression," *American Historical Review* 59 (1954):543–67; also her *Scholarly Privileges in the Middle Ages* (Cambridge, Mass., 1962), chap. 1; and Etienne Delaruelle, "De la croisade à l'université. La fondation de l'université de Toulouse," in *Les universités du Languedoc au XIIIe siècle* (Toulouse, 1970), pp. 23–24.

3. James A. Weisheipl, O.P., "Classification of the Sciences in Medieval Thought," *Mediaeval Studies* 27 (1965):54–62; James Westfall Thompson, *The Literacy of the Laity in the Middle Ages* (Berkeley, 1939), especially chap. 1; Lynn Thorndike, *A History of Magic and Experimental Science*, vol. 1 (New York, 1923), chap. 21 and especially pp. 485–87, on "Christianity and Natural Science"; also Pearl Kibre, "The Christian: Augustine," in *The Educated Man: Studies in the History of Educational Thought*, ed. Paul Nash, Andreas M. Kazamias, and Henry J. Perkinson (New York, 1965), p. 98.

4. M. L. W. Laistner, *Thought and Letters in Western Europe*, A.D. 500 to 900 (Ithaca, 1957), chap. 4; Pierre Riché, *Education et culture dans l'occident barbare, VI–VIIIe siècles* (Paris, 1962), pp. 108–11, 146–56, 237–38, 247–49; William H. Stahl, *Martianus Capella and the Seven Liberal Arts* (New York, 1971), pp. 154–70; Pearl Kibre, "The Boethian *De Institutione arithmetica* and the Quadrivium in the Thirteenth-Century University Milieu at Paris," in *Boethius and the Liberal Arts*, ed. Michael Masi (Nashville, forthcoming); and Lowrie J. Daly, S.J., *The Medieval University* (New York, 1961), pp. 8–10.

5. A. Forest, Fernand van Steenberghen, and Maurice de Gandillac, "Le mouvement doctrinal du XIe au XIVe siècle," in *Histoire de l'église*, vol. 13 (Paris, 1951), 1ff.; M. L'Abbé A. Clerval, *Les écoles de Chartres au moyen âge (du Ve au XVIe siècle)* (Paris, 1895; reprinted Frankfurt a.M., 1965), pp. 108–30, 235–48. For a recent study of the twelfth-cen-

tury approach to natural science see Brian Stock, *Myth and Science in the Twelfth Century* (Princeton, 1972); also M.-D. Chenu, O.P., *Nature, Man and Society in the Twelfth Century*, trans. Jerome Taylor and Lester K. Little (Chicago, 1968).

6. Van Steenberghen, *Histoire de l'église*, 13:182ff.; Pearl Kibre, *The Nations in the Mediaeval Universities* (Cambridge, Mass., 1948), pp.; 98–99; also *Chartularium universitatis Parisiensis*, ed. Heinrich Denifle and Emile Chatelain, 4 vols. (Paris, 1889–97), 1:nos. 180–182, for the efforts of Pope Innocent IV, in 1248, to establish and promote the study of Arabic and other non-Western languages.

7. Astrik L. Gabriel, *Skara House at the Mediaeval University of Paris* (Notre Dame, Ind., 1960); Gordon Leff, *Paris and Oxford Universities in the Thirteenth and Fourteenth Centuries* (New York, 1968), pp. 113–15.

8. For the situation particularly in Paris in the fifteenth century, see Pearl Kibre, *Scholarly Privileges*, chap. 6; and for Oxford, ibid., especially pp. 317–19; and Anna M. Campbell, *The Black Death and Men of Learning* (New York, 1931), chap. 6.

9. Astrik L. Gabriel, *Garlandia: Studies in the History of the Mediaeval University* (Notre Dame, Ind., 1969), chaps. 1 and 2.

10. Rashdall, *Universities*, 1:275–82; 3:37, 41–47; Pearl Kibre, *Nations in the Mediaeval Universities*, pp. 160–63.

11. Gabriel, *Garlandia*, chap. 4; "Preparatory Teaching in the Parisian Colleges during the Fourteenth Century."

12. Louis J. Paetow, *The Arts Course at Medieval Universities with Special Reference to Grammar and Rhetoric* (Urbana, Ill., 1910), chaps. 1 and 2; also Philippe Delhaye, "La place des arts libéraux dans les programmes scolaires du XIIIe siècle," in *Arts libéraux et philosophie au moyen âge. Actes du Quatrième Congrès International de Philosophie Médiévale* (Montreal, 1969, henceforth cited as *Arts libéraux*), pp. 168–70; *Statuta antiqua universitatis Oxoniensis*, ed. Strickland Gibson (Oxford, 1931), pp. 25–26; and James A. Weisheipl, O.P., "Curriculum of the Faculty of Arts at Oxford in the Early Fourteenth Century," *Mediaeval Studies*, 26 (1964):168–70.

13. Hugh of St. Victor, *Didascalicon: A Medieval Guide to the Arts*, trans. Jerome Taylor (New York, 1961), p. 59.

14. Pearl Kibre, "The Quadrivium in the Thirteenth-Century Universities (with special reference to Paris)," in *Arts libéraux*, pp. 178–79.

15. Ibid., p. 178.

16. Frederick M. Powicke, *Ways of Medieval Life and Thought* (London, 1949), pp. 189–94; Heinrich Roos, "Le trivium à l'université au XIIIe siècle," in *Arts libéraux*, p. 196.

17. Pearl Kibre, "The Quadrivium in the Thirteenth-Century Universities," in *Arts libéraux*, pp. 176–83; and for Oxford, Weisheipl, "Curriculum of the Faculty of Arts at Oxford," pp. 170–73, cover this and the remaining quadrivial arts.

18. For the prohibitions see James A. Weisheipl, *The Development of Physical Theory in the Middle Ages* (Ann Arbor, 1971), pp. 26–27; also Daly, *Medieval University*, pp. 81–83.

19. Chartularium universitatis Parisiensis, vol. 1, no. 246; Weisheipl, "Curriculum of the Faculty of Arts at Oxford," pp. 173–76; Thorndike,

History of Magic, 1:657–61.

20. Weisheipl, "Curriculum of the Faculty of Arts at Oxford," p. 175.

21. On lecturing, see Daly, *Medieval University*, pp. 151–56.

22. On the technique of the disputation, see ibid., pp. 156–58; also P. Glorieux, *La littérature quodlibétique*, 2 vols. (Kain-Paris, 1925–1935).

23. *Chartularium universitatis Parisiensis*, vol. 1, no. 441, statute of the faculty of arts, for April 1, 1272.

24. Thomas Aquinas, *Summa theologiae*, ed. and trans. William A. Wallace, O.P. (Dover, Mass., 1967), 10:xx–xxi; appendices 3–9, and especially pp. 213–16.

25. Albertus Magnus, *Commentarii in secundum librum Sententiarum*, dist. 13C, art. 2, in *Opera omnia*, ed. Auguste Borgnet, vol. 27 (Paris, 1894), p. 247; *De vegetabilibus et plantis*, bk. 6, tract. 1, in *Opera omnia*, ed. Borgnet, vol. 10 (Paris, 1891), pp. 159–60.

26. For license requirements see *Chartularium universitatis Parisiensis*, vol. 1, nos. 444, 451–56; vol. 2, nos. 921, 922, 996; vol. 4, no. 2659. See also Alexander Birkenmajer, "Le rôle joué par les médecins et les naturalistes dans la reception d'Aristote au XIIᵉ et XIIIᵉ siècles," *Studia Copernicana*, vol. 1 (Warsaw, 1970), pp. 73–97, especially p. 77ff.

27. Carlo Malagola, ed., *Statuti delle università e dei collegii dello Studio Bolognese* (Bologna, 1888), pp. 261–64; *Statuta dominorum artistarum academiae patavinae* (Padua, n.d.; Hain 15015), statutes of 1465, fol. 25r–v.

28. Kibre, *Scholarly Privileges*, pp. 42–51, 61–63.

29. Nancy G. Siraisi, *Arts and Sciences at Padua* (Toronto, 1973), chap. 1.

30. Mario Tabanelli, *La chirurgia italiana nel'alto medioevo* (Florence, 1965), 1:198–210.

31. *Conciliator differentiarum philosophorum et praecipue medicorum* (Venice, 1496), diff. 1, fol. 3r.

32. Malagola, *Statuti*, pp. 274–77, 251–52, 247–48.

33. Ibid., pp. 425–26, 448.

34. George Fowler, *Intellectual Interests of Engelbert of Admont* (New York, 1947), pp. 21–22.

35. Regarding philosophy at Bologna, see Charles Ermatinger, "Averroism in Early Fourteenth Century Bologna," *Mediaeval Studies* 15 (1954):35–56; the same author's "Some Unstudied Sources for the History of Philosophy in the Fourteenth Century," *Manuscripta* 14 (1970): 67–87; and bibliography there cited.

36. Malagola, *Statuti*, pp. 261–64 (1405); *Statuta dominorum artistarum*, fol. 21v (1465).

37. See note 35, above.

38. Thorndike, *History of Magic*, vol. 2 (New York, 1923), pp. 874–947; vol. 3 (New York, 1934), pp. 386–97; and Silvio A. Bedini and F. R. Maddison, *Mechanical Universe: The Astrarium of Giovanni de'Dondi* (Philadelphia, 1966); Lynn White, Jr., "Medical Astrologers and Late Medieval Technology," *Viator* 6 (1975):295–308.

39. Antonio Favaro, "I lettori di matematiche nella Università di Padova," *Memorie e documenti per la storia della Università di Padova*, vol. 1 (Padua, 1922), p. 25; *Statuta dominorum artistarum*, fol. 21v.

40. Malagola, *Statuti*, p. 432 (doctoral college statutes, 1378), pp. 274–75 (student university statutes, 1405), p. 489 (fragment of doctoral college statutes, undated).

41. *Expositiones . . . in subtilissimum Joannitii Isagogarum libellum* (Venice, 1527), fol. 343r–v.

42. John H. Randall, *The School of Padua and the Emergence of Modern Science* (Padua, 1961).

43. In his commentary on the *Isagoge* of Johannitius, Taddeo habitually broke down his material into (1) the opinion of the philosophers and (2) the opinion of the physicians, and then attempted reconciliation (see, for example, fol. 344v of the cited edition). The title of Peter's principal work, the *Conciliator*, indicates the same purpose.

44. Malagola, *Statuti*, p. 438. One may contrast with this the contemporary situation in England, where a substantial majority of educated physicians were still clerics. See Huling E. Ussery, *Chaucer's Physician: Medicine and Literature in Fourteenth-Century England* (New Orleans, 1971), pp. 29–30, 35–38, 40.

45. Malagola, *Statuti*, pp. 248–49.

46. Of the many examples that could be cited, two will suffice: Bartolomeo da Varignana, who taught medicine for many years at Bologna, attended the Emperor Henry VII on his ill-fated expedition to Italy; Gentile da Foligno (d. 1348), author of copious commentaries on Avicenna and a professor in various university centers, was the personal physician of Ubertino da Carrara, ruler of Padua.

47. Thorndike, *History of Magic*, 2:827–28.

48. Malagola, *Statuti*, p. 264.

5 Michael S. Mahoney   **Mathematics**

The development of mathematics in medieval Europe from the sixth to the fifteenth century shows clearly how mathematics depends on the cultural context within which it is pursued.[1] The barbarian cultures that succeeded Roman rule in the fifth century had no indigenous mathematical traditions; in mathematics, as in most other intellectual activities, they followed the Roman lead. But the Romans themselves had had little interest in mathematics beyond its practical application to business and surveying. Roman thinkers who wished to learn the theoretical mathematics of Euclid, Archimedes, or Apollonius did so in the same way they learned the philosophy of Plato or Aristotle—in its original Greek from Greek teachers. But Greek theoretical mathematics received no reinforcement from native Roman intellectual traditions, with the result that those few Romans who learned the subject made no contributions to it. Greek habits of mathematical thought made little or no impact on Roman culture, and Greek mathematics remained in Greek down to the end of the Empire.[2]

In mathematics, then, to succeed the Romans was to succeed to nothing beyond the rudiments of computational arithmetic on the abacus and of mensurational geometry immediately applicable to surveying and architecture. The fragments of Greek mathematics that Boethius (d. 524/25) tried to preserve by translating Nicomachus of Gerasa's *Arithmetic* and a portion of Euclid's *Elements* constitute a measure not of how little mathematics survived the fall of the Empire, but of how little had been present there before. Not only did the new cultures lack the mathematical texts of classical antiquity; more important, they were heirs to no intellectual tradition that would make them aware of what they lacked.

When, then, the texts did become available in translations from Greek and Arabic in the twelfth and thirteenth centuries (see above, chap. 2), they posed fundamental problems of comprehension and assimilation. Unprepared to read them, medieval scholars had to learn almost from scratch and without teachers.[3] In one sense, they learned quickly and well. In another sense, they never did quite learn. For they read the texts with their own intellectual concerns in mind: the union of faith and reason (a difficult problem exacerbated by the simultaneous introduction of Aristotelian philosophy), the education of the clergy, the effective governance of church and state. Hence, medieval mathematics assumed its own peculiar form in the period from the mid-eleventh to the mid-fifteenth century. Rarely pursued for its own sake, it served philosophical, pedagogical, and practical ends, and the internal technical development it underwent was largely dictated by those ends. To appreciate the medieval achievement in mathematics, one must bear this context in mind. Doing so will save one the feeling of disappointment experienced, say, when seeing how little impact the translation of Archimedes into Latin had on the subject.

Doing so will also make clear when the medieval period of the history of mathematics ended, to wit, in the later fifteenth and early sixteenth centuries, when the humanist translation of the Greek mathematical corpus into Latin stimulated the reunion of form and content; that is, when European mathematicians started thinking in Greek mathematical terms again.[4] The context explains, moreover, why mathematics had no single institutional locus and the disciplinary unity that comes from it, but, rather, was pursued in several different settings by a variety of practitioners.

Abacus and Agrimensor: The Early Centuries

A student in a monastic or cathedral school of, say, the ninth century who wanted to learn the arithmetic and geometry of the quadrivium had at his disposal little by way of learning materials.[5] With Boethius's *Arithmeticae institutiones*[6] and the Venerable Bede's (d. 735) *De computo vel loquela digitorum*,[7] he would have exhausted the arithmetical literature of his day, though not all the techniques. From the first work, a paraphrased translation of Nicomachus of Gerasa's *Arithmetica*,[8] he could learn the most elementary Pythagorean number theory: even and odd numbers; primes; perfect, abundant, and deficient numbers; figurate numbers; names and classes of numerical ratios (see below); arithmetic, geometric, harmonic, and other means. For the most part, Boethius simply set forth propositions, offering little by way of proof and even less by way of techniques for further investigation. Bede, by contrast, taught a technique, one that had existed for centuries as an oral tradition before he committed it to writing.[9]

Finger-counting, still used today by merchants in the Near and Far East, represents numbers by configurations of the fingers and the palm of the hand; in Bede's system, one can count up to 9999. Although one can also carry out the basic arithmetical operations for results up to that limit, the most common medieval use of the technique involved neither large numbers nor complicated operations. Primarily, Bede taught it as part of his procedure for determining the movable dates of Easter and of the feasts dependent thereon; it was the simplest way for monks to keep track of their religious calendar. Secondarily, finger-counting and computing supplemented the most common and effective means of calculation in antiquity and much of the Middle Ages, the abacus.

Our ninth-century student had no text from which to learn the abacus. Until the turn of the eleventh century it, like finger-counting, had formed a part of a body of orally communicated techniques practiced largely by merchants and government administrators.[10] Long after the first texts, which were addressed to a scholarly audience, the day-to-day use of the instrument remained an oral tradition. It existed side by side with the written techniques of the Hindu-Arabic algorism (see below) right up to the modern period and, indeed, is still used in many parts of the world.[11]

The Roman abacus used in the early Middle Ages overcame the unwieldiness of the Roman numeral system by employing a strictly additive, decimal place-value representation of numbers. In both the smaller hand format and the larger board form, it consisted of several columns containing successive powers of 10 counted from the right and beginning with 1. The particular digit within each column was represented by small stones (*calculi*, whence *calculare*, "to calculate"), each counting as 1 or, in some versions, as 5 when placed above a horizontal line drawn through the columns. Addition required nothing more than laying down the numbers and regrouping columns in which the total exceeded 9; subtraction followed an equally simple inverse pattern. Multiplication demanded two supplementary steps: first, calculation of the individual, digit-by-digit subproducts (by continued addition on the board or by reference to a 9 x 9 multiplication table or even by finger-reckoning); second, determination of the proper column into which to place the product of two other columns (for example, 10s columns x 100s column = 1000s column). Once laid down, the subproducts directly yielded the final result, needing regrouping at most. For example, 496 x 23 = 11,408 followed from the sequence of operations presented in figure 2. Division was only slightly more complicated and followed two basic patterns: "golden division" computed directly with the divisor, while "iron division" supplemented the divisor to the next multiple of 10 or 100, adding the product of subquotient and supplement to the remainder before each successive subdivision.[12]

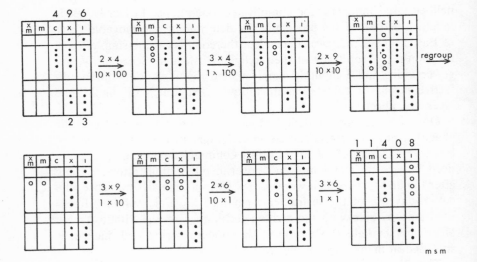

Fig. 2

By expressing all fractions in terms of the basically duodecimal system of Roman weights and measures,[13] ancient and early medieval abacists could treat the fractions as integers. Some boards had separate columns headed by the various subdivisions of the *as* (the basic unit of measure), but computation with all boards required supplementary tables of products and equivalents;[14] regrouping, of course, took place duodecimally rather than decimally.

The first texts on the abacus were written toward the end of the tenth century for monastic schools like the one our student had attended. Those texts reveal both by omission and by emphasis where the main difficulties lay in learning the instrument and in understanding its principles.[15] The rules of decimal place-value arithmetic—for example, regrouping and carrying over to the next column, products of powers of ten, borrowing for subtraction—constituted the biggest hurdle, particularly in the absence of any clear concept of the underlying structure of the system itself. That hurdle grew larger when the texts turned to the duodecimal system of fractions. For the most part, the treatises on the abacus remained collections of rules and examples with little or no attempt at justification.

Many of those treatises include the most original medieval contribution to the abacus. Sometime in the 970s or 980s, Gerbert of Aurillac, later Pope Sylvester II (d. 1003), invented or copied from Spanish Arabs an abacus with *apices*, or counters.[16] Each *apex* bore a number from 1 to 9 in the West Arabic form and was used as a single token in place of the corresponding number of *calculi*. Thus, the number 7683, for example, which required on the old abacus either 24 or 12 stones (the latter for abaci that grouped at five), now took only four *apices*. But the greater simplicity of representation had its price in greater complexity of operation. Sums no longer emerged directly from laying down the addends; one now had to know the sums of the numbers symbolized and to replace *apices* in a column by an *apex* bearing that sum.

In the realm of geometry, too, our student had little to go on. Cassiodorus recorded that Boethius translated Euclid's *Elements* into Latin.[17] By 800 only fragments—limited mostly to enunciation of the definitions, postulates, axioms, and theorems of books 1–4—still existed, and they had been grafted onto a collection of surveying texts known as the *ars gromatica* (from *groma*, "surveyor's pole").[18] Written during the late Empire, but certainly containing techniques known for centuries, the texts focused on the measurement of plane areas of varying regular and irregular shapes, along with the simpler solid figures. Reflecting a technological rather than mathematical tradition, they freely employed approximations and made no attempt at derivation or demonstration. Some of the texts simply cataloged systems of weights and measures.

It was the *agrimensores* (surveyors) rather than Euclid who determined the pattern of early medieval geometrical thought. The very etymology of "geometry" emphasized its metric purpose, and not enough of the Greek model had survived to set an example of a more abstract, theoretical enterprise. So, at the beginning of the eleventh century, Gerbert tried to explain to his student Adelbold, later Bishop of Utrecht, why the area of an equilateral triangle of side 30 differed from the triangular number of the same side;[19] Franco of Liège considered cutting up a parchment circle and rearranging the pieces as one means of squaring the circle; and two men named Ragimbold of Cologne and Radulf of Liège exchanged opinions on what could be meant by the "exterior angle" of a triangle, whether 7:5 or 17:12 was the correct ratio of the diagonal of a square to its side, what was meant by "linear feet," "square feet," and "solid feet" (*pedes recti, pedes quadrati, pedes solidi*).[20] From the Lotharingian cathedral and monastic schools to which these men belonged came also the composite edition of Euclid, *ars gromatica*, and Gerbertian abacus that circulated for centuries as the *Geometria* of Boethius;[21] its compiler showed the same insensitivity to the nature of Greek geometry characteristic of the period.

From the sixth to the eleventh century, there was little mathematics known and even less pursued. More important than inactivity, however, was the absence of any historical tradition carrying back to Greek mathematics. Whatever mathematics was going on was headed in a different direction. The introduction of new Greek texts in translation during the twelfth century might change that direction, but it could not do so entirely. Latin Europeans had already begun to develop their own habits of thought.

New Texts, New Problems

Three works translated from Arabic into Latin in the twelfth century gave new impetus to mathematics and determined its primary content for the next centuries: the arithmetic and algebra of al-Khwarizmi, and the *Elements* of Euclid. Other works translated during the twelfth and thirteenth centuries contributed to the investigations provoked by these three, but none had quite the same impact.

Al-Khwarizmi's *Treatise on Calculation with the Hindu Numerals* (ca. 825)[22] may have been the first Arabic arithmetic to employ separate symbols for the numbers from 1 to 9 (plus a symbol for 0 as a placeholder) in a decimal place-value system and to set forth written arithmetical techniques consonant with the system. A version of it was certainly the first to introduce the system into Latin Europe, where it was christened *algorismus* ("algorism," whence the modern "algorithm"), after a corrupted form of the author's name. Neither the original Arabic text nor the original Latin translation (done before 1143) survives, but three Latin versions produced soon after that translation give a good idea of its contents.[23] It began by presenting the symbols and offering an explanation, with examples, of the place-value system; then it provided a detailed exposition, also with examples, of the combinatory operations: addition, subtraction, doubling, halving, multiplication, division, squaring, extraction of the square root.

As has been shown above, the concept of number-symbols employed in a place-value system was not entirely new to European mathematicians. It underlay Gerbert's abacus with *apices*, where the columns of the board gave tangible meaning to the notion of place-value. To one familiar with the abacus, the operations of written arithmetic would not seem completely strange. Gerbert's abacus, for example, had already trained its users to do the mental arithmetic of the addition and multiplication tables, one of the fundamental hurdles encountered when individual symbols replace groups of units. Moreover, al-Khwarizmi and his immediate Latin successors began the transition from abacus to paper by merely translating the techniques of the former onto the latter. For example, doubling and multiplication both began with the highest digits, adding to already computed values on the left when subproducts exceeded 10; halving began with the lowest digit, adding 5 to the right when an odd digit was encountered.[24] Those were the habits of the abacist. So, too, was that of wiping out, or later crossing out, not only intermediate results, but the multiplicand and the dividend, which in the original algorism were gradually replaced by the emerging product and quotient, respectively.[25] So, for example, 496 x 23 proceeded according to the following pattern:

$$\frac{496}{23} \to (2\cdot 4 = 8) \to \frac{8496}{23} \to (3\cdot 4 = 12) \to \frac{9296}{23} \to$$

$$(\text{shift multiplier}) \to \frac{9296}{23} \to (2\cdot 9 = 18) \to \frac{11096}{23} \to$$

$$(3\cdot 9 = 27) \to \frac{11276}{23} \to (\text{shift multiplier}) \to \frac{11276}{23} \to$$

$$(2\cdot 6 = 12) \to \frac{11396}{23} \to (3\cdot 6 = 18) \to \frac{11408}{23};$$

$$496 \times 23 = 11408$$

Al-Khwarizmi's text was the prototype for the two most popular medieval algorisms, the *Carmen de algorismo* of Alexander de Villa Dei (d. ca. 1240) and the *Algorismus vulgaris* of John of Holywood (Sacrobosco) (d. 1244 or 1256). The first, composed around 1202, apparently aimed at the same sort of audience as Bede's work; it taught the new algorism as a tool of the computus of movable feasts, casting al-Khwarizmi into verse for easier memorization.[26] The second, written around 1240 for the arts curriculum of the new universities, struck a more theoretical note by adding some material from Boethius's *Arithmetica*, revising the order of presentation of the combinatory operations, and expanding the descriptions and illustrative examples. It became, especially with the commentary on it by Peter of Dacia (1291), the standard university text for several centuries.[27] Only in the later fifteenth and early sixteenth centuries, when separate schools of mercantile arithmetic proliferated, did al-Khwarizmi's techniques gradually give way to the modern forms of written computation, which retain and build on intermediate results without the need for erasure, though with increased reliance on mental computation.[28]

The abacus had offered no direct method for handling common fractions, and the early algorists inherited the resulting preference for conventional systems that reduced fractions to integers. With the introduction of Greco-Arabic astronomy, the sexagesimal system of degrees, minutes, seconds, and so forth, replaced the Roman system as the most common convention among scholars. Nonetheless, common fractions could not be avoided, especially as increased use of the new algorism encountered fractions with no finite sexagesimal equivalents. Somewhat more slowly than for integer arithmetic, but by the time that Johannes de Lineriis wrote his *Algorismus de minutiis* (ca. 1340), a succession of medieval writers worked out the now common notation of numerator over denominator (the terms themselves are medieval) separated by a line (*virgula*) and the now familiar procedures for combining fractions.[29]

Where al-Khwarizmi's arithmetic encountered broad understanding and provoked further research fully consonant with it, Euclid's geometry met a different reception. As noted above, medieval scholars were not well prepared for it by anything they possessed previously. Their first task was to understand it, and, despite its reputation, Euclid's *Elements* is not an elementary, self-explanatory textbook, especially in the later books on ratio and proportion, number theory, and incommensurable magnitudes. Not only did medieval (and, for that matter, classical) Latin lack the technical terms requisite for accurate translation; it occasionally lacked fairly common terms, such as that for rhombus (a Greek transliteration that replaced the Arabic *elmuayn* in the sixteenth century). Add to this two distinct Arabic versions of the *Elements*, three different Latin translators, and several variant versions by one of them, and one has some notion of the problems that stood in the path of assimilation of Greek geometry in the twelfth and thirteenth centuries.[30]

Assimilation, rather than original research, characterizes the theoretical geometry of the Middle Ages. It has, indeed, become a commonplace that no new theorem dates from the period, and more than one historian has recorded disappointment over the meager results achieved with such a rich heritage.[31] But medieval scholars faced a difficult task in merely trying to understand Greek geometry. It was foreign to them in more than a linguistic sense and in ways that arithmetic was not. As a result, although medieval scholars did learn to carry out geometric constructions and to prove geometric theorems, they did not always appreciate the subtlety of Euclid's *Elements* nor the reasons for it. Hence, on occasion, while paying lip service to Greek mathematical canons, medieval mathematicians violated them (at times quite creatively) in practice.

Moreover, the primary purpose for learning geometry in the Middle Ages was not to carry out further research in the area, but, rather, to understand the geometrical references of Aristotle[32] and the Church Fathers or to be able to do the mathematics demanded by astronomy or optics or to improve mensurational practice and the instruments designed for it.[33] The level of geometrical knowledge required by these applications was relatively low; they called, rather, for ingenuity of a logical or practical sort. It is, then, either in the didactically couched effort of assimilation or in ingenious application to other disciplines that one must seek the medieval geometrical achievement.

Although both Gerard of Cremona's accurate and complete translation of the *Elements* from the Arabic version of Ishaq ibn Hunayn as revised by Thabit ibn Qurra and Adelard of Bath's not quite so excellent translation (Adelard I) from the Arabic version of al-Hajjaj existed from the mid-twelfth century on, medieval geometers on the whole took their Euclid from less complete sources: an abridged version of Adelard's translation, containing the definitions, axioms, postulates, and theorems without proofs (Adelard II), plus an expanded version of that with quasi-original proofs (Adelard III; Roger Bacon called it the *editio specialis*). In the late 1250s Campanus of Novara edited and expanded these latter versions into what soon became and remained until the sixteenth century the authoritative medieval text of the *Elements*.[34]

In a recent study, John E. Murdoch has signaled the main characteristics of the Adelard-Campanus tradition.[35] For all its minor variances, it remained remarkably close to the original text. Few theorems were omitted, and the omissions were soon noted and remedied where necessary in the proofs of other theorems. The order of Euclid's theorems remained essentially the same; the few changes made affected the elegance of the work, not its logical coherence. With one major exception, to wit, the definition of equal ratios in book 5 (see below), no important misunderstandings blocked the original intent of the propositions or confused the thrust of the argument.

Indeed, it was the argument rather than the mathematical content that seems most to have interested Adelard and Campanus. For the *Elements* was accompanied into scholastic thought by Aristotle's *Posterior Analytics*, which held it up as a model of scientifically demonstrated knowledge. Hence, the variant proofs and proof-schemes of the Adelard III and Campanus versions, together with many of the additions made to both first principles and propositions, concentrated on the logical structure of the *Elements*: the form of individual theorems and their proofs; their relation to other theorems as converses, corollaries, and counterparts in another domain; their place in the work as a whole; their pertinence to issues outside geometry. In similar fashion, the first principles underwent careful scrutiny both for their own sake and for the sake of understanding first principles in general. Commentaries focused on the adequacy, mutual independence, and epistemological priority of Euclid's definitions, axioms, and postulates and on their occurrence in the proofs that followed.

What the Arabic mathematicians discovered by trying to apply the *Elements* to new mathematical situations (for example, the postulates of continuity necessary for quadrature by the method of exhaustion), medieval Latin writers often found by examining its logical structure and by revising it for use in the university classroom. Indeed, the initial and primary purpose of the logical analysis was to make the *Elements* understandable to students. The very language of the proofs, comments, and additions in the Adelard-Campanus tradition reveals this concern. For most propositions, the student learns of their relation to the ones immediately preceding, of any special problems in the construction, of any subsidiary lemmas or postulates demanded by the proof but not previously supplied, or peculiar characteristics of the proof itself (for example, the use of unusual forms of inference), of similarities the proposition or its proof might have to other propositions or proofs (for example, a comparison of books 5 and 7), and of interesting or important corollaries. The use of the first or second person abounds, the latter often in an imperative mood, as if guiding a construction, and occasionally the tone becomes almost conversational. Adelard and Campanus were not research mathematicians talking to colleagues; rather, they were teachers sharing their knowledge with students.

Beyond Euclid's *Elements*, very little Greek geometry was available to medieval scholars. Of the few mathematical works of Archimedes translated into Arabic, only the *Measurement of the Circle* and some theorems of *On Spirals* were, in turn, transmitted to Europe, the latter by means of a mensurational treatise (see below) which, because of its authorship by the Banu Musa, gained the Latin title *Verba filiorum.* A paraphrase of book 1 of *Sphere and Cylinder*, taken possibly from some Byzantine source, circulated under the title *Liber de curvis superficiebus Archimenidis* by one Johannes de Tinemue.[36] Although in 1269 William of Moerbeke translated almost the entire extant Archimedean corpus from the Greek, his effort had little effect, and most medieval authors took their knowledge of Archimedes from the earlier versions. Those authors treated Archimedes as they had Euclid; indeed, they used Euclid to do so. That is, they concentrated on the logical structure of Archimedes' methods and proofs, often supplying from the *Elements* the intermediate steps and formal justifications that Archimedes had expected his readers to know.[37] Hence, perhaps the most sophisticated treatment, a *Circuli quadratura* written in 1340 perhaps by Jean de Meurs, used fourteen of the first eighteen propositions of *On Spirals* to provide the straight line equal to the circumference of a circle called for in proposition 1 of *Measurement of the Circle*.[38] But the sophistication lay in seeing that proposition 1 presupposed that construction and that *On Spirals* provided it. If, by the mid-fourteenth century in Paris, some medieval mathematicians knew about Archimedes' works and understood them, none used them creatively.

In discussing the impact of Archimedes on European mathematics, Clagett has pointed to the conscious differentiation between empirical and theoretical techniques made after the introduction of the Greek material.[39] Yet, even before then, medieval scholars were beginning to distinguish between the subject's abstract structure and its practical applications. In the early twelfth century, Hugh of St. Victor, building on the distinction in his *Didascalicon* between theoretical sciences and practical sciences (see below, chap. 14), divided geometry into a theoretical, speculative part which proceeded "by rational consideration alone" (*sola rationis speculatione*) and a practical, active part which employed instruments.[40] In his *Practica geometriae*, which introduced the traditional generic title for works of the sort, he treated the measurement of heights (*altimetria*), areas (*planimetria*), and spherical volumes (*cosmimetria*). The geometry he applied to these problems scarcely went beyond what the *agrimensores* had taught and, like the writings of the eleventh-century Lotharingian authors, revealed a fundamental innocence of the most elementary theoretical knowledge beyond the measure of right triangles.[41] Nonetheless, Hugh did add to the instruments of the *agrimensor* the multipurpose Greco-

Arabic tool Gerbert was credited with having introduced into Europe, namely, the astrolabe.

About a century later, Robertus Anglicus of Montpellier (fl. 1270) introduced another instrument with Arabic forebears, the quadrant.[42] Neither the description of the instrument's construction nor the applications made of it represented a marked advance in mathematical sophistication over Hugh's treatise. For example, in marking off the curved lines that represent the twelve "artificial" hours of the day between sunrise and sunset, Robert seems to have assumed a knowledge of books 1 and 3 of the *Elements* in calling for the construction of an isosceles triangle, given the base, one (indefinitely extended) side, and the angle between them. But he gave no hint of the actual construction, and his constant references to manipulating the compasses give the impression that he actually worked by trial and error. That impression is strengthened when the reader is instructed to divide the circular edge of the quadrant first into two halves, then each half into thirds, then each third into thirds, and finally each subdivision into five parts. Here one finds not only a lack of constructions but the demand for an impossible construction given the tools available; Robert showed no awareness that the division of a given angle into thirds, not to say into fifths, poses rather special problems for the geometer.

It was not until the first half of the fourteenth century that Euclid's *Elements* influenced the style and content of practical geometry. Dominicus de Clavasio's *Practica geometriae* of 1346, besides introducing yet a third instrument, the gnomon, or geometrical square, explicitly tied its subject to its theoretical foundations.[43] Since, for example, all the instruments and techniques of practical geometry rested ultimately on similar triangles and, thus, on proportions, Dominicus's treatise began with four "suppositions" about proportions and about solving for unknown terms in them. Each application of an instrument to a problem in measurement—Dominicus maintained the traditional division into altimetry, planimetry, and stereometry—was accompanied by a "proof of this practice" founded on Euclid's *Elements*. The style of the whole is reminiscent of the Adelard-Campanus tradition; Dominicus seems to have been most concerned with emulating a logical model of exposition. Occasionally this concern got in the way of his material, as when in construction 15 of book 2 he confused the determination of the area of a regular polygon, given one of its sides, with the construction of a similar polygon by *Elements* VI.18. He was not confused about the circle, however. Although by *Elements* XII.2, circles are as the squares on their diameters, their absolute measure is theoretically unattainable: "The ratio of the circumference of any circle to its diameter is a triple sesquiseptimate [ratio], or thereabouts, . . . because there is no definite demonstrated ratio of the circumference to the diameter. Therefore, when I speak of measuring a circle with a square, I do not mean to speak demonstratively but only to teach how to find the area such that no sensible error remains."[44]

Dominicus's treatise represents the culmination of the tradition of practical geometry deriving from the *ars gromatica*. His techniques all but exhaust the possibilities of measurement by use of similar triangles and proportions alone. The further development of practical geometry required assistance from two quarters, first from the Arabic art of algebra as introduced in the twelfth century and, second, from the emergence of trigonometry, a largely original contribution of fifteenth-century mathematicians.

In addition to writing the first arithmetic using Hindu numerals, al-Khwarizmi also compiled a text that became the prototype for all Arabic and medieval Latin works on algebra. Indeed, as he supplied his own name to arithmetic (algorism), so the title of his algebra text gave a name to that art. Called in the original *Kitab al-mukhtasar fi'-l-hisab al-jabr wa'l-muqabalah* ("Compendious Book on Calculation by Completion and Balancing"), it became in Latin simply "algebra." Of its original three parts, only two were transmitted to Europe, and those two arrived separately.[45] Hence, it took Latin scholars some time to realize that the two parts fit together.

The first part, translated by Robert of Chester in 1145[46] and again by Gerard of Cremona some decades later, contained solution paradigms for six types of problems expressed in modern symbolism by the equations (1) $ax^2 = bx$, (2) $ax^2 = b$, (3) $ax = b$, (4) $ax^2 + bx = c$, (5) $ax^2 + c = bx$, and (6) $bx + c = ax^2$, where a,b,c are all positive rational numbers. Although al-Khwarizmi employed no symbols (in fact, he even wrote out the numbers in words), his standardized vocabulary of *shay'* ("thing") or *jidhr* ("root") for the unknown, *mal* ("wealth") for its square, and *dirham* (a common coin) for the known unit permitted him to state the paradigm in general terms before illustrating it by examples. Here is not the place to discuss al-Khwarizmi's sources, or the continuity of his algebraic tradition since Babylonian times, but his specific examples both at the beginning and later on in the text have long, cross-cultural histories in themselves.[47] After setting forth rules for solving all six forms, al-Khwarizmi supplied geometrical demonstrations of the rules for forms (4)-(6); his rules reveal the influence of book 2 of the *Elements*.

After a substantial discussion of binomial multiplication, both for numbers explicitly written as the sums of tens and unit (to reinforce the concept of decimal arithmetic, or to show that algebra is a form of arithmetic?) and for sums and differences of numbers and "things," he took up a number (it varies among the different versions) of examples solved by use of his six rules. A short presentation of the "rule of three" for solving problems by finding the fourth proportional to three given values—for example, "If ten cost six, how much do eight cost?"—completed the first part.

The second part, which belonged to the Arabic tradition of ʿilm al-misaha, or "science of measure," treated problems similar to those of the later Latin practical geometries but did so in a mathematically more sophisticated way.[48] In particular, it added algebra to the tools of the *mensurator* and, hence, added to the standard problems of determining heights, areas, and volumes those of dividing areas and volumes and of determining lengths from various combinations of dimensions. Al-Khwarizmi's original text remained in Arabic, but an expanded version of it by Abraham bar Hiyya (known in Latin as Savasorda) was translated by Plato of Tivoli in 1145 as the *Liber embadorum* ("Book of Areas"). Quite similar material, including a more extensive use of algebraic techniques, entered Europe through Gerard of Cremona's translation of the *Liber in quo terrarum et corporum continentur mensurationes* ("Book in which Are Contained the Measures of Lands and Bodies"—*Liber mensurationum*, "Book of Measures," for short) by an Abu Bakr, whose exact identity is uncertain. Part 3 of al-Khwarizmi's algebra, dealing with the division of inheritances, also remained untranslated, though portions of it may perhaps be found in the works of Leonardo Fibonacci (see below).

As impressive as all this algebraic material may seem to modern eyes, it apparently attracted little attention during its first two or three centuries in Latin. Not only do very few manuscripts of the works exist, but, more importantly, even fewer mathematicians of the twelfth, thirteenth, or early fourteenth centuries show any familiarity with them. Jean de Meurs's *Quadripartitum numerorum* and *De arte mensurandi*, written in the mid-fourteenth century, may represent important exceptions to this rule. But, in the knowledge of al-Khwarizmi and Fibonacci shown by part 3 of the *Quadripartitum* and in the familiarity with Abu Bakr and Archimedes displayed by the second part of chapter 5 and chapters 6-12 of the *De arte mensurandi* (according to Clagett, the portions actually by Jean), they are exceptions, quite uncharacteristic of the other texts being produced by masters of the arts curriculum in the universities.[49]

Al-Khwarizmi's algebra, with its emphasis on the art's utility to merchants and surveyors, failed to attract the attention of the masters of the arts faculty and appears to have remained in the marketplace as a largely oral tradition. Not until the late fifteenth century at Leipzig did it appear in a university lecture and in new texts.[50] Despite the learned traditions of practical geometry since Hugh of St. Victor and Dominicus de Clavasio's effort to give the subject a theoretical basis, the mensurational literature seems largely to have shared algebra's nonuniversity circulation. So it is, for example, that Dominicus's treatise appeared in a German translation as early as 1400, most likely in response to the demands of a practical audience.[51]

The Limits of Originality

The two outstanding mathematicians of the early thirteenth century, Leonardo Fibonacci of Pisa and Jordanus de Nemore, illustrate how much Greco-Arabic mathematics was available at the time and how original Europeans could be in mastering it. By the same token, the striking lack of successors to their work over the next two centuries emphasizes the peculiar orientation of medieval mathematics.

To speak of Leonardo is to use superlatives; to attempt to capture in brief scope the breadth and variety of his mathematics is futile.[52] Born around 1180 to a wealthy Pisan merchant, Leonardo learned the commercial trade in ports from Algeria to Byzantium; during that apprenticeship he also learned practical and theoretical mathematics from most of the available Greek and Latin sources. In his *Liber abbaci* (1202, revised 1228), he introduced Hindu-Arabic arithmetic into Europe by another route than the translations of al-Khwarizmi.[53] But there he went far beyond the mechanics of computation to present a vast and varied array of arithmetical problems, both practical and recreational, solved by ingenuity and by algebra. The later chapters of his book focus on algebra as a systematic problem-solving technique and contain the results not only of al-Khwarizmi, but of the less well-known al-Karaji as well.[54] Leonardo combined in his *Practica geometriae* (1220) the practice of the *ars gromatica* and the *'ilm al-misaha* with the theoretical foundations of Euclid and the imaginative manipulations of Archimedes.[55] In this one work, determining the height of a tower by use of the quadrant was presented side by side with the duplication of the cube as set forth by Archytas, Plato (as reported by Eutocius via the Banu Musa), and Philo of Byzantium (through a source also available to Jordanus); the division of areas by classical means accompanied the inscription of a rectangle and a square in an equilateral triangle by means of algebra; and the Euclidean theory of irrationals (book 10) was set next to the arithmetic of common weights and measures.

It was in his *Flos* and his *Liber quadratorum* (both ca. 1225), however, that Leonardo showed his true mettle.[56] In the former, he examined the equation (here in modern symbols) $x^3 + 2x^2 + 10x = 20$ (the same problem appears in al-Khayyami's *Algebra*), showed by skillful use of Euclid's theory of quadratic irrationals that the solution was not among them, as it was also neither integer nor rational, and then set forth without derivation or demonstration the sexagesimal solution $x = 1;22,7,42,33,4,40$, which is too large by 1½ parts in 60^6. In the second work, he showed himself the equal of Diophantus and without peer until the seventeenth century in his use of number theory to derive a general solution in rationals of the system $x^2 + 5 = y^2$, $x^2 - 5 = z^2$, or $y^2 - x^2 = x^2 - z^2 = 5$, where the particular value 5 was replaced by a parameter on which Leonardo set conditions of solvability.

Where Jordanus (fl. 1220) learned the mathematics that made him the second most literate European of his day in the subject may never be known, since he himself is known only through his works. Of these the largest and perhaps most original is the *De numeris datis*.[57] Treating there the linear and quadratic algebraic techniques of the Arabic writers and of Leonardo, Jordanus turned from numerical example and rhetorical precept to a literal symbolism which, though neither technically nor conceptually as complete as the later system of François Viète (*In artem analyticen isagoge*, 1591), nonetheless represented a step in that direction.[58] An example from Jordanus's text best illustrates his achievement:

> (I,7) If a number be divided into two [numbers], of which only one is given, but from the [number] not given times itself and times the given there results a given number, the number that was divided will also be given. Let the number be divided into a and b, and let b be given; and let d, which is given, result from a times itself and times b, i.e. times the whole ab. Let c be added to ab, and let it [c] be equal to a, so that the whole abc is divided into ab and c. Then, because ab times c yields the given d, and the difference of ab to c, namely b, is given, abc and c will be given; similarly also a and ab [will be given]. For an example of this operation, let 6 be one of the numbers, and from the other times itself and times 6 let 40 result, of which the double, i.e. 80, be doubled, and [the result] will be 160, to which is added the square of 6, i.e. 36, and they make 196, of which the root is 14, from

which, when 6 is taken away and the remainder halved, 4 results, which is the other number. And the whole divided will be 10, 6 and 4 added together.[59]

Though clearly original, it was a mixed achievement. The symbolism would seem to have derived from the use of single letters to denote line segments usually described by their endpoints. Note, for example, that c, though equal to a, would be a different line segment and, hence, would require a different symbol. Here at least it was still unnecessarily cumbersome. c is not needed to carry out the reduction of the original problem to one of the standard forms of mixed simultaneous systems that underlay traditional algebra; indeed, in a more suitable symbolism the desired result would follow immediately from a statement of the problem. Finally, Jordanus did not operate on the symbols, or with them. They were essentially a shorthand. Where modern operational symbolism would move from $xy = a$, $x - y = b$, via $(\frac{x-y}{2})^2 + xy = (\frac{x+y}{2})^2$, to explicit values for x and y in terms of a and b, Jordanus had to turn to a specific numerical example handled by a recipe. That is why the second part of the demonstration did not illustrate the first part but, rather, complemented it.

Nonetheless, Jordanus's symbolic treatment was original at the time, and that same originality shows clearly through his other works. In addition to an *Algorismus* along the lines of al-Khwarizmi's, he wrote a book, *De triangulis*, in which, among material also found in Leonardo's *Practica geometriae* and ultimately derived from Greco-Arabic sources, he set forth (IV.20) the trisection of the angle (two solutions taken from the Banu Musa and his own based in part on a proposition from Ibn al-Haytham's *Optics*), Hero's Theorem for the area of a triangle from its three sides, two solutions of the problem of finding two mean proportionals (IV.22), and a quadrature of the circle differing essentially from the Archimedean approach.[60] Jordanus's *Arithmetica*, perhaps the best known of his works, brought the substance of Euclid's arithmetical books (*Elements* VII–IX) to the subject, then dominated by Boethius's much less technical presentation.

Yet, neither Leonardo nor Jordanus had a following, either in his own lifetime or over the next century or more. Leonardo became the master of Hindu-Arabic arithmetic as taught in the Italian *scuole d'abaco* from the fourteenth century on, and several of the problems in the *Liber abbaci* and his other works found their way into the myriad problem-texts that circulated. But the more sophisticated techniques and problems apparently lay unused until resuscitated by Italian algebraists of the fifteenth century. When revived, they were translated into the "art of the coss" or *arte della cosa*, which had turned the abbreviations of the technical terms *res* ($\curlyvee$), *census* ($\varkappa$), *cubus* (σ), and so forth, into quasi-symbols (the algebraic term *cosa*, like the Italian word itself, derived from *causa*, which Leonardo had used as a synonym for *res*). That symbolism overwhelmed Jordanus's system of algebra, which few seem to have known and even fewer to have used.[61] Leonardo and Jordanus shone in the decades just after the introduction of Greco-Arabic mathematics. Owing to special circumstances, clear in the case of Leonardo, unknown in the case of Jordanus, they had all of that mathematics at their disposal, they learned it from its masters, and they pursued it in its own tradition. But few others shared those special circumstances. For most medieval thinkers, mathematics was something to be puzzled over, to be edited and revised for use in the classroom, to be examined for insights it might provide into logic and philosophy. In short, it was something to be talked about. Those who actually did it had no audience.

On the Way toward Autonomous Development

During the fourteenth century, medieval mathematics moved from a stage of assimilation of the Greco-Arabic texts to one of independent development of the material in those texts. As has been emphasized several times above, that development did not take place along the same lines as those followed by Greek and Arabic mathematicians; rather, it followed the interests and demands peculiar to medieval Europe. Those interests and demands dictated not only the particular topics pursued but also the sort of treatment they received. As John E. Murdoch has suggested, the medieval theory of ratio and proportion may serve as a representative example.[62] It was born of the peculiar circumstances in which Greco-Arabic mathematics was introduced into Europe, it took a turn possible only in the absence of a continuous tradition, and it was aimed ultimately at a philosophical point.

Euclid's *Elements* contains two separate theories of ratio and proportion reflecting two historical stages of development and two domains of applicability. The earlier theory, contained in book 7, derives from early Pythagorean doctrine and applies only to ratios of integers. It rests on the notion that all numbers have a common measure, that is, the unit, and, hence, that one can determine the equality of two ratios, or proportionality, by a finite counting procedure. To be precise: "(*Elements*, VII. def. 20) Numbers are *proportional* when the first is the same multiple, or the same part, or the same parts, of the second that the third is of the fourth."[63] That is, $(A,B) = (C,D)$ if and only if when (for some integers m and n) $mA = nB$, then $mC = nD$.[64] Because A and B are numbers, there will always exist some pair m, n, as Euclid shows in *Elements* VII.4.

The later theory represents Eudoxus's ingenious solution to the problem posed by the existence of geometrical magnitudes, such as the diagonal and side of a square, that have a ratio not expressible in integers, that is, for which no pair of integers m, n exists. Using a variation of the principle that underlay his famous "method of exhaustion," he made a useful definition out of the fact that the Pythagorean method for determining equality of ratio would involve an endless search for a common measure: "(*Elements*, V. def. 5) Magnitudes are said to *be in the same ratio*, the first to the second and the third to the fourth, when, if any equimultiples whatever be taken of the first and third, and any equimultiples whatever of the second and the fourth, the former equimultiples alike exceed, are alike equal to, or alike fall short of, the latter equimultiples respectively taken in corresponding order."[65] That is, $(A,B) = (C,D)$ if and only if, for all m and n, $mA \gtreqless nB$ if and only if $mC \gtreqless nD$. By this definition and its employment in the theorems of book 5, Eudoxus overcame in one fell swoop the difficulties of the infinite, the continuous, and the incommensurable by using a potentially infinite set of discontinuous elements to narrow in on a common measure. It was no mean achievement, and Euclid fully appreciated it.

However, neither Arabic nor European mathematicians appreciated it to the same extent. Some Arabic writers complained of its difficulty and sought to replace it, others defended it, but none completely understood it. European mathematicians simply ignored it, led astray in part by a spurious fourth definition in book 5 that not only failed to replace the genuine definition 4, but also skewed the meaning of definition 5.[66] Unable to see the subject through Euclid's or Eudoxus's eyes, medieval Latin writers viewed it through the distorting lenses of Boethius's *Arithmetica*; harking back to the earlier Pythagorean concept, they declared that "those ratios are equal of which the denominations are equal." And how did one determine the denomination of a ratio? According to Jordanus's brief tract, "the denomination of a ratio of this to that is what results from dividing this by that"; Campanus of Novara agreed word for word in his *De proportione et proportionalitate*.[67]

As the term "denomination" implies, each ratio received a name by this division, and the name depended on the class into which the ratio fell. If, in the ratio (A,B), $A = B$, it was a "ratio of equality"; if $A > B$, a "ratio of greater inequality"; if $A < B$, a "ratio of lesser inequality." Within ratios of greater inequality, if $A = mB$ for some integer m, then (A,B) was, in general, a "multiple ratio," in particular an "m-tuple ratio" (for $mA = B$, (A,B) was a "sub-m-tuple ratio"). If A contained B once with a remainder of 1, (A,B) was a "super-particular ratio"; specifically, $(3,2)$ was a "sesquialterate ratio," $(4,3)$ a "sesquitertian," $(5,4)$ a "sesquiquartan," and so on. If A contained B once with a remainder greater than 1, (A,B) was a "superpartient ratio"; specifically, for example, $(5,3)$ was a "superbipartient thirds," $(8,5)$ a "supertripartient fifths," and so on. When A contained B more than once with a remainder, (A,B) fell into the class of "multiple superparticular" or "multiple superpartient" ratios with specific names formed in the corresponding manner. Most medieval texts on ratio and proportion devoted a major section to this classificatory scheme, and throughout the literature of the late Middle Ages right into the late seventeenth century, ratios were expressed by their denomination rather than by pairs of numbers.[68]

If the concept of denomination embodied on the one hand the idea that a ratio is a special entity, a relation (*habitudo*) between two quantities, rather than a quantity itself, it acted on the other hand to erase that very distinction operationally. For, when the quantities in a ratio were viewed primarily as numbers, and when those numbers were divided by one another, the result had to be a number. Thus, in *De numeris datis*, II.2, Jordanus multiplied one term of a ratio by the ratio's denomination in order to determine the other term, and in II.3 he divided 1 by the denomination of a ratio in order to determine the denomination of the inverse ratio. In short, via the procedure of denomination, ratios came to be manipulated by the arithmetic of fractions. The short-term result was an arithmetization of the theory of ratio and proportion that evaded or ignored the subtler aspects of that theory. As a long-term result, denomination abetted the extension of the concept of number to include fractions as ratios to the unit and, ultimately, to include any quantity linked to a unit quantity by a ratio or relation definable in specified ways.[69]

But the procedure of denomination rode roughshod over the problem of irrational ratios, or ratios of incommensurables. Some authors argued that such ratios had no denomination, others that they were *mediately*, rather than *immediately*, denominated. The latter distinction, apparently introduced by Thomas Bradwardine (d. 1349) and later developed by Nicole Oresme (d. 1382), seems from the texts to have lacked specific, commonly agreed on, operational meaning; thus, although the ratio of the diagonal of a square to its side was called "half a double ratio," Bradwardine and Oresme disagreed on what is the mediate denomination and what the immediate. Hence, though here medieval writers could name things, their procedure for doing so neither rested on, nor provided insight into, the mathematical structure of the entities named.[70]

Moreover, by the time Oresme was writing his *De proportionibus proportionum*, two sorts of ratios were being denominated. Medieval writers had combined two minor definitions in book 5 of the *Elements* with traditional music theory and with a theorem transmitted through Arabic sources but dating back to Menelaus to develop a theory of the compound ratio, or ratio of ratios.[71] Here, perhaps more than anywhere else in mathematics, they showed both their ability to explore concepts at a profound level and their predominant concern with nonmathematical issues.

They began with inadequate information. Although Euclid had mentioned the compounding of ratios, noting in particular that in a continued proportion $(A,B) = (B,C) = (C,D)$ the ratio (A,C) is the *duplicate* of ratio (A,B), and (A,D) its *triplicate* (V. defs. 9, 10), and that "equiangular parallelograms have to one another the ratio compounded of the ratios of their sides" (VI.23), he did not use or develop the concept further.[72] Mathematical astronomers found more use for it, as for example in Menelaus's transversal theorem, which states that in figure 3 $(AG,AE) = (GD,DZ)(BZ,BE)$ and $(GE,AE) = (GZ,ZD)(DB,BA)$. The theorem appears in this form in Ptolemy's *Almagest*, and in a commentary on that work Theon of Alexandria added four more cases involving different combinations of the six line segments. Arabic writers brought the total number of cases up to eighteen; while studying the theorem for its own sake, however, they did not lose sight of its trigonometric context. Only Leonardo

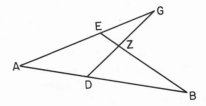

Fig. 3

Fibonacci among the early European writers recognized and recorded
that context. The far more influential Jordanus and Campanus, though
working from Arabic sources, turned, rather, to the already bastardized
Euclidean-Boethian theory of ratios for their language and style of
presentation. For Jordanus, "to produce or compound a ratio from
ratios is to produce the denomination of the ratio by multiplying the
denominations of the ratios by one another."[73] The conceptually vague
notion of "compounding" two relations to produce a third had become the operationally clear and simple matter of multiplying two
fractions, and, through the notion of denomination, compound ratio
followed simple ratio out of the domain of geometry into that of
arithmetic, not to return until the emergence of trigonometry in the
fifteenth century.

But, as has been repeatedly pointed out, medieval thinkers liked to
push ideas to their limits, and Bradwardine's application of compound
ratios to the science of motion (see below, and chapter 7) provoked
a development of the mathematical theory that soon made compound
ratios more than simply an application of the arithmetic of fractions.
To translate the new theory into arithmetic would have required the
concept of the rational exponent, for which Oresme and his contemporaries completely lacked the symbolism and had only the most rudimentary grasp of the basic notions. But Oresme did not have to
translate it into arithmetic. Both his mathematical and his nonmathematical purposes made possible a more abstract treatment. In the
following account, only the symbolism is anachronistic, and it has
been carefully designed not to import alien concepts into Oresme's
theory.[74]

Oresme began his presentation in *De proportionibus proportionum*
by assuming that his reader knew how to determine the equality of
two ratios (A,B) and (C,D) and knew also from Euclid that two
ratios (A,B) and (B,C) formed the compound ratio (A,C); in symbols, $(A,B) + (B,C) = (A,C)$.[75] Compounding, therefore, has an
inverse operation of division or decomposition, since, for any value C
between the terms of a ratio (A,B), $(A,B) = (A,C) + (C,B)$; that
C exists, at least in the realm of magnitude, was explicitly assumed
by all writers since Campanus.[76] Clearly, any ratio may be increased:
given (A,B), simply take a value $C > A$ or $D < B$; then $(C,D) >$
$(C,B) > (A,B)$. (Though Oresme did not say so explicitly, any
ratio may be similarly decreased.) To compound, that is "add," any
two ratios (A,B) and (C,D), one determines the quantity E, such
that $(C,D) = (B,E)$, or the quantity F, such that $(A,B) = (F,C)$

(the existence of the fourth proportional is another explicit assumption of medieval mathematicians);[77] then $(A,B) + (C,D) = (A,B) + (B,E) = (A,E) = (F,C) + (C,D) = (F,D)$. "Subtraction" takes place by the obvious inverse procedure.

Since any two ratios may be added together, a ratio may be added to itself any number of times, so that one may define $n(A,B) = (A,B) + (A,B) + \ldots + (A,B)$ for n terms. Similarly, by finding $m - 1$ mean proportionals between the terms of any ratio (A,B), one may give precise meaning to the inverse multiple $\frac{1}{m}(A,B)$. Hence, too, the expression $\frac{n}{m}(A,B)$ has precise meaning.[78] But n and m themselves stand in a ratio. If, therefore, $(C,D) = \frac{n}{m}(A,B)$, (n,m) may be called the "ratio of the ratios (C,D) and (A,B)." According to Oresme (in Grant's interpretation), (C,D) or $\frac{n}{m}(A,B)$ is immediately denominated by (A,B) and mediately denominated by (n,m).

With this apparatus Oresme already has enough to carry out the main purpose of chapter 4 of his treatise, to wit, an exposition of Bradwardine's Rule to show that it is the only internally coherent version of Aristotle's statement about forces, resistances, and speeds. If, that is, the ratio (V_2,V_1) is the ratio of the ratios (F_2,R_2) and (F_1,R_1), or $(F_2,R_2) = V_2/V_1(F_1,R_1)$, then Aristotle's various assertions in the *Physics* are satisfied.[79] Yet, two chapters intervene between the apparatus and the physical discussion, and Oresme clearly intended two more chapters, since separated from the text, to follow.[80] His purpose there went well beyond Bradwardine's Rule. He tried to establish for compound ratios criteria of commensurability analogous to those of simple quantities, with the ultimate goal of providing a dialectical basis for rejecting astrology.

Throughout the discussion in chapters 2 and 3, Oresme plays on a comparison of quantities under addition and ratios under composition that he has set up in the second half of chapter 1. Because continuous quantity is infinitely divisible, so, too, is any ratio of continuous quantities; because number is finitely divisible, so, too, is any ratio of numbers. Ratios may be rational or irrational, the former if their terms are commensurable, the latter if not. So, too, ratios of ratios may be rational or irrational, depending on the commensurability of the ratios constituting them. Finally, just as Euclid provided in books 7 and 10 a set of criteria for determining the commensurability of quantities, both discrete and continuous, so, too, one may establish similar criteria for ratios of ratios, as Oresme then undertakes to do.

The concept of commensurability rests on that of "part" and ultimately of "unit." A quantity A is said to be a *part*, or an *aliquot part*, of quantity B if and only if for some integer n, $nA = B$. A is *parts* of B if and only if there exists some quantity C that is a part of A and a part of B; that is, C is a *unit* common to both A and B. For Euclid, A and B are commensurable if and only if one of them is part or parts of the other. Clearly, compound ratios allow the same definitions: (A,B) is part of (C,D) if and only if, for some n, $n(A,B) = (C,D)$, or if and only if the ratio of ratios (C,D) and (A,B) is a multiple ratio. (A,B) is parts of (C,D) if and only if there exists a unit ratio common to them both, that is, if and only if there is some (E,F), such that $(A,B) = m(E,F)$ and $(C,D) = n(E,F)$, whence $(A,B) = \frac{m}{n}(C,D)$. Two ratios are commensurable if and only if one is part or parts of the other.

The various propositions of chapters 2 and 3 establish first the ways in which rational and irrational ratios can be decomposed[81] and then the conditions under which rational ratios are commensurable with one another.[82] The discussion leans heavily on Jordanus's *Arithmetica* and Euclid's *Elements* VII–IX to provide criteria for the existence and number of proportional means between integers. Although a similar set of conditions might have been possible for irrational ratios, Oresme did not attempt the task beyond some asides in chapter 4. One reason is that the conditions for commensurability among rationals already supplied the ammunition he needed. Those conditions seemed so to limit the number of commensurable ratios—for example, $(3,1)$ is incommensurable with $(2,1)$, as is $(5,1)$, $(6,1)$, $(7,1)$, and so forth—that "(Proposition III.10) Given two unknown ratios, it is likely (*verisimile*) that they are incommensurable; and if many unknown [ratios] are proposed it is most likely (*verisimilimum*) that one is incommensurable with another." Therefore, as Oresme pointed out here and in his other writings, it is most unlikely that the as yet unknown exact ratios of planetary motions will be commensurable.[83] But astrology rests on the commensurability of those motions, else the cycles of conjunction and opposition are destroyed. Hence, astrology is at best scientifically suspect.

Oresme's original mathematical theory had little impact or influence; only two writers, Alvarus Thomas and Georg Lockert, seem to have followed him, and Lockert may have got his information second-hand.[84] But that should not obscure the sophistication and depth of thought evidenced by this treatise. Nor should that sophistication and depth obscure the ultimately nonmathematical goal of the theory. For Oresme, as for the majority of his medieval contemporaries, mathematics served other purposes than its own.

Oresme was probably the best mathematician of his own day and ranks with Leonardo and Jordanus as among the best of medieval Europe. His contemporary Jean de Meurs may have matched him in breadth of learning but not in originality. Nor would someone as good or better, for example, Regiomontanus, appear for another century. But Regiomontanus belonged to another age, one in which Apollonius, Diophantus, and Pappus began to circulate in Greek, one in which increased interest in technical astronomy fostered the development of trigonometry out of Greek and Arabic sources, and one in which the arithmetical and algebraic practices of merchants and engineers, long part of an oral tradition, were being set down in textbooks for wide circulation. The astronomer (who was often called a "mathematician"), the *maestro d'abaco*, the *Rechenmeister*, the *cossista* (practitioner of the *arte della cosa*, the *ars rei et census*, or algebra) replaced the university master of arts as the focal figure of mathematics, and their needs for improved techniques and the demands placed on them for solutions to new problems shifted mathematics away from the didacticism and philosophical application of the schools and toward internal development. That is the mathematics of the Renaissance, and it would culminate in the brilliance of the seventeenth century. As such, it marks the end of medieval mathematics.

Notes

1. Most general histories of mathematics have a section devoted to the Middle Ages. The most extensive of these, not yet entirely superseded (especially when supplemented by the additions and corrections published in the third series of *Bibliotheca Mathematica*), is Moritz Cantor's *Vorlesungen über die Geschichte der Mathematik* (Leipzig, 1907), vols. 1 and 2. A. P. Yushkevich's *Istoriia matematiki v srednie veka* (Moscow, 1961), available in German translation as A. P. Juschkewitsch, *Mathematik im Mittelalter* (Leipzig, 1964), sets European mathematics of the period in the context of Chinese, Indian, and Arabic developments. J. Tropfke's *Geschichte der Elementar-Mathematik*, 3d ed., 4 vols. (Berlin, 1930–40); 2d ed., 7 vols. (Berlin, 1921–24), now undergoing revision by Kurt Vogel and others, contains a wealth of detail organized by subject; by its nature the work lacks any consideration of medieval mathematics in its cultural context.

2. On Greek mathematics see Thomas L. Heath, *History of Greek Mathematics* (Oxford, 1921) and B. L. van der Waerden, *Science Awakening* (Groningen, 1954).

3. Indeed, one can get a good sense of the content and nature of early medieval mathematical thought from observing the difficulties scholars had in understanding what the Greek texts were about.

4. Of course, they could not do so completely, and the peculiar combination of medieval, Renaissance, and Greek motifs found in the sixteenth century constitutes the major factor in the emergence of modern mathematics at the turn of the seventeenth. But that is another story.

5. Regarding the place of mathematics and its subdivisions in the medieval curriculum, see below, chap. 14.

6. Ed. G. Friedlein (Leipzig, 1867).

7. *Patrologiae cursus completus, series latina*, ed. J.-P. Migne, vol. 90 (Paris, 1850), cols. 295–98.

8. *Introduction to Arithmetic, Translated Into English by Martin Luther d'Ooge; With Studies in Greek Arithmetic by Frank Egleston Robbins and Louis Charles Karpinski* (London, 1926).

9. See the full account in Karl Menninger, *Zahlwort und Ziffer* (Göttingen, 1958; English trans. Cambridge, Mass., 1969), 2:3–25, esp. 11ff., where he discusses the many allusions to the technique in the writings of classical authors and the Church Fathers.

10. On the age of the abacus and its early forms, see Menninger, 2: 102–28; in the remainder of vol. 2, chap. 2, he sets out its subsequent development and various techniques.

11. A famous illustration in Gregor Reisch's *Margarita philosophica* of 1503 (reproduced in Menninger, 2:162) shows Lady Arithmetic looking over the shoulders of Boethius and Pythagoras using written symbols and a form of abacus, respectively (they were mistakenly thought in the Middle Ages to be the inventors of the respective systems). By the early sixteenth century, neither method had a clear advantage over the other, and both were widely employed. In particular, the early abacus had been transformed into the form of a table and its mode of operation altered. The English Exchequer takes its name from the peculiar form of abacus used at Westminster to keep tax accounts, and *Rechnen auf den Linien*, a German variation which placed jetons on the lines separating the columns to represent values of 5×10^n, became extremely popular in the fifteenth and sixteenth centuries. For further detail, see Menninger.

12. To carry out 1081:23 by "iron division," one uses as divisor 30 ($= 23 + 7$). Then, 1081:30 = 36 with 1 remaining (note that only division by 3 is involved; the factor of 10 is accounted for by knowing what columns to place the results in). To that 1, one adds $7 \times 36 = 252$; 253:30 = 8 with a remainder of 13. To 13 add $8 \times 7 = 56$; 69:30 = 2 with a remainder of 9. To 9 add $2 \times 7 = 14$; 23:23 = 1. Then 1081:23 = $36 + 8 + 2 + 1 = 47$.

13. 1 *as* = 12 *unciae* = 288 *scripuli* = 6·288 *siliquae* = 18·288 *oboli*. All the integral multiples from 1 to 12 of the *uncia* ("ounce") had individual names, as did the major submultiples from 1/2 to 1/144; cf. the treatise cited in the following note.

14. The most extensive listing is perhaps the anonymous *De minutiis* (tenth century) published by Nicolaus Bubnov in his *Gerberti . . . opera mathematica* (Berlin, 1899), pp. 225–44; the tract contains the products of all named units of weight taken two at a time and must have been written with an enormous abacus in mind.

15. For some of the earliest texts, see Bubnov, pp. 197ff. G. R. Evans, in his "Duc oculum: Aids to Understanding in Some Mediaeval Treatises on the Abacus," *Centaurus* 19 (1976):252–63, cites other texts (many of them edited) and thoughtfully analyzes the pedagogical and conceptual problems their authors faced.

16. Cf. Bubnov, appendix 1; and Menninger, 2:133ff.

17. Menso Folkerts, *"Boethius" Geometrie II: Ein mathematisches Lehrbuch des Mittelalters* (Wiesbaden, 1970), pp. 69ff. Two groups of texts circulated under the title *Boethii geometria*. The first consisted of five books, of which the first and fifth contained excerpts from the *ars gromatica* (see below); the second excerpts from Boethius's genuine *Arithmetica*; and the third, fourth, and beginning of the fifth the definitions, postulates, axioms, and most propositions of *Elements* I–IV. *Geometria* II had only two books, of which the first contained a preface, excerpts from the *Elements* for the most part similar to those of *Geometria* I, proofs of *Elements* I.1–3, excerpts from the *ars gromatica*, and a section on the abacus; the second book added further gromatic texts and a section on fractions. For greater detail, see ibid., pp. xi ff. Folkerts provides a new critical edition of II, but I remains unedited. Following the lead of Bubnov and Paul Tannery (cf. the former's *Gerberti opera*, the latter's "La géométrie au XIe siècle" cited below, n. 20), Folkerts firmly establishes that II is the product of some Lotharingian compiler of the early eleventh century; perhaps he was the same one who linked Boethius's name to the *apices* introduced by Gerbert.

18. On the *ars gromatica* and *agrimensores* (surveyors), see Moritz Cantor's *Die römischen Agrimensoren und ihre Stellung in der Feldmesskunst, eine historisch-mathematische Untersuchung* (Leipzig, 1875) and Bubnov, appendices 4 and 7. Karl Lachmann's *Gromatici veteres* (Berlin, 1848) omitted a central text later supplied by Cantor and, in a different version, by Tannery ("Un nouveau texte des traités d'arpentage et de géométrie d'Epaphroditus et de Vitruvius Rufus," in *Mémoires scientifiques*, vol. 5 [Paris, 1922], pp. 29–78).

19. According to the Pythagorean doctrine of figurate numbers, which can be found in Boethius's *Arithmetica*, a triangular number is one of the series 1,3,6,10,15, . . . , $\dfrac{n(n+1)}{2}$ of successive sums of the integers from 1 to *n*. The numbers can easily be disposed in an equilateral triangular array of side *n* as follows:

20. Paul Tannery's studies remain the best source for the mathematical efforts of the Lotharingian schools; see in vol. 5 of his *Mémoires scientifiques*: "La géométrie au XIe siècle" (pp. 79–102), "Une correspondance

d'écolâtres du XIᵉ siècle" (pp. 103–11, 229–303), and "Notes sur la Pseudo-Géométrie de Boèce" (pp. 211–28). These articles contain full documentation for the examples cited above, except for the Gerbert-Adelbold correspondence, for which see Bubnov, pp. 41–45.

21. See above, n. 17.

22. The original title is not certain, but it must have been something of this order; cf. G. J. Toomer, "al-Khwarizmi," *Dictionary of Scientific Biography*, 7:360. See also J. Ruska, *Zur ältesten arabischen Algebra und Rechenkunst* (Heidelberg, 1917) and A. P. Yushkevish, "Arifmeticheskii traktat Muchammeda ben Musa Al-Chorezmi," *Trudy instituta istorii estestvoznaniia i techniki 1954*, 1:85–127, and "Über ein Werk des Abu ʿAbdallah Muhammad ibn Musa al-Huwarizmi al-Magusi zur Arithmetik der Inder," *Schriftenreihe fur Geschichte der Naturwissenschaften, Technik und Medizin*, Beiheft 1964 (Leipzig, 1964), pp. 21–64.

23. Of these the oldest seems to be the unique Cambridge MS text beginning, "Dixit Algorizmi . . . ," a thirteenth-century copy of an original predating 1143; it has been twice edited, by Baldassarre Boncompagni (*Trattati d'arithmetica, I. Algoritmi de numero indorum* [Rome, 1857]) and by Kurt Vogel (*Mohammed ibn Musa Alchwarizmi's Algorismus: Das früheste Lehrbuch zum Rechnen mit indischen Ziffern* [Aalen, 1963]). The second version, *Liber Ysagogarum Alchorizmi a Magistro A. compositus*, exists in five MSS and has been edited by M. Curtze ("Über eine Algorismusschrift des XII. Jahrhunderts," *Abhandlungen zur Geschichte der Mathematik* 8 [1898]:2–27) and by A. Nagl ("Über eine Algorismusschrift des XII. Jahrhunderts und uber die Verbreitung der indischarabischen Rechenkunst und Zahlzeichen im christlichen Abendland," *Zeitschrift für Mathematik und Physik*, Hist.-Lit. Abt., 34 [1889]:129ff., 161ff.). The third version, *Liber algorismi de practica arismetrice*, generally ascribed to John of Spain, was edited by Boncompagni as the second of the *Trattati* cited above; at least seven MSS copies are extant.

24. An especially simple operation when using an abacus that groups at 5.

25. Early algorists appear to have worked on sand or wax, drawing symbols where they once dropped stones into columns. Erasure on such surfaces was easy and, hence, emulation of the techniques of the abacus followed naturally. On parchment or paper, using ink, it was less simple to eradicate symbols; so one crossed out, often trying to array the intermediate results in a pleasing pattern, as in the "sail" for multiplication (using the example given below):

$$
\begin{array}{c}
4 \\
\cancel{3} \\
\cancel{2}0 \\
11\cancel{0}\cancel{0} \\
\overline{\cancel{9}2\cancel{7}8} \\
8\cancel{4}9\cancel{6} \\
\cancel{2}\cancel{3} \\
\cancel{2}\cancel{3} \\
\cancel{2}\cancel{3}
\end{array}
$$

496 x 23   = 11408

26. J. O. Halliwell, ed., *Rara mathematica* (London, 1838; 2d ed. 1841). The Latin text served as the basis for an English exegesis, "The Crafte of Nombryng," published by R. Steele, *Earliest Arithmetics in English* (London, 1922).

27. Maximilian Curtze, ed., *Petri Philomeni de Dacia in Algorismum vulgarem Johannis de Sacrobosco commentarius. Una cum algorismo ipso* . . . (Copenhagen, 1897). On the relation of Sacrobosco's text to Alexander's and on the use of the works as school texts, see Guy Beaujouan's enlightening article, "L'enseignement de l'arithmétique élémentaire à l'Université de Paris aux XIIIᵉ et XIVᵉ siècles," *Homenaje à Millas-Vallicrosa* (Barcelona, 1954), 1:93–124.

28. Cf., among other sources, Louis C. Karpinski, *The History of Arithmetic* (Chicago and New York, 1925).

29. For a brief survey of earlier texts and techniques, together with an edition of the *Algorismus*, see H. L. L. Busard, "Het rekenen met breuken in de Middeleeuwen, in het bijzonder bij Johannes de Lineriis," *Mededelingen van de koninklijke Vlaamse Academie voor Wetenschappen, Letteren en schone Kunsten van België, Klasse der Wetenschappen*, vol. 30, pt. 7 (Brussels, 1968). Johannes's work which treated both sexagesimal and common fractions was originally intended as a prologue to a large collection of astronomical tables.

30. On the translations themselves, see Marshall Clagett's still standard "The Medieval Translations from the Arabic of the Elements of Euclid, with Special Emphasis on the Versions of Adelard of Bath," *Isis* 44 (1953):16–42. For a brilliant account of the problems posed, both textually and conceptually, see John E. Murdoch, "The Medieval Euclid: Salient Aspects of the Translations of the *Elements* by Adelard of Bath and Campanus of Novara," *Revue de Synthèse* 89 (ser. 3, nos. 49–52) (1968):67–94. Much of what follows is derived from Murdoch's study. Of the medieval Latin versions, only that of Hermann of Carinthia has been published in a modern version; see H. L. L. Busard, *The Translation of the Elements of Euclid from the Arabic into Latin by Hermann of Carinthia* (Leiden, 1968). Murdoch has found an early Greek translation: "*Euclides graeco-latinus*: A Hitherto Unknown Medieval Latin Translation of the *Elements* Made Directly from the Greek," *Harvard Studies in Classical Philology* 71 (1967):249–302. A complete *Corpus Euclidis medii aevi* remains a major desideratum.

31. See, for example, the comments of Edward Grant on Marshall Clagett's "Archimedes in the Late Middle Ages," in *Perspectives in the History of Science and Technology*, ed. Duane H. D. Roller (Norman, Okla., 1971), p. 260.

32. On this point, see A. G. Molland, "The Geometrical Background to the 'Merton School': An Exploration into the Application of Mathematics to Natural Philosophy in the Fourteenth Century," *British Journal for the History of Science* 4 (1968–69):108–25; esp. p. 112. The many investigations of the incommensurability of the diagonal of a square to its side, and the almost as popular problem of squaring the circle, owe their inspiration more to Aristotle's *Posterior Analytics* and *Physics* than to Euclid or Archimedes.

33. Cf. the emphasis on practical measurement in the preface to one MS version of the so-called Adelard III tradition (see below) in Clagett, "Medieval Translations," p. 33.

34. Campanus's text served as the basis for the first printed edition by E. Ratdolt (Venice, 1482) and was republished several times in the first half of the sixteenth century until replaced by translations made directly from the Greek. Medieval scholars realized they were not dealing with an exact translation. They often referred to Campanus's text, in which the proofs bore his name and in which his *additiones* stood out clearly, as a "commentary" (*commentum*).

35. See above, n. 30.

36. For fuller information, together with editions and translations of the pertinent texts, see Marshall Clagett's magisterial *Archimedes in the Middle Ages*, vol. 1 (Madison, Wis., 1964); vols. 2–3 (Philadelphia, 1976–). Although portions of Apollonius's *Conics* entered medieval Europe via Arabic treatises on optics, it remained part of the optical tradition and underwent no mathematical development until Greek manuscripts of the first four books began circulating in the fifteenth century; see Sabetai Unguru, "A Very Early Acquaintance with Apollonius of Perga's Treatise on *Conic Sections* in the Latin West," *Centaurus* 20 (1976): 112–28; and "Witelo and 13th-Century Mathematics: An Assessment of His Contributions," *Isis* 63 (1973):496–508. On Pappus of Alexandria's *Mathematical Collection*, particularly valuable for the information it contains about the Greek mathematicians, see Unguru's "Pappus in the 13th Century in the Latin West," *Archive for History of Exact Sciences* 13 (1974):307–24.

37. For example, Archimedes' use of the method of exhaustion (cf. Thomas L. Heath, *The Works of Archimedes* [Cambridge, 1897/1912; New York, 1956], chap. 7) makes tacit assumptions about continuity, existence of proportionals, and convergence of series that medieval commentators made explicit by reference to *Elements* V and X; cf. Clagett's discussions in the works cited in notes 31 and 36.

38. Clagett, "Archimedes in the Late Middle Ages," pp. 248–49.

39. *Archimedes*, 1:562.

40. R. Baron, "Sur l'introduction en Occident des termes 'geometria theorica et practica,'" *Revue d'histoire des sciences* 8 (1955):298–302; also Baron, "Hugonis de Sancto Victore Practica geometriae," *Osiris* 12 (1956):176–224, republished in *Hugonis de Sancto Victore Opera Propaedeutica* (Notre Dame, 1966), pp. 3–64. For an edition, see Maximilian Curtze, "Practica geometriae," *Monatshefte für Mathematik und Physik* 8 (1897):193–224.

41. Paul Tannery, "Practica Geometriae," in *Mémories scientifiques*, 5: 204–10; esp. p. 205: "The interest provided for the history of geometry by these few pages results from their marking a singular moment in its teaching. Euclid had already been translated from the Arabic (by Adelard of Bath) at least a half-century earlier; but, if our Hugh is no longer ignorant, as were the scholars of the eleventh century, that there is a theoretical geometry, it must have been for him something like, in our day, higher geometry for instructors in primary schools. He shows, in fact, no trace of Greco-Arabic science; he does not even know, for example, how

to put letters on his diagrams to denote points. He represents, then, the pure Latin tradition at the instant when its normal evolution in our West is going to be brusquely redirected by the introduction of eastern doctrines."

42. Tannery, "Le traité du Quadrant de Maître Robert Anglès (Montpellier, XIIIᵉ siècle). Texte latin et ancienne traduction grecque," *Mémoires scientifiques*, 5:118–97; see also Maximilian Curtze, "Der Tractatus Quadrantis des Robertus Anglicus in deutscher Übersetzung aus dem Jahre 1477," *Abhandlungen zur Geschichte der Mathematik* 9 (1899): 41–63.

43. H. L. L. Busard, ed., "The Practica Geometriae of Dominicus de Clavasio," *Archive for History of Exact Sciences* 2 (1965):520–75.

44. Ibid., p. 556.

45. H. L. L. Busard, "L'algèbre au Moyen Age: Le 'Liber mensurationum' d'Abû Bekr," *Journal des Savants* (1968):65–124.

46. Louis C. Karpinski, *Robert of Chester's Latin Translation of the Algebra of al-Khowarizmi* (New York, 1915; repub. in Karpinski and J. G. Winter, *Contributions to the History of Science* [Ann Arbor, 1930; reprint ed., New York and London, 1972]). For the Arabic text and a direct translation of it, see F. Rosen, ed. and trans., *The Algebra of Mohammed ben Musa* (London, 1831).

47. For some idea of the sources and the continuity of tradition, see the concordances of problems in the texts of the Babylonians, al-Khwarizmi, Abu Bakr, and Abraham bar Hiyya in Busard's article cited in n. 45; see also S. Gandz, "The Sources of al-Khowarizmi's Algebra," *Osiris* 1 (1936). For present purposes, I should like to avoid the hotly contested issue of whether the tradition to which these problems and solution techniques belong is, in fact, "algebraic" under some suitable definition of the term. It suffices here to note that certain problems and techniques found in the Babylonian tablets have a continuous tradition down through the work of al-Khwarizmi and into medieval and Renaissance Europe. Al-Khwarizmi, along with his contemporaries and successors called this material "algebra"; so do I.

48. See *Encyclopedia of Islam*, under ᶜilm al-misaha.

49. The *Quadripartitum* remains for the most part unedited; see A. Nagl, "Das Quadripartitum numerorum des Johannes de Muris," *Abhandlungen zur Geschichte der Mathematik* 5 (1890):135–46, for extracts of book 2; for the verse preface and portions of book 3 see Louis C. Karpinski, "The Quadripartitum numerorum of John of Meurs," *Bibliotheca mathematica*, 3d ser., vol. 13 (1912–13), pp. 99–114. Portions of the *De arte mensurandi* containing Archimedean material appear in Clagett's *Archimedes*, vol. 3; see also his "Johannes de Muris and the Problem of Proportional Means," in L. G. Stephenson and R. Multhauf eds., *Medicine, Science, and Culture* (Baltimore, 1968), pp. 35–49; and H. L. L. Busard, "The Second Part of Chapter 5 of the *De arte mensurandi* by Johannes de Muris," in *For Dirk Struik* (Dordrecht, 1974), pp. 147–67. According to Clagett (*Isis* 60 [1969]:383–84), the treatise *Commensurator* commonly attributed to Regiomontanus and published in 1956 by W. Blaschke and G. Schoppe (*Akademie der Wissenschaften und Litteratur* [*Mainz*], *Abhandlungen der mathematisch-naturwissenschaftlichen Klasse*

[1956], no. 7) is, in fact, a series of propositions (without proofs) from the *De arte mensurandi*.

50. MS Dresden (Staatsbibliothek) C.80 fols. 301v–305r, contains notes for a lecture course on the "rule of three" given by one Gottfried Wolack in the summers of 1467 and 1468 at the University of Erfurt. The codex, which may have been the property of Johann Widmann of Eger (d. 1460, professor at Leipzig and a leading algebraist of his day), offers a good sample of the state of mathematics at the end of the period under study here. For a partial inventory, see Karpinski, *Robert of Chester's Latin Translation*, pp. 53–55.

51. H. Mendthal, *Geometria culmensis. Ein agronomischer Tractat* (Leipzig, 1886).

52. Two major studies are Baldassarre Boncompagni's *Della vita e delle opere di Leonardo Pisano* (Rome, 1852) and J. and F. Gies, *Leonardo of Pisa and the New Mathematics of the Middle Ages* (New York, 1969), though the latter is largely nontechnical. An admirable analytic précis of Fibonacci's work is Kurt Vogel's article in *Dictionary of Scientific Biography*, 4:605–13.

53. Baldassarre Boncompagni, ed., *Scritti di Leonardo Pisano* (Rome, 1857–62), vol. 1.

54. On al-Karaji, see *Dictionary of Scientific Biography*, 8:240–46.

55. *Scritti*, 2:1–224.

56. *Flos*, ibid., pp. 227–247; *Liber quadratorum*, ibid., pp. 253–79.

57. Barnabas B. Hughes, "The *De numeris datis* of Jordanus de Nemore, a Critical Edition, Analysis, Evaluation and Translation" (Ph.D. dissertation, Stanford University, 1970), contains an extensive introduction setting out what is known of Jordanus's life. For Jordanus's other work, see Edward Grant's article, "Jordanus de Nemore," *Dictionary of Scientific Biography*, 7:171–79, and below, chap. 6.

58. I do not mean to imply that Viète worked from Jordanus's example, but merely that the two systems have a common characteristic. Jordanus's system was, as will be seen, stillborn. For the extent of Viète's originality, see Mahoney, "Die Anfänge der algebraischen Denkweise im 17. Jahrhundert," *Rete* 1 (1971):15–30.

59. Hughes, "The *De numeris datis*," p. 133 (my translation).

60. Grant, "Jordanus de Nemore," passim.

61. Oresme, for one, cited the *De numeris datis* in his *De proportionibus proportionum* (see below). Interestingly, Jordanus's more traditional texts on algorism, fractions, and arithmetic (number theory) were apparently quite popular and well known.

62. See his "Medieval Euclid" (n. 30) and especially his "The Medieval Language of Proportions: Elements of the Interaction with Greek Foundations and the Development of New Mathematical Techniques," in *Scientific Change*, ed. A. C. Crombie (London, 1963), pp. 237–71. These two articles are the source for much of the following discussion and contain the pertinent bibliography.

63. Thomas L. Heath, *The Thirteen Books of Euclid's Elements* (Cambridge, 1908), 2:278.

64. Here, and throughout the discussion to follow, I shall use the symbol (x,y) to denote the ratio of x to y. As the most general concept, the

notion of an ordered pair seems least likely to import anachronistic and unwanted assumptions into the discussion; it may also be closest to the medieval view (see below). In this convention I follow the lead of E. J. Dijksterhuis, *Archimedes* (Copenhagen, 1956), p. 51. I shall also follow classical usage by referring to (A,B) as the *ratio* of A to B, and to (A,B) $= (C,D)$, or the equality of ratios, as a *proportion* among the *proportionals A, B, C, D*. Medieval writers, following the Arabic terminology *nisâb, tanâsub*, called ratios "proportions" and proportions "proportionalities."

65. Heath, Euclid's *Elements*, 2:114.

66. "Quantities that are said to have continuous proportionality are those of which equimultiples either are equal or equally exceed or fall short of one another continuously (*sine interruptione*)"; see Murdoch, "Language of Proportions," p. 241.

67. H. L. L. Busard, "Die Traktate *De proportionibus* von Jordanus Nemorarius und Campanus," *Centaurus* 15 (1971):193–227; esp. pp. 205, 213.

68. See Isaac Newton, *Principia* (London, 1687), bk. 3, hyp. 7: "The periodic times of the five primary planets and (either of the sun about the earth or) of the earth about the sun are in sesquialterate ratio of the mean distances from the sun." Here, of course, Newton was not only using the medieval terminology, but using it in its extended application to ratios of ratios (see below).

69. For example, by means of a finite algebraic equation with rational coefficients.

70. Nor were medieval writers alone in their quandary; cf. the discussion by H. L. Crosby of Murdoch's "Language of Proportions" (n. 62) in the same volume, pp. 324–27; Murdoch's reply, pp. 340–42; and the more recent discussion by Edward Grant in Nicole Oresme, *De proportionibus proportionum and Ad paucas respicientes*, ed. Edward Grant (Madison, Wis., 1966), pp. 31–35.

71. See Molland, "Geometrical Background" (n. 32), pp. 116ff.; Busard, "Die Traktate *De proportionibus*" (n. 67), pp. 193–98; Murdoch "Language of Proportions" (n. 62), pp. 263–69; Grant's introduction to Oresme, *De proportionibus*.

72. It appears in a spurious definition added to book 6, perhaps to cover its presence in VI.23; cf. Heath, Euclid's *Elements*, 2:189–90. Pappus of Alexandria's use of compound ratios to overcome the dimensional difficulties of expressing geometrically the extension of Apollonius's 3- and 4-line locus problem to more than 6 lines was, of course, unknown to medieval writers; cf. Pappus d'Alexandrie, *La collection mathématique*, trans. P. Ver Eecke (Paris and Bruges, 1933), 2:510.

73. ⟨*De proportionibus*⟩, ⟨def. 4⟩, ed. Busard, p. 205.

74. What follows was suggested by Molland's article (n. 32) but represents further development stimulated in part by Oresme's *De proportionibus proportionum*. The major danger in the use of symbolism to express the structure of Oresme's theory is that it may make obvious and thereby suggest to the modern eye lines of development that Oresme never imagined. I hope to counterbalance that danger by setting forth symbolically only what Oresme actually wrote in words. Yet, it is worth noting that

Oresme says enough to make the precise limits of the danger unclear. The treatise is by far the most abstract and venturesome piece of mathematical thinking in the period from Greek antiquity to the seventeenth century.

75. The "+" is not anachronistic in any but a symbolic sense. Oresme frequently speaks of compounding as "addition" (note his use of "subtraction" to denote the inverse operation), and his theory, especially in establishing the relations of "greater" and "less," trades on the parallels between adding numbers and "adding" ratios.

76. On medieval assumptions of continuity, see Murdoch, "Language of Proportions" (n. 62), pp. 240–51.

77. Campanus, for example, added to the "common notions" of the *Elements*, "As much as some quantity is to some other of the same sort, by so much is some third to another fourth of the same kind" (ed. Paris, 1516, fol. 4r). An alternate definition of compounding takes the form $(A,B) + (C,D) = (AC,BC) + (BC,BD) = (AC,BD)$.

78. One may easily verify that the laws of reflexivity, transitivity, associativity, distributivity of "multiplication" over "addition," and commutativity hold for compound ratios if they hold for the terms of the ratios. In addition, one may use those laws to establish that $n(A,B) = (A^n,B^n)$ for all n, integer or fraction, and, hence, that Oresme's system indeed yields a theory of exponential ratios when translated into more modern symbolism.

79. See below, chap. 7.

80. Grant, *Oresme* (n. 71), 72–81.

81. For example, a rational ratio may be divided into two equal rational ratios, as $(4,1) = (2,1) + (2,1)$; into two unequal rationals, as $(16,1) = (8,1) + (2,1)$ or $(3,2) = (6,5) + (5,4)$; and so on.

82. For example, for $(A,B) = n(C,D)$, there must exist $n-1$ mean proportionals between A and B in the ratio (C,D), that is, $(A,A_1) = (A_1,A_2) = \ldots = (A_{n-1},B) = (C,D)$.

83. See Oresme, *De proportionibus*, ed. Grant, pp. 111ff.

84. Ibid., pp. 69ff.

6

Joseph E.   **The Science of Weights**
Brown

The medieval Schoolmen called it *scientia de ponderibus*, the science of weights. To a modern reader this designation might suggest a scientific interest in metrology—the definition of standard units of weights with their multiples and correlations—and it is true that the development of a technical metrology was an important element in the European economic expansion of the twelfth and thirteenth centuries. However, such development proceeded, as far as we can tell, independently of the medieval science of weights. The latter was a highly theoretical study of problems now treated in the science of statics, the branch of mechanics which investigates the arrangements of forces that produce equilibrium.

In modern mechanics, where a single set of principles embraces both dynamics and statics, equilibrium is treated as a special or limiting case of motion, that is, motion of zero or constant speed. During the Middle Ages, however, the general scholarly consensus was that rest and motion are fundamentally opposed conditions, each governed by its own distinctive principles. Given that consensus, we might expect to find one science of bodies in motion (as described in the following chapter of this book) and another for bodies at rest (to be described here). In fact, the situation was more complex than that, and many medieval writers on the science of weights deliberately invoked dynamic principles to explain equilibrium. By so doing they reflected and perpetuated an ambivalence in the ancient sources of their discipline. To understand how this came about, we must begin with a brief description of the ancient sources and the two independent approaches to the problem of weights developed in antiquity.

The Ancient Traditions

Although Aristotle did not formulate a science of weights, his works were a source of fundamental notions that would be employed by those who did. His *Physics*, in particular, contained a store of rudimentary ideas and suggestions awaiting fuller development. In Aristotle's view, there were two ways to refer to a heavy body—represented by the Latin terms *pondus* and *grave*. *Pondus* (weight) was more likely to designate a stationary body, thus emphasizing its situation as still-weight, whereas *grave* (heavy body) denoted an object descending toward its natural place. Neither Aristotle nor his followers made the distinction explicit or observed it scrupulously, but it was implicit in their writings and suggested that heavy bodies could be studied under two modes—statically at rest or dynamically in motion. Aristotle concentrated on the latter, finding in *gravitas* (heaviness) an irreducible natural attribute which caused the downward motion of a heavy body. Despite the differing connotations of *pondus* and *grave*, the two words retained a close affinity of meaning, thus opening the possibility that the analysis of a *pondus* (still-weight) might incorporate elements drawn from the study of a descending *grave*. This occurred in what we shall call the "Aristotelian dynamic tradition" of the science of weights.[1]

This dynamic tradition began with a Greek author, usually thought to have been among the first few generations of Aristotelians, in a seminal work entitled *Mechanical Problems*. Here we find considerable attention devoted to the balance, whose principles, it was asserted, illumine a wide range of mechanical phenomena. From this point onward, the study of the balance and the science of weights would be closely associated. The series of questions and case studies that followed a brief introduction contained several important and original contributions: an appreciation of compound motion, for which the author developed a primitive sort of vector analysis containing the first known statement of the parallelogram of forces; an attempt to geometrize the constrained movement of a rotating balance beam; and an incipient recognition of the principle of virtual velocities, from which the earliest known statement of the law of the lever is deduced. These contributions consolidated some of the general notions of Aristotle into a core around which a dynamic approach to the science of weights could be organized.

The author of the *Mechanical Problems* began by examining some
physical and mathematical properties of circular motion, and from
these developed an analysis of the behavior of the lever or balance.
A circular or circumferential path, he explained, was a resultant of
two component motions related to each other according to a constantly changing ratio. One component was "natural" and directed
along the tangent at every point of the circumference. The other component, "unnatural" in the sense that it interfered with the natural
tangential movement, was directed toward the center of the circle
and, thus, at a right angle to the tangential component. In the author's
view, the ratio between the natural (tangential) and the unnatural
(radial) components changed continuously, and the resultant was,
therefore, a curved arc. By contrast, when there is a *constant* ratio
between components acting at an angle, the resultant will be a straight
line—namely, the diagonal of a parallelogram.

The author's discussion included a geometric demonstration which,
although more bold than sound, was nevertheless important and por-
tentous as the first effort to geometrize constrained movement. In
connection with his examination of compound circular motion, he inquired why the endpoint on a rotating radius traveled more quickly
than some intermediate point on the same radius—"more quickly"
being defined as traversal of "a greater distance in equal time."[2] Having just established that the curved figure of the circle was a resultant
of a natural (tangential) and an unnatural (radial) component, he
drew the further conclusion that the ratio of natural to unnatural
movement increases in direct proportion to the length of the radius.

We may recast his geometric argument as follows. Let the equal
perpendiculars ZT and YO (fig. 4) represent the natural tangential
components (n_1 and n_2) of arcs XT and BO. Let the vertical projec-
tions XZ and BY of the same arcs represent their respective unnatural
components (u_1 and u_2). Take $n_1 = n_2$ ($ZT = YO$), so that arc XT,

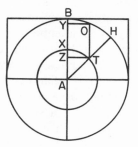

Fig. 4

described by the shorter radius, can be compared with arc BO, a *part* of arc BH described in the same time by the longer radius. Since $XZ > BY$, it follows that $n_2/u_2 > n_1/u_1$. Or, if we put the argument in terms of physical objects, a weight on the endpoint of a rotating radius moves more rapidly than a similar weight at an intermediate point because it is less interfered with.

This analysis is deficient in several respects. YO represents the tangential component, even when the radius is no longer at B on the circumference, but at O, where YO is by no means tangential. In order to express the greater velocity of the radius on the outer arc, the author compares radius AX when it is at point T to AB, an extension of the same radius, when it is at point O, where it was formerly but is no longer. He asserts, but cannot prove, that the inequality between the ratios n_2/u_2 and n_1/u_1 when the radii are at T and O holds everywhere on the circumferences. Clearly, the idea of geometrizing constrained movement was a pregnant one for mechanics, but it awaited medieval mechanicians to bring it to term.

This analysis raised another problem, concerning the proper way to define natural motion. In Aristotle's *Physics* natural motion (of a heavy body) was defined as motion directed toward the center of the universe, that is, vertical fall. Motion in any other direction was to some extent unnatural. In the *Mechanical Problems*, however, it was the tangential motion that was natural. This natural, outward component did not, of course, coincide with Aristotle's natural, downward component and could even nullify it. Medieval authors, while following the method of the *Mechanical Problems*, reverted to Aristotle's center of the universe as their point of reference for natural motion. That modification not only removed an inconsistency but, more importantly, made it possible for them to compare natural components of constrained motion to pure natural motion of a freely falling body.

Building on these initial conclusions, the author made his final contribution to the science of weights by proposing a theoretical explanation to account for equilibrium on a balance of unequal arms. His demonstration is noteworthy as an adumbration, for the first time, of the principle of virtual velocities. This principle, one of the most powerful in rational mechanics, appeared in a primitive form, which rested on further development of the Aristotelian doctrine of the "quicker." We have seen how the endpoint of a long rotating radius moves "more quickly" than the endpoint of a shorter radius rotating with the same angular velocity. But even when motionless, a longer radius can be conceived as "quicker" than a shorter radius in proportion to its greater length—although the advantage is only potential as long as the radius does not actually move. Now if we recall that according to Aristotle speed of descent (for a freely falling body) is directly proportional to weight, we can counterbalance the advantage of the longer, potentially quicker arm by giving the shorter arm the advantage of a heavier weight. If the weights are inversely proportional to the arm lengths, the advantages, one from potentially quicker arcal displacement and the other from quicker descent on account of greater weight, are offsetting, and equilibrium is maintained. The above reasoning seems evident in the following passage:

> Now since a longer radius moves more quickly than a shorter one·
> under pressure of an equal weight; and since the lever requires
> three elements, viz. the fulcrum—corresponding to the cord of a
> balance and forming the centre—and two weights, that exerted
> by the person using the lever and the weight which is to be moved;
> this being so, as the weight moved is to the weight moving it, so,
> inversely, is the length of the arm bearing the weight to the length
> of the arm nearer to the power. The further one is from the
> fulcrum, the more easily will one raise the weight; the reason being
> that which has already been stated, namely, that a longer radius
> describes a larger circle. So with the exertion of the same force the
> motive weight will change its position more than the weight which
> it moves, because it is further from the fulcrum.[3]

Although its lack of precision, generality, and mathematical conciseness must make us wary of reading this passage in the light of later and better articulated mechanics, it does contain significant insights. The author mentions both muscular fatigue and measured weight as indicators of the lever's effectiveness, but clearly prefers to deal with weights that can be expressed as quantities. He considers a special case of equilibrium rather than a general law relating forces, weights, and their distances from the fulcrum, but his case is based on a kind of deduction. While the coexistence of the variable factors, weight and radius length, does not permit us to attribute to the author a fully worked out principle of static moment (*force* $\times$ *distance*), his description does point to an intuitive groping toward that principle and an implicit recognition of the possibility that it could be expressed in the language of proportions. Although the passage was unsatisfactory because of its looseness, it was provocative for the same reason.

If the Aristotelian *Mechanical Problems* gave rise to a dynamic approach to the science of weights, an alternative static approach appeared under the aegis of Euclid and Archimedes. What distinguished the static approach from the dynamic was the presence of formal elegant geometrical demonstrations in the Euclidean mold and the total absence of any reference to movement. The contrast between the two traditions is nicely revealed by a brief analysis of Archimedes' demonstration of the law of the lever.

Archimedes based his proof on two powerful mathematical abstractions—the ideal balance and centers of gravity. He obtained the former simply by reducing the material beam of the balance to a weightless mathematical line, arbitrarily long and pivoted about a point, from which weights represented by geometrical magnitudes can be suspended. His center of gravity is an imaginary geometrical point to which the entire weight of an extended body has been reduced. In addition to these abstractions, Archimedes makes the crucial assumption, based on the principle of symmetry, that equal weights at equal distances from the pivot are in equilibrium. The proof of the law of the lever, then, aims simply to reduce the general case in which the lever arm lengths are inversely proportional to the suspended weights to the special case of equal arms and equal weights. A paraphrased version of his proof follows.[4]

Let commensurable magnitudes A and B (fig. 5) represent two weights suspended by their centers of gravity at points E and D on

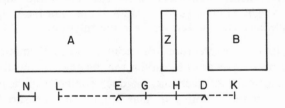

Fig. 5

lever ED, such that $A/B = DG/GE$; prove, then, that the system is in equilibrium about G. Now add to the lever segments DK (equal to GE) and EL (equal to DG), and locate H in such a way that $EH = DG$. The effect of this step is to double the length of the lever, permitting it to be analyzed as two separate suspension systems, LEH and HDK, each pivoted about its center, respectively E and D. Let Z represent some common measure of the unequal commensurable magnitudes A and B, such that $n \cdot Z = A$ and $m \cdot Z = B$. That is, for magnitude A we can substitute n magnitudes, each equal to Z, and for magnitude B we can substitute m magnitudes, each equal to Z. Because magnitudes A and B retain their proportionality to the doubled arm lengths ($LEH = 2\ GD$ and $HDK = 2\ GE$), there is also some corresponding common measure of linear distance N, such that $n \cdot N$

$= LEH$ and $m \cdot N = HDK$. Now at the midpoint of each of the n segments in LEH suspend one of the n magnitudes (each equal to Z) that make up A; similarly, at each midpoint of the m segments in HDK suspend one of the m magnitudes (each equal to Z) that make up B. This step redistributes the two original unequal magnitudes into a series of equal magnitudes suspended at equal intervals along the beam. Then, by the principle of symmetry, this whole system LGK is in equilibrium about its center, G. Further assumptions about substituting and combining centers of gravity permit the suspension system EGD with magnitudes A and B suspended at E and D to be equated with system LGK. Therefore, it, too, must be in equilibrium. In the succeeding proposition Archimedes proves that equilibrium also holds when the magnitudes and distances are incommensurable, thus making the law for equilibrating a lever of unequal arms completely general.

The Translations and their Impact

To put the ancient sciences into the public scholarly domain in western Europe required translations of the Greek authorities and their Arabic successors into Latin. This process of translation was beset with difficulties. Not only did translators work with technical treatises in unfamiliar subjects requiring some knowledge of mechanics, but their hand-copied documents were frequently defective and difficult to read as well. Moreover, translators using more than one manuscript often had to cope with multiple versions of the same passage. Nevertheless, translators in the generations on either side of 1200 A.D. overcame these difficulties and produced a succession of competent translations. Their new translations showed all too clearly that the science of Greece and Islam was far superior to that of Latin Europe. Manifest inferiority of the native materials predating the translations no doubt goaded the Schoolmen to master the alien sources.

Their scant, native resources did include some bare definitions and tables for units and subunits of weights that were given by Isidore of Seville (d. A.D. 636). They knew also the dramatic story of Archimedes' discovery of the respective amounts of gold and silver in the sacred crown of King Hiero of Syracuse, recounted by several Latin authors. In the anonymous *Carmen de ponderibus (Song of Weights)* of ca. 500, which accompanied the influential *Grammar* of Priscian, two methods were provided for solving the "crown problem." However, readers of this or other tales about Archimedes, attracted by accounts of his feats as a wonder worker, were little interested in the mathematical details of hydrostatics. This is unfortunate, for with even a modest understanding of hydrostatic theory it would have been possible to extract from the poetic description the basic formulas for determining the component weights or volumes. But such a step was well out of reach of Priscian's early medieval readers.

In the twelfth and thirteenth centuries, enthusiasm to locate and to translate as many of the Greek and Islamic sources as possible produced an embarrassment of riches, whose assimilation would require patient, comparative study. This study was complicated because the novelty and conspicuous superiority of the translated materials concealed certain internal problems. Some treatises were fragmentary and mutilated: even when intact, a single work was not always correct or consistent; separate ancient traditions appeared to their Latin discoverers to belong to a single unified corpus. It would be some time until they realized, reminiscent of King Hiero contemplating his crown, that not all the constituent elements were of equal value. Even when aware of discrepancies, the Latin Schoolmen shied away from isolating the rival schools and choosing between them. They preferred instead to "harmonize" the ancient authorities by ignoring or ingeniously glossing the discontinuities, difficulties, and discrepancies of their sources—much as biblical scholars were accustomed to do. Their disposition was to find complementarity, rather than rivalry, in the ancient dynamic and static traditions.

By the year 1200, Aristotle's *Physics, On the Heavens,* and *Meteorology*, the major sources of his cosmological and dynamical ideas, had each been translated at least once. As for the *Mechanical Problems*, with its analysis of the balance, the record of transmission is obscure. The first positively authenticated translation into Latin was not until 1497, almost three centuries after the formative period of the medieval science of weights. Yet it is difficult to read some of the early thirteenth-century authors without detecting an influence of some of its ideas. In particular, there is a close similarity of argument between the *Mechanical Problems* and crucial sections of the two treatises called *Liber de ponderibus*, described below, and *De motu*, a work on the kinematics of motion by Gerard of Brussels. Moreover, what seems to be a citation of the *Mechanical Problems* occurs in Emperor Frederick II's *De arte venandi*, a famous work on falconry; and two additional apparent references to it occur on title lists that itemize the holdings of medieval libraries, one at the University of Pavia and another belonging to a Spaniard.[5] Another influential translated work belonging to the dynamic tradition was *On the Heavy and the Light and the Comparison of Bodies to Each Other*. Hellenistic in origin and deductive in method, it assembled Aristotle's scattered correlations between force, volume, and speed and made them more precise.

Translations also included treatises by Arabic successors of the Greek mechanicians. Most influential by far was Thabit ibn Qurra's ninth-century work on the theory of the steelyard, *On the Kariston* (the Arabic name for a balance of unequal arms). Resembling the *Mechanical Problems* in some respects, this treatise contains a doctrine of virtual velocities or virtual displacements built on the understanding that a longer radius sweeps out a greater arc than a shorter radius in the same time. This doctrine, here elaborated in formal geometric proofs, provided a theoretical basis for the equilibration of an ideal balance. Thus, it was a statement of the law of the lever in the dynamic tradition. Thabit's approach was clearer and more sure than that of his Aristotelian predecessor, of whose work he probably had no firsthand knowledge. Since Thabit made no attempt to analyze compound circular motion, his treatise contained none of the inept geometry or the confusion about how to specify the component of natural motion that marked the Greek *Mechanical Problems*.

An even stronger reason for Thabit's superiority was that he had blended significant parts of the Archimedean static tradition into his treatment. Archimedes, as we noted, simply assumed in his proof of the law of the lever that distributed weights could be represented as if concentrated at point centers of gravity. Thabit, using the doctrine of virtual displacements, first proved the law of the lever and then, from this basis, demonstrated the validity of Archimedes' assumption —showing that the distributed weight of a uniform homogeneous balance arm was equivalent to the same weight concentrated at the midpoint of the arm. Where Archimedes had juxtaposed geometric magnitudes on either side of a horizontal bar, Thabit juxtaposed "powers of movement"; Archimedes' appeal to symmetry as the explanation of equilibrium was thus buttressed by Thabit's argument that the causes of motion on each side were equal.

Thabit's ninth-century effort to integrate the two traditions indicated the direction to be followed by Latin Schoolmen. The static tradition, with its superior mathematics, would make its impact on them through the dynamic tradition rather than as an independent entity. This fact is illustrated by the relative obscurity that fell upon the translations of Archimedes' works. Latin versions of Archimedes' *Equilibrium of Planes* and his hydrostatic *On Floating Bodies* were made in 1269, but the inherent difficulty of understanding the unfamiliar mathematics, augmented by unsatisfactory texts, severely restricted their circulation. Consequently, these treatises did not find a place with the standard works that collectively comprised the science of weights.

The mathematics of Euclid was another story. Euclid's famous geometrical text, the *Elements*, already translated in several versions in the twelfth century, was ubiquitous and paradigmatic for mathematical demonstrations, including those of the science of weights. Euclid's authority in statics was linked to his standing in geometry and enhanced by the report that he had written a work on the balance, even though that work was unknown to the Latin West. His reputation was such that anonymous works written in much later times were sometimes credited to him.

A final source of the static tradition in the West was an anonymous Hellenistic work, *On the Balance*, translated under the title *De canonio* early in the thirteenth century. It was an important work whose proofs depended throughout on the validity of the ideal balance and centers of gravity, as we shall see.

With the translation of Thabit's *On the Kariston*, the purely static tradition came to be known in partnership with the dynamic tradition. Even so, as we shall see, the ancient disjunction between motion and rest was recognizable in later efforts to maintain the independence of the science of weights from the science of motion. Meanwhile, by the tenth century the initiative of Thabit and others in studying weights on a balance had led to a recognition that such subject matter constituted a distinct *scientia*. This acknowledgement first appears in al-Farabi's *Catalogue of the Sciences*, a classificatory work written in the tenth century and translated from Arabic to Latin in the twelfth, which enumerates the various sciences without elaboration. In the thirteenth century, writers on this *scientia* attained a collective identity as *auctores de ponderibus*, though they have left no record of the process by which this common association was formed. In all probability, they were members of university faculties, but we know very little about them and often cannot even identify them by name.

Early Harmonizers of the Two Traditions

One of these anonymous medieval authors—identified sometimes as Archimedes!—grafted a fragment of the newly received Archimedean tradition onto an introductory section containing conspicuous elements drawn from early, native Latin sources. This work went by several titles, of which the earliest seems to have been *De incidentibus in humido (On Bodies Thrust Down in Liquid)*. Its structure is decidedly heterogeneous. The initial section, written in nontechnical language, urged in a prologue the importance of reliable weights and measures for merchants, and then proceeded to explain the taking of weights and measurements. A second feature was an unmistakable, if somewhat fuzzy, Aristotelian definition of naturally heavy and naturally light bodies. Its high point was the distinction between gross heaviness (*gravitas secundum numerositatem*) and specific heaviness (*gravitas secundum speciem*), which were both clearly defined for the first time in Latin.

The concluding section contained a distinctive vocabulary suggesting that it was based in some way on an Arabic original. It attempted to deduce a formula for finding the two component volumes making up a mixture—an obvious echo of the famous crown problem—from six postulates and eight theorems. The proofs, however, are so defective and unsatisfactory that one is tempted to conclude that they did not originally belong with the theorems at all. A further illustration of the heavily eclectic character of the treatise was its inclusion, in this concluding section, of two definitions, one contradicting and the other reiterating earlier definitions. The conspicuous patchwork of *De incidentibus* showed intellectual curiosity, inventiveness, and, above all, the hand of a harmonizer. As such, it invited further refinement, and later commentators would attempt unsuccessfully to clarify or remedy its faulty proofs.[6]

There was a more successful remedial effort to fill gaps in the argument of the Hellenistic treatise on the Roman balance, *De canonio*. This was, as we noted, an anonymous work belonging to the Euclidean-Archimedean static tradition. The Roman balance, or steelyard, was familiar in the marketplace as a versatile instrument that gave its name to a section of fourteenth-century London. Vendors hung a commodity to be weighed from its short arm and read the weight from a scale on the long arm by adjusting a movable counterpoise. Since the arms were unequal, a compensating weight had to be added to the short arm before the beam of the steelyard was balanced. A major aim of *De canonio* was to establish the formula by which this equilibrium would be established on an unloaded beam. The formula, given correctly, can be expressed algebraically as

$$\frac{W}{W_2 - W_1} = \frac{L}{2L_1},$$

where W stands for the weight that must be added to obtain equilibrium, W_1 and W_2 for the weights of the short and the long arms, and L and L_1 for the lengths of the full beam and the short arm.[7]

All four theorems included in this work depended for their validity on two assumptions made by the author. These assumptions which were not defended in his treatise, were (1) a material beam whose weight is uniformly distributed may be replaced by an ideal weightless beam of the same length that supports at its midpoint a weight equivalent to that of the material beam; and (2) an ideal lever with unequal weights at unequal distances is in equilibrium when $W_1/W_2 = L_2/L_1$. The author notes that the first assumption "has been demonstrated in the books which speak of those matters," and that the second "has been proved by Euclid, Archimedes, and others." Assumption 2 is the law of the lever, and he adds for emphasis that it is "the foundation on which all [the propositions] depend."[8]

Since the cited sources were no longer available in medieval times, early students of the science of weights were left with a small crisis. *De canonio* was by far the most popular work in the medieval literature of statics: more than thirty manuscript copies remain extant. As long as there was no proof for its assumptions, the validity of its applied theorems rested on an insecure foundation, and to provide a secure foundation required working out a basic theory of statics. That was the undertaking of the most famous of the medieval *auctores de ponderibus*, to whose work we now turn.

Jordanus de Nemore and the Foundation of Latin Statics

Jordanus de Nemore, a versatile but obscure mathematician of the early thirteenth century, presented a basic theory of statics in a work appropriately called *Elementa super demonstrationem ponderum* (*Elements for the Demonstration of Weights*). This tract contained seven suppositions and nine theorems, the last two of which were the pivotal theorems assumed but not proved in *De canonio*. Almost immediately the four propositions of the ancient *De canonio* were directly attached to those of Jordanus's *Elementa*, making it appear that the earlier work disappeared into the later one. We can trace two steps in the process of joining the four old propositions to the nine new ones. First, an alert editor-commentator replaced assumption 2 of *De canonio*, with its telltale reference to "Euclid, Archimedes, and others," by a reference to the proof in the immediately preceding treatise: "as was shown in the penultimate proposition of the above demonstrations."[9] A later commentator, unaware that the two works had originally been independent, was understandably perplexed as to why the eighth proposition was designated "penultimate" when five propositions followed it. He therefore changed "in penultima" to read "in potentia."[10] With the references to external authorities and foundations removed and the two sets of propositions numbered continuously (as they soon were), the Hellenistic and Latin treatises fitted together neatly, if not altogether unobtrusively.

About Jordanus de Nemore, author of the *Elementa*, we know very little. In the *Elementa* he cited his own earlier work on geometry, but his writings are otherwise frustratingly silent about his professional and personal life. Nor do surviving contemporary records yield any information about this man who was so highly regarded by his contemporaries and successors that his name joined those of Euclid and Archimedes as an authority on the science of weights.

The seven suppositions of the *Elementa* were the basis for an original doctrine of "positional gravity" that opened a novel and promising way to compute the effectiveness of a weight on an incline. The suppositions were in the tradition of Aristotelian dynamics and the *Mechanical Problems*, but Jordanus had extensively rewritten the conditions for geometrizing constrained movement—not, however, without some residual ambiguity. Positional gravity, *gravitas secundum situm*, was the component of a body's natural heaviness that governed its speed of descent on a constrained path. It was distinct from *gravitas in descendendo*, heaviness in free fall, which measured the efficacy of unconstrained weight and set an upper limit for positional heaviness. The suppositions, given immediately below, show some trace of the notion, drawn from the *Mechanical Problems*, that the greater the ratio of natural (tangential) motion to unnatural (radial) motion, the quicker the movement. Perhaps they also extend the notion introduced by *De incidentibus* that there are various kinds of gravities: whereas the author of *De incidentibus* had differentiated gross heaviness from specific heaviness, Jordanus assigned distinct gravities to bodies on the basis of whether their descent was free or constrained.

1. The motion of every weight (*ponderosus*) is toward the center [of the world].

2. That which is heavier descends more rapidly.

3. It is heavier in descending (*gravius in descendendo*), when its motion toward the center is more direct.

4. It is heavier positionally (*secundum situm gravius*), when, at a given position, its path of descent is less oblique.

5. A more oblique descent is one in which, for a given distance, there is a smaller component of the vertical.

6. A weight is less heavy positionally (*minus grave secundum situm*) than another, which is caused to ascend by the descent of the other.

7. The position of equality is that of equidistance [that is, parallelity] of the beam to the plane of the horizon.[11]

To compare motions (understood as paths or distances traversed in the same time), the demonstrations of Jordanus included diagrams in which a vertical line represented the simple motion of free fall and a nonvertical line represented compound motion. A vertical line indicated that the body traversing it possessed its full heaviness, while the nonvertical line—taken sometimes as an oblique and at other times as a curved line—showed the descending body to have less heaviness than in free fall. According to supposition 4, loss of positional gravity is inversely proportional to the obliqueness of the path. But when obliqueness is defined in supposition 5, there is an ambiguity in the phrase "for a given distance." If the given distance is taken along a straight incline, as by the dotted line HB (fig. 6), its obliqueness could be expressed by the ratio HL/HB or, in trigonometric terms, as the sine of the angle HBL.[12] That ratio correctly gives the effective force of weight b directed downward along the straight incline HB. If, however, the ambiguously specified distance is taken as an arcal path, such as AX, the analysis by positional gravity will be erroneous. In the first seven propositions of the *Elementa*, Jordanus adopts this erroneous approach, comparing arcs with their vertical intercepts. He explains, for instance, that the reason why the longer arm AC on a balance of unequal arms ACB, loaded with equal weights a and b (fig. 6), has greater positional gravity is because when $LM = XY$, $\dfrac{\text{arc } BM}{LM} > \dfrac{\text{arc } AX}{YX}$; therefore, the weight on the longer arm descends. This choice of arcal paths as the measure of obliqueness perpetuated the mistake initiated by the *Mechanical Problems* and continued in Thabit's treatise *On the Kariston*. However, in his final two propositions Jordanus rectifies his mistake and measures positional gravity by the ratio between vertical component of the arcal path and distance from the pivot point (HE/EC in fig. 7).

The definition of obliqueness was not the only ambiguity in the *Elementa*; nor were Jordanus's incompatible analyses of positional gravity his only inconsistency. Supposition 6, as given, would lead the reader to understand "caused to ascend" as the elevation of a resisting weight in a connected mechanical system rather than as the free ascent of a naturally light body. But this would be a misunderstanding. Proposition 1, in which the supposition figured, was an awkward

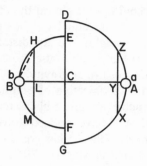

Fig. 6

but quite recognizable rendering of Aristotle's law for free fall and free ascent.[13] Its demonstration and diagram support the interpretation that Jordanus meant to speak of inherently light bodies freely rising upward and not of bodies forced upward by the action of a heavier body. Thus, proposition 1 simply failed to consider the effect of a mechanical connection between weights; the next six propositions (examined below), considered the connection but analyzed it improperly. But then we find once again that Jordanus corrects his mistake in the final two propositions.

To understand the dynamic factors in equilibrium, Jordanus invented a very peculiar ideal balance whose components behave independently rather than as members of a connected system. When the ideal beam is rotated from the horizontal, it no longer behaves as a rigid diameter, becoming, instead, two disconnected radii. The pivot, acting as a hinged joint, now allows each weight to swing freely as an ideal pendulum. Demonstrations in the first seven propositions assume that both weights are dropped simultaneously, fall simultaneously, and behave independently. Thus, while weight *a* is approaching *X* (fig. 6), weight *b* is approaching *M* instead of *H*—as if supposition 6 had never been set down. In the fourteenth century Blasius of Parma, himself one of the *auctores de ponderibus*, described this strange mechanical arrangement, noting that "sometimes a balance is of two arms, sometimes of one."[14]

These two incorrect choices—measuring positional gravity by arcal displacement and treating the arms as two independent pendulums—made the first section of the *Elementa* generally unsound in theory and led to particular errors as well. According to Jordanus, an ideal balance of equal arms and equal weights is unstable in every position except the horizontal; in fact, the system is everywhere in indifferent equilibrium. By his account, a weight on the end of a bent radius has less positional gravity than an equal weight on the end of a straight radius, even though the weights are equidistant from a vertical line through the fulcrum; in fact, their static moments are the same, and there is no difference in positional gravity.

Worst of all, Jordanus's objective of equilibrating a beam of unequal arms cannot be achieved, given his assumptions and the mathematics at his disposal. We have noticed Jordanus's conclusion that weight a on longer arm AC (fig. 6) has greater positional gravity than an equal weight b on the shorter arm BC. To find by Jordanus's arcal analysis the exact weight that must be added to weight b to offset that advantage and thus balance beam BCA would require three steps: (1) Precisely determine the arc lengths AX and BM. (2) Find k, such that weight k plus weight b has the same ratio to weight a as arc $\dfrac{BM}{LM}$ has to $\dfrac{\text{arc } AX}{YX}$. (3) Increase weight b by the amount k. But any such calculation was doubly foredoomed. With the mathematics available to him, the rectification of arc AX was impossible, and the rectilinear and curvilinear line segments were therefore incomparable. But even had Jordanus possessed the mathematics to rectify a curved line segment, it would still have been impossible to equilibrate BCA by his method of comparing arcal displacements. The fallacy of Jordanus's method becomes clear if expressed in terms of the work principle—although he had as yet neither stated nor used that principle in any form. If the concept of work, taken as the product of weight and distance, is accommodated to his erroneous method, it would entail multiplying the suspended weight by its arcal displacement instead of its vertical displacement.

Whether or not he recognized this impasse, Jordanus suddenly abandoned arcal displacements, disconnected weights, and virtual velocities in the last two propositions. Instead he introduced a kind of reasoning that approximates the principle of virtual work. In the first of these two propositions, he presents an indirect proof for the law of the lever, which may be summarized as follows. Suppose that beam ACB (fig. 7) is not in equilibrium when $a/b = CB/AC$. Since one weight must descend, let b on the shorter arm BC descend to E (where it is designated e). For this to happen, it is necessary that weight e be able to lift a lighter weight a on AC through the vertical distance GD. By substitution, it will also be necessary that weight e be able to lift weight l, equal to itself, on arm CM, equal to BC, through the vertical distance LM. But it was previously proven that equal weights at equal distances from the fulcrum are in equilibrium. There-

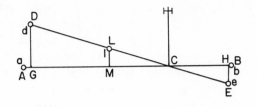

Fig. 7

fore, since b cannot lift l to L, neither can it do the equivalent and raise a to D. Proportionality is established by similarity of triangles, and the proof is short and concise. Jordanus does not point out that he has introduced a new method of measuring obliqueness, but arcs are conspicuously missing from the figure.

This change also introduces, implicitly, the principle of work. The text reads: "what suffices to lift a to D, would suffice to lift l through the distance LM. Since therefore l and b are equal, and LC is equal to CB, l is not lifted by b."[15] Jordanus does not set down the principle of work as the product of force and distance; nor does he exploit the full range of its applicability. Apart from these limitations, however, his statement carries an essential recognition of the factors and relationships constituting the work principle as it applies to cases of equilibrium. Jordanus understands that a given force can accomplish equivalent operations against different loads when weights and distances from the pivot are varied simultaneously. He expresses that procedure mathematically and shows that any potential increase in elevation is purchased at the expense of a proportional decrease in weight. He recognizes that elevation is to be taken along a vertical line and that the amount of elevation is governed by the length of the arm. Thus, his germinal expression of the work principle converted the Aristotelian doctrine of virtual velocity into a far clearer, sharper, and more effective principle that would be crucial in the subsequent history of mechanics.

A final proposition justifies the substitution of a concentrated weight on an idealized arm for the uniformly distributed weight of a material beam. These correct propositions, eight and nine, were exactly the ones that would supply the assumptions needed by *De canonio* and give that work a foundation which, although not altogether satisfactory, was a very promising attempt.

One final slip comprised Jordanus's corrected version of positional gravity. In the final proposition he carelessly transposed a set of ratios and immediately repeated the error. The erroneous ratios gave the formula for balancing unequal weights on unequal arms as $W_1/W_2 = L_1/L_2$ instead of $W_1/W_2 = L_2/L_1$. That arrangement, placing both the heavier weight and the longer arm on the same side of the fulcrum, would, of course, throw the beam violently out of balance.[16]

A treatise already strained from lack of internal consistency did not need a further, confusing inconsistency at its culminating point. Copyists, transcribing the *Elementa* faithfully, reproduced the error that made it seem that Jordanus misunderstood his own theory. The two editors who successively fused the nine propositions of the *Elementa* to the four propositions of *De canonio* also let the error stand—if, indeed, they recognized it. One anonymous commentator, known through an Oxford manuscript (Corpus Christi College 252), did recognize the error, corrected it, and modified the demonstration in some details.

Another commentator, whose Aristotelian ties were more obvious than those of Jordanus, discussed positional gravities and levities in terms of physical causes, an aspect not developed in the geometrical *Elementa*. His *Liber de ponderibus*, in its long and interesting preface, noted that the science of weights had a philosophical as well as a geometrical aspect. This Peripatetic Commentator, as we shall call the anonymous author, elaborated the former in a way that strongly resembled the *Mechanical Problems*. Motion, he explained, was the key to understanding heaviness or lightness. Motion was sometimes simple, as in free vertical descent, and sometimes mixed. In the case of mixed motion, an unnatural component reduced a body's original heaviness, while deflecting it from the line of natural descent. This interaction of natural and unnatural components determined effective heaviness and the resulting velocity of the body. Along a more curved arc, where there was a greater impediment to descent than along a less curved one, there was a consequent reduction of positional gravity.

The Peripatetic Commentator's remarks illustrate and illumine the general philosophical milieu in which Jordanus's formal suppositions and proofs had emerged. After him, there was a pronounced scholastic flavor to some of the points debated—for example, how the balance was affected when one of two equal weights was suspended closer to the center of the world, where (in one opinion) it weighed more; and whether the arbitrarily long, imaginary lines by which weights were suspended at both sides of an ideal balance were parallel or converged toward the center of the world. Some commentators wrote in both the philosophical and geometrical modes, as Roger Bacon in the thirteenth century and Blasius of Parma in the fourteenth century, but the work of Jordanus, despite its internal weaknesses, was plainly superior to these attempts.

The Redactor of Jordanus and the Originality of Latin Statics

A larger and better work, *De ratione ponderis (On the Theory of Weight)*, following soon after the *Elementa*, made up for many of the inadequacies of the earlier treatise. Modeled on Jordanus's work, it followed the original very carefully but also introduced needed qualifications, made subtle linguistic changes, and expanded the original nine theorems to forty-five, while adding two completely new sections. The author, whom we shall refer to as the Redactor, puzzlingly deleted all three references to Jordanus, but the book was frequently attributed to Jordanus nonetheless.[17] *De ratione* was beyond question the finest work of medieval statics.

Whoever was responsible for the revision—whether a more mature Jordanus or a modest disciple with gifts equal to his master's[18]—was someone whose command of the principles of static moment and of work outdid that of any ancient or medieval predecessor. Besides contributing the first correct solution to the inclined plane problem, he elegantly corrected errors in the *Elementa* about the bent lever and the conditions for stable equilibrium in material and ideal beams. Throughout those new demonstrations he was faithful to the two unstated principles, work and static moment.

At the same time he did not, as we might have expected, simply expunge the erroneous section of the *Elementa*. He retained much of that introductory section with its unproductive interpretation of positional gravity. Even when, with great skill and economy, he gave the first correct demonstration of the bent lever, he appended to the new theorem a "more subtle variant" into which the erroneous theorem of the *Elementa* seems to be subsumed as a refined, theoretical concept helpful for understanding unstable equilibrium.

Once again, as with *De incidentibus* and the *Elementa*, we find a work containing two imperfectly matched sections. In *De ratione* it is just possible that the Redactor, unlike his predecessors, recognized the discrepancy that he had left in the work. What may be a hint of such recognition is his introduction of slight but systematic changes into the technical terminology used in the retained fallacious theorems, changes permitting the early doctrine of positional gravity to be differentiated from the revised doctrine of positional force. He also redrew the diagram of fig. 6, deleting the upper quadrants *BHE* and *AZD*, so that it illustrated not an ideal balance, *BCA,* but two ideal pendulums, *BC* and *CA*. Such changes in a work otherwise very faithfully copied may indicate that the Redactor, whether Jordanus or his disciple, meant to describe two types of ideal "balances." In this connection, it should be noted that the overwhelmingly theoretical interests of the *auctores de ponderibus* put a certain distance between their intellectual habits and our own. The imaginations of these academic men, who speculated about beams at the rim of the universe or its center and about ideal suspension cords thousands of miles long, were not likely to be strained by a queer, hybrid contraption such as Jordanus's ideal balance that behaved like two pendulums.

The Redactor's original theorems avoided comparisons of force and resistance, as if the weights suspended from the ends of the balance were a pair of heavy bodies released simultaneously. His insistence on properly connected weights moving on opposite arms of the balance in accord with supposition 6 of the *Elementa* was an instance of the sound intuition and use of the work-principle that marked the original sections of *De ratione*. Likewise, his grasp of static moment, evident in the demonstration of the mechanical action of the bent lever, was surer than any predecessor's. In that demonstration he showed correctly for the first time that the effective force exerted by a weight applied to the end of a bent lever depended on the *horizontal* distance between the weight and the vertical axis of the system. Thus if the weight *a* is attached to the bent arm represented by *ZYC* (fig. 6) its effective force is exerted over the distance *YC*. The crowning achievement of the Redactor was his explication of the ambiguous doctrine of positional force, converting it into a correct measure of the effective component of a body's weight on a straight incline, now symbolized by the formula $F = W \sin a$. No one in antiquity had solved the inclined plane problem. The *Elementa* made a promising, but partly false, start. The Redactor's solution, summarized below, holds its place beside the much later ones of Simon Stevin (1548–1620) and Galileo (1564–1642).

Proposition 9 of *De ratione* proves that along a plane whose inclination is constant, the effective force of a supported body is everywhere the same. Thus, since the equal segments *EL* and *OC*, on plane *DC* (fig. 8), have equal vertical intercepts, *ER* and *OP*, they also have equal positional force (*virtus in hoc situ*). In proposition 10 the Redactor uses this conclusion to show that connected unequal weights W_1 and W_2 on oppositely inclined planes of lengths L_1 and L_2 are in equilibrium when $W_1/W_2 = L_2/L_1$. The indirect proof begins by

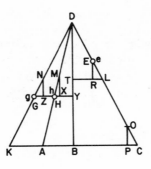

Fig. 8

assuming that when $e/h = DA/DC$, equilibrium is not achieved, and weight e descends through the vertical distance ER to L. This requires that the connected weight h on the plane AD be raised through the vertical distance XM to M. But raising h to M is equivalent to raising g to N (where $g = e$ and $DK = DC$), so that any force sufficient to do the former could equally do the latter. But the resistance of g to ascent is the same as the force of e for descent, since by proposition 9 their positional forces are everywhere the same on their respective inclines. Therefore, contrary to the initial assumption, equilibrium is maintained, and there is no motion.[19]

The Uncertain Synthesis

Although the medieval *auctores de ponderibus* combined the Archimedean and Aristotelian traditions with great skill, they were not able—even in the brilliant Redactor—to produce a completely satisfactory synthesis. A misconception on which the dynamic tradition itself rested was the source of their failure. This misconception was the unqualified proposition that speed of fall was directly proportional to the free weight of a heavy body, a proposition that is correct only for vertical descent within a constrained system where a weight is paired against a resistance. As a description of free fall, it was a historic mistake that confused and misdirected the science of dynamics until Galileo. The propositions of the *Elementa* appeared to begin with the mechanics of free fall and then move with some uncertainty to those of released pendulums and the equilibrated balance beam. The Redactor did nothing to limit the meaning of the initial proposition that he inherited from the *Elementa*; in his version it remained a description of movement in free fall. Under those circumstances it is not surprising that some Aristotelians came to view the entire science of weights as no more than an adjunct to understanding the motion of free descent, into which certain modifications were introduced by the presence of a balance. To these Aristotelians, resistance encountered in depressing the end of a lever was analogous to that encountered by a body falling through a resisting medium. The English mathematician and natural philosopher Thomas Bradwardine put such a construction on the first proposition of Jordanus in his influential *Treatise on Proportions* of 1328.

Bradwardine's interpretation seems to have prompted the composition of an anonymous major treatise known only as *Aliud commentum (Another Commentary)*. Its gifted author, whom we will call Commentator, argued vigorously and plausibly that Jordanus's demonstrations were invalid except when limited to displacements in a constrained system. Working in ignorance of the Redactor, he argued that proposition 1 of the *Elementa* correlating weights to speeds of descent and ascent could not be interpreted in isolation from the remaining propositions. The Commentator understood Jordanus's first proposition as a comparison of the movements of two heavy bodies connected across an ideal pulley. He pointed out, quite correctly, that the proof of the law of the lever in proposition 8 required that when weight *b* (fig. 7) descends, weight *g* ascends. Since any ascent of weights is exclusively the subject matter of proposition 1, he declared that proposition 1 must be implicit in proposition 8. Arguing backwards, he then reasoned that proposition 8, which obviously excludes free fall, dictates the proper meaning of descent and ascent in proposition 1. The intention behind Jordanus's dense and murky prose in that first proposition, Commentator said, was to limit himself to the interactions between connected bodies. He argued that the citation of proposition 1 to support any form of Aristotle's general law of movement was a misappropriation, which invalidated the remaining theorems of the treatise.

Within this argument we find the most explicit medieval statement of the work principle. Assume a lever *BACD* pivoted at *A* (fig. 9), with a given resisting weight on arm *BA*. The Commentator then says:

> A weight *g* at position *C* is related to the same weight *g* at position *D* according to the ratio of the total descent which it can have at position *C* to the total descent which it can have at position *D*.
> Therefore, from the fact that it cannot descend farther than the length [of the radius of the circle] whose circumference it describes, it follows, from this exposition, that the weight *g* at position *C* is related to the same weight *g* at position *D* according to the ratio of

CA to *DA*. Thus, if the weight *g* at position *C* (with another weight on the other arm) suffices to descend to the vertical position, the same weight *g* at position *D* (with another weight on the other arm of the balance) would suffice to descend to the vertical position with equal ease and would be related to the weight first raised as the ratio of *CA* to *DA*.[20]

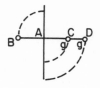

Fig. 9

This rather cumbersome statement can be reformulated as follows: "What suffices to lift a weight W through a vertical distance H will suffice to lift a weight kW through a vertical distance H/k, and it will also suffice to lift a weight W/k through the vertical distance kH."[21]

Although the fourteenth-century Commentator's appeal to the logic of the *Elementa* was impressive argumentation, it could not remove the inconsistencies in Jordanus's treatise. Another writer of the fourteenth century resorted to a more radical solution to deal with these inconsistencies. Where Commentator had called for a restricted interpretation of the dynamical first proposition, this later writer abandoned the dynamic tradition altogether. He thus represents a return to the isolated study of bodies at rest that constituted the static tradition in its original Archimedean form. From Jordanus he retained the bare statements enunciating the final two propositions, but discarded everything else, including all the proofs. For his demonstration, he introduced a horizontal bar suspended from one arm of an ideal balance and a weight suspended vertically from the other arm. His proof for the law of the lever, applied to a balance of unequal arms, depended on the imaginary, symmetrical redistribution of weight segments on each side of the pivot—a technique reminiscent of Archimedes' *Equilibrium of Planes*, which may have been available in Paris, where the unknown author wrote this commentary.[22]

The tendency to anthologize and harmonize—smoothly or crudely —continued into the age of printing. This tendency, which imposed an external unity on the science of weights from its beginnings, may be seen in two dissimilar but complementary works of different centuries, the Peripatetic commentary (*Liber de ponderibus*) and the treatise referred to as *Another Commentary*. These two works were published together at Nuremburg in 1533, the former misattributed to Jordanus, called "the Woodsman (*Nemorarii*)" rather than "de Nemore." Niccolò Tartaglia, in his *Questions and Diverse Discoveries* of 1546, brought together the multiple medieval sources, much as Blasius of Parma had done late in the fourteenth century.[23]

More than three centuries of the *scientia de ponderibus* lie between the appearance of *De canonio* and the *Questions* of Tartaglia. Throughout these three centuries the scattered *auctores* and commentators produced more than a dozen treatises of varying lengths and value. A thirteenth-century program to recover and reproduce the ancient materials, to harmonize sources, and to reconstruct missing elements had produced a recognizably new science. Its overall strategy of uniting statics and dynamics proved visionary and premature, but it achieved a limited success. Aside from serious unresolved problems in the science of motion that stood in the way of a unified mechanics, aside, too, from the limitations of Euclidean geometry, the ecumenicism of the new science worked to its disadvantage in the long run. Ingenious at accommodation and reluctant to exclude traditional material, the *auctores* remained insufficiently critical of their subject. Put another way, the medieval science of weights met the organizational requirements for a *scientia*, but never fully became a system in a theoretical sense.

The trouble was not in its banalities, but in its eclecticism. We find and admire mechanical insights and solutions that were at best dimly appreciated by medieval readers in the science of weights. It is hard to conclude otherwise when we consider the following. One edition after another of the *Elementa* left uncorrected the transposed ratios that invalidated or confused everything that had gone before. The remarkable and unique statement of *Aluid commentum* that expressed the work principle was consistently garbled in copying. The author of that excellent statement used the work principle in polemics, but not to correct the erroneous propositions in the work on which he commented. Blasius of Parma, the man most conversant with the literature on weights in the fourteenth century, essentially buried the work principle in his *Questions*. Even Leonardo da Vinci, who was strongly influenced by the *auctores de ponderibus*, was unaware of his self-contradiction when he made the effective force of a weight along an incline a function both of the sine of the angle of inclination and of the tangent of the same angle.[24]

All of this is only to say that the founders of the medieval science of weights did not transcend the historical conditions of their times.

They demonstrated thoroughness and skill in collating scattered, diverse materials. They achieved a notable clarification and unification of theory leading to original solutions of particular problems that had resisted the ablest mechanicians of the past. And they bequeathed to the founders of modern mechanics a discipline that combined skillful insights with nagging inconsistencies demanding removal before statics could be integrated with dynamics.

Notes

1. In the terrestrial region of Aristotle's cosmos, two elements (earth and water) were absolutely heavy and moved naturally toward the center of the universe; two others (air and fire) were absolutely light and moved naturally toward the periphery. The result was a set of concentric spheres about the center of the universe (in ascending order, earth, water, air, and fire), and these spheres defined the natural place of each of the elements. Only when it was in its natural place was a body naturally at rest. Therefore, unless constrained, all heavy bodies above their proper spheres descended toward their natural places, and all suspended heavy bodies retained a potential downward tendency. This Aristotelian cosmos, with its emphasis on potential movements toward four different natural places, offered obstacles to the formulation of a science of weights. For instance, water beneath its natural place would flow naturally upward toward that sphere. And bodies such as smoke, which were endowed with absolute levity, would naturally ascend toward the outer rim of the terrestrial region.

2. E. S. Forster, in his translation into English, used the title *Mechanica* (Oxford, 1913), 848b7.

3. Pseudo-Aristotle, *Mechanica* 850a36–850b6.

4. Marshall Clagett, *The Science of Mechanics in the Middle Ages* (Madison, Wis., 1959), pp. 34–36. See also Clagett, *Greek Science in Antiquity* (New York, 1955), appendix 4. The diagram in figure 5 is slightly modified from Clagett's.

5. For possible influence of the *Mechanical Problems* on the *Liber de ponderibus*, see Ernest A. Moody and Marshall Clagett, eds., *The Medieval Science of Weights* (Madison, Wis., 1952), pp. 147, 150–53. The same work, pp. 402–4, also considers possible influence on the treatise *De ratione ponderis*. Clagett, *Science of Mechanics*, pp. 182–83, notes the similarities between *De motu* and the *Mechanical Problems* and, pp. 71–72, considers the citations by Frederick II and the possible link with the Pavian library. The Spaniard, Henricus de Villena (1384–1434), lists under Aristotle two books of *Mechanicas*; see Emilio Cotarelo y Mori, *Don Enrique de Villena: su vida y obras* (Madrid, 1896), p. 157. Interestingly, it is the first two chapters (that is, questions 1 and 2) of the *Mechanical Problems* that seem to be reflected in the medieval sources. The comparative silence about the thirty-three remaining chapters raises the possibility that only a fragment of the original work circulated in medieval times. A fragment containing two detached Aristotelian questions

might have been recorded by Henricus's cataloguer as two distinct mechanical treatises.

6. Moody and Clagett, *Science of Weights*, pp. 35–53, give text and
editorial comment. To this should be added the recent evaluation in
Clagett, *Archimedes in the Middle Ages*, vol. 3 (forthcoming), pt. 3,
chap. 6, sec. 2 and the separate appendix. Sources for the medieval works
discussed below are the editions of Moody and Clagett and of Joseph E.
Brown, "The *Scientia de Ponderibus* in the Later Middle Ages" (Ph.D.
dissertation, University of Wisconsin, 1967).

7. For the proof of this theorem, see Moody and Clagett, *Science of
Weights*, pp. 64–67.

8. Ibid.

9. Florence, Biblioteca Nazionale, MS J.IV.29, fol. 65r; see also the
discussion in Brown, "*Scientia de Ponderibus*," pp. 60–63.

10. Cf. MS Vat. Lat. 2975, fol. 169v.

11. Moody and Clagett, *Science of Weights*, pp. 128–29. Moody's Latin
text and translation into English of the *Elementa* form the basis for this
section. Jordanus is the most important of the known medieval *auctores
de ponderibus*. The "de Nemore" frequently but not always given with
his name perhaps links him with the city of Nemours. For details of his
other work and his significance, see the article by Edward Grant, "Jordanus de Nemore," *Dictionary of Scientific Biography*, 7:171–72; and
Ron B. Thomson, "Jordanus de Nemore: Opera," *Mediaeval Studies* 38
(1976):97–144.

12. Jordanus's supposition, as given, makes the *loss* of positional gravity
of weight b directly proportional to $\dfrac{HB}{HL}$, that is to the cosecant of angle
HBL. I have rewritten the proportion, making the gain in positional gravity directly proportional to $\dfrac{HL}{HB}$, the sine of *HBL*.

13. The first proposition reads in literal translation: "Among any heavy
bodies there is direct proportionality between velocity in descending and
weight. But the proportionality between descent and the contrary motion
is the same but inverse." (Translation changed slightly from Moody's
rendering, *Science of Weights*, pp. 127–28.)

14. Blasius of Parma, *Tractatus de ponderibus*, Clagett text and translation in *Science of Weights*, pp. 246–47.

15. Moody and Clagett, *Science of Weights*, pp. 139–40.

16. Moody inadvertently *corrected* these obviously incorrect ratios in
his English translation, *Science of Weights*, pp. 140–41, but his Latin text
on the facing page maintains them as they appear in the manuscripts.

17. Jordanus's name occurs in the title of a copy owned by Richard
Fournival, who was dead by 1260; see Moody and Clagett, *Science of
Weights*, pp. 171–72.

18. On the authorship of this work, see Brown, "*Scientia de ponderibus*," pp. 63 ff., 228–32.

19. Moody and Clagett, *Science of Weights*, pp. 188–91. Fig. 6 combines the diagrams of propositions 9 and 10.

20. Brown, "*Scientia de ponderibus*," p. 212.

21. Clagett, *Science of Mechanics*, pp. 78–79.

22. Clagett, *Archimedes*, vol. 3 (forthcoming), pt. 1 chap. 8. This work is known only through a single manuscript, Paris, Bibliothèque Nationale, MS Latin 7377B, fols. 93v–94r.

23. For a translation and commentary, see Stillman Drake and I. E. Drabkin, eds., *Mechanics in Sixteenth-Century Italy* (Madison, Wis., 1969).

24. Marshall Clagett, "Leonardo da Vinci (Mechanics)," *Dictionary of Scientific Biography*, 8:217.

7

John E. Murdoch and Edith D. Sylla

The Science of Motion

The Role and Scope of Motion in Medieval Natural Philosophy

Much of the conceptual setting for the subject of this chapter has already been described in chapter 3. In one sense, to speak of the "science of motion" in the Middle Ages is to speak of all physics or even of all natural philosophy. Aristotle had identified or defined nature as a principle (that is, a source or cause) of motion.[1] Medieval authors frequently cited this definition and, as a consequence, often used motion and its divisions as a basis in their discussions of the organization of Aristotle's *libri naturales*. Thus, the introduction to Geoffrey of Haspyl's thirteenth-century commentary on Aristotle's book *On Generation and Corruption* says:

> Since therefore the subject of natural philosophy is mobile bodies considered in general, it follows that the division of natural philosophy reflects the distinctions between mobile bodies. This division is made in one way as follows: mobile bodies may be considered (1) in a strict sense or (2) by analogy. Mobile bodies taken in a strict sense may be considered (a) in general or (b) under a limitation. The book *On Physics* considers mobile bodies in general. Mobile bodies under a limitation (*contractum*) may be divided further into (i) locally mobile, (ii) generable and corruptible, (iii) alterable, and (iv) augmentable. The book *On the Heavens and the Earth* deals with locally mobile bodies. That book deals with circular motion [of the heavens] and also with rectilinear motion, which belongs primarily to the elements, but secondarily indeed to mixed bodies according to Aristotle's statement in *On the Heavens and Earth* that all mixed bodies are moved by the element dominant in

them. The book presently under consideration, that is, *On Generation and Corruption*, deals with generable and corruptible bodies.
The book *On Meteors* deals with alterable bodies, for when vapors
are transmuted into hail, rain, and so forth, no proper generation
occurs, but only alteration, for the subject remains the same in
species. The [pseudo-Aristotelian] book *On Plants* deals with
augmentable bodies, for all plants can be augmented, this being the
proper operation of the vegetative [soul] as is determined in the
second book of Aristotle's *On the Soul*. Mobile bodies taken by
analogy [category 2 above] are either bodies with a sensitive soul
(*sensibile*), on which there is the book *On Sensible Creatures*
(*De sensibilibus*) or *On Animals*, or bodies with intellective souls,
about which, because of their nobility, there are two natural
sciences. Aristotle's *On the Soul* deals with the science of the human
soul, and the [pseudo-Aristotelian] book *On Medicine* deals with
the human body.[2]

This tidy division of Aristotle's works is a medieval creation, but it
does indicate the importance and pervasiveness of the consideration
of motion throughout all of medieval natural philosophy. To the modern reader, motion may be identifiable with local motion (that is,
change of place), but this is a product of the Scientific Revolution of
the seventeenth century, which reduced all other Aristotelian motions
to secondary manifestations of local motion, and it is not applicable
to medieval science, within which motion includes alteration and augmentation, and, in some cases, generation and corruption, as well as
local motion. This chapter deals with the science of motion in its
broad, medieval sense.

Because of the pervasiveness of the consideration of motion throughout all of medieval natural philosophy, it is extremely difficult to identify a medieval "science of motion" somehow standing on its own apart
from the tradition of natural philosophy. Other areas of science, such
as optics or statics, were regarded as disciplines separable from natural
philosophy, and explanations of just how they were separable were
given. But this did not occur for a "science of motion." It was viewed
as part and parcel of natural philosophy. We cannot, however, deal
with all of natural philosophy in this chapter. Fortunately, other chapters of this book deal with the medieval science of motion as it was
related to Aristotle's *On the Heavens, On Generation and Corruption*,
and *Meteorology*. We will therefore restrict ourselves in this chapter to
dealing, first, with medieval commentaries on Aristotle's *Physics* and,
second, with other independent works devoted to motion, discussing
the subcategories of motion only as these are found in works devoted

to motion in general. In restricting our scope in this way, however, we are imposing a modern distinction upon the medieval materials, and this fact must not be forgotten.

Motion and Change in Commentaries on Aristotle's Physics

The text of Aristotle's *Physics* was known to the Middle Ages from the twelfth century, when it was first translated both from the Arabic and from the original Greek. It was by far the most influential work in medieval discussions of motion. Like Aristotle's other works, the *Physics* was above all else a *teaching* text within medieval intellectual life, particularly during the thirteenth and fourteenth centuries. Although Aristotle's works may not have been ideally suited for this role because of their difficulty, and although they were sometimes held in suspicion and even banned by Christian authorities because of their doctrinal content, nevertheless, because of their superiority to the available alternatives they became the set texts for many areas of the arts curriculum within the medieval universities and remained so for centuries.

Treating the Aristotelian works as teaching texts, then, medieval authors felt that their prime duty was to clarify Aristotle's meaning and to make his works understandable to his new Latin audience, while their secondary task was to qualify or to revise, perhaps even correct, his views, should that appear necessary. Some topics that attracted the attention of medieval commentators on the *Physics* arose primarily out of the text itself, while others also reflected concerns generated elsewhere, perhaps by Aristotle's logical works or by theological considerations. But since the medieval contributions, whatever their motivation, mainly took the form of variations on the Aristotelian base, the necessary first step toward an accurate view of the medieval work must be to look at the *Physics* itself—a work which, although neatly arranged into eight books, does not as a whole present its contentions in a finished and polished form, but, rather, reveals Aristotle's ongoing dialectical analysis of the conceptions and problems he regarded as most basic to an investigation of nature.

Books 1 and 2 of the *Physics* provide the foundations of all natural philosophy. It is here, for example, that Aristotle establishes that motion is involved in the very definition of nature or the natural and introduces the four causes (the material, efficient, formal, and final) that are to be used in explaining just how and why motion and change come about.[3] Here, too, he rejects the view that things come about by chance or necessity and argues for a teleological interpretation of nature, according to which things occur for the sake of something, for some purpose or end. But for the investigation of motion, perhaps the most important part of this introductory section of the *Physics* is that in which Aristotle sets forth his doctrine of matter and form and, in terms of it, elicits a "model" according to which motion and change always involve an underlying something (a substratum) that remains constant throughout change, the change itself consisting in the successive presence in the substratum of contrary or opposite forms. For example, if something undergoes a motion of alteration from white to black, there is a substance or substratum that survives, while it possesses the contrary forms first of white and then of black. Aristotle presents this account of motion and change in response to the views of his predecessors and, in particular, to that of Parmenides, who had argued that no change at all is possible within nature.

In the opening sentences of book 3, Aristotle tells us that we must inquire not just into motion itself, but also into all those things that are involved in it.[4] Thus, the whole latter part of book 3 is concerned with an intensive investigation of the infinite, and book 4 is devoted entirely to an examination of place, time, and the void, all of which are involved in motion (or, in the case of the void, thought by others to be involved in motion and so to be considered, if only to prove that the void does not exist). Similarly, parts of book 5 and all of book 6 are consumed in discussing continuity and the various problems and ramifications that one must deal with, given the view that all motion is absolutely continuous. The *Physics* ends by proving in book 8 that motion has always existed and always will exist, and that the only proper way to account for this eternity of motion is by the assumption of an immobile and unmoved first mover, without parts or magnitude, acting at the circumference of the universe, something that was naturally assimilated by medieval Aristotelians to their Christian God.

The intervening sections of books 3–8 deal more directly with motion (indeed, already in antiquity, books 5, 6 and 8 were referred to as Aristotle's work "on motion").[5] Thus, in the first three chapters of book 3 Aristotle sets forth his view on the nature of motion, how it should be defined, and what relation it has to the mover and the moved body. In book 5 he explains the three types of motion narrowly defined, namely, locomotion, alteration, and augmentation, and discusses the properties that any motion must have.[6] And in the apparently incomplete book 7 he considers at length the relations that must obtain between mover and moved body (that they always must be distinct and yet in contact) and sketches out things of essential importance in dealing with the comparison of motions (circular versus rectilinear, for example) and with the proportion of motions in terms of the forces and resistances giving rise to them.

The Relations of Movers and Mobiles

An excellent example of how Aristotle's text raised problems for medieval commentators may be taken from several of these concerns of book 7. Aristotle begins book 7 by arguing for the truth of the proposition that everything that is in motion is moved by something else.[7] This proposition—so basic to Aristotle's natural philosophy that it might even be regarded as an axiom—followed closely from Aristotle's notions of actuality and potentiality and from his ideas of how the potential was brought to actuality.[8] As far as local motion is concerned, everyday observations seem by and large to support the proposition. Carts do not move forward, loads are not lifted, and work is not done unless there is a force acting continuously.

There were, however, two areas in which commentators had a difficult task defending this Aristotelian "axiom." The most obviously difficult area concerned projectile motion. In book 7 Aristotle had argued not only that everything that is in motion is moved by something else; he had argued also that this mover must be in contact with the thing moved. In the case of projectile motion, the only thing in contact with the mobile seemed to be the medium, usually the air, through which it moved. How could the air plausibly be assumed to move a projectile? One theory, with few if any supporters, was that in projectile motion the air ahead of the projectile moved out of the way (since two bodies could not be in the same place at the same time) and rushed around behind the projectile, filling the incipient vacuum there (since there can be no vacuum or void in Aristotle's world), at the same time pushing the projectile forward.[9] Aristotle himself, although mentioning this solution—usually called the theory of *antiperistasis* or "mutual replacement"—did not seem to favor it, but argued rather vaguely that the original propellant of the projectile gives the air immediately surrounding it the power to move the projectile further and that this power is passed on through the medium with the projectile.[10] In fact, it was difficult for anyone to believe that the medium served as the moving force in projectile motion, since in most cases it appeared that the medium resisted rather than caused the motion.

A second difficulty with the proposition that everything in motion is moved by something else is a less obvious one, but becomes apparent when we consider the Aristotelian distinction of local motion into natural and violent. In natural motion, heavy bodies, in which the elements water or earth predominated, moved downward, and light bodies, in which the elements air or fire predominated, moved upward (the fifth element, ether, of which the heavenly bodies were composed, moved naturally in a circular motion around the center of the cosmos). In violent motion, heavy bodies moved upward, and light bodies downward, these motions taking bodies away from their natural places (the natural place of earth being in the center of the cosmos, surrounded, in order, by the natural places of water, air, and fire). In the case of violent motion, it was not difficult to see that an external mover was involved; but in the case of natural motion, things were more problematic. If one had a mixed body undergoing natural motion, one element might serve as a force and another as a resistance, but in an elemental or simple body, how were mover and moved to be distinguished as Aristotle's proposition required? In fact, this question was almost impossible to answer in a proper Aristotelian way unless one took things as far back as the Prime Mover, or at least back to the generation of the simple body in question, and identified the mover with whatever generated the body and the resistance with the resistance of the body from which it was generated. Then the subsequent natural motion of the body had to be taken as a result of the action of the Prime Mover or as an accidental consequent of the original generation of the body, perhaps delayed by some impediment to the motion.[11] Many commentators passed over this problem and simply assumed that in natural motion the mover, in the sense of the force involved, was the body in question, and that the moved, in the sense of the resistance, was the medium through which it moved.[12]

Much more time and effort, however, seems to have been spent by medieval scholars in deliberating over the continuation of projectile motion and its "fit" with the standard Aristotelian principles about movers and moved bodies. One drastic resolution of this problem was that of William of Ockham (d. ca. 1349): "I maintain that in such motion as occurs through the separation of a movable object from its first projecting body, the moving agent is the very thing that is moved and not some power in it, whether absolute or relative, so that this mover and the thing it moves are absolutely indistinguishable."[13] Ockham supported this view by using the Aristotelian analysis of how one body acts on another to show that the original mover could not transmit any power to the projectile—or else the same power would be transmitted if the mover and projectile were simply placed side by side at rest, which does not occur. To the objection that the motion, as a new effect, had to have a cause, Ockham replied that "a local motion is not a new effect, neither an absolute nor a relative one . . . , for motion is nothing more than this: the movable body coexists with different parts of space."[14] Where there was no new effect, no cause was necessary.

As rigorous as Ockham's solution may be, another answer to the problem was more popular in the fourteenth century and, as a result, has been more noticed by modern historians: it is that of "impetus theory."[15] Ideas related to this theory appeared already in late Greek commentators on Aristotle and in several Islamic writings as well,[16] but these works were not available in translation, and it seems fair to conclude that medieval Latin "impetus theorists" created their conceptions on their own. According to their theory, projectile motion is to be explained by saying that when a ball is thrown or an arrow shot, the projector transmits an *impetus* to the ball or arrow, which then continues to act as an internal cause of its continued motion (making unnecessary the implausible use of the medium as a mover). In the fourteenth-century, the Parisian commentator Jean Buridan (d. ca. 1358)—to cite the best-known proponent of this theory—treated impetus as a quality inherent in the mobile, proportional both to the mass of the mobile and to its velocity.[17] Furthermore, Buridan believed impetus to be a quasi-permanent quality and, consequently, inferred that, once the mobile is set in motion, it would tend to continue to move under the action of its impetus until some counteracting cause or resistance intervened. In the case of the heavenly spheres, where there is no resistance, Buridan suggested that impetus given them by God at the time of creation could continue to move them *ad infinitum*.[18]

Furthermore, impetus was utilized in explaining yet another puzzle in Aristotle's physics: that of the obvious acceleration of a body undergoing natural motion, especially a natural motion of "free fall." How could acceleration occur if both the moving body, here functioning as the moving force, and the medium, functioning as a resistance, remain the same, since in Aristotle's view a constant force and a constant resistance lead to uniform motion? The answer of the impetus theorist was that, as a body in free fall moves, it gains impetus (presumably owing to the continual action of its natural heaviness or gravity), and this ever-growing impetus added to the original force or heaviness that moves it is what causes the acceleration.[19]

As a whole, impetus theory was a departure from Aristotle, but one which did not seriously conflict with the general Aristotelian system, while at the same time giving an elegant solution to a problem which had plagued successive generations of commentators. Therefore, it tended to be widely accepted, although with variations from author to author.[20] And, in general, commentators on book 7 worked within the framework that Aristotle had established. Other examples could be given of medieval innovations that were improvements upon Aristotle, in the sense of agreeing more nearly with observation or of helping to form a more complete and comprehensive system.[21] But, like impetus theory, such innovations could be, and were, incorporated into revised versions of Aristotle, differing from his views in detail but not in general approach.

The Nature of Motion

In studying projectile motion, medieval scholars addressed themselves to a problem that continued to be of concern during the sixteenth and seventeenth centuries. In fact, the history of the impetus theory has often been written as a background to the study of Galileo's theories of *impeto* and inertia.[22] In contrast, other problems dealt with in medieval *Physics* commentaries lack such modern continuations and are, as a result, less treated in histories of science.

One of these was the problem of the nature of motion, usually addressed in commentaries on the first three chapters of book 3. Here, the section of Aristotle's text attracting the most attention was that in which he raises the question of the relation of motion to the various categories of things that exist (where these categories were those described by Aristotle in his work called the *Categories*: substance, quantity, relation, quality, activity, passion, time, place, position, and state). Labeled "text 4" of book 3 by the scholastics, the crucial passage reads: "Now no motion exists apart from things; for that which changes always does so either with respect to substance or with respect to quantity or with respect to quality or with respect to place, and there can be no thing common to these which is not, as is our manner of speaking, a *this* or a quantity or a quality or some one of the other categories. Thus neither a motion nor a change can exist apart from these [categories] if nothing else exists but these."[23] Thus, Aristotle asserts that motion is not a separate type of being unto itself, but belongs to the already established categories.

When the medievals came to discuss this passage, one of the problems they had to face lay in the fact that in the *Categories* Aristotle had included some motions (for example, heating) in the category of passions or affectations,[24] whereas here in the book 3 (and also in book 5)[25] of the *Physics* he might be interpreted as putting motion variously into the categories of substance, quantity, quality, and place. Elsewhere, in the *Metaphysics*, book 5, chap. 13,[26] Aristotle seemed to say that motion as a continuous magnitude belonged to the category of quantity.

It was this issue of Aristotle's apparent inconsistency, then, that animated a good deal of the medieval discussion devoted to text 4 of book 3 of the *Physics*. To appreciate what was involved at closer hand, it will be helpful to begin by noting three of the most standard questions devoted to this problematic text 4 in the thirteenth and fourteenth centuries: (1) To what category or categories does motion belong? (2) Is motion essentially the same as its terminus? (Or how does motion differ from what is gained by it?) (3) Is motion a thing apart from permanent, absolute things? It is apparent that all three of these questions deal with the same issue, namely, what kind of entity is motion?

In treating this central issue and in reconciling the Aristotelian texts behind it, most Latin commentators made use of two influential distinctions initially set forth by Averroes (1126–98). According to the first distinction, Aristotle sometimes (in the *Categories*, for example) spoke of motion according to the more common or "famous" opinion, whereas at other times (as in text 4 of book 3) he spoke according to the "truer" view. Thus, in Averroes' way of looking at the issue, to say that motion is a process (*via*) toward some terminus or perfection yet different from that perfection implies that it is either a category *per se* or belongs to the category of passions or affectations, and this is to speak of motion according to the more common or famous view. On the other hand, to say that motion is nothing other than a part-by-part generation of the terminus or perfection toward which it tends implies that it belongs to the category of this terminus and is to speak of motion according to the more correct view.[27]

The second distinction made by Averroes was found in his comments on book 5 and afforded yet another way of expressing this same difference concerning the nature of motion. If one speaks of motion according to its *matter*, then it belongs to the same category as does the terminus toward which it tends (since the "matter of a motion" belongs to one or another of the categories of quantity, quality, or place). But if one considers motion insofar as it is a *form*, then one must view motion as constituting a category *per se*.[28]

These two distinctions of Averroes served to focus attention upon the difference between viewing motion as a process and viewing it as the loss or acquisition of various termini or forms. As such, Averroes' ideas entered into almost all later questions on the nature of motion. One way of broaching the question in these later discussions was to ask whether motion was merely the terminus or form achieved by the motion (*forma fluens*) or whether it was an additional flux or transformation (*fluxus formae*) distinguishable from the terminus or form acquired. This terminology seems to have originated with Albertus Magnus (ca. 1200–1280), but, as Albertus acknowledged, the distinction behind it was already present in Averroes.[29]

Up through the thirteenth century, then, using distinctions like those of Averroes and Albertus, commentaries on this section of book 3 arose pretty much from the text itself, or from closely related texts, and used distinctions such as that between matter and form that were common throughout Aristotelian commentaries. Late in the thirteenth century or early in the fourteenth century, however, the issues to be treated became much clearer and more distinct. By this time many commentators believed that nothing existed in the outside world except individual substances (objects) and their individual qualities, or nothing except individual substances, qualities, and quantities. If this were so, motion, insofar as it was real, had to fall not just into one of the ten categories, but specifically into the two or three categories of really existing things. Equally clearly, however, the term 'motion' as usually understood did not simply denote substances, quantities, qualities, or even places: it also connoted a successive acquisition of such termini. For this latter reason it seemed wrong to some to claim that motion was nothing except the mobile and the forms or termini it acquired, since this could not account for the relation the mobile had to these termini and, especially, for the "successiveness" of this relation.[30] The problem then was whether a sparse ontology containing only individual permanent things could be shown to be adequate to account for motion, or whether it was necessary to postulate something else beyond.

How did this issue come to seem so important to the commentators on the first part of book 3? Admittedly, Aristotle's text, which said that there is no motion over and above real things, did contain a basis for pushing in such a direction, but influences from other texts and concerns are also striking. Most commentators discuss in some form at least one of the three questions listed above, but just how they approach and treat the question very often reflects the main direction of their philosophical position irrespective of the issue of motion in particular.

William of Ockham provides a very clear, if extreme, case of a commentator who "follows through" on his philosophical convictions in general when discussing motion. Ockham's conception of motion was almost exactly the same as the *forma fluens* conception then very common among his contemporaries. Yet, concerned as he was to show that it is not necessary to posit the existence of any real entities except individual things falling under the categories of substance and quality,[31] Ockham exerted a good deal of effort and care dealing with Aristotle's assertion that no motion exists apart from things. Before Ockham, many Aristotelian commentators had agreed with Averroes that, indeed, the "truer" conception of motion was the one that associated it with the termini gained rather than with some supposed special category of passions or motions. Thus, Ockham and many others agreed that in alteration, for example, there is a body altered, there are the qualitative forms (say, the degrees of heat) it successively and continuously takes on, and there is nothing else, no "motion" as a separate entity. And the same was also often held with respect to local motion and augmentation and diminution. Ockham was unique, however, in the logical rigor with which he drove this point home. Although Ockham's view and the way he presented it had their dissenters, his resolution of the problem of the nature of motion quickly became one that was considered at length by everyone who came after him, clearly indicating the historical impact of what he had maintained.

In his *Exposition of the Physics*,[32] Ockham makes it abundantly clear that the successful resolution of this problem of motion lies in focusing one's attention on the abstract noun 'motion.' It and other abstract nouns, he claims, have wrongly led many thinkers to "imagine that, just as there are distinct nouns, there are distinct *things* corresponding to them." Ockham maintains that this is where the real difficulty lies concerning the assumption of motion as some real thing apart from the individual mobiles and the individual places and forms they successively occupy. That nothing more than these individual mobiles, places, and forms is needed to account for motion can be shown, he argues, because any proper proposition about motion in which the abstract term 'motion' occurs can always be reduced to another proposition, or set of propositions, that contains nothing but "absolute" terms which refer only to individual, concrete things. This is all, he says, that we need in order to account adequately for motion, and, as always with nature, "it would be futile to accomplish with a greater number of things what can be accomplished with fewer" —an instance of Ockham's famous "razor" or principle of parsimony. For example, the proposition 'Every motion derives from an agent' can be reduced to the proposition 'Each thing that is moved, is moved by an agent', where the abstract 'motion' of the first proposition is replaced by 'Each thing that is moved', which does and can refer only to individual things.[33]

The motivations behind Ockham's "razor" or principle of parsimony may have come from logic or from theology or elsewhere, but whatever combination of considerations may have led Ockham to his views on motion, it is reasonably certain that they did not come primarily from this section of Aristotle's *Physics* or even from the *Physics* as a whole. In the commentary of Jean Buridan on book 3 of the *Physics* we have another example of the influence of concerns or positions not arising from the text itself on the questions asked and the conclusions drawn, but with quite a different result.

Like many commentators before him, Buridan treats the major categories of motion, in particular alteration and local motion, separately. For alteration, he comes to the (by then quite standard) conclusion that the reality of alteration only involves the mobile and the *forma fluens*, or the forms successively acquired. In this (and in giving a like answer with respect to augmentation and diminution), he was at one with Ockham; indeed, many of the considerations he employed in establishing these conclusions were themselves quite Ockhamist. In the case of local motion, however, he breaks completely with his English predecessor by introducing a new consideration and, therewith, a new problem into the debate. If, he says, local motion were nothing but the mobile and the places successively acquired, then what would one say about motion of the entire cosmos?[34] Place, for Aristotle and the scholastics, usually meant the innermost surface of the surrounding medium.[35] The entire world could have no place in this sense because there was no body, not even a void, surrounding it. In reply, some commentators supposed that the world was in a place by reason of its center. Revolutions of the last sphere, for instance, could be judged by reference to the stationary earth at its center.[36] But what would happen, Buridan asks, if God were to rotate the whole cosmos *as one solid body*? In support of this hypothetical question, he appeals to the condemnation of propositions by Stephen Tempier, the Bishop of Paris, in 1277, where it was held as heretical to maintain that God cannot move the heavens in a straight line.[37] On the basis of this, Buridan infers that there is also no reason why He could not cause the whole universe (the earth included) to rotate. If He did, the rotation of the cosmos as a single, solid whole could be referred to no fixed place.

In response to this special case, Buridan inferred that, unlike alteration or augmentation, local motion does involve something other than the mobile and the forms or places acquired. If local motion were only the mobile and the places successively acquired, then there would be no motion in this case, since there are no places successively acquired. But the case assumes that there is motion. Therefore, local motion must involve something else. This something else, according to Buridan, is a "purely successive thing" (*res pure successiva*) intrinsic to the rotating body. And if local motion involves such an intrinsic successive thing in this case, so it does in natural cases as well.[38] Albert of Saxony, Marsilius of Inghen, Blasius of Parma, and other later authors, especially at Paris and at the central and eastern European universities which absorbed Parisian natural philosophy so quickly and thoroughly, all followed Buridan in his view of local motion as a varying or successive quality inhering in the mobile as in a subject.[39]

It is worthy of note that, in appealing to the counterfactual possibility that God could rotate the entire cosmos as a single body, Buridan was doing something that was quite common among his contemporaries. References to God's absolute power—most often without any mention of the 1277 condemnations—were a frequent characteristic of fourteenth-century natural philosophy, indeed, of all philosophy and theology in that century. Since for the late scholastics God's absolute power was limited only by the law of noncontradiction, to inquire what could be the case under God's absolute power (*de potentia Dei absoluta*) was simply to ask what could possibly be the case without logical contradiction. Inquiry could thus go beyond the confines of Aristotelian natural philosophy with its particular set of physical and metaphysical doctrines into a broader area in which Aristotelian laws might be violated. The result was that Aristotle's arguments and conceptions were put through an unprecedented process of logical testing. Because many of Aristotle's most basic principles were considered (rightly or wrongly) to be logically unassailable—so that to violate them would, indeed, lead to logical contradiction—almost all of Aristotelian natural philosophy survived. But there were some cases in which this logical testing—particularly when combined with specific theological doctrines that led in anti-Aristotelian directions—did result in a fundamental reformulation of Aristotelian conceptions.

It is apparent, then, in the two discussions just considered of the kind of entity motion is, that Ockham presents an example of a more general philosophical position serving as an "outside influence" in the resolution of this problem, while Buridan qualifies the scope of this outside influence by uncovering what he took to be a decisive "counter instance" (involving the place of the universe) to Ockham's conclusions in the case of local motion. At times, however, the outside influence that affected the treatment of some issue concerning motion was not as general as an overall philosophical persuasion, but derived instead from a more specific conception or conviction. A rather striking example of the latter kind of influence can be seen in the impact upon conceptions of quality and alteration of the theological doctrine of transubstantiation—that is, the view adopted by the Fourth Lateran Council in 1215 that the Eucharistic bread and wine are actually transformed or transubstantiated into the body and blood of Christ, while the superficial appearances of bread and wine remain.[40] The proper understanding of the Eucharist had, of course, been pondered since the early days of Christianity, but after 1215 theologians were given something much more definite to explain in the doctrine of transubstantiation. Just as they were given this problem, they were also given the translations of the Aristotelian *libri naturales*, including the *Physics*, in which they might have hoped to find, not the explanation of the Eucharist, but perhaps a vocabulary and set of conceptions useful in formulating such an explanation.

Were it not for the Church's determination that transubstantiation did, in fact, occur, the principles of Aristotelian physics would most likely have led to the conclusion that transubstantiation and the results of transubstantiation fell into the class of occurrences that were "logically impossible" and, therefore, impossible even under the assumption of God's absolute power. Thus, Aristotle taught in his logical works as well as in his *Physics* that all things capable of separate existence fell into the category of substance. Accidents—or everything that was not a substance, including quantities, qualities, relations, and so forth—could exist only insofar as they inhered in, or were properties of, substances. Thus, for example, it would be impossible to have whiteness unless there was some subject or thing that was white. Similarly, quantities were necessarily always the quantities of some thing or things. Such assertions were considered as not only physically true, but also logically necessary, since it was of the very essence or definition of an accident that it be an accident *of something*.[41] According to the doctrine of transubstantiation, however, it was asserted that the properties or accidents of the bread—for example, its whiteness, size, taste, smell, texture, and so forth—were present, but did not inhere in any substance. They were not in the bread, because that no longer existed after transubstantiation. Neither were they in the body of Christ, which had taken the place of the bread, because Christ was thought to retain his own properties, invisible though they might be.

In the face of this dilemma, some philosophers did follow Aristotle and conclude that the understanding of the Eucharist in terms of transubstantiation is impossible.[42] Even God could not make an accident that did not inhere in any substance, they concluded, because to do so would involve a logical contradiction. Such a view, however, was never very popular during the thirteenth and fourteenth centuries, and, moreover, it was among the propositions condemned in 1277.[43] For most philosophers and theologians in the late thirteenth and fourteenth centuries it was necessary, therefore, to assume that there could be accidents not inhering in any substance, or that this was at least not logically impossible or unachievable under the the assumption of God's absolute power. Some thinkers, among them St. Thomas, argued that after transubstantiation the quantity of the bread or wine served the role that had previously been served by the substance, so that the qualities did at least have the quantity to inhere in and were not entirely without subject.[44] Others, however, including the likes of Godfrey of Fontaines, Walter Burley, and Ockham, faced the anti-Aristotelian conclusion at full strength and proceeded to try to explain just how it could be that there were quantities, qualities, and accidents of all sorts that did not inhere in any subject or substance.[45]

All of this became relevant to the problem of motion when it was considered further that the transubstantiated elements could be moved, altered, and augmented and diminished without undoing the results of transubstantiation. If medieval philosophers and theologians were anything, they were sticklers for detail and completeness, and there was no detail of any theological doctrine which they willingly allowed to remain vague. It was obvious that the Eucharist might, for instance, be warmed or cooled between the time it was transubstantiated and the time the faithful partook of it—especially if, as was the custom, some of the Eucharist was reserved to be given to the sick or others at a later time. The body and blood of Christ continued to be present (and, hence, partaking of the Eucharist continued to be efficacious), it was believed, as long as the Eucharist still had the essential properties of the original bread and wine even if slight alterations had occurred. How, then, did these motions of the Eucharist take place?

Earlier philosophers who had sought to explain alteration had argued that qualitative forms such as whiteness and hotness in themselves or in their essences had a single definite degree. It was because
material bodies participated to varying extents in these forms, they
thought, that the varying degrees of whiteness or hotness resulted. A
motion of alteration, then, might occur when a body participated more
and more fully in heat or whiteness.[46] In the Eucharist, however, there
was no such matter or body which could participate in the qualitative
forms to produce the observed alteration. Similarly, there was no
matter which could take on the observed quantities to produce augmentation or diminution. In whatever way these motions were explained, they had to be explained without assuming that the participation of matter or substance played an absolutely necessary role. As
Buridan had reasoned from the hypothetical case of the rotation of
the whole cosmos to a conclusion about all local motions, similarly
some philosophers reasoned from the case of the Eucharist to a conclusion about all alterations, augmentations, and diminutions. Thus,
one of the two most prominent later medieval explanations of alteration, the so-called succession of forms theory, was originally taken
seriously and worked out carefully as an explanation of the alteration
and augmentation or diminution of the Eucharist. The theory appears
in this context in the work of Godfrey of Fontaines (d. after 1303),
who was regularly cited as the originator of the theory; and Walter
Burley, who developed the theory to its fullest extent, also began from
this context.[47] In the succession of forms theory, alteration is explained as the result of a subject's taking on a continuous series of
forms of varying degrees, each form being corrupted as the next form
is introduced. In the alteration of the Eucharist, according to Godfrey
of Fontaines, one simply has the continuous series of forms without
any underlying subject.[48]

The succession of forms theory, then, is a theory of motion which
one frequently finds presupposed or discussed in later commentaries
or questions on book 3 of the *Physics* when nothing whatsoever about
the Eucharist is at issue. The fourteenth century saw an increased
emphasis upon looking at what was happening in motion at every
point along its path rather than considering only the mobile and its
starting and finishing points. Many factors may have contributed to
this emphasis, but probably among them was the concentration on the
problem of the motion of the Eucharist, since in that case, as many
understood it, there was *no* mobile, and one was therefore forced to
concentrate on the places, sizes, or qualities taken on.

In this chapter there is not space to give further examples of the treatment of motion in medieval *Physics* commentaries. Perhaps enough has already been said, however, to give a fair indication of the interaction that occurred between the Aristotelian text and other medieval concerns, leading the medieval commentators at times to support or clarify Aristotle and at times to diverge from him. As we have seen, medieval *Physics* commentaries shared some concerns, such as the problem of projectile motion, with the science of the seventeenth century, while other concerns, such as the problem of the nature of motion, are much more prominent in medieval than in early modern science.

Motion and Change in Other Medieval Works

Although the bulk of medieval discussions of motion occurs within Aristotelian commentaries, other kinds of works were also significant, especially as the carriers of some of the more original medieval ideas concerning motion and change. Particularly in the fourteenth century, many of the questions that were dealt with in commenting on Aristotle's *Physics* also appeared as separate, independent works, usually, but not always, in a more elaborate form. Thus, an issue such as whether motion could occur in a void is frequently found alone or together with other assorted questions on motion instead of as a question on book 4 of the *Physics*; or the problem of the nature of motion might appear apart from its usual place as a question on book 3 of the *Physics*.[49] Similarly, these and other questions on motion are found, not separately, but within other kinds of philosophical, and even theological, works. The quodlibetal questions of the theological faculties often included the discussion of motion and other matters of natural philosophy; and commentaries on Peter Lombard's *Sentences*, which was the required first work for all bachelors of theology, also frequently contained elaborate questions dealing with motion and related issues.[50]

In the fourteenth century, however, there occurred another phenomenon: a remarkable increase in the utilization of logic and mathematics in the treatment of motion. This utilization can be found in commentaries on the *Physics* and on Lombard's *Sentences*, but it is developed in an unprecedented way in independent works dealing with motion.

An examination of some of these works will constitute the major portion of the remainder of this chapter, but first brief mention should be made of a very different sort of mathematical work: Gerard of

Brussels's *Book on Motion.*[51] Written in the thirteenth century, it
bears the axiomatic format of Greek mathematical works, such as
those of Euclid and Archimedes. Its topic is that of the rotational
motion of geometric lines, surfaces, and solids, specifically, the determination of the average or mean velocities involved in their motion.[52] The procedure is relatively simple for the rotation of a line,
where only one point of the line corresponds to each velocity, and
the average velocity of the line is therefore equal to the velocity of its
midpoint. The problem is more difficult for the rotation of a circular
surface in its own plane, because then a larger proportion of the surface moves with faster velocities than with slower. Such problems
would today be solved by integrating to find the relevant average
velocities, and the mathematical ingenuity that Gerard exhibits in resolving these problems without such techniques at his disposal is, by
medieval standards, most impressive.[53]

His work is quite different in character and subject, however, from
fourteenth-century works that applied mathematics extensively (and
at times with equal ingenuity) in treating motion—works such as
those of Thomas Bradwardine, Richard Swineshead, or Nicole Oresme.
Gerard deals with the motions of geometrical entities and not, as do
these fourteenth-century works, with motions of physical subjects (no
matter how mathematicized the treatment of these subjects might be).
But such internal differences are also confirmed by external factors.
The first is that Gerard's work consistently appears in medieval manuscripts together with other works on mathematics, statics, and optics,
and not with questions or treatises in natural philosophy, something
that was characteristic of fourteenth-century works on motion, even
those which were most mathematical in character. Secondly, although
Thomas Bradwardine does cite Gerard's work, it appears to have had
little influence on other fourteenth-century treatments of motion.[54] It
seems clear, then, that the *Book on Motion* belonged to a different
tradition in the Middle Ages from that in which these later treatments
fell.

We can conveniently discuss these later treatments in terms of three
commonly treated problem areas—the problem of relating changes in
the velocity of a mobile to changes in the forces and resistances causing the velocity; the theory of the latitude of forms; and the question
of the continuity and limits of change and motion.

The Proportions of Motions

In a number of places in his works, Aristotle had considered the relations that obtain between speeds or velocities and the forces and

resistances involved in the determination of these velocities. At times Aristotle used these relations in order to disprove doctrines that he held to be false, such as that maintaining the existence of a void (*Physics*, book 4, chap. 8) or the existence of an infinite magnitude or weight (*On the Heavens*, book 1, chaps. 6–7), but in at least one instance, the final chapter of book 7 of the *Physics*, his attention to the matter was more direct. Yet even here we should not assume that he wished to generalize the particular relations that he asserted (by way of example) to hold between velocities, forces, and resistances, or even that his primary goal was to set up any relation of this sort.[55] For if we note that the seventh book of the *Physics* was most likely an early draft of the beginning of what we now have as book 8, it would seem that Aristotle's purpose in considering these velocity-force-resistance relations was to elucidate what was involved in the strength of movers, directing his attention in advance, as it were, to the discussion of the Prime Mover that would form the climax of the *Physics*.[56]

It was not long, however, before the fundamental burden of this particular chapter of book 7 was taken as that of formulating the appropriate relations between velocities (or motions) and the forces and resistances determining them. Such occurred already in the Greek commentators of late antiquity and again in Aristotle's Islamic expositors, and it can be found clearly stated in the scholastic *expositiones* of the *Physics* in the thirteenth-century Latin West. St. Thomas, for example, tells us that Aristotle's words allow us to infer general rules for the comparison of motions to one another and that the reason behind these rules is that they preserve the same proportion between the speed of a motion and a changing resistance (assuming a constant force) or between the speed of a motion and a changing force (assuming a constant resistance).[57] Although St. Thomas speaks of "general rules," it is notable that the rules he cites do not go beyond the particular examples of force-resistance changes used by Aristotle and that he also follows Aristotle in varying the relevant forces and resistances one by one, always holding the other constant. Furthermore, his tendency to remain close to Aristotle on these points seems to have been characteristic of all thirteenth- and early fourteenth-century commentators on the *Physics*.[58]

The situation changed rather dramatically, however, in 1328. In that year Thomas Bradwardine (d. 1349) wrote his *Treatise on the Proportions of Velocities in Motion* and in so doing removed the whole problem of relating velocities, forces, and resistances from the context of an exposition of Aristotle's words and investigated it in its own right.[59] After an appropriate, and rather detailed, opening chapter setting forth the required mathematical preliminaries, Bradwardine began his investigation by considering and rejecting various theories purporting to resolve this problem[60] and then proposed his own solution: "The proportion of the velocities in moving bodies follows the proportion of the moving powers to the resistive powers and vice versa. Or to express the same thing in other terms: the proportions of the moving powers to resistive powers and the velocities in moving bodies are, taken in the same order, proportional and vice versa. And this is to be understood in the sense of geometric proportionality."[61] Given merely this enunciation of his solution or theory, it is to the modern reader far from clear precisely what Bradwardine meant or how one should, if one wished, put his answer into symbols. However, the examples he appeals to in proving his theory make it clear that in our terms his solution is that velocities will vary arithmetically when the proportions of force to resistance determining these velocities vary geometrically. Thus, if a given proportion of force to resistance produces a given velocity, then when that proportion is squared, the velocity will be doubled (and if the proportion were cubed, the velocity would be tripled, and so forth).[62]

One can, however, obtain a much more adequate appreciation of Bradwardine's solution and of why he proposed it if attention is paid to how he expressed it. The crux of the matter is that when we, in modern terms, say that a proportion is "squared" or has its "square root taken," the medieval scholar would have said that the proportion is, respectively, "doubled" or "halved" (more literally, "subdoubled"). Given this medieval convention of expression, Bradwardine's solution appears supremely simple and straightforward: double the velocity, double the force-resistance proportion; halve the velocity, halve the proportion; and so on.

Since Bradwardine was here following the standard medieval mathematics of proportions, it will be helpful to inquire into that mathematics in slightly greater detail. In general terms, it tells us that the only way to increase or decrease proportions is to do so by (in *our* terms) multiplying or dividing proportions *by proportions*. For example, to increase the proportion 3 to 1 to its double would be to multiply it by itself (yielding 9 to 1 as its double); to decrease the proportion 16 to 1 to its half would be to divide it by the proportion 4 to 1 (yielding 4 to 1 as its half); or to increase the proportion 3 to 2 to the proportion of 2 to 1 would be to multiply it by the proportion 4 to 3 (yielding 12 to 6, which equals 2 to 1). One should note a difference between the first two of these examples and the third. In the first two, we have a multiplication or a division which amounts, respectively, to taking powers or roots.[63] Thus, for the first example we have (in modern notation): $9/3 = 3/1$, and thus multiplying $3/1$ by itself gives $9/3 \times 3/1 = 9/1$. For the second example, we have: $16/4 = 4/1$ and thus dividing $16/1$ by $4/1$ gives $16/1 = 16/4 \times 4/1$.[64] However, in the third example, we do not have equal proportions, and, therefore, we do not have a problem of powers or roots. But the multiplication of $3/2$ by the *unequal* proportion $4/3$ is still what is involved in the "increase" of $3/2$. Thus: $3/2 \times 4/3 = 12/6$.[65]

In *medieval* terms these operations would have been expressed, not as the multiplication or division of proportions, but as the "compounding" or, more commonly, the "adding" or "subtracting" of proportions, precisely as one adds or subtracts other quantities, be they lines or numbers. There are even marginal diagrams in the manuscripts likening a line and its aliquot parts to a proportion and its parts. Thus, the medieval would say that if the proportion 9 to 8 is "added" (*additur*) to the proportion 4 to 3, one obtains the proportion 36 to 24 (or 3 to 2). Similarly, one obtains the proportion 4 to 3 when the proportion 9 to 8 is "subtracted" (*subtrahatur*) from the proportion 36 to 24. Alternatively, it was often said—to use the same example—that the proportion 36 to 24 is compounded out of (*componitur ex*) the proportions 9 to 8 and 4 to 3. In modern terms, all of this amounts to $4/3 \times 9/8 = 36/24 = 3/2$.[66] Thus, if one was appropriately versed in the mathematics of proportions and then generalized what Aristotle had said in book 7 of the *Physics* about the particular relations of velocities, forces, and resistances and concluded that variations in velocity must be tied to variations in the *proportion* of force to resistance, then Bradwardine's answer to the problem was, mathematically speaking, the natural one.[67] It is true that "exponential-type functions," such as that contained in his solution, had been previously utilized in pharmacology in attempting to set forth the relation between the effects of a compound medicine and its constituent *virtutes*, and this may have had some influence upon Bradwardine.[68] But it certainly need not have been so; the realization of just how to increase and decrease proportions would have given the answer without any such influence.

Unlike Gerard of Brussels's *Book on Motion*, Bradwardine's *Treatise on Proportions* very clearly had an influence on his contemporaries and immediate successors. Numerous manuscripts of the *Treatise on Proportions* are still extant, as well as abbreviations of it, commentaries on it, and references to, uses of, and reworkings of it in other works.[69] Some subsequent scholars, like Blasius of Parma and John Marliani, disagreed with Bradwardine's "function," basing their objections largely on what they felt to be its erroneous way of dealing with the doubling of proportions,[70] but most later medieval authors accepted Bradwardine's theory.[71] Its popularity became such as to allow its exposition and application in theological as well as physical works.[72]

Most specific treatments of Bradwardine went little beyond what is to be found in his *Tractatus*, frequently confining their attention merely to the specific cases of the doubling and halving of velocities and proportions as Bradwardine himself had done, but two subsequent thinkers did go well beyond Bradwardine: Richard Swineshead and Nicole Oresme.

Writing his *Book of Calculations* about 1350,[73] Swineshead set down what clearly seems to be both the most brilliant application and the most brilliant development of Bradwardine's function that the Middle Ages was to see. The application occurs in treatise 11 of the *Book of Calculations*, where Swineshead brings Bradwardine to bear on the resolution of a problem concerning the motion of a heavy body near the center of the universe: whether, assuming motion in a void, a heavy body in free fall will ever reach the center of the universe (in the sense that the center of the body and the center of the universe coincide). If one regards the heavy body as a thin rod and assumes that it acts as the sum of its parts (rather than as a unitary whole), then the relevant variables of the problem are evident: as soon as any part of the rod *passes* the center of the universe, that part may be regarded as a resistance against its continued motion, while the part of the rod that has not yet passed the center of the universe will function as a force causing its continued motion. This means that lengths or segments of the rod function both as measures of the distances traversed as the rod approaches the center and as the forces and the resistances involved in determining its motion. If one agrees, with Swineshead, that Bradwardine's function must govern the velocities that result from varying force-resistance proportions, the crux of the problem is to discover a way to apply this function to the particular changes in force and resistance provided by the falling rod. Swineshead establishes how this can be done, and through the application of a series of quite sophisticated lemmas dealing with the mathematics of changing force-resistance proportions, he proves that if the rod acts as the sum of its parts, its center will never come to coincide with the center of the universe. Although Swineshead spends by far the major portion of treatise 11 establishing this result, he concludes by rejecting it as impossible, holding instead that heavy bodies act as single units and not as the sum of their parts.[74]

Treatise 14 of the *Book of Calculations*, entitled "On local motion," constitutes Swineshead's second important development of Bradwardine. Here he comes closer than any other medieval scholar to realizing the full potential (within the limitations of medieval mathematics) of Bradwardine's function for dealing with variations in force and resistance. Beginning with the function as a supposition, Swineshead deduces approximately fifty *regulae* or rules specifying precisely what *kinds* of changes in velocity correspond to what *kinds* of changes in force and resistance, and vice versa. In his opening rules he formulates the fundamental mathematics of force-resistance changes, basing his analysis on the tradition of compounding proportions. If, for example, a resistance R is held constant while a Force F_1 acting upon it is allowed to increase to F_2, then the increase of the final force-resistance proportion F_2/R over the initial force-resistance proportion F_1/R is equivalent to the proportion F_2/F_1.[75] If we then turn our attention to the velocity V_1 resulting from F_1/R and the velocity V_2 resulting from F_2/R, we can ascertain that the difference between V_2 and V_1 corresponds to the proportion F_2/F_1 (that is, the increment of velocity between V_2 and V_1 is equal to that which would result from a proportion F/R equal to F_2/F_1 standing alone).[76] But this means that to discover just what velocity changes correspond to changes in force and resistance, we need pay attention *only* to the relevant F/F proportions (or to the relevant R/R proportions if the force is held constant while the resistance is varied).[77]

This provides Swineshead with the key to all of his succeeding rules. When, and only when, the F/F (or R/R) proportions are equal will the corresponding positive or negative increments of velocity be equal. This situation will occur, for example, if (for the case of constant resistance) a given force increases proportionally (that is, if $F_2/F_1 = F_3/F_2 = F_4/F_3 = \ldots$). On the other hand, if a given force increases uniformly (that is, if $F_2 - F_1 = F_3 - F_2 = F_4 - F_3 = \ldots$, so that $F_2/F_1 > F_3/F_2 > F_4/F_3 > \ldots$), then the corresponding velocity increments will successively decrease (or, in medieval terms, motion will be "intended more and more slowly").

Throughout roughly the first half of his rules, Swineshead establishes his results on the basis of changes in force or resistance that are independently given. But he then breaks new ground by ascribing changes in *resistance* to the medium through which a mobile moves. Hitherto all increase or decrease in resistance has been considered merely *relative to time*; but to ascribe variations in resistance to a medium is to speak of increase or decrease *relative to space*. Swineshead's problem, then, was to find a way to connect variations in resistance over time with variations in resistance over space, something that he accomplished in a series of rules that are striking for their ingenuity as well as their complexity.[78] For to the medieval supporter of Bradwardine's function (or, for that matter, to anyone supporting an "Aristotelian" view maintaining velocity to be *some* function of force and resistance), motion in a nonuniform medium was exceedingly problematic. As soon as the resistance in a medium was allowed to vary, one had to face the difficulty that the degree of resistance of the medium determined the velocity of the motion, while at the same time the velocity determined where in the medium the mobile would be and, hence, the resistance it would encounter. One seemed caught in a situation involving a double dependency of the relevant variables on each other. But Swineshead's "translation" of spatial increments of resistance into temporal ones automatically rendered the resistance encountered time-dependent and thus circumvented the troublesome double dependency.

Swineshead had taken Bradwardine as far as he was ever to be taken in the Middle Ages, at least in terms of relating changes in velocities, forces, and resistances. There was, however, another ingenious "working out" of Bradwardine in the later fourteenth century, one that developed the mathematics of the "exponential-type function" he had provided. It was carried out by Nicole Oresme (ca. 1325–82), chiefly in his work *De proportionibus proportionum*.[79] Viewing Bradwardine's function as a "proportion of proportions," Oresme set aside for the moment its relevance to such physical variables as force, resistance, and velocity and concentrated his attention upon the relations of proportions *per se*, in particular upon which proportions can be said to be an aliquot part, or aliquot parts, of other proportions. Thus, if we take the set of proportions 2/1, 4/1, 8/1, 16/1, 32/1, we can say that 2/1 is a *fifth part* of 32/1 (precisely as a one-foot length is a fifth part of a five-foot length). In our terms, of course, what is meant is that $(2/1)^5 = 32/1$. Similarly, 2/1 is a *third part* of 8/1. But if we now wish to express the "parts relation" of, say, 32/1 and 8/1, we can say that since there are five 2/1 parts in 32/1 and three 2/1 parts in 8/1, it follows that 32/1 is *five third parts* of 8/1 (in our terms $32/1 = (8/1)^{5/3}$).[80] There are cases, however, where one proportion is not a part or parts of another proportion. 3/1 is, for example, neither part nor parts of 32/1. In such instances, we have, Oresme tells us, an "irrational proportion of proportions."[81] Although Oresme could not deal with such "irrational" cases in the sense of calculating what would be for us the exponent relating the two proportions involved, he did make claims about the likelihood of our meeting such cases. One can show, he maintained, that, given any unknown set of proportions, it is more probable that any two of the set would constitute an irrational proportion of proportions than a rational proportion of proportions. Indeed, the more proportions there are in the set, the more probable it will be that any two proportions picked will give us an "irrational case."[82]

Now if one applies what Oresme has developed to the proportions involved in motions, in particular the motions of the heavenly bodies, the consequences are considerable. For the proportions of force to resistance for the heavenly motions are unknown, and, hence, constitute a set of proportions to which his conclusions about the probability of irrational proportions of proportions are applicable. If, then, it is likely that any two F/R proportions governing the motion of the heavenly bodies are such as to constitute an irrational proportion of proportions, it follows, from Bradwardine's function, that it is also likely that the velocities of the heavenly bodies are mutually incommensurable.[83] Oresme then brought this result to bear in a criticism of astrology. For, as previous writers had shown, an incommensurability of the motions of any two heavenly bodies would entail that the bodies never twice return to particular locations in the heavens at the same time, an event that is absolutely crucial to astrological prediction.[84] Consequently, if Oresme's analysis is correct, it follows that predictive astrology cannot, with any reliability, predict.[85]

Bradwardine, Swineshead, and Oresme are surely the outstanding examples of the development and treatment of *proportiones motuum* in the later Middle Ages. But they are only examples. We should not ignore the fact that they, and Bradwardine's exponential-type function in general, constitute only a part of a broader tradition of calculating with proportions, a tradition that extends into medicine and theology as well as into all other corners of philosophy.[86] Such was a natural consequence of the fact that, within the Greek mathematics and exact science that the Middle Ages had inherited, "measure" was accomplished, not by the utilization of constants and standard units, but by an appeal to the theory of proportion.[87] This meant that wherever "measurement" or quantitative comparison was of concern— whether in the velocities, forces, and resistances of Bradwardine and his followers or in the "distances" between God and His creatures, no matter—proportions were likely to be utilized.

There is, moreover, another difference between "measure" as we find it in modern science and as it appears in the treatises on the proportions of motions that have just been discussed. It is that in no one of these works was it ever the case that one of the goals was to calculate a velocity from a single given proportion of force to resistance, or a force from a given velocity and a given resistance, and so forth. At times the velocities of two separate mobiles were compared with the force-resistance proportions causing these velocities. But in most instances the aim was to compare *changes* in velocity (expressed as a proportion of velocities) to the relevant *changes* in proportions of force to resistance (expressed as a proportion of proportions). Here the fourteenth-century natural philosopher was following what he believed he found in Aristotle himself.

The Intension, Remission, and Latitude of Forms

A second problem area treated in fourteenth-century works arose out of the theory of the "latitude of forms." Some of the most important issues in this problem area find their clearest treatment in Nicole Oresme's *On the Configurations of Qualities and Motions*, written about the middle of the fourteenth century, and in the pseudo-Oresmian treatise *On the Latitudes of Forms*.[88] But these works are relatively late in the history of the theory of the latitude of forms and represent the personal twist that Oresme gave to that theory, so it will be better to start with earlier discussions.

It would appear that the doctrine of the latitude of forms was first nourished within the medical and pharmacological traditions. The earliest known use of the concept appears in Galen, where it is applied to the range within which the human constitution can vary, going from perfect health at one extreme to death at the other. Thus, according to Galen, there are various states of less than perfect health, and a neutral zone in which one is neither definitely healthy nor definitely sick, before one reaches the point at which sickness clearly sets in.[89]

This idea of the latitude of human health was then combined with the doctrine of the four humors. The perfect mixture or balance of humors or qualities, the temperate complexion, represented health. There might be slight variations in the humors or qualities without sickness, so Avicenna later argued, but human life cannot exist beyond certain limits.[90] These limits were the extremes of the latitude within which human life could exist. In medieval pharmacology, the latitude of human health and sickness was then quantified using the four Aristotelian qualities—hot, cold, wet, and dry. The healthy body was temperate, neither hot nor cold, neither wet nor dry to excess. Moving away from the temperate, one could become hot or any of the other three qualities in the first degree, the second degree, the third degree, or the fourth degree (with the fourth degree causing obvious injury and, beyond that, death). Aside from specific medicines for specific complaints, the general purpose of pharmacology was to correct such imbalances of the qualities, administering a cold medicine to someone who had become too hot, a drying medicine to someone in whom there was an excess of wetness, and so forth.

Thus, within medicine, and within pharmacology in particular, there developed a clear concept of a latitude or range of degrees of qualities, often with numerical values assigned. Frequently these latitudes were for pairs of qualities like hot-cold, wet-dry, white-black, with a neutral point or zone at the center, the degrees of one quality increasing in one direction and the degrees of the contrary quality increasing in the other. Depending upon the width of the neutral zone, the entire latitude might have eight or ten degrees.[91]

By the thirteenth century the term 'latitude' had become a technical term also in philosophy and theology, possibly transplanted from medicine to this new, broader context. Here, in its more general use, "having a latitude" was the opposite of "indivisible."[92] Thus, since qualities were generally assumed to have latitudes, there would not be, for example, only one indivisible whiteness, but varying degrees of whiteness. In another context, the concept of latitude was adapted to refer to the whole "Great Chain of Being" from prime matter (or even nothingness) at its beginning up to God at its "supreme" limit, with all existing things or "creatures" ordered in between.[93] Here latitude became a "scale of perfection" and was imagined as a dense series of indivisible perfections corresponding to the species of all things in the universe. In similar fashion, some authors, Walter Burley in the fourteenth century, for instance, visualized the degrees of a single quality as a dense series of indivisible perfections. The ontology behind this view was the so-called "succession theory" of qualitative intensification, according to which, when a quality was made more intense, each successive degree of the quality was destroyed and replaced by a new higher degree.[94]

Of perhaps greater interest, however, are the contexts in which philosophers and theologians treated both latitudes and degrees not as a series of closely packed, but discrete, individuals, but as continua, very like lines in their mathematical properties. The ontology that allowed this representation of latitudes and degrees was the so-called "addition theory" of qualitative intensification. According to this theory, put into circulation by Duns Scotus[95] early in the fourteenth century, a quality is made more intense by the addition of a new part of form which combines with the old form to produce a higher degree. Perhaps contrary to expectations, this theory did not assume that the previous form survived as such, but that, within the new higher degree, there was a part equal to the old lower degree. Thus, degrees, like latitudes, came to be imagined as lines, rather than points, and higher degrees contain lower degrees, just as longer lines contain shorter ones. That is, within the latitude of quality, the degrees were not only ordered on a scale, but, like the lines imagined to represent them, they were also additive. The physical conception that helped make plausible the additive nature of degrees was the addition of illuminations in a given medium: it could be further assumed that agents producing other qualities act in a similar manner. Thus, it was thought that degrees of heat could be added physically by causing two heating agents to act upon the same body.[96] In the conceptual realm, however, once a single degree came to be imagined as identical to the latitude of that quality from zero degree up to a given point, then physical theory was no longer required, and the additive nature of intensities could be worked out mathematically—that is, by analogy to line segments alone.[97]

This use of latitudes and additive degrees is exemplified in the work of various fourteenth-century Oxford philosophers. One of the most common concerns in formulating such measures was the problem of "measuring" qualities as they were distributed in various ways over or through given subjects, and here these philosophers developed what was soon to be a standard terminology. Qualities, they said, could be *uniformly* distributed in a subject, or they could be *difformly* (that is, nonuniformly) distributed.[98] Among difform qualities, a special class was picked out in which a quality increased or decreased at a constant rate from one end of the subject to the other. These qualities were called *uniformly difform*. Aside from being one of the simplest cases of difformity, this uniformly difform distribution was probably also of special interest because it was believed that it represented the distribution of the initial effect of a single agent. Thus, it was assumed that the intensities of illumination caused by a single light source were uniformly difform, decreasing at a constant rate as they were more distant from the light source.[99] By analogy, then, it was believed that the effect, for example, of a heating agent was also uniformly difform.

Difform distributions of quality that were not uniformly difform were called *difformly difform*. A special class of difformly difform distributions were what might be called "stair-step" distributions, with two or more segments each in themselves uniform; but such stair-step distributions were less often discussed than uniformly difform ones. The same terminology of uniform, uniformly difform, and difformly difform was also applied to velocities in local motion, but then the distribution of the velocity could be not only "over space" (as in the case of a quality like heat in an extended subject), but also "over time."

Let us turn now to several specific medieval works and sample their use of the latitudes and degrees of qualities and motions. In the works of Oxford authors, the use of latitudes and degrees is typically combined in a single work with discussions using other techniques, such as the Bradwardinian approach to the proportions of velocities in motion discussed in the previous section. A good first example is the final section of the *Rules for Solving Sophisms* by William Heytesbury (fl. 1335),[100] where the sophisms treated concern local motion, augmentation, and alteration. In casting his examination of these three standard types of motion in terms of sophisms, Heytesbury was following a rapidly growing fourteenth-century tradition. In the two preceding centuries, sophisms had been almost totally the property of writings on logic and grammar, where they functioned as deceptive or puzzling propositions or examples that were to be resolved by the appropriate (but often very difficult and complicated) application of logical and grammatical rules.[101] In "solving" such sophisms, the student presumably became much more familiar with these rules and adept at using them. In the later thirteenth and fourteenth centuries the same process was transferred to natural philosophy, the deceptive or puzzling example now dealing with some aspect of motion, the relevant rules being both logical and physical ones, and the main purpose being the development of a similar expertise in the understanding and application of such rules and techniques.

Heytesbury's application of this sophism tradition to the treatment of motion begins with motions uniform in time if not in space, specifically with the motions of bodies not all of whose points are moving with the same velocity. In such cases, he asserts, a local motion should be measured by the distance traversed by the fastest moving point in the body, if there is one, and by the distance that would be traversed by a point moving indivisibly faster than any point in the body if there is no fastest moving point.[102] He applies this rule to such imaginative cases—or, in his term, sophisms—as the rotation of a body of which the outermost part continually is destroyed or undergoes corruption while the remaining, inner parts expand or are rarefied. If the inner parts are rarefied more slowly than the outer parts are corrupted, so that the body as a whole decreases in size, then, since at any instant there is always a different point furthest from the center of rotation, one has the result (by the "fastest moving point rule") that the body as a whole moves continuously more and more slowly, although every moving point within it is moving continuously faster and faster.

Turning to mobiles moving difformly over time, Heytesbury holds that velocity at an instant is measured by the space that the fastest moving point in the body would have traversed had it continued to move uniformly with that degree of velocity for a period of time.[103] He then distinguishes motions difform over time into those that accelerate or decelerate uniformly over time and those that accelerate or delecerate difformly, and defines a uniform acceleration or deceleration as one which in any equal parts of time gains or loses an equal latitude of motion.[104] In this context Heytesbury states the rule that a uniformly accelerated or decelerated motion corresponds to its mean degree, insofar as the space traversed is concerned, a rule that has come to be known among historians as the "mean-speed theorem."[105] A given latitude of motion gained or lost uniformly in a given time always causes a mobile to traverse a space equal to that which would be traversed if the mobile moved with the middle degree of the latitude for the whole time. If the acceleration starts from zero velocity, the middle degree will be equal to half the maximum degree gained, but if the acceleration begins from a finite velocity, then the middle degree will be greater than half the maximum degree. Using the mean-degree theorem, Heytesbury and others could thus correlate uniform accelerations with the spaces traversed, showing, for instance, that in a uniform acceleration from zero velocity, three times as much distance is traversed in the second half of the time period as in the first half.[106] Some historians have compared Galileo's later analysis of the acceleration of free fall to the medieval mean-speed theorem and argued that Galileo's work may reflect a knowledge of these medieval ideas.[107]

Having related velocities to distances, Heytesbury goes on to relate accelerations to velocities. He imagines a latitude of motion as a linear scale of velocities starting from zero velocity at one end and extending infinitely at the other end to higher and higher degrees of velocity. Accelerations should be measured by the latitudes of motion acquired, he says, just as velocities are measured by distances acquired. Thus, for example, an acceleration from one degree to three degrees of velocity is equal to an acceleration from two degrees to four degrees of velocity, because in both cases the latitude of motion acquired is two degrees. Heytesbury uses this rule to solve sophisms which presuppose that accelerations or decelerations should be measured proportionately, so that in the previous example the accelerations would be considered unequal, since in one case the velocity is tripled (from one to three degrees), whereas in the other it is doubled (from two to four degrees).[108] Heytesbury concludes his treatment of local motion by mentioning briefly that one can imagine a latitude of accelerations containing different rates of acceleration, just as one has imagined a latitude of velocities.[109]

In the second and third parts of his chapter Heytesbury devotes himself to determining the proper measures involved in motions of augmentation and alteration. Augmentations, Heytesbury decided, should be measured not by the absolute amount acquired, but by the proportion of the quantity acquired to that previously possessed.[110] On the other hand, he argued that alterations should be measured by the maximum latitude gained, whether that was uniformly gained in the whole subject or only in some part or point.[111]

The same extensive application of latitudes and degrees to motion appears in the writing of Heytesbury's Oxford colleague Richard Swineshead. As is perhaps evident from what has been said above about its discussion of local motion in connection with Bradwardine's rule, Swineshead's *Book of Calculations* everywhere emphasizes the mathematical development of the rules and topics with which it deals, and does so as exhaustively as the techniques at Swineshead's disposal allow. His treatment of latitudes and degrees, of their distribution over specific subjects or parts of a subject, and of the possibilities of changing one distribution into another is no exception. Here, and throughout the *Book of Calculations*, he operates in a thoroughly *secundum imaginationem* ("according to the imagination") fashion, developing variant cases involving latitudes, degrees, and distributions to show that the rules he has elicited are capable of dealing with all such cases. Indeed, when it is a question of competing opinions or theories of how one should measure a given distribution of a quality, Swineshead sometimes seems to leave it a relatively open question as to which theory is to be accepted or even preferred, his main objective apparently being to work out, by means of logic and mathematics, just which results follow from *each*. The exhaustiveness of results and the rigorous demonstration that they are results seem to be more important to Swineshead than the particular conceptions about latitudes and degrees from which they follow.[112]

But to be accurate here, we should note that, like Heytesbury's *Rules for Solving Sophisms*, Swineshead's *Book of Calculations* is a book of logico-mathematical techniques, with examples of their use in solving sophisms or in deducing other variant cases of the phenomena under investigation. Since its purpose is to demonstrate the efficacy of these techniques, it is understandable that it does not place first importance upon choosing between theories if both theories are internally consistent and capable of handling a diversity of cases. The point of the book is to show how a multiplicity of consequences can be elicited from any given theory so that its weaknesses or strengths become more manifest. But this does not tell us what Swineshead might have done had he devoted himself to a commentary on Aristotle's *Physics* or even to a commentary on Lombard's *Sentences*, in which a decision about what was true would naturally have had greater importance.[113]

It was against this English background that Nicole Oresme developed his own system for measuring qualities and motions. The term 'latitude' as used at Oxford had connoted a single geometrical dimension or line, but in its root geometrical meaning, of course, latitude or width generally denotes a second geometrical dimension after longitude or length. Oresme took advantage of this linguistic situation to propose a system of measuring qualities or velocities in two "dimensions" of latitude and longitude, latitude referring to the intensity of a quality or motion and longitude to its extension either in the qualified body or mobile or in time.[114]

Like his predecessors, Oresme considered uniform, uniformly difform, and difformly difform distributions of intensity or latitude over space or longitude. Two characteristics specifically distinguish his work from that of his predecessors. First of all, he begins by systematically describing his method for "plotting" intensity versus extension in a way that has led some historians to see his work as a step in the direction of graphing. Although the term 'graphing' is anachronistic, the comparison is not fully unfounded. Oresme uses his system to treat the qualities or motions of points, lines, planes, and solid figures. The quality or motion of a point is represented by a single line indicating its intensity. For the quality or motion of a line, Oresme represents the extension of the subject by a horizontal baseline (see fig. 10). On this baseline he draws vertical lines to represent the intensity of the quality or the velocity of the motion at each point of the subject.

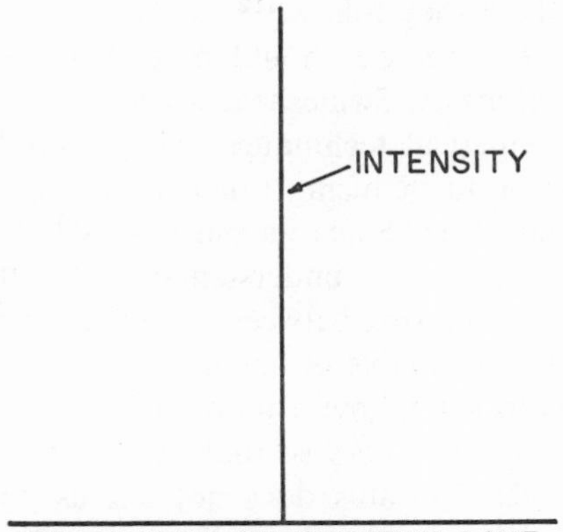

EXTENSION OR SUBJECT LINE

Fig. 10

For a uniform quality or motion, all the verticals are equal and the resulting figure is a rectangle. For a uniformly difform quality or motion, the resulting figure is a triangle or a trapezoid. Difformly difform qualities may be represented by stair-step figures, semicircles, or other irregular figures—it being understood that every line representing intensity must start from a point on the horizontal baseline corresponding to a point of the subject. (See fig. 11 for some possible and impossible configurations.) For the qualities or motions of planes, the configuration of quality or motion is easily represented by a solid figure with a base representing the subject; and the system can also be extended to three-dimensional subjects with the configurations of quality or motion for the infinite planes within the subject imagined as interpenetrating solids.[115] Apart from representing the variation of intensity or velocity from one part of a subject to another, Oresme's system was also used to represent variation over time, in which case the baseline (fig. 10) represented time. Thus, the configuration of velocities of a uniform motion was a rectangle, and the configuration of velocities for a uniform acceleration was a triangle if the acceleration started from zero velocity, or a trapezoid if the acceleration was from one velocity to another.

Aside from his emphasis on geometrical representation, Oresme also differed from his Oxford predecessors in that his primary measure of qualities and motions became not intensity, pure and simple, or velocity, pure and simple, but the so-called "quantity of quality" or "quantity of motion," where the quantity of a quality or motion was equal to its intensity times its extension. For a linear subject, the quantity of quality or motion was easily represented by the area of the imagined configuration: by the area of the rectangle for a uniform quality or motion, by the area of the triangle or trapezoid for a uniformly difform quality or motion, and so forth. To consider such a quantity of quality or motion was an important step away from the ideas of the Oxford authors, for whom the product of an intensity times an extension had no real ontological significance. In the case of Oresme's configurations of velocities over time rather than over an extended subject, however, the product of the intensity of velocity times extension in time does have an obvious ontological significance —it corresponds to the total distance traversed. Thus, if one equated the "quantities" of two velocities with respect to time, this was the same as equating the distances traversed. The seemingly obvious rightness of saying that two motions are equal if the distances they traverse are equal made the concept of "quantity of quality" seem right by analogy. On the basis of this analogy, then, Oresme adopted the mean-degree measure for uniformly difform distributions of all sorts. A uniformly difform motion is equal to its mean degree because it traverses as much space as would be traversed by a uniform velocity at its mean degree in the same time. This could be proved geometrically by showing that the area of the triangle or trapezoid representing

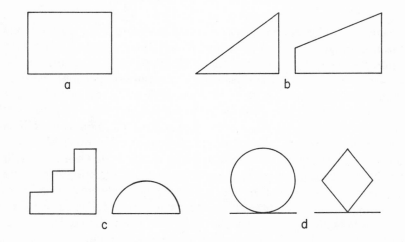

Fig. 11   Configurations of qualities or motions: (a) uniform quality or motion; (b) uniformly difform qualities or motions; (c) difformly difform qualities or motions; (d) impossible configurations (since each vertical line within the plane figures must touch the baseline).

the uniformly difform motion is equal to the area of the rectangle representing the uniform velocity at the mean degree (see fig. 12).[116] Similarly, then, a uniformly difform quality was said to equal a uniform quality at its mean degree, because both qualities contain the same quantity of the quality.

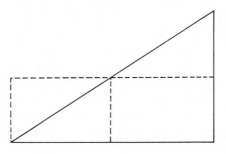

Fig. 12   Proof of the mean-speed theorem. The area of the triangle (representing the distance traversed by a body in uniformly difform motion) equals the area of the rectangle (representing the distance traversed by a body moving uniformly with a speed equal to the mean speed of the uniformly difform motion).

Beyond the sorts of cases already considered, Swineshead and Oresme often dealt with cases involving infinite intensities and even infinite total qualities. If we allow the intensity of a quality to increase to infinity over the succeeding decreasing proportional parts of a finite subject (that is, if a qualified subject has an intensity of a certain degree over its first half, twice that intensity over its next fourth, three times over its next eighth, and so on), what can we then say of the appropriate quantitative measure of this quality?[117] The amount of time spent by these authors in deliberating such cases tells us that the consideration of the infinite in their attempts to measure qualities was not an incidental extension of the application of latitudes and degrees, but was integral to their enterprise. It fits rather well, moreover, with the penchant that has already been noted for analyzing sophisms, since variant cases involving the infinite can often be considered, and at times are even explicitly labeled, *sophismata*.[118]

In the decades after the appearance of the works of Heytesbury, Swineshead, and Oresme, discussions of the intension, remission, latitudes, and degrees of forms were quite common, and many rather elementary handbooks of the basic concepts of their works were compiled.[119] But although the technique of "measuring" qualities and motions by latitudes and degrees remained popular for a good while, it had reached its high point with Swineshead and Oresme.

The Continuity and Limits of Change and Motion

The proportions of forces and resistances producing motion were sometimes called by medieval scholars the measures of motion with respect to cause. By contrast, the distances gained in local motion and the latitudes or degrees gained in alteration were sometimes called the measures of motion with respect to effect. But there was also a third kind of measure connected with motion, and that was the measure in terms of *limits*—either the temporal limits of the beginning and ceasing of a motion, or the maxima and minima of the powers or actions involved in motion. As our third general problem area, we will consider several aspects of the problem of beginning and ceasing.

This problem area derived its great interest from the fact that Aristotle, and most medievals following him, contended that all motions, as well as the times and distances and the ranges of powers and resistances that might be involved in them, are absolutely *continuous*, in the sense of being infinitely divisible. Any number of issues crucial to the continuity of motion were very much in Aristotle's mind in the later books of the *Physics*: the composition of continua in general and the relation of their parts to one another, the impossibility of infinite change in a finite time or finite change in an infinite time, and the impossibility of the motion of an indivisible.[120] And the fourteenth-century natural philosopher closely followed Aristotle in this regard, expending considerable effort in treating such issues in questions on the *Physics* and in separate treatises devoted specifically to such problems.[121]

But perhaps the most important segment of this preoccupation with the continuity of motion concerns the problems of relating the indivisible degrees of a motion or a quality to the whole continuum of that motion or quality. Here the central passages in the *Physics* were chap. 5 of book 6 and chap. 8 of book 8. In the former, Aristotle attacked the problem of the first moment of a continuous change (or, in his terms, "that in which first a thing has changed"), concluding that if the change in question is truly continuous, there is no such moment.[122] In the latter chapter from book 8 he came back to the same problem, this time viewing it in a different way, namely, from the standpoint of the necessity of there being an intervening pause, a *quies media*, when a mobile in rectilinear motion reverses its course.[123]

These two passages in Aristotle formed the point of departure for a considerable body of fourteenth-century literature dealing with the issues they presented, the major portion of this literature occurring not in commentaries (of whatever fashion) on the *Physics*, but in separate *questiones* or treatises entitled *De primo et ultimo instanti* ("On the First and Last Instant [of a thing's being or nonbeing]") or *De incipit et desinit* ("On the [terms] 'it begins' and 'it ceases' [to be such and such]").[124]

The definitions, distinctions, and rules found in these treatises on temporal limits vary somewhat from author to author, but those presented in what seems to have been the most popular work *De primo et ultimo instanti*—that of Walter Burley—became relatively standard for most natural philosophers who devoted themselves to the problem.[125] In this treatise and in most others, the criteria for ascribing limits seem to have been based on a rigorous distinction between the kinds of things being limited: permanent things, all of the parts of which exist simultaneously (such as a stone, or Socrates' being white); and successive things, whose parts necessarily exist one after the other (such as a given motion or a given stretch of time). This distinction having been made, Burley then sets up his rules for limits. A successive thing (such as a run that Socrates makes) is limited at its beginning by having a last instant of nonbeing and at its end by having a first instant of nonbeing. That is, Socrates' run occurs within the continuum of time and, with respect to its inception, there is no first instant at which he *is* running, only a last instant at which he is *not* running. Similarly, with respect to the termination of Socrates' run, there is no last instant in which he *is* running, only a first instant in which he is *not*. Socrates' run, then, is *extrinsically* limited at both ends, a conclusion in which Burley was following Aristotle.[126]

On the other hand, the requisite ascription of limits for permanent things is quite different and not so simple. To begin with, if the permanent thing exists for only an instant, then it naturally has both a first and last instant of being, namely, the sole instant for which it exists. Among the examples Burley gives of such instantaneously existing permanent things are the occupation of a particular place (*ubi*) by a body in local motion, or of a particular size or particular degree of quality by a subject undergoing a motion of augmentation or alteration. On the other hand, if the permanent thing exists for an interval of time, then its inception is limited intrinsically by a first instant of being. The termination of the existence of such a permanent thing is more complicated, however. If its existence is determined by an individual, indivisible form (such as the existence of Socrates), then it is terminated intrinsically by a last instant of being. If, alternatively, its existence is tied to a form having a latitude of degrees (such as a subject's being hot), then the termination is extrinsic, a first instant of nonbeing.[127]

One should realize that it was often not the change itself that created the problem in ascribing a proper limit, since the changes or motions involved could be either continuous or discontinuous. But in almost every example (save those, of course, dealing with something that existed for but an instant) the change in question crossed some fixed point as viewed against a continuous "background," usually that of the continuum of time. Thus, the changes involved in the coming into existence of some entity, in the inception of some motion, or in some subject becoming white immediately after it was not white were problematic because they necessarily occurred within an absolutely continuous time.[128]

More important than this distinction of what kinds of limits for what kinds of things was something one finds in Burley and almost all other authors: an absolute insistence that for all things, be they successive or permanent, a given limit, whether relevant to the beginning of the thing or relevant to its termination, must be either intrinsic or extrinsic, never both. To return to the example of the inception of Socrates' run, if there is a last instant in which Socrates is not running, there could not possibly also be a first instant in which he is running, since this simultaneously extrinsic and intrinsic limiting would imply the existence of two instants immediately next to one another, which would, in turn, imply the composition of time out of indivisible instants and, hence, deny its continuity.

Rules similar to those set down by Burley were accepted by most late medieval authors who wrote *questiones* on first and last instants or even treatises *De incipit et desinit*. In the latter treatises, however, the notions of first or last instants of being or of nonbeing were traditionally expressed in a different manner. Where Burley would speak, for example, of there being a last instant of nonbeing as the beginning limit of Socrates' run, a typical *De incipit et desinit* approach would be to speak of Socrates' run beginning through the removal or negation of the present and the positing or affirmation of the future (*incipere per remotionem sive negationem de presenti et positionem sive affirmationem de futuro*). This meant that Socrates' run is such that it does not exist in the present instant (since that instant is removed or negated and is, of course, equivalent to Burley's last instant of nonbeing for the beginning of the run), but immediately after this same present instant will exist.[129] The purpose of this was not simply to translate rules and distinctions like those of Burley into a different terminology, but, rather, to make way for the metalinguistic treatment of the physical problem of ascribing limits.[130]

To see how this was so one need only note that in *De incipit et desinit* treatises any "limit problem" was interpreted as a *proposition* in which one of the two "limit words" 'incipit' or 'desinit' occurred. Secondly, the positing or removal of the present, past, or future of which these treatises spoke could be interpreted, respectively, as an affirmative or a negative *proposition* stated in the present, past, or future tense. All elements were, therefore, on the "propositional level." This meant that one solved the "limit problem" at hand by giving a proper logical exposition of the proposition in which the term 'incipit' or 'desinit' occurs in terms of a conjunction of affirmative and negative propositions stated in the present, past, or future tense. Thus, to return to the example of the onset of Socrates' run, the limit problem involved can now be stated as 'Socrates begins to run' and solved as soon as one realizes that the proper logical exposition of this proposition amounts to resolving it into the conjunction of the two propositions 'Socrates is not running at this instant' and 'Immediately after this instant Socrates will run'. What we then have is, again, Burley's results, but now metalinguistically expressed. The elements in the analysis of the particular limit problem in question are now propositions that speak about Socrates' run. Further, even the distinction Burley made between permanent and successive things received a metalinguistic equivalent in speaking of propositions composed of *terms* dealing with permanent things or with successive things.[131]

Taken this far, the *De incipit et desinit* treatises of the fourteenth century seem to have added little, if anything, to what was accomplished in works, like that of Walter Burley, on first and last instants. But if one goes more deeply into this *De incipit et desinit* literature, it is an historical fact that it is just this literature, and not that on first and last instants, that carried the analysis of the limits of continuous change further. A great deal further. The key fourteenth-century treatises here are the *Sophisms* of Richard Kilvington (d. by 1362) and the fourth chapter (expressly titled "De incipit et desinit") of William Heytesbury's *Rules for Solving Sophisms*, together with the subsequent literature that these two works engendered.[132] The complexity and ingenuity of the new cases invented in this literature for analysis is such as to make their description impossible in a short compass. There is one general technique, however, that was often adopted in constructing these new cases: the authors saw fit to compare several instances of continuous change with one another and then investigate just what effect this "multiplication of variables" might have upon their rules and distinctions for the ascription of limits. Indeed, often the central rules paid little attention to the kind of basic distinctions that were the concern of Burley and the more elementary *De incipit et desinit* "translations" of him; they were, instead, keyed to explaining the complex cases that compared changes.[133]

What is more, the comparison frequently had to do with instances of continuous change that were stipulated to occur at different *rates*. One did not, for example, have to do simply with a comparison of Socrates' growing in size with Plato's growing in size, but of Socrates' beginning to grow from a size smaller than that from which Plato begins to grow, yet allowing Socrates to grow at a rate faster than that of Plato. If both are allowed to grow for an indefinite period of time, Socrates will obviously reach a size larger than that of Plato. But what can we say of the sizes had by both Socrates and Plato if we also stipulate that, at the instant Socrates would have been the same size as Plato, both of them cease to exist? There is then no instant at which the proposition 'Socrates is the same size as Plato' will be true. But since in his process of growth Socrates will "run through" every size Plato will run through and none greater, can we not now, before both begin their growth, say that 'Socrates will be of such a size as Plato will be'? The answer to this question varied from author to author, but the important point to observe is that, whatever the answer, those who deliberated on this and similar puzzles or (to be more exact) sophisms came to a clearer understanding of just what was involved in a limiting process. By comparing continuous changes such as those exemplified by the sophism of Socrates' and Plato's growth, they were often not merely able to assert that these changes did or did not have the same limit (be it intrinsic or extrinsic), but they were also often able to explain just why this was so.

As unfitting as this kind of deliberation might initially seem to a "science" of motion, one can hardly overstate how very pertinent it was to the investigation of motion and change in the eyes of fourteenth-century thinkers. Not only was it a natural outcome of Aristotle's concerns with motion, but the amount of time and effort that was spent in treating the problem of the continuity of change and its limits exceeded that devoted to many other particular aspects of motion. It was greater, for example, than that expended in discussing the proportions of motions, forces, and resistances. In addition, the techniques and results that were developed in this investigation of limits had an attraction and applicability that went beyond the particular concerns from which they grew. Many other areas of natural philosophy and even theology were seen to benefit from their employment as well.[134]

Conclusion: The Evaluation of the Medieval Investigation of Motion

The present chapter began with an indication of how very central motion and change were to the whole of Aristotelian natural philosophy in the Middle Ages, a fact that makes it only too evident that what has been here described is only a fragment of the medieval deliberations regarding motion and change. We have, however, attempted to set forth the most important and most basic features of those deliberations and to provide a fairly accurate map of the areas of major concern to medieval, especially fourteenth-century, thinkers. Still, what has been said of these major concerns is itself necessarily incomplete, for almost all of them were far more multifaceted, and harbored developments that were far more complex and comprehensive, than we have been able to indicate. This was especially true of the investigations into the proportions of motions, the latitudes of forms, and the various problems of the continuity and limits of change.

In each of these cases, we find the late medieval natural philosopher pushing his ingenuity to its very limits in inventing a never-ending series of new examples of this or that kind of motion or change to be treated and resolved by his new conceptions and techniques. The more complicated and difficult the example, the better, for such would assure that one truly had expertise in operating with proportions or latitudes or limits, and would at the same time prove that the conceptions and rules that had been devised in each of these types of investigation could adequately cover all variant cases of the kinds of change that were their charge. As we have seen, these variant cases often involved coming to grips with infinite intensities and values of one sort or another, but there, too, the results were almost always satisfactorily dealt with by applying no more than the usual conceptions and rules at the medieval scholar's disposal (although, admittedly, the application frequently required more than an average fourteenth-century know-how).

We have also seen that there was a long-standing tradition in medieval logic of inventing complex, and often seemingly outlandish, examples that were called sophisms. That this predominantly logical tradition should have been brought to bear in many late medieval works dealing with motion and change should occasion no surprise if we think for a moment of the common base and unity of all that was learned in a medieval faculty of arts. But several other general things should be noted about these new and complicated examples provided by such *sophismata*. They were, to begin with, not experiential; they were not meant to be tested by an appeal to nature. This goes without saying when the invented instance dealt with some subject that was infinitely intense in some part, but it was also true in instances where the infinite was nowhere present. Everything was formulated and analyzed *secundum imaginationem*, not *secundum cursum naturae* (following one's creative imagination and not what occurs according to the order of nature). This is even the case when, as with some of Heytesbury's later commentators, some concrete object is specified to illustrate a specific example of motion or change: to clarify what is involved in something having a uniform motion of rotation while at the same time suffering an overall motion of rarefaction, yet undergoing corruption at its fastest moving point, by referring to a wheel of ice revolving in an oven may assist one in understanding the text, but it is still an assistance that is derived *secundum imaginationem*.[135] One is surely no closer to "things being that way in fact in nature" than one was without the clarifying example.

Secondly, even the most cursory examination of the variant cases and examples that were analyzed in the areas of proportions, latitudes, and limits makes it apparent that either logic or mathematics was being applied in the analysis at hand. However, although it is undeniable that mathematical and logical conceptions and methods are constantly being used, one should avoid inferring too much from such a distinction. To begin with, it seems clear that, as we find them applied within medieval natural philosophy, one should not distinguish mathematics from logic too rigidly, since fourteenth-century scholastics frequently moved from one to the other, and even amalgamated them, with ease. In place of speaking of the application, then, of logic or mathematics, it might be better to speak of a number of specific techniques or "languages" of analysis, each one tied to some specific kind of investigation but incorporating both logical and mathematical elements as required. Thus, one might distinguish one "language" of analysis for dealing with the proportions of motions, another for latitudes and degrees, a third for limits, and so forth.[136] What is more, even when we can discriminate something that appears to be relatively straightforward mathematics, it sometimes turns out that the kind of mathematics being applied is not to be found in the mathematics inherited from Greek antiquity. Such was particularly the case when the infinite was being treated. But even when the mathematics at stake was drawn from the likes of Boethius or Euclid, it would be wrong to classify those fourteenth-century works concerned with the measurement of motion that utilized this mathematics as part of the quadrivium. Evidence for this can be found in the fact that the treatises of Bradwardine, Heytesbury, Swineshead, and Oresme were almost never copied together with other mathematical works such as Euclid's *Elements* and Jordanus de Nemore's *Arithmetic*. They did not form, it seems fair to infer, part of that mathematical corpus.[137] The codicological evidence is that these and similar works were put together with other physical and logical writings and were hence regarded every bit as much a part of natural philosophy as some treatise or question dealing with motion that applied no mathematics at all.

Nevertheless, however one may categorize these late medieval writings on the proportions, latitudes, and limits of motion and change, it seems plausible that the persistent application of logic and mathematics—or logic mixed with mathematics—that we find within them, and the endless multiplication of variant cases as well, are somehow related to another, much more general, late thirteenth- and early fourteenth-century phenomenon: a growing concern with evidence, certitude, and precision, with the characteristics requisite for evident knowledge, and the modes of analysis involved in obtaining it. And it is not merely the attempts to measure change that bespeak this concern with precision. It would seem that Ockham bears testimony to the same thing when he worries the problem of what motion is through one kind of "test" after another.[138]

Furthermore, the way in which Ockham and most other fourteenth-century thinkers approached the question of what motion is appears to have something else in common with the discussions of the measurement of motion: both exhibit a marked tendency to move away from the analysis of motion in terms of mover and moved, agent and patient, something that was so prevalent in much of what Aristotle himself had to say about motion and, consequently, in the works of his thirteenth-century commentators. In the treatises on the measurement of motion and change, in particular, the focus was usually on measuring the change or motion itself—its embodiment in a mobile or its being due to a mover for the most part dropping from view. Even when the causes of a motion were present and were measured as the forces and resistances determining that motion, the concern was not with these forces and resistances as dwelling in some particular mover, mobile, or medium, but with the forces and resistances in abstraction from concrete agents and patients. Alternatively, the measure may have treated what Aristotle termed the "consequents" of motion: the times, places, and distances involved in a motion and the continuity and infinity necessarily connected with it.[139] Yet in all of these instances of the measurement of motion, the ontology of mover-mobile and agent-patient is noticeably in the background.[140] Is this not why, perhaps, the efforts of Bradwardine, Heytesbury, or Swineshead appear to be somehow newer and more divergent from Aristotle than, for example, Buridan's efforts to deal with the question of impetus, since the latter remains within the Aristotelian matrix of mover and mobile?

It would be an error, however, to regard these new and distinctive fourteenth-century efforts as moving very directly toward early modern science. The history of science to date has tended to view much of the material covered here mainly as background to the Scientific Revolution of the seventeenth century; but to confine one's perspective of medieval deliberations about motion to that of "anticipations" of what came later, or even to center one's perspective on such anticipations, is misleading and incomplete. To be sure, Galileo was aware of medieval works on motion, although research still remains to be done to determine just which works and how he came to know them.[141] And he may well have put, for example, the medieval mean-speed theorem and even its proof to his own use in his investigation of naturally accelerated motion. But what was then being used was but one part, one fragment, of the medieval "science of motion," a part removed from its context and, in Galileo's hands, made to perform quite different duty. And the same thing must be said of other pieces and parts of medieval thinking about motion that were mentioned or used in seventeenth-century mechanics.

Put in a more general way, the fact remains that the goal, indeed, the whole enterprise, of many a medieval scholar who treated motions was worlds away from that of Galileo and his confreres. Thus, although the mean-speed theorem or, more generally, mean-degree measure, appears in and about any number of fourteenth-century writings on motion, it is seldom assigned a special significance setting it apart from other "rules of measure"; and the uniformly difform motion with which it is so often connected is never (save once, almost by accident) put together with the motion of free-fall, as was the case with Galileo.[142]

The medieval thinker was intent upon carrying as far as possible each of his separate areas of investigation. To give as exhaustive a development as he could of the proportions or of the latitudes of motions, for example, was his primary goal. Such a goal explains the absolute plethora of examples and variant cases. Moreover, it is also strong evidence against another way the Middle Ages is often compared to the Scientific Revolution. If only, this comparison runs, the medievals had gone further, had taken the necessary additional steps, they might have, perhaps even would have, come much closer to what we find in the seventeenth century. The error in this is that they did take the "additional steps" that were available to them under the mathematical and logical techniques at their disposal—but they did so within the logic and spirit of their own particular enterprise, their own particular kind of investigation of this or that aspect of motion and its concomitants and variables. Yet carrying all of this out did not lead them to Galileo or the seventeenth century; it even seems fair to say that it did not move them in that direction.

In fact, in contrast to those medieval scholars who developed and dealt with proportions, latitudes, and limits in their analysis of motion and change, it is the late scholastics who approached motion and related topics in a more traditionally Aristotelian manner and context who can be more closely associated with Galileo and the Scientific Revolution. This is the case, for example, with what we find in the commentaries of Buridan, Nicole Oresme, or Albert of Saxony on Aristotle's works, or even with what we see in separate questions that grew out of such commentaries. When scholars such as these treat such issues as impetus or the possible motion of the earth, and so forth, they seem to have a greater kinship with Galileo and the seventeenth century than do the more distinctive and original medieval works on motion associated with the Merton School at Oxford. And this is so not primarily because of the particular conceptions and considerations they employ in investigating these issues, but rather because the whole framework of their approach is, like that of Galileo, more consonant with Aristotle, howsoever much they—or Galileo—may have been critical of him on particular points.

The fate, then, of medieval discussions of motion should not be viewed solely as providing some kind of background from, or against which, early modern thinking about motion developed—and this applies in particular to those discussions of the measurement of motion in which such a background has often been held to exist. It is more accurate to locate its fate in at least three other developments. First, in the veritable pushing to their limits of the separate kinds of investigation of motion and change. Second, in the wholesale application of the techniques and results of each of these investigations within other areas of medieval learning, especially within the commentaries on Peter Lombard's *Sentences* that formed the backbone of the faculties of theology, and within commentaries on Avicenna's *Canon*, Galen's

Ars parva, and other medical writings.[143] And lastly, in the repetition of these techniques and results—perhaps rephrased and with a few new examples and connections noted—but without the creation of anything new that was of substance. This last phenomenon, largely of the fifteenth century, may still have been within the logic and spirit of fourteenth-century discussions, but it seems to have been a static one, no longer part of a productive development in philosophy and theology.

Notes

1. Aristotle, *Physics* 2.1, 192b13–23.
2. Oxford, Merton College, MS 272, fol. 176rb. On Geoffrey of Haspyl, whose scholastic activity probably took place at Oxford some time between the years 1243 and 1263, see A. B. Emden, *A Biographical Register of the University of Oxford to* A.D. *1500*, vol. 1 (Oxford, 1957), under "Geoffrey de Aspale"; Enya Macrae, "Geoffrey of Aspall's Commentaries on Aristotle," *Medieval and Renaissance Studies* 6 (1968):94–134; and Charles Lohr, "Medieval Latin Aristotle Commentaries, Authors G–I," *Traditio* 24 (1968):150–52 ("Galfridus de Aspall").
3. *Physics* 2. 3. See also above, pp. 99–102.
4. *Physics* 3. 1, 200b15–25.
5. By Porphyry, Philoponus, Simplicius, and others; see W. D. Ross, *Aristotle's Physics: A Revised Text with Introduction and Commentary* (Oxford, 1936), pp. 1–6.
6. Generation and corruption, although sometimes included as motions, are not here considered as motions narrowly defined because, as Aristotle argues, they occur instantaneously and are therefore "mutations," whereas all motions in the strict sense take time.
7. *Physics* 7. 1, 241b34.
8. For the appearance of these ideas in the work of Thomas Aquinas, see above, pp. 97, 101–2.
9. Cf. *Physics* 4. 8, and 8. 10. Also Simplicius, *Physics* 8. 10. See Anneliese Maier, *Zwei Grundprobleme der scholastischen Naturphilosophie*, 3d ed. (Rome, 1968), p. 117 n. 6.
10. *Physics* 8. 10, 266b25–267a20. Cf. Maier, *Zwei Grundprobleme*, pp. 117–18.
11. Cf. Maier, *Zwei Grundprobleme*, pp. 115–16.
12. Aristotle himself had done this in the *Physics* 4. 8, 215a24–215b12, 210a11–17; and in *On the Heavens* 1. 6, 273b30–274a2.
13. Ockham, *Philosophical Writings*, ed. and trans. Philotheus Boehner, O.F.M. (Edinburgh, 1957), p. 140. Cf. Maier, *Zwei Grundprobleme*, pp. 154–60.
14. Ockham, *Philosophical Writings*, p. 140. One should note that this is a brief passage within a discussion of quite a different matter and that Ockham does not spell out his assumptions in detail. Although Ockham says here that, in the case at hand of the continued motion of a projectile, the local motion is not a new effect, thus refuting the argument that as a

new effect it requires a cause, he generally assumes that when a body is moved locally there is a mover.

15. Maier, *Zwei Grundprobleme*, pp. 113–314. Clagett, *The Science of Mechanics in the Middle Ages* (Madison, Wis., 1959), pp. 505–40.

16. Maier, *Zwei Grundprobleme*, pp. 120–34. Clagett, *Science of Mechanics*, pp. 509–15, 543–45.

17. Maier, *Zwei Grundprobleme*, pp. 201–28. Clagett, *Science of Mechanics*, pp. 522–23, 534–35. The main source is Buridan's *Questions on the Physics* (Paris, 1509; reprint ed. Frankfurt, 1964), bk. 8, quest. 12.

18. Maier, *Zwei Grundprobleme*, pp. 222–23. Clagett, *Science of Mechanics*, pp. 523–25, 536–37. Because Buridan considered impetus to be permanent and not self-expending and because he suggested that impetus could explain the everlasting motion of the heavens, some historians (for example, Clagett, p. xxviii) have suggested his impetus to be an "analogue of inertia." But there is a most important ontological difference between the two: impetus causes a motion to continue because it is involved as a continuously acting force, whereas in inertia a force is needed only to change, and not to continue, motion. Similarly, one might be tempted to see a prefiguration of modern inertial ideas in Ockham's solution to the problem of projectile motion: his explanation is ontologically similar to one involving inertia, since in his view no continuously acting force is required to explain continued motion. But he does not further say that this motion would continue indefinitely unless or until a counteracting force intervened.

19. Maier, *Zwei Grundprobleme*, pp. 219–21. Clagett, *Science of Mechanics*, pp. 525, 535–36.

20. Maier, *Zwei Grundprobleme*, pp. 228–90.

21. Thus some medieval commentators, considering that natural motion was motion toward natural place and that violent motion was motion away from natural place, came to recognize a third type of motion in which a body remained equidistant from its natural place, as when a ball rolled on a spherical surface concentric with the center of the universe. They then reasoned that since such a motion was not violent, it could be produced by any force, however small, if the accidental impediments of friction and the like were removed; cf. Albert of Saxony, *Questions on the Physics* (Paris, 1518), bk. 4, quest. 12, fol. 51r; also Marsilius of Inghen, *Questions on the Physics* (Lyon, 1518), bk. 4, quest. 12, fol. liv verso, and bk. 7, quest. 7, fol. lxxix recto. For somewhat similar arguments in later authors, see Jerome Cardan, *Opus novum de proportionibus*, prop. 40, in Cardan's *Opera Omnia* (Lyon, 1663), 4:480; Simon Stevin in *The Principal Works of Simon Stevin*, vol. 1, ed. E. J. Dijksterhuis (Amsterdam, 1955), pp. 185, 187. A somewhat similar idea, but overlaid with an emphasis on the special nature of the sphere, is found in Nicholas of Cusa, *De ludo globi*, in *Opera* (Paris, 1514), vol. 1, fol. cliv verso. See Allan Franklin, *The Principle of Inertia in the Middle Ages* (Boulder, Colo., 1976), chap. 5.

22. Pierre Duhem, who was the first to discover the existence of impetus in the medieval sources, saw in it the kernel of the modern principle of inertia. See Duhem's *Etudes sur Léonard de Vinci*, 3 vols. (Paris, 1906–13), and the correction of Duhem's views by Anneliese Maier, *Zwei*

Grundprobleme, pp. 113–314, and *Ausgehendes Mittelalter*, vol. 1 (Rome, 1964), pp. 353–79. For a recent brief treatment see Franklin, *Principle of Inertia*.

23. 200b32–201a3.

24. *Categories* 9, 11b1–8.

25. *Physics* 5. 1, 225a34–b9 (text comment 9). Book 3 puts motion into the categories of substance, quantity, quality, and place, but book 5 rejects substance and accepts only the other three categories.

26. *Metaphysics* 5. 13, 1020a26–32 (text comment 18).

27. Averroes, *Commentary on the Physics*, bk. 3, comment 4. Cf. the discussion in Anneliese Maier, *Zwischen Philosophie und Mechanik* (Rome, 1958), pp. 62–64.

28. Averroes, *Commentary on the Physics*, bk 5, comment 9. Cf. Maier, *Zwischen Philosophie und Mechanik*, pp. 64–67.

29. Maier, *Zwischen Philosophie und Mechanik*, pp. 68, 74–77.

30. This is one of the points made, for example, by Walter Chatton in his *Commentary on the Sentences*, bk. 2, dist. 2, quest. 2 (Florence, Biblioteca Nazionale Centrale, MS Conv. Soppr. C.5.357, fols.185v–186v; and Paris, Bibliothèque Nationale, MS Latin 15887, fols. 90r–91v).

31. Ockham, *Philosophical Writings*, pp. 137–39.

32. This is Ockham's first, and most complete, treatment of the *Physics*. Still unedited—for MSS see Charles Lohr (above, note 2), pp. 206–7—substantial portions of the sections on motion (and on place and time as well) circulated in the fourteenth century as a separate work entitled *De successivis*. This has been edited by Philotheus Boehner (St. Bonaventure, N.Y., 1944) and provides the most easily available source for Ockham's contentions about motion. Note should also be made of the fact that Ockham treated the problem of motion in his two other works on the *Physics* (*Questiones* and a *Summule*, both still awaiting a modern edition) and, earliest of all, in his *Commentary on the Sentences* (Lyon, 1495; reprint ed. London, 1962), bk. 2, quest. 9.

33. *De successivis*, pp. 37–38. Ockham speaks of "mutation" rather than motion in this example, but he later (p. 48) makes it clear that everything he has said about mutation applies, *mutatis mutandis*, to motion as well.

34. Bk. 3, quest. 7 (Paris 1509; reprint ed. Frankfurt, 1964). Cf. Maier, *Zwischen Philosophie und Mechanik*, pp. 121–31. See also below, chap. 8.

35. Aristotle, *Physics* 4. 4, 212a20–21.

36. Buridan ascribes such an opinion to Averroes in his questions on the *Physics*, bk. 4, quest. 6, conclusion 4, fol. 72v. Cf. Maier, *Zwischen Philosophie und Mechanik*, p. 125, n. 84.

37. H. Denifle and A. Chatelain, eds. *Chartularium Universitatis Parisiensis*, vol. 1 (Paris, 1889), item 473, proposition 49. Translated in *A Source Book in Medieval Science*, ed. Edward Grant (Cambridge, Mass., 1974), p. 48.

38. Buridan, *Physics*, bk. 3, quest. 7, conclusion 6, fol. 50v. Cf. Maier, *Zwischen Philosophie und Mechanik*, pp. 126–131. Also Anneliese Maier, *Metaphysische Hintergrunde der spätscholastischen Naturphilosophie*

39. See Maier, *Zwischen Philosophie und Mechanik*, pp. 139–43. (Rome, 1955), pp. 347–53.

40. See "Transubstantiation," in *New Catholic Encyclopedia*, 14:259–61; also "Lateran Councils, Fourth (1215)," 8:407–9. See also Edith Sylla, "Autonomous and Handmaiden Science: St. Thomas Aquinas and William of Ockham on the Physics of the Eucharist," in *The Cultural Context of Medieval Learning*, ed. John E. Murdoch and Edith D. Sylla (Dordrecht, 1975), pp. 361–72.

41. See Thomas Aquinas, *Commentary on the Sentences*, bk. 4, dist. 12, quest 1, art. 1, in *Opera omnia* (Parma, 1852–73; reprint ed. New York, 1948), 7:652, and Aristotle, *Metaphysics* 7:1, 1028a10–34; *Topics* 1. 5, 102b4–26.

42. See Sylla, "Autonomous and Handmaiden Science," p. 361 and n. 43.

43. Denifle and Chatelain, *Chartularium*, item 473, props. 138–41. Prop. 140 says: "To make an accident exist without a subject is impossible and implies a contradiction."

44. St. Thomas Aquinas, *Commentary on the Sentences*, bk. 4, dist. 12, quest. 1, art. 2, pp. 657–58; and *Summa Theologiae*, 3a, quest. 77, art. 2.

45. For Godfrey of Fontaines, see his *Quodlibet Undecimum* in *Les Philosophes Belges*, vol. 5 (Louvain, 1924), quest. 3, pp. 12–22, esp. p. 19: "And it should be said that it is necessary to posit motion without a mobile in the proposed case [of the Eucharist]." Walter Burley expresses his view in his *First Treatise* and *Second Treatise on the Intension and Remission of Forms*. See the article on Burley in the *Dictionary of Scientific Biography*, 2:608–12. For Ockham, see his *Commentary on the Sentences*, bk. 4, quest. 7; and *Quodlibet* 4, quest. 31; cf. also Sylla, "Autonomous and Handmaiden Science," pp. 371–72.

46. See Anneliese Maier, *Zwei Grundprobleme*, pp. 3–35. Maier traces the participation theory back to Simplicius.

47. See the sources for Godfrey of Fontaines and Burley cited in n. 45; also Maier, *Zwei Grundprobleme*, pp. 36–73. Maier, however, does not seem to pay enough attention to Godfrey's *Quodlibet XI*, quest. 3, and, therefore, erroneously claims that Godfrey apparently did not advocate the view widely ascribed to him by later medieval authors.

48. See n. 45. Godfrey goes on to say, p. 20, that there is not one accident, but a continuous chain of different accidents (*accidens aliud et aliud*) of the same species following each other without interruption.

49. Thus, Otto of Merseburg includes the question *Utrum supposito vacuo aliquis motus possit fieri in ipso* among assorted *questiones naturales* (for MSS see Jan Pinborg in *Bulletin de philosophie médiévale*, 15 [1973]: 148–49), and the anonymous Magister Ricardus (probably Richard Kilvington) does the same in his four *Questiones de motu: Utrum aliquod corpus simplex possit moveri equevelociter in vacuo et in pleno* (Venice, Biblioteca Nazionale Marciana, MS VI, 72 (2810), fols. 101r–107v). The nature of motion was treated, for example, by the early fourteenth-century Franciscan William of Alnwick among a series of *determinationes* devoted to subjects of all sorts (see Guillelmus Alnwick, *Questiones disputatae de esse intelligibili et de quodlibet*, ed. A. Ledoux (Quarrachi, 1937), pp. xxiii–xlvi).

50. For the investigation of motion in quodlibetal questions see the indices (under *motus*) in Palémon Glorieux, *La Littérature quodlibétique*,

2 vols. (Kain-Paris, 1925–1935). For the treatment of motion in *Sentence Commentaries*, see Maier, *Zwei Grundprobleme*, pp. 142–200.

51. See Marshall Clagett, "The *Liber de motu* of Gerard of Brussels and the Origins of Kinematics in the West," *Osiris* 12 (1956): pp. 73–175; and *Science of Mechanics*, pp. 187–97.

52. One should note that all of the motions that Gerard treats are uniform over time and that when he speaks of the rotational motion of, say, a line, he has in mind what we could call its linear velocity, not its angular velocity. Furthermore, when Gerard compares the motion of one line to that of another line and says that they are "equally moved," he means that the lines have an equal average or mean velocity. The fact that he speaks merely of the motion of a rotating line and not of its mean velocity explains why Thomas Bradwardine could later disagree with him (see reference in note 59) and instead opt for the fastest moving point of the rotating line as the proper measure of its motion.

53. Thus to find the average velocity of a circular disc *ABC* (fig. 13)

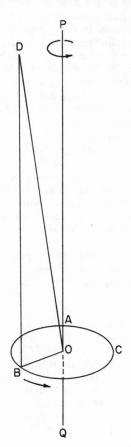

Fig. 13

rotating in its own plane about an axis PQ through its center O, Gerard finds a substitute geometrical figure that can be rotated with velocities perpendicular to its plane so that the paths of no two points of the figure overlap. The figure he substitutes for the circular disc is a right triangle DBO with base DB equal to the circumference of the circular disc and height BO equal to its radius. If this triangle is rotated in three dimensions so that the base of the triangle (representing the circumference of the circular disc) traces out the surface of a right cylinder about an axis through the vertex of the triangle, and if the base of the triangle moves with the same linear velocity as any point on the circumference of the original circular disc, then for every point or line of the original disc there is a corresponding point or line of the triangle moving with equal velocity; the triangle is therefore moved neither more nor less than the circular disc, and they have equal motions. But the motion of the triangle can now be compared to the rotations of rectangles around the same axis, showing that the motion of the triangle (and, hence, of the original circular disc) is equal to 4/3 of the motion of its radius. Since the average velocity of the radius is equal to the velocity of its mid-point, the average velocity of the circular disc is in 4/3 proportion to the velocity of the midpoint of its radius. For further explanation of the particular proposition being discussed see Edith Sylla, "The Oxford Calculators and the Mathematics of Motion, 1320–50: Physics and Measurement by Latitudes" (Ph.D. dissertation, Harvard University, 1970), appendix B, pp. B1–B14; or Wolfgang Breidert, *Das aristotelische Kontinuum in der Scholastik (Beiträge zur Geschichte der Philosophie und Theologie des Mittelalters, n.s.,* vol 1) (Münster, 1970), pp. 47–67. The interpretation offered here is intended as a correction of that found in Clagett's works.

54. Bradwardine cites the work by another title ("On the Proportionality of Motions and Magnitudes"); see the work cited in n. 59, pp. 128–29. Marshall Clagett has speculated that Gerard's first proposition equating the motion of a radius to that of its midpoint may have influenced later Oxford scholars in their equation, with respect to space traversed, of a uniformly accelerated motion with its mean velocity. Although such a connection is a logically attractive one, there is no solid historical evidence to support it.

55. *Physics* 7. 5, 249b27–250b7. When speaking of local motion in this chapter, Aristotle treats only violent motion, so that all resistances are weights of the mobiles in question. The "rules" Aristotle gives can be be symbolized as follows:

If F moves R a distance D in time T,

a. F moves $R/2$ a distance $2D$ in time T,
b. F moves $R/2$ a distance D in time $T/2$,
c. F moves R a distance $D/2$ in time $T/2$,
d. $F/2$ moves $R/2$ a distance D in time T,
e. F *need not* move $2R$ a distance $D/2$ in time T,
f. $F/2$ *need not* move R a distance $D/2$ in time T.
g. If F_1 moves R_1 a distance D in time T and
   F_2 moves R_2 a distance D in time T, then
   $F_1 + F_2$ moves $R_1 + R_2$ a distance D in time T.

Note that, with the exception of cases d and g, Aristotle holds the force constant while varying the resistance, something that is explicitly noted by his medieval interpreters (see below). Furthermore, the specification of what *need not* occur in cases e and f clearly indicates that Aristotle did not have in mind the formulation of a general rule relating velocities, forces, and resistances. Indeed, his instantiating example for these two cases, of one man not being able to haul a ship which a number of men might be able to haul, suggests that Aristotle had strength in mind and not an abstract force or power, a conclusion supported by the fact that he used the term 'strength' itself as well as 'power' in these passages.

56. For opposing views on whether there is a mechanics of velocity-force-resistance in Aristotle, see Israel Drabkin, "Notes on the Laws of Motion in Aristotle," *American Journal of Philology* 59 (1938):60–84, and Henri Carteron, *La notion de force dans le système d'Aristote* (Paris, 1923). The latter maintains that any such mechanics is "prétendue." Part of the introductory chapter of Carteron's book is now available in a revised translation in *Articles on Aristotle*, vol. 1, *Science*, ed. Jonathan Barnes, Malcolm Schofield, and Richard Sorabji (London, 1975), pp. 161–74.

57. Thomas Aquinas, *In octo libros physicorum Aristotelis expositio*, ed. P. M. Maggiolo (Turin and Rome, 1954), bk. 7, lectio 9, pp. 492–93.

58. Anneliese Maier, *Die Vorläufer Galileis im 14. Jahrhundert* (Rome, 1949), pp. 85–86.

59. H. Lamar Crosby, ed. and trans., *Thomas of Bradwardine, His Tractatus de Proportionibus: Its Significance for the Development of Mathematical Physics* (Madison, Wis., 1955).

60. Bradwardine, *Tractatus de proportionibus*, pp. 86–110. To cite but two of these erroneous theories, Bradwardine considers and rejects, first, the possibility that the proportion of velocities might "follow the excess of the power of the mover over the power of the thing moved," that is, in modern symbols, $V \propto (F - R)$. Historians have noted that this rejected alternative corresponds to that advocated by the sixth-century Greek commentator on Aristotle, John Philoponus (whom Bradwardine could not have known, since his relevant work was not translated into Latin) and have also, questionably, seen this alternative as that advocated by the Islamic commentator Ibn Bajja (Avempace, d. 1138) whose views would have been known to Bradwardine through their citation by Averroes in his comments on book 4 of the *Physics*. Historians have also noted that this first rejected theory corresponds to that held by Galileo in his youth. On all of this see Ernest Moody, "Galileo and Avempace: The Dynamics of the Leaning Tower Experiment," *Journal of the History of Ideas* 12 (1951), 163–93, 375–422. Secondly, Bradwardine also considers and rejects the alternative that the proportion of velocities follows the inverse proportion of resistances when the forces are held constant, and the proportion of forces when the resistances are held constant. Considering the procedure found in *Physics* 7. 5 (see note 55), it is not surprising that he finds that this erroneous theory has support from Aristotle, although in the final analysis he regards his own theory as the proper interpretation of Aristotle.

61. Bradwardine, *Tractatus de proportionibus*, p. 112. Note that in *our* terms the Latin *proportio* is a ratio, but we have throughout rendered this as "proportion" to remain closer to the medieval mathematics involved.

62. Ibid. Anachronistically, Bradwardine's resolution of his velocity-force-resistance problem can be generalized symbolically as follows: If some force-resistance proportion F_1/R_1 gives rise to a velocity V_1, then increasing this proportion to F_2/R_2 will yield V_2, such that the following relation holds: $F_2/R_2 = (F_1/R_1)^{V_2/V_1}$. *Equally* anachronistically, "Bradwardine's law" has often been symbolically expressed as $V \propto \log F/R$.

63. This is always the case when one is dealing with equal proportions expressed as a series of continuous proportions (that is, where the equal proportions have a common term joining them, such as 8 is to 4 as 4 is to 2 as 2 is to 1, giving 8 to 1 as the cube or third power of 2 to 1, or 2 to 1 as the square root of 4 to 1, and so on).

64. In the first example, the operation of multiplying (= squaring) amounts to taking 9 as the continuous proportional to 1 and 3, while in the second example, the operation of dividing (= taking the square root) amounts to taking 4 as the mean proportional between 16 and 1. The operations of "increasing" and "decreasing" proportions in these two examples are, incidentally, sufficient for what one finds in Bradwardine, since he limits his own examples to "doubling" and "halving," which are equivalent to squaring and taking square roots.

65. Note should also be made of the fact that the above manner of increasing and decreasing proportions—particularly when unequal proportions are involved, as in the third example above—was especially prevalent and standard in medieval music theory (cf. Boethius, *De institutione musica*, ed. G. Friedlein [Leipzig, 1867]). In fact, the last example above amounts, in music theory, to "adding" an interval of a fourth (*diatesseron*, in medieval terms) to that of a fifth (*diapente*) and obtaining the octave (*diapason*).

66. Note that although medieval scholars frequently spoke of the addition, subtraction, or compounding of proportions in terms of specific examples similar to those given above, they also frequently did so in general, abstract terms, saying (for example) that the proportion of A to C is compounded of the proportions A to B and B to C. This is done, for instance, by Richard Swineshead when he deals with Bradwardine's "proportions of velocities" (see below).

67. This is clearly so for Bradwardine and his followers, who are rigorous in interpreting the increasing and decreasing of proportions to mean, in modern terms, the taking of powers and roots. Thus, for Bradwardine the modern meaning of $V \propto F/R$, which historians have used to interpret Aristotle, would have been impossible. He would have regarded $V \propto F/R$ as equivalent to his own function, since it speaks of the increase and decrease of a proportion. He could state something like the traditional interpretation of Aristotle only by splitting it into two parts (which is, after all, what Aristotle had done): that when force is constant, velocity varies inversely as the resistance, and when resistance is constant, velocity varies directly as the force. In this version of the Aristotelian relation, the variation of a proportion does not come into play. In fact, this is the only ver-

sion of the Aristotelian relation that Bradwardine considers among his erroneous theories.

68. Michael McVaugh, "Arnald of Villanova and Bradwardine's Law," *Isis* 58 (1966):56–64.

69. Clagett, *Science of Mechanics*, pp. 440–44, 465–503.

70. Ibid., p. 443.

71. Ibid., pp. 440–44; 465–503; also Maier, *Die Vorläufer Galileis*, pp. 95–104.

72. See John E. Murdoch, "From Social into Intellectual Factors: An Aspect of the Unitary Character of Late Medieval Learning," in *Cultural Context of Medieval Learning*, ed. Murdoch and Sylla, pp. 289–97.

73. The most easily available text of the whole of this work is *Liber calculationum* (Venice, 1520). For manuscripts and other editions, see the article cited in n. 75.

74. M. A. Hoskin and A. G. Molland, "Swineshead on Falling Bodies: An Example of Fourteenth-Century Physics," *British Journal for the History of Science* 3 (1966):150–82.

75. That is, in modern terms, $F_2/R = (F_2/F_1) \times (F_1/R)$. Cf. John E. Murdoch and Edith D. Sylla, "Swineshead, Richard," in *Dictionary of Scientific Biography*, 13:184–213. Although anachronistic, for the sake of simplicity the modern equivalent of multiplication has been used here (and throughout the remainder of this section) for the medieval "addition" of proportions, and the modern notation for a ratio has been used in symbolizing the medieval proportions themselves. In the fourteenth-century literature treated in this section, there is no notation for a proportion (although the quantities involved in it may be expressed by either numbers or literal variables), a proportion always being expressed simply in words: "the proportion of A to B" (*proportio* A *ad* B).

76. This is what Swineshead intends by his fourth rule: "Whenever a force increases . . . with respect to two equal or unequal, but constant, resistances, it will intend . . . motion with respect to each [of these resistances] with equal swiftness."

77. Reference in n. 75.

78. Ibid.

79. Nicole Oresme, *De proportionibus proportionum and Ad pauca respicientes*, ed. Edward Grant (Madison, Wis., 1966).

80. Ibid., pp. 208–9.

81. In our terms, the exponent x relating the two proportions in the function $(3/1)^x = 32/1$ would be irrational.

82. Oresme, *De proportionibus*, pp. 246–55.

83. Ibid., pp. 384–87.

84. Ibid., pp. 111–21.

85. Ibid., pp. 422–29.

86. See, for example, Jeanne-Marie Dureaux-Lapeyssonie, "L'oeuvre d'Antoine Ricart, médecin catalan du XVᵉ siècle. Contribution à l'étude des tentatives médiévales pour appliquer les mathématiques à la médecine," in *Médecine humaine et vétérinaire à la fin du moyen âge*, ed. Guy Beaujouan et al. (Geneva and Paris, 1966), pp. 171–364; and Murdoch, "From Social into Intellectual Factors," pp. 329–30.

87. Cf. John Murdoch, "The Medieval Language of Proportions: Elements of the Interaction with Greek Foundations and the Development of New Mathematical Techniques," in *Scientific Change*, ed. A. C. Crombie (London, 1963), pp. 261–65.

88. See Marshall Clagett, *Nicole Oresme and the Medieval Geometry of Qualities and Motions* (Madison, Wis., 1968); and *Science of Mechanics*, pp. 331–418.

89. Galen, *Microtegni*, in *Opera* (Venice, 1490), vol. 1, fol. 10v. Cf. Edith Sylla, "Medieval Concepts of the Latitude of Forms: The Oxford Calculators," *Archives d'histoire doctrinale et littéraire du moyen âge* 40 (1973):226–27, and Michael McVaugh, ed., *Arnaldi de Villanova Opera medica omnia II: Aphorismi de gradibus* (Granada and Barcelona, 1975), pp. 92–93.

90. Avicenna, *Canon* (Venice, 1507), fol. 2. Cf. McVaugh, *Arnaldi de Villanova Opera medica omnia II*, pp. 21–22; Sylla, "Medieval Concepts of the Latitude of Forms," pp. 227–28.

91. For the development of medieval pharmacy in general, see McVaugh, *Arnaldi de Villanova Opera medica omnia II*, pp. 3–143.

92. Sylla, "Medieval Concepts of the Latitude of Forms," pp. 228–29.

93. It is used in this way, for example, by Jean de Ripa. See Jean de Ripa, *Quaestio de gradu supremo*, in *Textes philosophiques du moyen âge*, ed. André Combes and Paul Vignaux, vol. 12 (Paris, 1964), pp. 143–222. For other fourteenth-century literature applying latitudes and degrees to a "scale of perfections," see John E. Murdoch, "*Mathesis in philosophiam scholasticam introducta*: The Rise and Development of the Application of Mathematics in Fourteenth Century Philosophy and Theology," in *Arts libéraux et philosophie au moyen âge* (Actes du quatrième congrès international de philosophie médiévale, Montréal, 27 August–2 September 1967) (Montreal, 1969), pp. 238–46.

94. See Sylla, "Medieval Concepts of the Latitude of Forms," pp. 233–38. See also the discussion of this theory above and the references in notes 45, 47, and 48.

95. Maier, *Zwei Grundprobleme*, pp. 54–55; Sylla, "Medieval Concepts of the Latitude of Forms," pp. 230–32.

96. See Sylla, "Medieval Quantifications of Qualities: The 'Merton School,'" *Archive for History of Exact Sciences* 8 (1971):12–15.

97. The best exposition of the ontological background of the Oxford Calculators' uses of latitudes and degrees is found in parts 2 and 3 of the *Sum of Logic and Natural Philosophy* by John Dumbleton (fl. ca. 1340). See Edith Sylla, "Medieval Concepts of the Latitude of Forms," pp. 251–64.

98. For typical uses of these terms, see Clagett, *Science of Mechanics*, pp. 210–12, 243–50, 382–401, 445–62.

99. They believed, in modern terminology, that the intensity of light produced was proportional to $1/r$ where r was the distance from the light source. In fact, intensities of light are proportional to $1/r^2$.

100. Printed at Venice in 1494 along with Heytesbury's *Sophisms*. On this work and the preceding, see Curtis Wilson, *William Heytesbury: Medieval Logic and the Rise of Mathematical Physics* (Madison, Wis., 1956).

There is also a shorter treatment of Heytesbury by Curtis Wilson in the *Dictionary of Scientific Biography*, 6:376–80.

101. See G. Wallerand, *Les oeuvres de Siger de Courtrai* (Louvain, 1913), pp. [20]–[33], and Martin Grabmann, *Die Sophismataliteratur des 12. und 13. Jahrhunderts mit Textausgabe eines Sophisma des Boetius von Dacien (Beiträge zur Geschichte der Philosophie und Theologie des Mittelalters*, vol. 36, pt. 1) (Munich, 1940).

102. Heytesbury, *Regulae solvendi sophismata* (Venice, 1494), fol. 37v. Bradwardine had earlier adopted the first half of this conclusion, measuring velocities of rotation by the distance traversed by the fastest moving point—thus rejecting the answer given by Gerard of Brussels (see above, note 53). See Bradwardine, *Tractatus de Proportionibus*, pp. 128–33.

103. Heytesbury, *Regulae*, fols. 38v–39v. Clagett has attached importance to the emergence of a concept of instantaneous velocity among these authors. See *Science of Mechanics*, pp. 212–15, 236–37, 243–44, 261.

104. Heytesbury, *Regulae*, fol. 39v.

105. Ibid., fol. 40v. See Clagett, *Science of Mechanics*, pp. 255–329.

106. Heytesbury, *Regulae*, fol. 40v.

107. Clagett, *Science of Mechanics*, pp. 255–69.

108. Heytesbury, *Regulae*, fols. 41v–44r.

109. Ibid., fol. 44v.

110. Ibid., fol. 45r.

111. Ibid., fol. 51r.

112. See Murdoch and Sylla, "Swineshead," in *Dictionary of Scientific Biography*, 13:187–95, 208.

113. In at least one place in his *Book of Calculations* Swineshead does reject the results he has elicited from a given theory in favor of what presumably obtains in nature; see above, pp. 227–28.

114. Clagett, *Oresme and the Medieval Geometry of Qualities and Motions*, pp. 168–71; also *Science of Mechanics*, pp. 348–49.

115. Clagett, *Oresme and the Medieval Geometry of Qualities and Motions*, pp. 208–11.

116. See Clagett, *Science of Mechanics*, pp. 342–43.

117. See Murdoch and Sylla, "Swineshead"; also Murdoch, "Philosophy and the Enterprise of Science in the Later Middle Ages," in *The Interaction between Science and Philosophy*, ed. Yehuda Elkana (Atlantic Highlands, N.J., 1974), pp. 67–68, and Clagett, *Oresme and the Medieval Geometry of Qualities and Motions*, pp. 412–35, 495–517.

118. See Murdoch, "From Social into Intellectual Factors," pp. 336–37, nn. 144, 152.

119. Such a work is included, for example, in Oxford Bodleian Library, MS Bodley 676. Cf. Clagett, *Science of Mechanics*, pp. 632–35. Similar glossaries of the terminology of latitudes and degrees, and so forth, were still included in the printed undergraduate handbooks at Oxford and Cambridge in the late fifteenth and early sixteenth centuries (*Libellus sophistarum ad usum Oxoniensium* [London, 1499–1500]; *Libellus sophistarum ad usum Cantabrigiensium* [London, 1524]; and so on).

120. These topics constitute the major portion, for example, of book 6 of the *Physics*.

121. For example, numerous separate questions and treatises were written in the fourteenth century on the problem of the composition of *continua*, the most notable being Thomas Bradwardine's *Tractatus de continuo*. On the late medieval literature dealing with this problem, see John Murdoch and E. Synan, "Two Questions on the Continuum: Walter Chatton (?), O.F.M. and Adam Wodeham, O.F.M.," *Franciscan Studies* 26 (1966):212–88 (esp. pp. 212–25).

122. *Physics* 6. 5, 236a10–15.

123. Ibid., 8, 262a12–263a3.

124. Several of these treatises will be discussed in what follows, but further references to the relevant medieval literature will be found in Wilson, *William Heytesbury*, pp. 29–56.

125. The text of Burley's work has been edited—unfortunately not without error—by Herman and Charlotte Shapiro in *Archiv für Geschichte der Philosophie* 47 (1965):157–73. For an analysis of part of this text see Wilson, *William Heytesbury*, pp. 32–35.

126. Aristotle was explicit (*Physics* 6. 5, 236a10–15) in denying the existence of a first moment in which something *begins* to change and in affirming the existence of a first moment in which its change is completed. Burley and other scholastics "sharpened" the latter affirmation into the ascription of a first instant in which the thing was *no longer* changing or in which the change *no longer* existed.

127. Burley, *De primo et ultimo instanti*, ed. Shapiro, pp. 166–72.

128. The same occurrence of a "continuous background" was operative in treatises on maxima and minima. Thus, in ascribing a maximum weight which Socrates could lift or a minimum which he could not, the "continuous background" was provided by the fact that the range of weights was considered to be continuous. Similarly, distance served as the "continuous background" when it was a question of ascribing maxima or minima for Socrates' power of vision.

129. Note should be made of the fact that a definite significance was attached to the order of the words in this last clause (specifically, to the position occupied by the term 'immediately'). Thus, for the fourteenth-century logician or—more to the present point—for the fourteenth-century natural philosopher treating issues metalinguistically through the application of logical techniques, the following two propositions were quite different: (1) 'Socrates' run will exist immediately after the present instant' and (2) 'Immediately after the present instant Socrates' run will exist'. The first proposition implies that there is a particular instant following the present instant between which and the present instant there is no other instant. Since there is no such instant "immediately in contact with" the present instant (for that would entail that time would be composed of indivisible instants), this first proposition is false. But the second proposition does not imply (by the relevant fourteenth-century rules) the existence of such an "immediate successor" instant, and, therefore, it is true. Indeed, it provides for what is in fact true: namely, that between the present instant and any other instant there is always another instant.

130. This metalinguistic approach is also that used by Ockham in the treatment of the nature of motion, discussed above. Yet more specifically relevant to the metalinguistic treatment of the problem of limits was the fact that the terms 'incipit' and 'desinit' had themselves long been considered as special logical terms which affected elements within the propositions in which they occurred. See William of Sherwood, *Treatise on Syncategorematic Words*, trans. Norman Kretzmann (Minneapolis, 1968), pp. 106–16; Peter of Spain, *Tractatus Syncategorematum and Selected Anonymous Treatises*, trans. Joseph P. Mullally (Milwaukee, 1964), pp. 58–65; Norman Kretzmann, "Incipit/Desinit," in *Motion and Time, Space and Matter*, ed. Peter Machamer and Robert Turnbull (Columbus, Ohio, 1976), pp. 101–36.

131. Thus, in his presently unedited treatise *De incipit et desinit*, Thomas Bradwardine distinguishes between *termini* or *dictiones* that stand, respectively, for either permanent or successive things and then goes on to establish such conclusions as: "Every affirmative proposition composed of terms standing for successive things in which the verb 'it begins' occurs must be expounded by the removal of the present and the positing of the future. For example, this proposition: 'motion begins to exist' must be expounded as follows: 'motion does not exist now and immediately after this will exist' . . ." (Biblioteca Vaticana, MS Vat. Lat. 3066, fols. 50v–52r). This is but another way of saying that the motion in question (Socrates' run, for example) does not have a first instant of being; the present "now" is "removed" and serves, therefore, as an *extrinsic* (beginning) limit to the motion.

132. Kilvington's *Sophismata* exists only in manuscript form, but is presently being edited and translated by Norman and Barbara Kretzmann. For the influence of Heytesbury's work in the later Middle Ages see Wilson, *William Heytesbury*, pp. 25–28.

133. Thus, in the chapter on 'incipit' and 'desinit' in his *Rules for Solving Sophisms*, Heytesbury sets forth principal rules that are too cumbersome and complicated to be applied easily to the usual simple cases of ascribing limits to a change; they are instead expressly formulated to be utilized in resolving the specific sophisms Heytesbury was dealing with. He even "glosses" these rules by pointing out how they fit with the kind of "comparisons" that his sophisms instantiate. See Wilson, *William Heytesbury*, pp. 48–50.

134. See Murdoch, "From Social into Intellectual Factors," pp. 289–97.

135. The example is taken from Gaetano da Thiene's (fl. ca. 1400) commentary on Heytesbury's *Rules for Solving Sophisms* (Venice, 1494), fol. 38rb.

136. See Murdoch, "From Social into Intellectual Factors," pp. 280–89.

137. Furthermore, it seems to be only in the fifteenth century that we find (for example, Munich, Staatsbibliothek, MS CLM 19850, fols. 7r–10v) the latitude of forms classified as a *scientia media*, falling between mathematics and natural philosophy.

138. See the references in n. 32.

139. Physics 3. 1, 200b15–20.

140. The major exception to this is the treatment of *reactio* in Swineshead, John Marliani, and others. On this, see Marshall Clagett, *Giovanni Marliani and Late Medieval Physics* (New York, 1941), pp. 34–58.

141. Research on this issue is still in progress, but for the present see: A. C. Crombie, "Sources of Galileo's Early Natural Philosophy," in *Reason, Experiment, and Mysticism in the Scientific Revolution*, ed. M. L. Righini Bonelli and William R. Shea (New York, 1975), pp. 157–75, 303–5; and William A. Wallace, "Galileo Galilei and the *Doctores Parisienses*," to appear in *New Perspectives on Galileo*, ed. R. E. Butts and J. C. Pitt, forthcoming.

142. For that lone instance, see William A. Wallace, O.P., "The Enigma of Domingo de Soto: *Uniformiter difformis* and Falling Bodies in Late Medieval Physics," *Isis* 59 (1968):384–401. One should keep in mind that a fair number of the medieval deliberations—indeed, some of the most ingenious among them—about "mean degree measure" were not concerned with uniformly difform distributions of a quality, but rather with "stair-step" distributions where various parts of the subject were uniformly qualified by various degrees. See, for example, the description of treatise 2 of Swineshead's *Book of Calculations* in Murdoch and Sylla, "Swineshead."

143. For theology see Murdoch, "From Social into Intellectual Factors," and the references in n. 93. For medicine, see, for example, J.-M. Dureaux-Lapeyssonie, "L'oeuvre d'Antoine Ricart" (above n. 86), and Michael McVaugh, *Arnaldi de Villanova* (above, n. 89), and the same author's "Quantified Medical Theory and Practice at Fourteenth-Century Montpellier," *Bulletin of the History of Medicine* 43 (1969):397–413, and "An Early Discussion of Medicinal Degrees at Montpellier by Henry of Winchester," *Bulletin of the History of Medicine* 49 (1975):57–71.

8 Edward Grant   **Cosmology**

It would be futile in the brief compass of this article
to attempt even a skeletal history of medieval cosmology. Despite the monumental ten-volume work
of Pierre Duhem (*Le système du monde: Histoire
des doctrines cosmologiques de Platon à Copernic*
[Paris, 1913–59]), much remains to be done before
a sound history of our topic can be presented with
any reasonable degree of confidence and thoroughness. But even our present, inadequate state of
knowledge encompasses much more than can be
properly incorporated here. Rather than attempt
more than is feasible and practicable, I shall confine myself to problems that were not only of major
concern in the Middle Ages, but which also convey
a reasonable picture of the cosmos as it was comprehended by medieval natural philosophers steeped
in the works of Aristotle. For although astronomy
established some of the boundary conditions of medieval cosmology, the latter was not fashioned and
shaped by technical astronomers but was the work
of natural philosophers, or physicists, in the Aristotelian sense of the term. Aristotelians, not Ptole-
maists, shaped the medieval world view that would
most nearly approximate to the concept of cosmology as "the theory of the universe as an ordered
whole, and of the general laws which govern it"
(*Oxford English Dictionary*).

With the translation of the works of Aristotle
from Arabic and Greek into Latin during the twelfth
and thirteenth centuries, the Latin West inherited
virtually all of the elements of a full-blown cosmology. Although Aristotle's *De caelo* (*On the
Heavens*) would emerge as the most fundamental

cosmological treatise in the Middle Ages, it was supplemented by significant sections in his *Metaphysics, Physics, Meteorology,* and *On Generation and Corruption.* If the works of Aristotle were seminal and dominant, they were augmented by commentaries on those works and independent treatises written by Arabic and Greek authors such as Avicenna, Averroes, Simplicius, Proclus, and others, whose works were also made available during the great age of translation,[1] as well as by works that had shaped medieval cosmological thought prior to the translations, especially Plato's *Timaeus,* in the partial fourth-century Latin translation and commentary of Chalcidius, and the hexaemeral treatises (commentaries on the six days of creation in Genesis).

Mention of commentaries on Genesis should serve to remind us that statements about the world and its structure in the Old and New Testaments remained ever present to medieval theologians and natural philosophers, who often enough were one and the same. In the Middle Ages, the structure of the world was never conceived solely in physical and metaphysical terms, but had to be made compatible with a variety of theological concepts which, in the end, transformed the Aristotelian cosmos into a Christian universe. Cosmological discussions in a theological context are found in many of the commentaries on the *Sentences* of Peter Lombard, a twelfth-century theological treatise on which theological students at the medieval universities were required to write commentaries. Since this requirement remained in effect for some three centuries, the hundreds of extant commentaries on the *Sentences* (especially the portion on the second book, which was concerned with creation and the nature of the world) are a significant source for a proper understanding of medieval cosmology. The impact of theology on cosmology is further evidenced by the 219 articles condemned in 1277 by the bishop of Paris, Etienne Tempier. A significant number of these condemned propositions concerned the eternity of the world, celestial movers, the possibility of a plurality of worlds, the impact of the heavens on terrestrial events, and other cosmological problems.[2] Since it is essential that we describe the structure and operation of the world as it was understood in the Middle Ages by scholastic natural philosophers, and not restrict ourselves to problems of interest to modern astronomers and cosmologists,[3] theological considerations must be included wherever relevant.

While the medieval cosmos is sometimes characterized by the metaphor of a Gothic cathedral whose separate elements flow together rationally and logically to form a splendid, compact, intelligible whole soaring upward toward the Deity himself, it was rarely, if at all, presented or described in a logically cohesive and systematic manner. To my knowledge, no genuine cosmological synthesis was developed in the late Middle Ages. The clue for comprehending this strange state of affairs may perhaps lie in the nature and form of medieval scholastic literature. In natural philosophy, which embraced cosmology, the most common mode of expression was by means of a commentary on a traditionally recognized authoritative text. This often took the form of a series of *questiones*, or specific problems, which followed the order of the commented text and developed from it; or it might take the form of a straightforward commentary in which the commented text was discussed systematically section by section. In the *questiones*, which furnished most of the interesting cosmological discussion, each *questio* was subjected to a reasonably thorough analysis by means of a series of pros and cons, followed by the commentator's solution. What emerged from all this was a series of distinct and often intensively considered problems that remained isolated from, and independent of, other related *questiones*, to which allusions and references were minimal. Since this approach was characteristic of the whole range of literature relevant to cosmology, it is important to recognize that the virtues of such an approach, which emphasized thorough and systematic analysis of distinct problems in a prescribed order, were offset by the absence of any cohesive integration of the many separately derived conclusions as well as a failure to detect inconsistencies between questions treated within the same treatise. Above all, however, the commentary form of literature wedded to authoritative texts may well have discouraged, or rendered superfluous, the conscious and explicit formulation of a coherent and consistent cosmology within which the disparate elements scattered throughout the *questiones* could be brought together, evaluated, and assessed as part of a larger whole. If an integrated world picture existed in the minds of scholastic natural philosophers, they failed to describe it in the treatises that have been left to us. By default, then, the burden of depicting the overall structure and operation of the medieval cosmos must fall to the modern scholar.

The ideas and conceptions essential to a comprehension of the medieval physical world lie at hand in the types of treatises already mentioned. They are embedded in commentaries and *questiones*, primarily those on the physical works of Aristotle (especially the *De caelo*), on the *Treatise on the Sphere* of John of Sacrobosco,[4] on the *Sentences* of Peter Lombard, and on the six days of creation in Genesis.[5] In such treatises, scholastic commentators formulated a wide and unusual range of questions. They inquired, for example, whether the world has a beginning or end; whether it was created from nothing; whether it is perfect; whether it is located in an immobile place; whether anything lies beyond it; and whether there is more than one world. Since it was unanimously agreed that the planets and stars were carried round on physical spheres, numerous questions were posed about the nature and motion of those spheres. How many are there? Does God move the *primum mobile*, or first movable sphere, directly and actively as an efficient cause, or only as a final or ultimate cause? Are all the heavens moved by one mover or several, and if by several, what kinds are they? Are the celestial movers conjoined to their orbs or distinct from them? Are the spheres moved by intelligences, angels, forms or souls, or by some principle inherent in their very matter? Do celestial movers experience exhaustion or fatigue? Does the celestial region form a continuous whole, or are the spheres contiguous and distinct? Are the orbs all of the same specific nature or of different natures? Are the orbs concentric with the earth as common center, or is it necessary to assume eccentric and epicyclic orbs? The nature of celestial matter was widely discussed. Was it like terrestrial matter in possessing an inherent substantial form and inherent qualities such as hot, cold, moist, and dry? Does it undergo changes involving generation and corruption, increase and diminution? On the assumption that celestial motion affected terrestrial, or sublunar, change, how was this achieved? Questions were also raised about the structure of the radically different region below the moon, as, for example, whether or not the four elements were continually proportional. And, finally, questions were proposed about the earth and its relation to the cosmos, among them: Is the earth spherical? Does it always rest in the center of the world? Is it as a point with respect to the heavens?

Medieval cosmology must be constructed from responses to these and other similar questions. Since these questions emerged in large measure from intense and uninterrupted study of Aristotelian treatises beginning in the thirteenth century, the cosmology described below represents a view of the world developed during the thirteenth to fifteenth centuries, the climactic centuries of medieval thought.

On the Eternity and Uniqueness of the World

With the reception of the works of Aristotle, a fundamental problem was immediately posed in the Christian West: Was the world eternal, without beginning or end? In Aristotle's opinion, the structure and processes of the world have always been, and will always remain, what they are at present. Left unchallenged, Aristotle's eternal world would have undermined, if not destroyed, one of the central themes of Christianity. The creation of the world by a God of infinite goodness and the end of that world in preparation for a final day of judgment would have to be repudiated. More than any other issue, perhaps, the doctrine of the world's eternity cast suspicion on the natural works of Aristotle through much of the thirteenth century. The doctrine itself was condemned in 1270 and again, in a variety of guises, in the massive condemnation of 1277.[6] Preoccupation with the issue is well illustrated by the fact that Thomas Aquinas concerned himself with it in at least seven different treatises.[7] Three major positions were developed during the thirteenth century.[8] At one extreme, the Averroists, following Aristotle, argued that the eternity of the world was rationally demonstrable; while at the other, Augustinians, for the most part Franciscans, like St. Bonaventure, insisted that the temporal origin of the world was capable of rational proof. In the middle was Thomas Aquinas, who, in agreement with Moses Maimonides, denied that reason could demonstrate the matter either way. Whatever the intrinsic merits of the numerous proofs and arguments and despite the intensity, acrimony, and subtlety of the debate, all scholastics were compelled to believe that as a matter of faith and revelation, God had indeed created the world in time and out of nothing (*ex nihilo*).

If the eternity of the world posed serious problems for the Church, its uniqueness did not. Here, at least, it seemed that Aristotle and Christianity were in agreement: there is only one world. For a variety of theological reasons, Christianity demanded a single world as the stage on which all the essential acts of its high drama could be played out. By contrast, Aristotle, appealing to natural principles and rational argument, concluded that it was demonstrably impossible that there be more than one world, since, among other reasons, all the matter in existence forms our single, finite world with no residue left to generate another.

Beneath the apparent harmony, however, lurked conflict and dispute; for although Aristotle's conclusion might be applauded, his derivation of it was found offensive. To argue, as Aristotelians often did, that the existence of more than one world was impossible implied that even if He wished, God could not create other worlds. A restriction was thus imposed on God's infinite and absolute power. Here, as in other areas of cosmic operation, the Aristotelian penchant for demonstrative conclusions clashed with the Christian sense of God's absolute power to do as He pleased in accordance with His inscrutable will. All through the 1260s and 70s, these conflicting tendencies were much in evidence. The triumph of theology over natural philosophy came with the condemnation of 1277, which had as one of its primary objectives the exaltation of God's absolute and unpredictable power and the damnation of attempts by natural philosophers to set limits to that power with appeals to rational demonstration.[9] It was with all this in mind that article 34 of the condemnation of 1277 declared it an excommunicable offense to hold "that the first cause [that is, God] could not make several worlds." As a consequence, it became respectable during the fourteenth century, as it had not been before 1277, to contemplate the possibility of simultaneously existing worlds beyond our own. Since no logical contradiction was involved in assuming that by His absolute power, God could create other worlds, the focus of attention centered on the effect such worlds might have on the validity of Aristotelian physical principles. For example, although Aristotle had argued that vacua of any kind were impossible, the existence of such spaces between these worlds now gained a degree of plausibility. Nicole Oresme, as we shall see, found the suggestion acceptable and was also prepared to argue that each world would function exactly as ours does, obeying the same physical laws. Each universe would have its own "up" and "down," "center" and "circumference."[10] Of great interest is Oresme's willingness to uphold, if only hypothetically, the existence of many co-equal world centers, a move that implied rejection of the kind of unique world center on which Aristotle based his physics and cosmology. Although Oresme sought to render the concept of a plurality of simultaneously existing worlds plausible and physically intelligible, his primary objective was apparently to undermine Aristotle's denial of the *possibility* of more than one world. For, in the end, Oresme acquiesced in the unanimous medieval belief that, in fact, "there has never been nor will there be more than one corporeal world."[11]

Infinite Void Space beyond the World

The same unanimity in favor of a single, unique world was also accorded to Aristotle's judgment that the world was finite in extent. Not all scholastics, however, were prepared to adopt Aristotle's conclusion denying the existence of bodies, places, time, or void beyond the farthest extremity of the world, namely, beyond the sphere of the fixed stars.[12] Following a tradition that derived from the Greek Stoics and which was transmitted to the Middle Ages in William of Moerbeke's 1271 Latin translation of Simplicius's sixth-century commentary on Aristotle's *De caelo*, some were led to inquire about the hypothetical status of a hand that was imagined thrust beyond the outer-most extremity of the world. And like the Stoics, a few scholars in the fourteenth century, such as Thomas Bradwardine, Nicole Oresme, and Johannes de Ripa, inferred that the hand would be plunged into a real infinite void space beyond the world.[13] But where the Stoics assumed a three-dimensional void space with no properties of its own, medieval proponents assumed that God himself was omnipresent in this infinite emptiness, which they often characterized as His "immensity" (*immensitas*). Although the origin of this concept may never be precisely known, it is probable that theological considerations were crucial. With the emphasis on God's absolute power remaining strong throughout the fourteenth century as a significant legacy of the condemnation of 1277, it must have seemed incongruous to some that the presence of an infinitely powerful God should extend no further than the finite world that He chose to create. But if His presence extended beyond the world, it seemed plausible to assume that He was infinitely extended, since no good reasons could be offered for supposing that His presence ceased at some finite distance beyond. In large measure, perplexing questions about the interrelationship between God's presence and the infinite void which He somehow filled were either ignored or left vague and obscure.[14] Extracosmic void did, however, become the locale for some interesting "thought experiments" concerning motion and place. Oresme argued that if God moved the entire world rectilinearly through an infinite void, the possibility of which had to be conceded by article 49 of the condemnation of 1277,[15] an absolute motion would result, since nothing existed outside the world to which its movement could be related; Buridan inquired whether a body located in this infinite void could be said to be in a place; and others asked about the hypothetical determination of distance measurements in such a void space. The medieval concept of a finite, spherical universe surrounded by an infinite, God-filled void space continued its influence into the seventeenth and eighteenth centuries, when it was discussed by such luminaries as Henry More, Isaac Newton, Otto von Guericke, and Samuel Clarke.

Is the World in a Place?

As for the finite world itself, to which we must now turn our attention, an unavoidable and peculiar problem emerged concerning the location of its outermost sphere. It was regularly asked whether the last sphere was in a place. While this question was occasionally equivalent to inquiring whether the whole world was in a place, queries about the place of the last sphere usually elicited different responses from queries about the place of the world. In the context of medieval Aristotelian physics and cosmology, the place of the world, conceived as a body, was not understood in terms of a separate space which penetrated the world and in which the latter was immersed. For, as Thomas Aquinas explained, such a space would either penetrate the world to a degree that made the two indistinguishable, in which event space was superfluous; or this separate space possessed its own three dimensions, from which the absurd consequence followed that two distinct three dimensional entities—space itself and the material world —could simultaneously occupy the same place.[16] At best, the world might be assigned a place only because its parts were in their respective places.

But the question posed most frequently by scholastics concerned the place of the last, or outermost sphere, of the world. In Aristotle's physics, the place of a body was defined as the innermost, motionless surface of the containing body in direct contact with the contained body.[17] From this definition, and a reasoned conviction that no material body could exist beyond the world to serve as its container, Aristotle concluded that the last sphere, or sphere of the fixed stars, could not itself be in a place, except, perhaps, in a secondary sense in which one part of the outermost celestial orb contains another part.[18] Paradoxically, then, although every body and part of the world had a place, so that the world could be conceived to be in a place by virtue of its parts, the outermost sphere, which contained the world, was not itself in a place. Denial of a place to the last sphere was a consequence forced upon Aristotle in order to avoid an infinite regress of material places; for if the outermost sphere of the world were contained by another sphere, the latter, in turn, would require a containing sphere or surface, and so on, *ad infinitum*, a process that would inevitably lead to the assumption of an infinite universe. Whatever its merits within the context of Aristotelian physics and cosmology, rejection of a place for the last sphere left many of Aristotle's followers uneasy, as evidenced by the numerous discussions of the problem from late antiquity through the Middle Ages. In the latter period, many sought to confer some sense of place on the outermost sphere of the universe.[19]

In the late Middle Ages, an interpretation by Averroes was widely accepted. Assuming that every body in motion presupposed the existence of a body at rest, Averroes declared "that the [outermost] heaven is in a place *per accidens*, namely because a certain part of it, namely its center, is in a place."[20] Since the motion of the heavens presupposed a body at rest, Averroes linked the outermost sphere, which is continually in motion, with the earth, which is perpetually at rest in the center of the universe, and, in effect, made a single entity of these ordinarily disparate and incommensurable bodies. He then concluded that the last sphere of the world is in a place *per accidens* because its center is in a place *per se*. The whole world, however, is not in a place by virtue of that same center but, rather, because all its parts are in a place.

Thus did Averroes and his Latin medieval followers invert Aristotle's conception of the place of a body from something determined by an external container to something determined from within. Indeed, despite the fact that each celestial orb was immediately circumscribed by another contiguous orb, a situation which fulfilled Aristotle's primary definition of place, Roger Bacon described these places as accidental. In his opinion, since the celestial orbs are already perfect, they require no real *per se* places which contain, limit, and perfect them. Aristotle's definition of place was thus restricted by Bacon to the subcelestial region.[21]

Numerous criticisms were formulated against Averroes' strange doctrine. As Aquinas observed, it subverted Aristotle's definition of a place *per accidens*, which was intended to apply only to instances involving, for example, the accidental change of place of a quality, say, whiteness, when the body in which it inhered actually changes place. Aquinas also objected to the assumption of the earth as the center and part of the outermost sphere, when, in fact, the latter lay wholly outside the former.[22] Moreover, as Albert of Saxony declared, the earth need not be immobile to serve as a reference point for celestial motion. It could function in this capacity as a body rotating in the same place, provided that its rotation was not in the same sense as, and with the same angular speed of, the heavens.[23] Indeed, the earth itself could be replaced as a fixed point of reference by a geometric point. But even the alleged necessity of a fixed point of reference for the determination of a motion was challenged by Albert of Saxony and Nicole Oresme, the latter, as we saw above, arguing that an absolute motion could be conceived without reference to anything at rest if God chose to move the entire world with a rectilinear motion.

The Place of the World Is an Immobile Sphere

Averroes' opinion represented only one of a number of competing views, most of which were transmitted to the Latin West by Averroes himself and all of which found at least an occasional supporter. Two of the major interpretations were drawn from the late Greek commentators Alexander of Aphrodisias (fl. 2d–3d c. A.D.), who firmly denied a place to the last sphere and the world (Roger Bacon, Johannes Canonicus, John Buridan, Albert of Saxony, and Marsilius of Inghen supported this opinion) and Themistius (fl. 4th c. A.D.), who, although denying a place *per se* to the last sphere, insisted that it was in place by virtue of its parts, which consisted of all the inferior celestial orbs (Aquinas was the most prominent advocate of this interpretation). It was also through Averroes that Avempace's rather unusual opinion became known, according to which the last sphere was assigned a place *per se* in the form of the convex surface of the next lower sphere, Saturn.

An opinion which Averroes did not pass on to the West sought to locate the place of the outermost celestial sphere in an all-encompassing, immobile sphere that met Aristotle's basic definition of place as a containing body whose innermost surface was both immobile and in contact with the contained body.[24] By identifying the immobile, containing orb with the theologically derived Empyrean sphere, whose characteristics will be described below, the popularity of this opinion was guaranteed.

Cosmologically, an outermost immobile sphere had a certain appeal. Campanus of Novara, in his rather widely read *Theory of the Planets* declared that "The empyrean's convex surface has nothing beyond it. For it is the highest of all bodily things, and the farthest removed from the common center of the spheres, namely the center of the earth; hence it is the common and most general 'place' for all things which have position, in that it contains everything and is itself contained by nothing."[25] Although the need for an immobile container for the last of the mobile spheres was the most popular argument, an outermost immobile sphere was defended on other grounds. The poles of the world could be fixed and immobile only in an immobile sphere; the absolute directions which Aristotle postulated as inherent in the very structure of the universe could be fixed only with respect to an immobile sphere; and observed differences on earth could be explained by assuming different powers in the various parts of an immobile sphere (for example, one part would dominate plants, another animals, another waters, and so on).[26]

Opposition to an outermost immobile sphere was largely based on Aristotelian principles. By their very natures, celestial spheres ought to be capable of motion. On the assumption that motion is nobler than, and prior to, rest, an immobile outer sphere would be less noble than the celestial spheres below it, which seemed absurd. But even if it rested, God could move it by His absolute power, so that, as Albert of Saxony explained, "it seems absurd that a power to act and a body whose perfection is motion should be prevented from doing so for an eternity." Thus, while most theologians accepted an immobile last sphere, others, such as John Duns Scotus, Buridan, and Albert of Saxony, were prepared to follow Aristotle and conclude that without the existence of matter beyond the world to serve as a container, the last sphere, and, therefore, the world itself, could not be in a place. Thus, an outermost sphere could be abandoned as contrary to the essentially mobile nature of all celestial orbs.[27]

On the Number of Heavens and Orbs

It was customary in the Middle Ages to distinguish a variety of heavens in the world. The term *caelum*,[28] or heaven, was very broadly conceived and could designate a cluster of particular spheres, a single sphere, and even part of a sphere. In a typical division, Thomas Aquinas distinguished three heavens in the celestial region:[29] the Empyrean; the crystalline, or aqueous; and the sidereal, which consisted of the traditional seven planetary spheres and sphere of the fixed stars. Thus, Thomas conceived of a total of ten separate celestial spheres distributed over three heavens. Thomas also mentions that the term *caelum* is sometimes used metaphorically, as when the Heaven of the Holy Trinity (*caelum sanctae Trinitatis*) "is sometimes referred to as heaven because of its spiritual sublimity and light. It is of this heaven that the text is interpreted where the devil said, 'I will ascend into heaven,' i.e., to equality with God."[30] But this fourth heaven, or eleventh sphere, was not thought of as a corporeal heaven.[31]

The Empyrean, or outermost, heaven had its roots in a distinction which the Church Fathers and early medieval commentators made between "the heaven" (*caelum*) created on the first day and the "firmament" (*firmamentum*) brought forth on the second day and made visible on the fourth. By the twelfth century, and perhaps much earlier, the firmament was designated as the visible heaven, with the heaven of the first day reserved as the abode of the angels and the place where they were created. This distinction appears, perhaps for the first time, in the *Glossa Ordinaria*, probably composed by Anselm of Laon (d. 1117), who incorporated material traceable to Walafrid Strabo (d. 849). Anselm (or possibly Walafrid) called the heaven of the first day "empyrean" (*empyreum*), a term drawn from Martianus Capella's *Marriage of Mercury and Philology*.[32] According to Anselm, *empyreum* is to be understood as "fiery or intellectual, which is so-

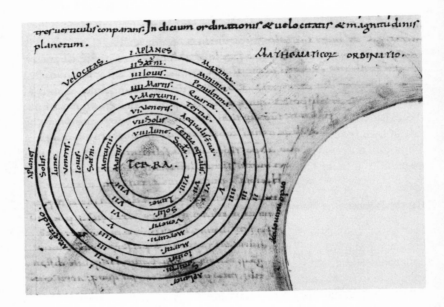

Fig. 14   The geocentric system of the universe in simplified
form. Paris, Bibliothèque Nationale, MS Lat. 6280, fol.
20r (12th c.).

called not by virtue of its burning [or heat] (*ardor*) but from its brilliance (*splendor*), since it is immediately filled with angels."[33] When
the description in the *Glossa Ordinaria* was quoted verbatim in Peter
Lombard's widely used *Sentences*,[34] the Empyrean sphere was assured
a permanent place in medieval cosmology.[35] Aquinas declared that its
existence and properties were arrived at by the intellect alone.[36] In
this manner, he concluded that the Empyrean sphere was "full of
light" (*lucidus*), immobile, and a worthy abode for the angels, who
were created there. Although, in his earlier works, Aquinas could not
conceive how the Empyrean sphere might influence the inferior celestial spheres, he concluded in later treatises that it conferred stability
on the first mobile heaven and permanence on all inferior spheres.

The heaven immediately below the immobile Empyrean sphere was
the starless crystalline, or aqueous, sphere. Centuries of commentary
on Genesis 1:7 concerning the meaning of the passage on "the waters
above the firmament" (that is, above the sphere of fixed stars), which
were separated from the waters below the firmament,[37] had produced
an interpretation of those uppermost "waters" as a solid, transparent
crystalline sphere. The commonplace opinion, as expressed by Aqui-

nas, was that "the waters above the heavens are not necessarily fluid, but rather are crystallized around the heavens in a state similar to ice."[38]

With the introduction of the works of Aristotle, the nature of the waters above the starry firmament was much discussed because in

Fig. 15   The eclipse of the sun (lower left) and of the moon (upper right). Paris, Bibliothèque Nationale, MS Lat. 6556, fol. 5v (13th c.).

Aristotle's cosmology water was one of four elements confined below the celestial region.[39] A customary resolution of the problem was to characterize the waters above the firmament as possessed of a nature wholly different from that of the elemental waters in the subcelestial zone.

Although the Empyrean and crystalline heavens were always taken to lie above the starry firmament, in time the number of orbs held to exist above the sphere of fixed stars became, to a considerable extent, a function of the number of astronomical motions that one assigned to the sphere of fixed stars. Throughout the late Middle Ages as many as three motions were assigned to it: (1) a daily motion from east to west; (2) a motion from west to east of one degree in 100 years as a consequence of the precession of the equinoxes[40] (at this rate, a complete revolution of the sphere of stars would be completed in 36,000 years); and (3) a progressive and regressive motion of the stars known as "trepidation," a theory proposed by the ninth-century pagan Arab astronomer Thabit ibn Qurra.[41]

Some scholastics, like Albert of Saxony, Pierre d'Ailly, and probably Roger Bacon,[42] assigned all three motions to the fixed stars, but, on the common assumption that a single orb could not be moved with more than one motion simultaneously, invoked a separate sphere for each motion. Albert of Saxony, for example, assigned the daily motion to a tenth orb, the *primum mobile*; motion of trepidation to a ninth sphere; and motion of precession to the eighth, or starry, sphere.[43] Since Albert rejected an outermost immobile, or Empyrean, sphere, it appears that he assumed a total of ten mobile celestial spheres, although which of the last two spheres would have functioned as the crystalline sphere is left unmentioned. Those who accepted the Empyrean sphere, like Pierre d'Ailly, were committed to an additional eleventh celestial sphere.

But there were those who assigned the three basic motions of the fixed stars to only two spheres, an eighth and ninth,[44] while others, ignoring trepidation, assigned the daily motion to a ninth sphere and precession to the eighth.[45] And occasionally, someone like Nicole Oresme confined himself to only eight mobile spheres by ignoring trepidation and assigning the daily and precessional motions to the eighth sphere.[46] In each case, the total number of celestial spheres would be determined ultimately by acceptance or rejection of an immobile Empyrean sphere. Thus, the seven planetary orbs, when taken along with the sphere of fixed stars, constituted for Aquinas, as we saw, the sidereal heaven. When added to the spheres already mentioned, they helped make up the full complement of eight to eleven heavenly orbs.

Throughout the late Middle Ages, the true order of the planetary spheres was assumed to be that assigned by Ptolemy, rather than that of Plato and Macrobius. In ascending order, Ptolemy placed the moon, Mercury, Venus, sun, Mars, Jupiter, and Saturn.[47] Unable to detect any planetary parallax, and, therefore, lacking criteria for arranging the planets by linear distances, Ptolemy used the sun as a natural divider "between those planets which can be any angular distance from the sun and those which cannot but which always move near it."[48] Themon Judaeus was expressing a widely held and traditional sentiment when he described the sun "as the wise king in the middle of his kingdom, and as the heart in the middle of the body and the fourth in order."[49] The sun's centrality, coupled with its vital role in the generation and perpetuation of life, was undoubtedly instrumental in evoking Nicole Oresme's characterization of it as "the most noble body in the heavens," even "more perfect than Saturn or Jupiter or Mars, which are all higher up . . ."[50] In thus glorifying the sun, Oresme avoided the path recommended by Averroes, who would have made greater nobility a direct function of the ascending order of the planets.[51] The sun's overwhelming significance, its centrality in the order of the planets, and the lavish praise heaped upon it in a variety of metaphors were perhaps helpful in paving the way for a new solar role in the Copernican heliocentric cosmology of the sixteenth century.

The concept of "heaven," or *caelum*, was not confined to the celestial region, but was also applied to the subcelestial realm below the concave surface of the moon. An alleged similarity in "actual or potential sublimity or luminosity" between the airy and fiery regions of the upper atmosphere just below the moon and the celestial region itself was probably the basis for distinguishing additional heavens.[52] Drawing on Rabanus Maurus's ninth-century treatise, *De universo*, Aquinas and others assigned four heavens to the sublunar realm by subdividing each of the two elemental spheres of fire and air.[53] The upper region of fire was designated the "fiery heaven" (*caelum igneum*), and the lower region the "Olympian heaven" (*caelum Olympium*), presumably because it was approximately the height of Mount Olympus. In the sphere of air, the upper part was called the "ethereal heaven" (*caelum aethereum*) and the lower part the "airy heaven" (*caelum aereum*). By adding these four "heavens" to the three distinguished in the celestial realm we arrive at a total of seven heavens distributed over some ten to thirteen spheres extending upward from the lower airy sphere to the immobile Empyrean orb embracing the entire world.[54]

The Dilemma of Medieval Cosmology (Aristotle's Concentric Spheres versus Ptolemy's Eccentrics and Epicycles) and Its Resolution

Up to this point we have spoken as if each planet had only a single sphere to which it was attached and by which it was carried around, rather than moving freely as a bird flies through air or as a fish swims through water.[55] But it was known to all that a single sphere could not account for the motion of any planet. While Aristotle and Ptolemy had assumed that each planet was attached to a single sphere, both employed a plurality of spheres to account for the resultant motion of each planet. Following statements and leads provided by the two great Greek authorities, it was generally believed that these many celestial spheres were contiguously nested one within the other from the lowest sphere of the moon to the outermost mobile sphere. Although the assumption of contiguity posed difficult problems about the possibility of vacua and waste space, the existence of which was routinely denied, the alternative, namely, a conception of the celestial region as one continuous body from the concavity of the moon's lowest orb to the convex surface of the outermost sphere, was even more unacceptable, since a single, continuous celestial body could not possibly account for the diversity of celestial motions.[56]

If general agreement could be found between Aristotle and Ptolemy on these, and other, issues, a fundamental difference was detected soon after the appearance of their major works in the twelfth century. With respect to the configuration and arrangement of the numerous celestial spheres, Aristotle had assumed a series of concentric orbs, each of which moved with a natural, uniform, circular motion in a fixed direction. Although some spheres moved in opposite directions and on different poles, they all shared a common center, which was the center of the universe, taken as coincident with the earth's center. In the most significant exposition and defense of this arrangement, Averroes insisted that every celestial orb required a physical center, the earth, for its motion.

The rival position proposed by Claudius Ptolemy in the *Almagest*, written in the 2d century A.D., represented a strong reaction to a grave deficiency in the Aristotelian system, the concentric spheres of which could not account for variations in the observed distances of the planets from the earth. To remedy this fatal flaw, Ptolemy, following an already well-established tradition, employed eccentric and epicyclic circles to represent planetary motions. In its simplest form, a planet's motion might be represented by an eccentric circle possessing a center other than the earth's center; or, if the earth's center was retained, an epicycle might be added to the circumference of the deferent circle; or, finally, some combination of eccentric and epicyclic circles could be employed. In all of these situations, however, the planet's distance from the earth varied—in stark contrast to the Aristotelian system. In a later treatise called *Hypotheses of the Planets*,[57] Ptolemy sought to formulate a physical model of the world that would reflect the geometry of eccentrics and epicycles formulated in the *Almagest*. Although the *Hypotheses* was apparently unknown in the Latin Middle Ages, its substantive content seems to have been incorporated into an Arabic astronomical treatise by Ibn al-Haytham (Alhazen) and translated into Latin under the title *Liber de mundo et coelo*.[58]

Late medieval scholastics were well aware that the Ptolemaic eccentric-epicycle astronomical representation was capable of saving the astronomical phenomena, while the Aristotelian concentric system was not. An attempt by al-Bitruji (Alpetragius), in his *On Celestial Motions*, to devise a scheme of concentric spheres that would account for variations in planetary distance and remedy the acknowledged deficiencies in the Aristotelian system was generally regarded as a failure.[59] But despite the superiority of Ptolemy over Aristotle on purely astronomical grounds, medieval natural philosophers were faced with a serious dilemma. Either they must reject the earth's centrality with its surrounding concentric spheres, and, thus, abandon a vital aspect of Aristotelian physics and cosmology, which might prove fatal to the whole, or accept a cosmology that was astronomically untenable. While a few in the thirteenth century opted for Aristotle unqualifiedly,[60] and yet others hesitated between the two,[61] a compromise solution, destined to gain wide acceptance by the fourteenth century, was drawn ultimately from Ibn al-Haytham's *Liber de mundo et coelo*.

The solution already appears, perhaps for the first time, in Roger Bacon's *Opus tertium*,[62] although it is by no means certain that Bacon himself accepted it.[63] With the occasional exception of Mercury, three separate orbs were assumed necessary to represent the motion of a planet.[64] In figure 16, which represents the moon, a planet that has both an epicycle and eccentric (unlike the sun which was only assigned an eccentric), let T be the center of the earth and world and also the center of the lunar orb. The entire sphere of the moon lies between the convex circumference $ADBC$ and the concave circum-

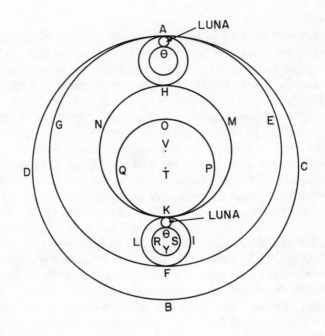

Fig. 16

ference *OQKP*, which are both concentric to *T*.[65] Between these two circumferences, three orbs are distinguished by assigning another center, *V*, toward the aux, or apogee, of the moon. Around *V* as center are two circumferences, *AGFE* and *HNKM*, which represent the surfaces that enclose the lunar eccentric deferent. Surrounding the eccentric deferent is the outermost orb, between surfaces *ABDC* and *AGFE*, and immediately surrounded by the eccentric deferent is the innermost orb, between *HNKM* and *OQKP*. Between the surfaces of the middle, or eccentric, orb is a concavity containing a spherical epicycle. The latter may be conceived in two ways: either as a solid globe, which Bacon calls a "convex sphere" (*spericum convexum*) because it lacks a concave surface; or as a ring with two surfaces, one convex (*KLFI*) and the other concave (*RYSθ*), where the central core belongs exclusively to the eccentric orb and forms no part of the epicyclic sphere itself. The solid spherical planet, or moon in the figure, which has only a convex surface, is located within a concavity of the epicyclic orb. The eccentric sphere is assumed to move around its center, *V*, carrying the epicycle with it; the epicyclic sphere, in turn, has its own simultaneous motion and carries the planet with it.

On the assumption that the outermost convex and innermost concave surfaces of every total planetary sphere were concentric with respect to the center of the earth, the foundation of the Aristotelian system was saved. With this achieved, the region between the convex and concave bounding surfaces of each planetary sphere could be subdivided into three, or more, orbs in accordance with the demands of Ptolemaic astronomy. That the middle orb, or eccentric deferent, did not have the earth as its physical center could now be ignored, since the sphere as a whole had the earth as center. In this qualitative manner, the most essential demands of Aristotelian cosmology and physics, on the one hand, and Ptolemaic astronomy, on the other, were met and a compromise effected. For now the variation in planetary distances, represented by eccentrics and epicycles, was incorporated into a system of concentric planetary spheres. Without technical competence in astronomy, scholastic natural philosophers could provide few details about the relationships and functions of the separate orbs. Such matters were better left to astronomers. It was enough to know that the planetary spheres were concentric, even though in other respects the system adopted differed radically from Aristotle's description of it. With Aristotle saved, however tenuously, it was possible to get on with the real business of cosmology, which had relatively little to do with technical astronomy.

It is natural to inquire, at this point, whether the eccentric and epicyclic orbs were considered physically real or merely convenient geometric devices for saving the phenomena? Moreover, was it possible to know the precise arrangement of the numerous celestial spheres? It is no exaggeration to claim that the existence of material celestial orbs was universally accepted in the late Middle Ages, whatever one's commitment to this or that astronomical or cosmological system. For those who adopted the compromise solution, it seemed reasonable to assume, as they did, that their system was the most plausible representation of the physical arrangement of the celestial spheres. Otherwise there would have been little reason to devise a compromise that incorporated the essential features of Ptolemaic astronomy and Aristotelian cosmology. Frequent denials that vacua or waste spaces would occur are sufficient testimony that the compromise was conceived as a physical system and not solely as a geometrical representation or device to save the phenomena. There is, however, no warrant to infer from this that proponents of the compromise believed they had derived the true and ultimate arrangement of the celestial orbs. Scholastic natural philosophers were too sophisticated for so naive a conclusion. They were fully aware that other combinations of spheres might be devised to save the same astronomical phenomena, combinations that would also conform to the requirements of cosmology;[66] or that new astronomical data might demand a revised arrangement of spheres.

Celestial Movers

Whatever the total number of mobile spheres or orbs in the compromise system—and this varied considerably—the cause of their motion was a problem of central importance. Despite the variety of instrumentalities and mechanisms invoked to explain celestial motions, it was always assumed in medieval cosmology that the ultimate source of all such motions was God, the Prime Mover. And like Aristotle's Unmoved Mover, God caused motion without in any way suffering motion or change. But how, it was frequently asked, did God move the first moving sphere, or *primum mobile*, whose most fundamental motion was associated with the daily movement in which the fixed stars and all the planets participated equally? Since the *primum mobile* was the only sphere which God moved directly, it was deemed vital to know whether He moved it as an active and efficient cause, as a passive and final cause, or as both. Most would have argued that God moved the first sphere in both capacities. His role as a passive mover and final cause was based on Aristotle's declaration that "the final cause . . . produces motion as being loved but all other things move by being moved."[67] The *primum mobile*, which embraces all the orbs involved in the daily motion, seeks to become as like the Prime Mover as possible and, thus, wishes to come to rest in imitation of the deity, for "all things desire and love the prime desirable thing (*primum appetibile*) and prime end, just as an inferior thing loves and desires a better thing."[68] Since it is impossible for any celestial orb to acquire such a state of perfection, the *primum mobile* remains in a continuous state of rotational motion as it strives for its unattainable goal.

But God's action as sole cause of motion in the *primum mobile* could also be interpreted as an efficient cause. For, as John Buridan explained, "He is called efficient [cause] insofar as He is understood to produce motion from not-being to being. But He is said to be an end [or final cause] insofar as the heaven that is moved seeks, as much as it can, to be assimilated to Him."[69]

For the rest of the celestial spheres, theological grounds compelled the scholastics to depart from the texts of Aristotle, who, after describing his seemingly unique Unmoved Mover, proceeded to explain the motions of the other celestial spheres, some 55 in all, by assigning to each of them an Unmoved Mover whose properties were in no way distinguishable from the first Unmoved Mover.[70] In another context, Aristotle was thought to have held a quite different, but equally repugnant, opinion that each celestial sphere was a composite of body and soul. Since souls are alive, Aristotle was held to believe that celestial bodies are animated beings moved by an intrinsic soul, an opinion that was condemned in 1277.[71] That celestial bodies could possess sensitive or vegetative life was highly objectionable,[72] as was the opinion that the source of natural celestial motion could reside within the magnitude of the orb itself and function in a manner akin to that of a natural terrestrial mover.[73]

To avoid these and other dilemmas, every celestial motion was attributed to a created intelligence conceived as immobile and completely separated from the celestial orb it moved. Although motive intelligences were frequently identified with angels, this identification was by no means unanimous.[74] Widely accepted, however, was the opinion that the separated intelligences moved their respective orbs by the intellect and will together.[75] In this context, celestial motions were conceived as voluntary, not natural or intrinsic. The Prime Mover was thought to have conferred upon each intelligence the potentiality for producing an infinitely intensive celestial velocity, the possibility of which was conceded because no internal or external resistance to its efforts could arise naturally in the heavens. Observation, however, made it obvious that celestial motions were finite. Scholastics explained that, as mere potentialities, infinite intensities could only be actualized by the intellect and will of the intelligences and need not necessarily be actualized, as would be the case if they operated from natural necessity.[76] As voluntary agents, celestial movers could, in principle, choose to operate intensively for a short time, and even move their spheres instantaneously by actualizing their full potential; or, they might choose to move with finite velocities for an infinite duration. That they had in fact "chosen" to operate with "forces" of constant, finite intensity to produce uniform regular motions was apparent by observation. In moving their orbs, celestial movers were considered *vires infatigabiles*, inexhaustible forces, either because the Prime Mover had initially conferred infinitely intensive power on each mover, or constantly supplied it from His own inexhaustible reservoir.[77]

Not all late medieval explanations of celestial motion relied on intelligences or angels. John Blund and Robert Kilwardby insisted that in creating the celestial orbs, God conferred upon each of them an intrinsic active principle which enabled it to incline naturally to a rotational motion in a given direction.[78] Another motive source was proposed by Buridan when he suggested that perhaps "one could imagine that it is unnecessary to posit intelligences as the movers of celestial bodies since the Holy Scriptures do not inform us that intelligences must be posited. For it could be said that when God created the celestial spheres, He began to move each of them as He wished, and they are still moved by the impetus which He gave to them because, there being no resistance, the impetus is neither corrupted nor diminished."[79] Oresme, for whom angels were the intelligences that moved the celestial orbs, also suggested, without elaboration, that when God created heavens, "He put into them motive qualities and powers just as He put weight and resistance against these motive powers in earthly things. These powers and resistances are different in nature and substance from any sensible things or quality here below. The powers against the resistances are moderated in such a way, so tempered and so harmonized, that the movements are made without violence; thus, violence excepted, the situation is much like that of a man making a clock and letting it run and continue its own motion by itself."[80] In the same treatise, Oresme also identified the celestial motive force as a "corporeal quality" within the orb itself which meets resistance only to prevent a more rapid motion.[81]

The Properties of Celestial Bodies

The properties assigned to the celestial region were in stark contrast with those attributed to the sublunary, or terrestrial, realm. Other than extension and motion, these disparate regions had little in common. The fifth element, or essence (*quinta essentia*, Aristotle's "ether"), of which the entire heavens was constituted, bore no resemblance to the four constantly changing sublunar elements, earth, water, air, and fire. The fundamental assumption from which so much else followed was that the celestial element suffered no alterations other than the positional changes of its parts associated with its incessant uniform circular motion. It lacked the primary contrary qualities that were indispensable for the manifold and continuous changes observed everywhere below the concave surface of the moon. The unalterable nature of the heavens also gained support from Aristotle's unchallenged claim that from earliest times no essential changes had been observed there.[82] For all these reasons, the celestial region was even said to lack matter, since, as Buridan explained, "the heaven is neither generable nor corruptible; and yet everything that has matter is generable by virtue of the fact that matter existing under one form is in potentiality, by its [very] nature, to other forms and naturally strives for them; but such potentiality and striving would be perpetually in vain" in the heaven "unless such forms could exist in this [celestial] matter at some time [or other]. Now it is absurd to assume something that is always in vain in nature."[83] Apparent changes, such as the sudden appearance of comets, haloes, shooting stars, and similar phenomena, were dealt with by locating them, as Aristotle did, in the upper reaches of the atmosphere just below the moon.

But what about the universally accepted qualities and properties assigned to the planets and stars in the centuries-old astrological tradition? How could the moon be described as wet and cold; Mars as hot and dry; Saturn dry and cold; and so on, if, as we have already said, the existence of the primary contrary qualities (namely, hot, cold, wet, and dry) in the heavens were denied? Moreover it was common to assign secondary contrary qualities, such as rare and dense, to the celestial zone. For example, the planets were considered denser and quite distinct parts of their orbs;[84] and the spot, or man, in the moon was sometimes explained as a consequence of a greater rarity of a part of the moon that was less able to reflect light than its denser parts.[85] Furthermore, if dense and rare occur in the celestial orbs and planets, then so must light and heavy.[86]

If all these primary and secondary qualities actually inhered in the celestial bodies and spheres, it would follow by generally accepted Aristotelian principles that generation and corruption would and must occur, since it was presumed that where contrary qualities existed, such as hot-cold, wet-dry, rare-dense, and so on, change must inevitably result. To avoid this disastrous consequence, many fourteenth-century scholastics assumed that the primary and secondary qualities in the celestial region did not formally (*formaliter*) inhere in the fifth celestial element, but existed there only virtually (*virtualiter*).[87] On this interpretation, Saturn, for example, would not be cold to our touch, since it does not actually possess the quality of coldness; rather, it has the capacity to produce the effect of coldness. Again, the sun is not actually hot but is capable of producing heat. For as Robertus Anglicus explained, "just as not all that blackens is black, because, if this were so, it would go on *ad infinitum*, so not all that heats is hot, but one must suppose first that which heats and is not hot, like a celestial body which by its movement inflames and heats bodies which are under it."[88] Thus, the qualities and properties assigned to the planets were based largely on their alleged capacities to produce certain effects on terrestrial bodies. And because of the different and unique effects which each planet exerted on the sublunar region, as well as the visible differences between planets, each planet was conceived as the sole member of a unique species. It was even

suggested that the orbs which carried the planets and stars formed
their own distinct species, since the sun, which illumined the planets
and stars, failed to illuminate the orbs.[89]

The all-pervasive medieval belief that the celestial and terrestrial
regions were necessarily of radically different natures led to the assignation of further startling properties to the heavens. The fifth celestial element, for example, despite its oft-proclaimed continuity, was
conceived as indivisible. It was not, many insisted, divisible into separate parts, since such divisibility would imply change. At best, one
might allow only that one part lies outside another in a fixed relationship. As support for such a view, it was assumed that the celestial
orbs were solid, although the status of the interior of each sphere—
whether solid or hollow—was left unclear.[90] The solid surfaces of the
spheres were usually thought of as transparent (except for the denser
portions representing the planets and stars), perfectly smooth, and
polished. Nested within each other in perfect contact (to avoid the
abhorrent vacuum), they moved without friction and produced no
heat, contrary to all experience with terrestrial motion. Similar con-
trasts might be mentioned, but we must now consider the manner in
which the celestial region was thought to affect behavior in the inferior and continually changing domain below the moon.

The Influence of the Celestial Region
on the Terrestrial

That the motions and natures of celestial bodies affected the behavior
of sublunar animate and inanimate bodies was accepted by all. Evidence for this belief was considered overwhelming. All agreed with
Aristotle that the sun's alternate approach and retreat every year was
essential for life on earth, since it produced heat, generation and corruption, and cyclical transformations of the four elements.[91] The moon
was also believed to exercise powerful influences on terrestrial events,
as its motion and phases brought on the ebb and flow of the tides,
influenced the increase and decrease of humors in animals, and affected the critical days in the course of diseases.[92] In the long astrological and alchemical traditions, the other planets were also assigned
important causative powers on terrestrial events, as in the formation
of metals, the spontaneous generation of lower and imperfect animals,
and the generation of plants and higher animals. Themon Judaeus
spoke for the Middle Ages when he declared that "every natural power
of this inferior, sensible world is governed by the heaven" and that

"to every active power in these inferior things, there corresponds a certain power in the heaven . . ."[93] Governance of the terrestrial region by the celestial was made possible by the contact of the concave surface of the lunar sphere with the convex surface of the sphere of fire, a contact that linked the two regions and made the world a continuous, uninterrupted whole.

Three major sources of terrestrial effects were distinguished in the heavens: motion, light, and "influence" (*influentia*).[94] The first of these, motion, was apparently conceived as a mechanical transmission in which some particular planetary influence was conveyed from rotating orb to rotating orb and ultimately from the innermost lunar sphere centerward into the terrestrial region. Unidirectional transmission was often premised on the conviction that by its motion any celestial sphere could drag its immediate inferior neighbor along with it.[95] Details of such transmission were not provided, very likely because mechanical transmission of effects was in complete conflict with the medieval concept of celestial motion that has been described above. Hard, smooth, highly polished surfaces that were moved without friction by separate immobile intelligences could hardly provide a mechanism for a chainlike transmission of an effect from the outermost to the innermost parts of the universe. The physical effect of one orb upon another by contact was precluded not only by the frictionless motion of each orb, but, more importantly, by the assumption that the motion of each orb was controlled exclusively by an incorporeal intelligence. On this problem, Scholastic natural philosophers muddled along with a conflict inherited from Aristotle but which neither they nor their mentor seem to have recognized.[96] Indeed, despite Aristotle's suggestion that celestial bodies could transmit effects by their motion, his system of concentric spheres was so constituted as to preclude the possibility of the transmission of effects from one set of planetary orbs to another.[97]

The sole feasible mode of transmission of celestial influences was by means of rays transmitted rectilinearly through the transparent and invisible orbs without mutual interaction. In this category, light was paramount because its rays produced both light and heat. On the common assumption that all planets and stars received light from the sun, all celestial bodies were deemed capable of transmitting light, or *lumen*, to the terrestrial region. As Roger Bacon explained, "This species [*lumen*] produces every action in the world, for it acts on sense, on the intellect, and on all matter of the world for the generation of things."[98] In this mode of transmission, Aquinas likened the role of the intermediate orbs through which the light passed to that of air in the attraction of a magnet for iron: they served as a medium of transmission without undergoing any alteration.[99]

That light, however, could not function as the sole instrumentality by which the heavens acted on inferior bodies was evident by the belief that metals were formed in darkness within the earth. To account for this, and a host of other phenomena in which light seemed uninvolved, a general, all-purpose celestial "influence" (*influentia*) was invoked. As described by Themon Judaeus, it is "a certain quality, or virtue, diffused through the whole world, just as the species of heat or light is multiplied."[100] By contrast with light, however, celestial influence was invisible and could pass through opaque and dense bodies. Thus, it could be pressed into service to explain a host of effects for which light was inapplicable, as, for example, magnetism, generation of metals, tides, and even the prevention of the formation of vacua.[101] Only the human intellect and will, which were immaterial faculties independent of material organs, were exempt from its power.[102] The supremacy of the human intellect and will over celestial influences was encapsulated in a popular medieval dictum that "a wise man will dominate the stars."[103]

The Possible Motion of the Earth and the Dimensions of the Universe

Much of interest and significance, which, unfortunately, cannot be described here, was proposed about the structure and nature of the sublunar region, including the arrangements, properties, and proportions of the four elements. Of major interest was the earth itself lying at the center of the universe. Despite the firm conviction of Aristotle and Ptolemy that the earth lay completely immobile at the center of the world, a few medieval natural philosophers of the caliber of Buridan and Oresme considered the possibility of its daily axial rotation. Although both ultimately rejected this rotation, they offered numerous arguments to "demonstrate" that terrestrial rotation, coupled with the assumed immobility of the heavens, could just as well save the astronomical phenomena as the traditional opinion. Oresme, indeed, was prepared to argue that it was also a physically tenable interpretation.[104] If the earth's axial rotation proved too much to uphold in the Middle Ages, its slight movement around the geometric center of the universe was quite acceptable.[105] Such a motion arose as a consequence of a continuous shift of the earth's center of gravity as its density perpetually altered under the influence of geological changes. In order to come to rest, the earth sought to bring its center of gravity into coincidence with the geometric center of the universe, an inclination which produced a slight motion that was endlessly repeated.

Although we have now ranged from the farthest reaches of the medieval universe to its geometric center, we have not yet mentioned its dimensions. In his widely known *Theory of the Planets* and the tables which accompany it, Campanus of Novara (ca. 1205–96) provided a representative estimate of the size of the universe and its parts. The table below incorporates certain dimensions of cosmological interest.[106] By setting the distance of the convex surface of a given planetary sphere equal to the distance of the concave surface of the next planetary sphere, a compact universe was assumed in which no waste space or vacua could occur. The modern reader is struck by the smallness of the medieval world as compared to the estimates of present-day astronomy. The visible universe from the earth to Saturn and the concave surface of the sphere of the fixed stars is little more than 73 million miles, with the sun approximately 4 million miles away from the earth and its inhabitants.

Small though it may appear to us, it would be rash to suppose that the universe was thought of in the same way during the Middle Ages. Quite the contrary. It was conceived as very large indeed, as witnessed by the oft-repeated belief that the earth is as nothing compared to the universe, and as a point when compared to the distance from the sphere of fixed stars.[107] In this regard, an absolute and relative sense were distinguished.[108] In the absolute sense, it was obvious that the earth was not a point when compared to the heavens, since it had size and magnitude, and its distances from the various celestial bodies were finite. In the relative sense, however, it could be so conceived, for with respect to the superior orbs, such as the fixed stars, Saturn, and Jupiter, the earth had an "insensible magnitude," although it acquired a "notable magnitude" when compared to the nearer spheres of Venus, Mercury, and the moon.

In no manner, then, is it a sense of size which distinguishes the medieval from the modern conception of the world. The fact that:

the height of the stars in the medieval astronomy is very small compared with their distances in the modern, will turn out not to have the kind of importance you anticipated. For thought and imagination, ten million miles and a thousand million are much the same. Both can be conceived (that is, we can do sums with both) and neither can be imagined; and the more imagination we have the better we shall know this. The really important difference is that the medieval universe, while unimaginably large, was also

Cosmological Table from Campanus of Novara's *Theory of the Planets*
(Distances given in miles)

Planet	Radius of planet	Circumference of planet	Distance of convex surface of planetary sphere from earth's center[a]	Distance of concave surface of planetary sphere from earth's center[b]	Thickness of planetary sphere[c]
Moon	948	5,958	209,198	107,936	101,261
Mercury	115	725	579,320	209,198	370,122
Venus	1,442	9,066[d]	3,892,866	579,320	3,313,545
Sun	17,850	112,200	4,268,629	3,892,866	375,762
Mars	3,786	23,800	32,352,075	4,268,629	28,083,446
Jupiter	14,820	93,160	52,544,702	32,352,075	20,192,626
Saturn	14,604	91,800	73,387,747	52,544,702	20,843,044
Fixed stars				73,387,747[e]	

[a]Subtracting a planet's radius from the distance of the convex surface of its sphere yields the planet's maximum distance from the earth's center.

[b]Adding the radius of a planet to the distance of its concave surface gives that planet's minimum distance from the earth.

[c]Since a planet is a body and not a point, Campanus added the value of each planet's diameter to the thickness of its sphere, which would otherwise be obtained by subtracting the distances of the convex and concave surfaces. See Benjamin and Toomer, *Campanus of Novara*, p. 56.

[d]Campanus actually gives 9,095, which is a calculating error (Benjamin and Toomer, p. 434 n. 56).

[e]This figure represents the distance of the concave surface of the sphere of fixed stars. Its thickness, and that of the spheres beyond, including the ninth, is not given or discussed.

unambiguously finite. And one unexpected result of this is to make
the smallness of Earth more vividly felt. In our universe she is small,
no doubt; but so are the galaxies, so is everything—and so what?
But in theirs there was an absolute standard of comparison. The
furthest sphere, Dante's *maggior corpo*, is quite simply and finally,
the largest object in existence. The word "small" as applied to
Earth thus takes on a far more absolute significance. Again, because
the medieval universe is finite, it has a shape, the perfect spherical
shape, containing within itself an ordered variety. Hence to look
out on the night sky with modern eyes is like looking out over
a sea that fades away into mist, or looking about one in a
trackless forest—trees forever and no horizon. To look up at the
towering medieval universe is much more like looking at a great
building. The 'space' of modern astronomy may arouse terror, or
bewilderment or vague reverie; the spheres of the old present us
with an object in which the mind can rest, overwhelming in its
greatness but satisfying in its harmony. That is the sense in which
our universe is romantic, and theirs was classical.[109]

Notes

1. For a brief discussion of the age of translation, see Edward Grant,
Physical Science in the Middle Ages (New York, 1971), pp. 13–19; and
above, chap. 2. The numerous works translated by Gerard of Cremona
and William of Moerbeke, the two greatest medieval Latin translators, are
listed in *A Source Book in Medieval Science*, ed. Edward Grant (Cambridge, Mass., 1974), pp. 35–41.

2. See Pierre Duhem, *Le système du monde*, 6:20–30, and Grant,
Source Book, pp. 45–50. All 219 propositions are translated by Ernest L.
Fortin and Peter D. O'Neill, in *Medieval Political Philosophy: A Source-
book*, ed. Ralph Lerner and Muhsin Mahdi (Toronto, 1963), pp. 335–54.

3. In this regard, it is worth noting that modern cosmologists who
speculate about the consequences of black holes in which there "would be
infinitely powerful gravity concentrated in an infinitely dense, infinitely
small spot where time and space have lost their meaning" (see Walter
Sullivan, "A Hole in the Sky," *New York Times Magazine*, 14 July 1974,
p. 26) would have no difficulty in recognizing their medieval predecessors
as kindred spirits in that endless quest for knowledge and understanding
of cosmic operations.

4. Composed in the thirteenth century, this elementary astronomical
treatise, which also contained a brief cosmological section, was used as a
text in medieval universities throughout the late Middle Ages. Because of
its enormous popularity, numerous commentaries on it were written that
far transcended it in sophistication and technical virtuosity. The text has
been edited and translated by Lynn Thorndike, *The Sphere of Sacrobosco
and Its Commentators* (Chicago, 1949). Treatises of the same title, though
often with additional or different topics, were written by Robert Grosseteste, John Pecham, and Nicole Oresme.

5. Although two famous encyclopedias of the thirteenth century, *De rerum proprietatibus* by Bartholomew the Englishman and the *Speculum maius* by Vincent of Beauvais, contain more cosmological information than did Sacrobosco's *Sphere*, they represent little more than unevaluated collections of passages drawn from a variety of sources and unintegrated into any coherent whole.

6. See articles 9, 87, 90, 91, 98, 101, 107, and 202 in Grant, *Source Book*, pp. 48–50.

7. See *St. Thomas Aquinas, Siger of Brabant, St. Bonaventure: On the Eternity of the World*, translated from the Latin with an Introduction by Cyril Vollert, Lottie H. Kendzierski, and Paul M. Byrne (Milwaukee, Wis., 1964), p. 14.

8. P. Dr. Zachary Hayes, O.F.M., *The General Doctrine of Creation in the Thirteenth Century with Special Emphasis on Matthew of Aquasparta* (Munich, 1964), p. 114.

9. For further discussion, see Grant, *Physical Science in the Middle Ages*, pp. 24–29.

10. Oresme's full discussion with commentary appears in Grant, *Source Book*, pp. 547–54.

11. Ibid., p. 554.

12. *De caelo* 1. 9. 279a13–24. For Aristotle, whatever might exist beyond the world would, of necessity, be eternal, changeless, and perfect.

13. The discussion of infinite extracosmic void space is drawn from the following of my publications: "Medieval and Seventeenth-Century Conceptions of an Infinite Void Space beyond the Cosmos," *Isis* 60 (1969): 39–60; *Physical Science in the Middle Ages*, pp. 76–82; and "Place and Space in Medieval Physical Thought," in *Motion and Time, Space and Matter*, ed. Peter K. Machamer and Robert G. Turnbull (Columbus, Ohio, 1976), pp. 137–67.

14. Although Oresme and Bradwardine conceived of the infinite void as co-eternal and co-extensive with God, and John de Ripa described it as in some sense created by God and circumscribed and contained by Him (John distinguished two infinites, an ordinary kind of which infinite void space was an example, and a single superinfinite which he equated with God's "immensity"), none considered explicitly whether infinite void space is an attribute of God, whether it is three-dimensional or only "transcendentally" dimensional, and so on. Consideration of such problems may not have begun until the sixteenth century.

15. Article 49 condemned the following proposition: "that God could not move the heavens [that is, the world] with rectilinear motion; and the reason is that a vacuum would remain." See Grant, *Source Book*, p. 48.

16. *S. Thomae Aquinatis In octo libros De Physico Auditu sive Physicorum Aristotelis Commentaria*, new ed. by A. M. Pirotta, O.P. (Naples, 1953), bk 4, lectio 7, par. 909, p. 202.

17. *Physics* 4. 4. 212a5–7, 20–21, 29–30.

18. *Physics* 4. 5. 212b6–22.

19. In much of what follows on this topic, I am indebted to Duhem, *Le système du monde*, vol. 7, chap. 3 ("Le Lieu").

20. *Commentary on the Physics* in *Aristotelis Opera cum Averrois commentariis* (Venice, Junctas ed., 1562–74; reprint ed. Frankfurt am Main, 1962), vol. 4, bk. 4, comment 45, fol. 144r, col. 1.

21. Roger Bacon, *Questiones supra libros octo Physicorum Aristotelis*, in *Opera hactenus inedita Rogeri Baconi*, edited by Ferdinand M. Delorme, O.F.M., with the collaboration of Robert Steele, fasc. 13 (Oxford, 1935), pp. 216–22. In *Le système du monde*, 7:164–168, Duhem summarizes Bacon's views.

22. *De physico auditu*, pp. 203–4. In Aquinas's opinion, the last sphere is in a place through its parts, each of which is potentially, though not actually, in a place. The similarity with Aristotle's view described above is obvious.

23. See Duhem, *Le système du monde*, 7:284–85.

24. See John Buridan, *Quaestiones super libris quattuor De caelo et mundo*, ed. Ernest A. Moody (Cambridge, Mass., 1942), bk. 2, quest. 6, p. 150; Albert of Saxony, *Questions on the Physics*, bk. 4, quest. 8, fol. 107r in *Questiones et decisiones physicales insignium virorum: Alberti de Saxonia: Octo libros Physicorum; Tres libros De celo et mundo; Duos lib. De generatione et corruptione; Thimonis in Quatuor libros Meteororum; Tres lib. De anima; Buridani in Aristotelis Lib. De sensu et sensato . . . Recognitae rursus et emendatae summa accuratione et iudicio Magistri Georgii Lokert Scotia quo sunt Tractatus proportionum additi* (Paris, 1518). Numerous references will be made below to Albert of Saxony's *Questions on the Physics* and *Questions on De caelo*, as well as to Themon Judaeus's *Questions on the Meteors* (or *Meteorology*). Another interpretation conceived the place of the world as the concave surface of the last celestial sphere. Apparently based on a misinterpretation of Gilbert de la Porrée's discussion of *ubi* in the *Liber sex principiorum*, this approach effectively denied that a surrounding body is required for a place. Instead the outermost portion of the last mobile or immobile sphere could be taken as the place of the world. See Duhem, *Le système du monde*, 3:197.

25. *Campanus of Novara and Medieval Planetary Theory: Theorica Planetarum*, edited with an Introduction, English translation, and commentary by Francis S. Benjamin, Jr., and G. J. Toomer (Madison, Wis., 1971), p. 183.

26. For these and other arguments, see Albert of Saxony, *Questions on the Physics*, bk. 4, quest. 8, fol. 107r; John Buridan, *Questions on De caelo*, bk. 2, quest. 6, pp. 149–53; and Pierre d'Ailly, *14 Quaestiones on the Sphere of Sacrobosco* in *Tractatus Ioannis de Sacro Busto Anglici Viri Clariss.: Gerardi Cremonensis Theoricae Planetarum Veteres . . . Petri cardin de aliaco episcopi camaracensis, 14 Quaestiones . . .* ([Venice] 1531), question 2, fol. 149v. While d'Ailly was a firm supporter of an immobile outer sphere, Albert of Saxony opposed it. By contrast, Buridan argued that one could accept the existence of an Empyrean sphere on faith alone, noting that Aristotle often assumed positions contrary to the faith "because he sought nothing unless it could be deduced from reasons derived from the senses and experience." Indeed, Buridan presented arguments in behalf of an immobile sphere. But, ultimately, he opted for

Aristotle's opinion and showed how one might counter opposing arguments.

27. John Duns Scotus, who denied every kind of place to the last sphere, held that the latter would rotate whether or not there were any bodies within or without. Celestial orbs rotate by virtue of their forms without regard to places or other bodies (Duhem, *Le système du monde*, 7:212, 282).

28. On the meanings assigned to the term *caelum* by Albertus Magnus, see P. Hossfeld, "Die naturwissenschaftlich/naturphilosophische Himmelslehre Alberts des Grossen," *Philosophia Naturalis* 11 (1969): pp. 329–30.

29. St. Thomas Aquinas, *Summa Theologiae*, Latin text, English translation, Introduction, Notes, Appendices and Glossary by William A. Wallace, O.P., vol. 10, *Cosmogony* (London, 1967), p. 89.

30. *Summa Theologiae*, p. 91. In his *Speculum Naturale*, Vincent of Beauvais also discusses "the heaven of the Trinity" (see *Vincentii Burgundi . . . Speculum Quadruplex: Naturale, Doctrinale, Morale, Historiale* [Douai, 1624], vol. 1, cols. 217–19).

31. *St. Thomae Aquinatis Scriptum super libros Sententiarum Magistri Petri Lombardi*, vol. 2, new ed. by R. P. Mandonnet, O.P. (Paris, 1929), bk. 2, dist. 14, quest. 1, art. 4, p. 356.

32. On the origin of the Empyrean heaven, see Benjamin and Toomer, eds. and trans., *Campanus of Novara*, p. 393, n. 52. Although the *Glossa Ordinaria* was regularly attributed to Walafrid Strabo during the Middle Ages, an attribution repeated by Benjamin and Toomer, it is now assigned to Anselm of Laon (see Wallace's note in Aquinas, *Summa Theologiae*, 10:40 n. 5).

33. *Campanus of Novara*, p. 393, n. 52. The translation is mine from the Latin text quoted by Benjamin and Toomer.

34. Peter attributes the passage to Strabo. See *Magistri Petri Lombardi Parisiensis Episcopi Sententiae in IV Libris Distinctae*, 3d ed., vol. 1, pt. 2, bks. 1 and 2 (Grottaferrata [Rome]: Editiones Collegii S. Bonaventurae Ad Claras Aquas, 1971), p. 340; in the older edition of the *Sentences* in J. P. Migne, *Patrologiae cursus completus, series latina*, see vol. 192 (Paris, 1855), col. 656.

35. To the brief list of discussants given by Thomas Litt, *Les corps céleste dans l'univers de Saint Thomas d'Aquin* (Louvain and Paris, 1963), pp. 255–61, we may add pseudo-Grosseteste (*Summa philosophiae*), Vincent of Beauvais (*Speculum naturale*), Bartholomew the Englishman (*De rerum proprietatibus*), and Michael Scot (*Commentary on the Sphere of Sacrobosco*).

36. Litt, *Les corps céleste*, pp. 257–60, has collected and summarized Aquinas's references to the Empyrean sphere.

37. For a few references, see Benjamin and Toomer, *Campanus of Novara*, pp. 393–94.

38. *Summa Theologiae*, p. 81.

39. Thomas Aquinas summarizes some of these opinions in *Summa Theologiae*, pp. 77–83.

40. On precession, which was discovered by Hipparchus in the 2d century B.C. and reported to the Middle Ages in Ptolemy's *Almagest*, see Benjamin and Toomer, *Campanus of Novara*, p. 378.

41. The theory of trepidation arose from discrepancies in the observation of precession (see Benjamin and Toomer, *Campanus of Novara*, pp. 378–79, and J. L. E. Dreyer, *A History of Astronomy from Thales to Kepler*, 2d ed., revised with a foreword by W. H. Stahl [N.Y., 1953], pp. 276–77). Ordinarily, either precession or trepidation should have been employed, but many scholastics assigned both motions to the stars.

42. For Albert, see *Questions on De caelo*, bk. 2, quest. 6, fol. 105v; for d'Ailly see *14 Quaestiones*, fol. 149r; and for Bacon, see his *Communia naturalium* in *Opera hactenus inedita Rogeri Baconi*, ed. R. Steele, fasc. 4 (Oxford, 1913), pp. 388, 449, 455.

43. In *Scriptum super libros Sententiarum*, ed. Mandonnet, 2:348, Aquinas assigned the daily motion to a tenth orb, precession to a ninth orb, which he identified as the crystalline heaven, but made no particular assignation to the eighth sphere.

44. As did Themon Judaeus (see Henri Hugonnard-Roche, *L'oeuvre astronomique de Themon Juif* (Geneva and Paris, 1973), pp. 104–5, and probably Robertus Anglicus and Michael Scot (see Thorndike, *The Sphere of Sacrobosco*, pp. 203, 283).

45. This was probably Campanus's position (see Benjamin and Toomer, *Campanus of Novara*, p. 183). That it was a common position is indicated by Nicole Oresme, *Le livre du ciel et du monde*, ed. Albert D. Menut and Alexander Denomy, translated with an Introduction by Albert D. Menut (Madison, Wis., 1968), pp. 488–91.

46. Oresme thought it superfluous to assign a ninth starless orb (see his *Le livre du ciel et du monde*, p. 491.)

47. *Almagest*, bk. 9, chap. 1 in the translation by R. Catesby Taliaferro in *Great Books of the Western World*, vol. 16 (Chicago, 1952), p. 270. The rival order proposed by Plato and Macrobius was moon, sun, Venus, Mercury, Mars, Jupiter, and Saturn. See W. H. Stahl, trans., *Macrobius: Commentary on the Dream of Scipio* (New York, 1952), pp. 162–64.

48. *Almagest*, p. 270.

49. My translation from the Latin passage quoted by Henri Hugonnard-Roche, *L'oeuvre astronomique de Themon Juif*, p. 75.

50. *Le livre du ciel et du monde*, p. 507. Much the same opinion was expressed by Richard of Middleton in bk. 2, distinction 14, question 1 of his *Sentence Commentary (Magistri Ricardi de Mediavilla . . . super quatuor libros Sententiarum Petri Lombardi Quaestiones subtilissimae* [Brixiae, 1591; reprint ed. Minerva G.m.b.H., Frankfurt am Main, 1963] vol. 2, p. 168, col. 1), and by Roger Bacon in bk. 4 of his *Questiones supra libros octo Physicorum Aristotelis*, in *Opera hactenus inedita*, ed. F. Delorme, fasc. 13 (Oxford, 1935), pp. 221–22.

51. In his Long Commentary on Aristotle's *Metaphysics*, bk. 12, comment 44, Averroes declares (*Aristotelis Opera cum Averrois commentariis*, vol. 8, fol. 327, F, G, and H) that the order of nobility of celestial movers depends on the order of their respective celestial orbs, arranged in descending order from the sphere of fixed stars. From this he concludes that the nobility of the planets themselves depends on their descending spatial order (*secundum locum*) with respect to the sphere of fixed stars with which the Prime Mover is associated. As a challenge to his own opinion, Averroes acknowledges that the commonly accepted assumption

that the velocities of the planets increase with proximity to the earth
might be explicable either by virtue of the greater nobility of celestial
movers in proportion to their proximity to earth, or because celestial
bodies nearer to earth have smaller magnitudes. Indeed, Averroes also
suggests the possibility that the sun might be the noblest planetary body,
since its great magnitude and profound influence on both animate and
inanimate beings perhaps indicate that its motion controls the other
planets. But Averroes dismisses all these possible interplanetary relationships as quite imprecise and opts for the view that greater planetary
nobility is in direct proportion to proximity to the sphere of the fixed
stars and its Prime Mover, a view he deems exact and comprehensible.
See also Harry A. Wolfson, "The Plurality of Immovable Movers in
Aristotle, Averroes, and St. Thomas," in *Studies in the History of Philosophy and Religion: Harry Austryn Wolfson*, ed. Isadore Twersky and
George H. Williams (Cambridge, Mass., 1973), 1: 13.

52. Thomas Aquinas, *Summa Theologiae*, p. 89.

53. Ibid.

54. The same seven heavens are also described by Bartholomew the
Englishman, *De rerum proprietatibus* (reprint ed. Frankfurt a. M., 1964
from Frankfurt edition of 1601), pp. 372–73.

55. These are the illustrations of free movement given, for example,
by John Buridan, *Questions on De caelo*, p. 210 and Pierre d'Ailly, *14
Quaestiones*, quest. 2, fol. 148v.

56. For a typical judgment in favor of contiguity, see Albert of Saxony,
Questions on De caelo, bk. 1, quest. 5, fol. 88r, col. 1. Some of the basic
issues involved in the debate over continuity or contiguity appear in
Thorndike, *The Sphere of Sacrobosco*, p. 282 (for views ascribed to
Michael Scot, who opted for contiguity), pp. 202–3 (for a defense of
continuity by Robertus Anglicus), and p. 353 (for Cecco d'Ascoli's rejection of both positions). The basic definitions of *continuity* and *contiguity*, which underlie much of the debate, were drawn from Aristotle's
Physics 5. 3. 227a9–19.

57. A German translation of its two books and the Greek text of the
first appear in Claudius Ptolemaeus, *Opera quae exstant omnia*, vol. 2,
Opera astronomica minora ed. J. L. Heiberg (Leipzig, 1907), pp. 70–145.
An additional Arabic part of the first book has recently been published
with an English translation by Bernard R. Goldstein, "The Arabic Version
of Ptolemy's *Planetary Hypotheses*," in *Transactions of the American
Philosophical Society*, n.s., 57, pt. 4 (1967):3–55. For a summary of
Ptolemy's physical model, see Duhem, *Le système du monde*, 2:86–99.

58. A Latin edition made from a single, untitled manuscript in the
Biblioteca Nacional de Madrid, has been published by José M. Millás
Vallicrosa, *Las truducciones orientales en los manuscritos de la Biblioteca
Catedral de Toledo* (Madrid, 1942), pp. 285–312. See also Duhem, *Le
système du monde*, 2:121–29, and A. I. Sabra, "Ibn al-Haytham," *Dic-
tionary of Scientific Biography*, 6:197–98, 210.

59. See F. J. Carmody, *Al-Bitruji De motibus celorum* (Berkeley,
1952) and J. L. E. Dreyer, *A History of Astronomy from Thales to
Kepler*, pp. 265–66. In his commentary on *Metaphysics* 12.8, comment

45, Averroes tells us that he had tried, and obviously failed, to explain all the essential astronomical phenomena, including variation in distances, by a system of concentric spheres. (See F. J. Carmody, "The Planetary Theory of Ibn-Rushd," *Osiris* 10 [1952]:572.)

60. For example, Michael Scot, in the commentary on Sacrobosco's *Sphere* ascribed to him (see Thorndike, *The Sphere of Sacrobosco*, pp. 248–342, where Michael makes no mention of Ptolemy, eccentrics, or epicycles) and William of Auvergne (see Duhem, *Le système du monde*, 3:249–60).

61. Robert Grosseteste (Duhem, *Le système du monde*, 3:286–87) and perhaps Thomas Aquinas (see Thomas Litt, *Les corps céleste*, pp. 346–65).

62. See Pierre Duhem, ed., *Un fragment inédit de l'Opus tertium de Roger Bacon précédé d'une étude sur ce fragment* (Quaracchi, 1909), pp. 128–31.

63. See Duhem, *Le système du monde,* 3:438–39.

64. In his *Sentence Commentary*, bk. 2, dist. 14, quest. 2, John Duns Scotus assigned five orbs to Mercury and three to each of the six other planets (see *B. Ioannis Duns Scoti Commentaria Oxoniensia ad IV Libros Magistri Sententiarum*, ed. P. Marianus F. Garcia, O.F.M., vol. 2 [Quaracchi, 1914], p. 547). Others who accepted this system, however, speak as if all planetary spheres consisted of precisely three distinct orbs.

65. The figure appears in Bacon's *Opus tertium*, p. 129. The best basis of the scholastic solution in Ibn al-Haytham's *Liber de mundo et coelo* is found in the opening lines of the lengthy descriptions of the individual planets, as when we are told (Millás Vallicrosa, p. 300) that "the lunar orb is a spherical body bounded by two equidistant spherical surfaces with their center [lying] at the center of the world. The upper [spherical surface] touches the concave [surface of the] orb of Mercury; the lower [touches] the sphere of fire. . . . " Similarly, we learn (p. 304) that "the orb of Mercury is a spherical body bounded by two equidistant surfaces with their center [lying] at the center of the world. The upper of these touches the concave [surface] of the orb of Venus; the lower [touches] the convex [surface] of the lunar orb . . . " The elaborate descriptions of the orbs and their operations within the bounding surfaces were, however, largely ignored.

66. Thomas Aquinas discusses the problem in his *Summa Theologiae* and *Commentary on De caelo* (for the passages, see James A. Weisheipl, "The Commentary of St. Thomas on the *De caelo* of Aristotle," *Sapientia* 29 [1974]: 25).

67. *Metaphysics* 12. 7. 1072b3–4.

68. John of Jandun, *Quaestiones in duodecim libros Metaphysicae* (Venice, 1553; reprint ed. Frankfurt a. M., 1966), bk. 12, quest. 11 ("Whether the Prime Mover moves [that is, causes motion] as an appetible thing, or end"), fol. 132v. See also Moses Maimonides, *The Guide of the Perplexed*, translated with an Introduction and Notes by Shlomo Pines (Chicago, 1963), p. 256.

69. *In Metaphysicen Aristotelis Questiones argutissimae Magistri Ioannis Buridani . . .* (Paris, 1518; reprint ed. Minerva G.M.B.H., Frankfurt a. M., 1964), bk. 12, quest. 6, fol. 68v, col. 2. Buridan and others

believed this was also Aristotle's opinion. But according to James Weisheipl, "The Celestial Movers in Medieval Physics," *The Thomist* 24
(1961):321 n. 97, Aquinas "frequently insists that those who interpret
Aristotle's God as a mere physical mover or a mere final cause are in complete error," since Aristotle also believed that God is the cause of the substance of the heavens. Elsewhere ("The Commentary of St. Thomas on
the *De caelo*," p. 31) Weisheipl argues that for Aquinas the *primum mobile*
is moved by a natural prime mover (*primum movens*), or angel, who is
not absolutely unmoved, but is rather a self-mover. "But Thomas insists
. . . that beyond such a self-mover there is another reality, whom we call
God." God, by contrast with the self-moving *primum movens*, is absolutely unmoved. Thus for Aquinas, God is not the immediate effective
mover of the *primum mobile*—as He seems to be for Buridan and others
—but is, rather, the ultimate efficient and final cause of all motions.

70. *Metaphysics* 12. 8. 1073a34–1073b2. Whether Aristotle conceived
all unmoved celestial movers to be of equal status, or whether he considered the unmoved mover of the sphere of the fixed stars as a unique
"first unmoved mover" superior to the immovable planetary movers is a
vexing, and perhaps insoluble, problem, which is brilliantly discussed by
H. A. Wolfson in his article ("The Plurality of Immovable Movers in
Aristotle, Averroes, and St. Thomas") cited above in n. 51. Since in
scholastic thought God is the "first unmoved mover," it is obvious that
medieval Christian commentators were compelled to proclaim the "first
unmoved mover" as incomparably superior to all the rest.

71. See Lerner and Mahdi, *Medieval Political Philosophy*, p. 344,
art. 73. The opinion described above was attributed to Aristotle by Albertus Magnus, Richard Kilwardby, and Thomas Aquinas (see Weisheipl,
"The Celestial Movers in Medieval Physics," pp. 306–7, 314, 321). The
problem of the relationship between souls and spheres is presented by
H. A. Wolfson, "The Problem of the Souls of the Spheres from the Byzantine Commentaries on Aristotle through the Arabs and St. Thomas to
Kepler," in *Studies in the History of Philosophy and Religion*, ed. Twersky
and Williams, 1:22–59.

72. See Litt, *Les corps céleste* pp. 108–9, where Aquinas allows only
an intellective life to celestial movers.

73. See A. Maier, *Zwischen Philosophie und Mechanik* (Rome, 1958),
p. 210, where Oresme argues that under these conditions instantaneous
motion would occur, since there is no resistance to motion in the celestial
region.

74. Whereas Aquinas, Richard Middleton, and Nicole Oresme, for
example, identified the intelligences with angels (see Weisheipl, "Celestial
Movers in Medieval Physics," pp. 321–22; Richard Middleton, *Sentence
Commentary*, bk. 2, dist. 14, quest. 6, p. 173, cols. 1–2; and Oresme, *Le
livre du ciel et du monde*, pp. 299–301), Albertus Magnus and Richard
Kilwardby did not (Weisheipl, "Celestial Movers," pp. 307, 315).

75. To hold that an intelligence moves its orb by will alone was condemned in 1277. See Lerner and Mahdi, *Medieval Political Philosophy*,
p. 344, art. 74.

76. On the problems of infinite celestial velocities and motion of infinite duration, see A. Maier, "Himmelsmechanik und allgemeine Bewegungsgesetze," in *Zwischen Philosophie und Mechanik*, pp. 189–236.
77. On the problem of *vires infatigabiles*, see ibid.
78. See Weisheipl, "Celestial Movers in Medieval Physics," pp. 313–18. Weisheipl notes the similarity of this position with that of Nicholas of Cusa and Copernicus.
79. Translation by M. Clagett, *The Science of Mechanics in the Middle Ages*, p. 561 from Buridan's *Questions on De caelo*, bk. 2, quest. 12 (edition of E. A. Moody). Despite his interesting suggestion, Buridan, in the very same treatise, seems to accept intelligences as celestial movers.
80. *Le livre du ciel et du monde*, p. 289.
81. Ibid., p. 299. In his much earlier *De proportionibus proportionum*, Oresme spoke analogically of "the ratio of a moving intelligence to its orb" but, nevertheless, concluded that "an intelligence moves by will alone and with no other force, effort, or difficulty, and the heavens do not resist it, as I believe were the opinions of Aristotle and Averroes" (see Edward Grant [ed. and trans.], *Nicole Oresme De proportionibus proportionum and Ad pauca respicientes* [Madison, 1966], p. 293). Oresme appears to have altered his opinion in *Le livre du ciel et du monde*, although he still speaks of angels as celestial movers.
82. *De caelo* 1. 3. 270b14–17.
83. *Questions on De caelo*, bk. 1, quest. 11, p. 51. Aquinas considered the fifth element as matter because, in his opinion, it possessed a single substantial form analogous to that of terrestrial matter, although it was an unalterable substantial form (see Litt, *Les corps céleste*, p. 59).
84. See Albert of Saxony, *Questions on De caelo*, bk. 2, quest. 20, fol. 114v, col. 2; John Buridan, *Questions on De caelo*, bk. 1, quest. 9, p. 41; and, for Robertus Anglicus, Thorndike, *Sphere of Sacrobosco*, p. 206.
85. See Albert of Saxony, *Questions on De caelo*, bk. 2, quest. 22, fols. 116r, col. 2–116v, col. 1.
86. Buridan, *Questions on De caelo*, quest. 9, p. 41.
87. This was the interpretation of John Buridan, *Questions on De caelo*, bk. 1, quest. 9, pp. 43, 44; Albert of Saxony, *Questions on De caelo*, bk. 1, quest. 2, fol. 87v, col. 1; and Themon Judaeus, *Questions on the Meteors*, bk. 1, quest. 3, fol. 157v, col. 2.
88. Thorndike, *Sphere of Sacrobosco*, p. 209.
89. Albert of Saxony, *Questions on De caelo*, bk. 2, quest. 19, fol. 114r, col. 2.
90. Their "hollowness" is proclaimed in Beniamin and Toomer, *Campanus of Novara*, p. 54 and implied by Weisheipl, "Celestial Movers in Medieval Physics," p. 303. But the evidence is unclear. Themon Judaeus, for example, declared that "the heaven is a hard non-fluid body, for otherwise there would be a mixing together of planets and stars . . ." (*Questions on the Meteors*, bk. 1, quest. 3, fol. 157v, col. 2). We are left to ponder whether Themon meant to include the interiors of the spheres in this description. Roger Bacon, however, was one who implied hollowness by suggesting that the interiors are gaseous or fluidlike (see *Liber se-*

cundus communium naturalium, in *Opera hactenus inedita*, ed. R. Steele,
fasc. 4 [Oxford, 1913], p. 403).

91. *De generatione et corruptione* 2. 10. 336a24–336b9 and 337a5–
7, 11–15.

92. See Themon Judaeus, *Questions on the Meteors*, bk. 1, quest. 1,
fol. 155v, col. 2.

93. Ibid., fols. 155v, col. 2–156r, col. 1.

94. These are discussed by Themon Judaeus, *Questions on the Meteors*,
bk. 1, quest. 1, fol. 155v; and Albert of Saxony, *Questions on De caelo*,
bk. 2, quest. 12, fol. 109v, col. 1.

95. Albert of Saxony, *Questions on De caelo*, bk. 2, quest. 6, fol. 105r,
col. 2.

96. See Joseph de Tonquédec, *Questions de cosmologie et de physique
chez Aristote et Saint Thomas* (Paris, 1950), pp. 41–43.

97. See Aristotle, *Metaphysics* 12. 8. 1073b1–5; and G. E. R. Lloyd,
Aristotle: The Growth and Structure of His Thought (Cambridge, 1968),
p. 160.

98. Translated by David C. Lindberg from Bacon's *Opus Maius*, 4
dist. 2, chap. 1, in Grant, *Source Book*, p. 393; see also pp. 385–86 for
Lindberg's translation from Grosseteste.

99. See Litt, *Les corps céleste*, p. 41 and also Buridan, *Questions on
De caelo*, bk. 1, quest. 10, p. 47.

100. *Questions on the Meteors*, bk. 1, quest. 1, fol. 155v, col. 2.

101. See ibid., and E. Grant, "Medieval Explanations and Interpretations of the Dictum that 'Nature Abhors a Vacuum,'" *Traditio* 29
(1973):329–31.

102. See Litt, *Les corps céleste*, p. 206 and Themon Judaeus, *Questions
on the Meteors*, fol. 156r, col. 1.

103. "Vir sapiens dominabitur astris," as Themon expressed it (ibid.).
For a further discussion of this saying, which was frequently attributed
to Ptolemy but does not appear in his works, see Litt, *Les corps céleste*,
pp. 207–8, n. 3.

104. See Grant, *Physical Science in the Middle Ages*, pp. 63–70. Some
of these medieval arguments would reappear in Copernicus's *De revolutionibus*.

105. Grant, *Physical Science in the Middle Ages*, pp. 70–71 and
Grant, *Source Book*, pp. 621–24.

106. The data for the table are derived from Benjamin and Toomer,
Campanus of Novara, pp. 356–63. For the underlying assumptions and
methods, see pp. 54–56, 146, 147, 189, and 399.

107. See Aristotle, *Meteorologica* 1. 14. 352a26–28 (and cf. 1. 3.
339b7–9), and Ptolemy, *Almagest*, bk. 1, chap. 6, p. 10 of R. Catesby
Taliaferro's translation in *Great Books of the Western World*, vol. 16.

108. This distinction appears in *Commentarii Collegii Conimbricensis
Societatis Iesu In Quatuor libros De coelo Aristotelis . . .* (Lyon, 1598),
bk. 2, quest. 2 ("An terra comparatione coeli instar puncti habeat"),
art. 2, pp. 379–80.

109. C. S. Lewis, *The Discarded Image: An Introduction to Medieval
and Renaissance Literature* (Cambridge, 1964), pp. 98–99.

9 Olaf Pedersen   **Astronomy**

From beginning to end the Middle Ages were imbued with the idea that astronomy, more than any
other science, was of immediate relevance to the
human situation. This had been the opinion of the
Pythagoreans, and Plato made it the central theme
of his *Timaeus*. Men of the early Middle Ages, however, were even more impressed by another work
composed early in the sixth century and widely
known thereafter. In the *De consolatione philosophie*, written in prison during the last months before his execution, Boethius (ca. 480–524) likened
himself to suffering mankind in general, remembering the days when he was free to contemplate the
sun, moon, and stars, and to comprehend their
movements by the use of numbers. This led him
to wonder why the Governor of the harmony above
refused to govern the unruly affairs of men here
below.[1] A century later Isidore of Seville pointed
out that astronomy could help men shift their attention from the mundane to the sublime: "This
succession of the seven secular disciplines was terminated in astronomy by the philosophers for this
purpose . . . , that it might free souls entangled by
secular wisdom, from earthly matters, and set them
at meditation upon things on high."[2] And finally,
in the fourteenth century Geoffrey Chaucer noted
in *The Knight's Tale* that experience of the heavens
teaches mankind about the eternal order. After describing the everlasting change and decay in the
earthly realm—"The broad river sometimes waxeth
dry; the great towns see we wane and wend"—
Chaucer calls attention to the immutability of the
world above, concluding: "Then may men by this

order well discern, that thilk [that is, the very] mover stable is and eternal."[3]

But there were also practical reasons for the importance of astronomy within medieval European scientific thought. The problem of orienting churches in the proper east-west direction may have been a factor—though it must be admitted that this was far easier than the analogous task facing Muslim astronomers of determining the *qibla* (the direction of Mecca from a given geographical locality)— so that the faithful would know what direction to face when praying.[4] Astronomy may also have served European sailors, plying the Mediterranean or the coastal waters of the continent, but again their Arabic counterparts faced the much more difficult task of finding their way across the vast Indian Ocean; the latter were forced to rely on astronomical methods of navigation not required in Europe until the discovery and exploration of the new world.

In many civilizations time-reckoning has been a major stimulus for astronomical studies. In medieval Christendom the civil (Julian) calendar was based on the motion of the sun, which is incommensurable with the motion of the moon, on which the equally important ecclesiastical calendar was founded. It might seem a difficult astronomical problem to bring the two systems into harmony and to create a calendar acceptable throughout Christendom; but, in fact, as we shall see, this problem was successfully solved early in the Middle Ages with a minimum of astronomical theory.

A stronger practical motivation for astronomical studies by the later Middle Ages was astrology. In late antiquity, belief in the possibility of astrological divination permeated every social stratum.[5] During the early Middle Ages it was largely held in check by the authority of the Church and the Fathers, and as late as the middle of the twelfth century John of Salisbury could ridicule it.[6] However, it soon began to gain ground as a result of Islamic influence. The thirteenth century saw the first university professorial chair in astrology at Bologna,[7] and princes began to consult astrologers on matters of state. The last barriers to astrology broke during the black death of 1348–50, when learned and unlearned alike placed themselves at the mercy of the local practitioner, often basing even the most intimate details of personal life upon his counsel. A scholar like Petrarch (1304–74) could unmask the astrologer as a greedy imposter,[8] while Nicole Oresme (ca. 1320–82) was exposing him as a public danger,[9] but all in vain. Astrology had come to stay, and many scholars came to regard astronomy principally as a theoretical introduction to astrological practice. Many of the astronomical manuals, tables, and computing instruments developed during the next few centuries owed their existence to their astrological utility.

Astronomy also benefited from its position as one of the seven liberal arts, which from the beginning (at least in theory) formed the educational framework of the medieval schools.[10] In antiquity these seven disciplines were regarded as "liberal," in that they represented proper occupations for free citizens, in contrast to the manual labor of slaves. An echo of this attitude is occasionally heard in medieval educational theory,[11] but already the humanists of the twelfth century preferred to regard the liberal arts as means of liberating man from his innate ignorance and darkness. Since these were consequences of the Fall, the study of the liberal arts, including astronomy, could be regarded as an aspect of the great work of redemption.

Finally, in the thirteenth century, the growth of natural theology led to the belief that the investigation of nature's visible operations could uncover their hidden causes and so reveal some of God's ways. Astronomy became particularly important because Aristotle's Prime Mover of the heavenly spheres could easily be associated with the Christian idea of God as upholder of all creation. The metaphor of the "Book of Nature"—as containing a revelation of its own parallel and supplementary to that of Scripture—began to provide motivation for the study of nature in all its aspects. Slightly damaged by the *devotio moderna* and pietist movements in the fifteenth century, it returned with full force in the seventeenth century to be utilized by Galileo and many others.[12]

Early Latin Astronomy

The beginnings of astronomy in Western Christendom were very modest. The works of Hipparchus, Ptolemy, and the other Greek astronomers disappeared or became useless when knowledge of Greek died out after the final collapse of the western half of the Roman Empire. Astronomical knowledge was limited to the fragments contained in Latin compilations, such as book 2 of Pliny's *Natural History*, Macrobius's neo-Platonic commentary on Cicero's *Somnium Scipionis*, and a treatise on the liberal arts by Martianus Capella entitled *The Marriage of Philology and Mercury*.[13] To these must be added a large body of patristic literature, which now became invested with scientific authority—especially the commentaries on Genesis of St. Ambrose and St. Augustine.[14]

Through such works as these, scholars of the early Middle Ages could get a vague inkling of the achievements of ancient astronomy.

However, since the encyclopedic sources available to them completely ignored both the methods by which the results had been obtained and the mathematical form in which they had been expressed, there was no possibility of extending or even fully understanding the Greek achievement, either observationally or theoretically. Even worse was the fact that the only original Greek work of relevance translated into Latin was the first half of Plato's *Timaeus*,[15] which across a gap of 800 years, transmitted to the Middle Ages the primitive Platonic cosmology, which Greek astronomers themselves had discarded long before. This is comparable to our losing access to all recent scientific literature and beginning anew from a textbook of pre-Copernican astronomy.

What scholars of the early Middle Ages were able to make of this poor heritage is best seen in the works of Isidore, Bishop of Seville (d. 636). His long encyclopedia, the *Etymologies*, gives in twenty books a survey of all the theoretical knowledge and practical skill known to its author, from science, law, and theology to agriculture and cookery.[16] The general level of information is low. In the astronomical section, most of the space is devoted to the constellations and their mythology. Among the merits of the work is Isidore's clear distinction between astronomy and astrology and his unambiguous rejection of the latter as impious superstition—an attitude, which, supported by the authority of St. Augustine, became normative for centuries to come.[17] Isidore's work became immensely popular and gave rise to imitations for 600 years—among them *De universo* of Rhabanus Maurus (9th c.), *De naturis rerum* of the English Benedictine monk Alexander Neckam (d. 1217), and the massive *Speculum naturale* of the French Dominican Vincent of Beauvais (d. ca. 1264).

Isidore wrote another treatise, *De natura rerum*, in which he expounded his astronomical, cosmological, and meteorological lore in greater detail and more systematically than in the *Etymologies*.[18] This work, too, inaugurated a whole genre of similar expositions, among which were a book of the same title by the Venerable Bede (673–735) and the very popular *De imagine mundi* by Honorius Inclusus, who flourished about 1100.

Most of these works of the early medieval period were written by Benedictine monks and used in the monastic schools or the cathedral schools created at Charlemagne's order in the ninth century. It is thus appropriate to refer to the four or five centuries after Isidore as the monastic period of Latin astronomy. Its history has not yet been sufficiently studied, but it is at least clear that the dominant concern was cosmology rather than mathematical astronomy.[19]

Fragments of mathematical astronomy do appear in the works of Isidore and his followers. Isidore himself knew of the axis of the world and its poles, and also of the zodiac as a belt of stars stretching 5 degrees to each side of the ecliptic, the sun's path through the stars; however, he had no notion of celestial coordinates and was unable to define precise positions of the stars. As for the phenomena, Isidore knew the direct (west to east) motion of the sun and planets against the stellar background, but his periods of revolution are correct only for Jupiter and Saturn; he mentioned retrograde motion but provided no further discussion. Bede was a little better informed. His periods of planetary revolution, taken from Pliny, were more nearly correct; he also knew about the periods of visibility of the planets and the maximum elongations of Venus and Mercury. It seems that he, like Pliny, assumed the planets to move around the earth on eccentric circles without epicycles and that he knew the signs of the zodiac within which the apogees and perigees of these orbits were situated. On the other hand, he was apparently unaware of the fact that a simple kinematic model of this kind is incapable of accounting for retrograde motion—the second anomaly of planetary motion.[20]

A more complicated model of planetary motion did not appear until the ninth century, when John Scotus Eriugena wrote his *De divisione nature*, the only original philosophical work of the entire period.[21] Here he revived the so-called geoheliocentric system, usually (but wrongly) attributed to Heraclides of Pontus, according to which Venus and Mercury move in circles around the sun, while the sun moves around the earth.[22] It is not clear whether Eriugena made a similar assumption for the remaining planets (Mars, Jupiter, and Saturn), thus anticipating the Tychonic system of the sixteenth century.

Despite their slender and faulty knowledge, early monastic astronomers successfully developed one discipline at least tenuously connected with astronomy, namely, the science of computing the date of Easter in accordance with the rules laid down by the Council of Nicea in 325. Early instructions on how to perform this computation were superseded when Bede published his large treatise *De temporum ratione*, which gave a careful, detailed, and correct exposition of all the problems of time-reckoning.[23] This book became the basis of the medieval science of *compotus*, an independent mathematical discipline of high standards and immediate practical relevance.[24] However, its connections with astronomy were only indirect. Bede had a firm grasp of the way in which the civil and ecclestiastical calendars were dependent, respectively, on the motions of the sun and moon. But his calendric methods were neither based on a geometrical theory of these motions nor dependent on astronomical observation. Rather, they were founded on purely arithmetic methods and parameters inherited from antiquity.

The Carolingian revival of the ninth century had no dramatic consequences for astronomy, although it did lead to small improvements in astronomical knowledge. On the one hand, the greater intellectual activity in the newly established schools put the teaching of astronomy on a sounder footing. An anonymous manual such as *De mundi coelestis terrestrique constitutione* reveals the greater care with which technical terms were now defined.[25] Such important distinctions as that between synodic and sidereal periods were made, and intricate problems were split into separate questions with their accompanying *oppositiones* and *solutiones*, in anticipation of the later Scholastic method. New mathematical methods for describing planetary motion were tried out: thus, we have a ninth-century diagram with a set of concentric circles, upon which both the direct and indirect motion in longitude and the motion in latitude could be crudely represented.[26] From the tenth century we have a rectangular grid of lines resembling Cartesian coordinates on which the motions of all the planets in longitude and latitude were plotted.[27] Such attempts might have led to interesting developments in both astronomy and geometry had they been followed up.

On the other hand, Carolingian astronomy still suffered from the lack of any significant connection between observation and theory. It possessed no instrument other than a crude sundial.[28] Observational records from this period are few in number and are usually restricted to brief notes of eclipses in calendars, obituaries, or historical annals. Ultimately the cause of this low level of astronomical practice must be traced to the limited and unreliable source material with which the early Middle Ages began. The fact that discussions of astronomy were filled with quotations from Vergil and Ovid rather than Ptolemy and Theon reveals better than anything else how little the early medieval period had to work with.

The Impact of Islamic Astronomy

While Latin scholars were doing what they could with their limited heritage, their Islamic counterparts succeeded in recovering a substantial portion of the Greek astronomical corpus. This body of material had been preserved and, indeed, improved and extended in Islam, and it was by acquiring Islamic astronomical works and Greek astronomical works preserved in Arabic that the Latin world would finally achieve a high level of astronomical knowledge. The first contact with Islamic astronomy was made in the last decades of the tenth century, when a few students from northern Europe crossed into Spain to study in monasteries on the southern slopes of the Pyrenees, where Islamic influence could already be felt. With the return of these students to northern Europe, the schools in which they taught became centers for the dissemination of Greco-Arabic science.[29] The best known of these pioneers is Gerbert of Aurillac (ca. 945–1003), who introduced the abacus, the armillary sphere, and, apparently, the astrolabe as teaching aids in the schools in which he taught before becoming Pope Sylvester II.[30]

It is hazardous and potentially misleading to attach decisive importance to a single event in the history of science; nevertheless, it appears that much of the development of Latin astronomy during the eleventh and twelfth centuries can be explained as a consequence of the introduction of the astrolabe. This beautiful and sophisticated little hand instrument dates from late antiquity and underwent further development in Islam. It made its entry into the Latin world through two brief treatises, *De mensura astrolabii* and *De utilitatibus astrolabii*, translated from Arabic to Latin in the second half of the tenth century and further diffused in the eleventh century through copies made by the Benedictine monk Hermann Contractus of Reichenau.[31] A specimen of the instrument itself was in the possession of Radulf of Liège about 1025.

The astrolabe (see fig. 17) consisted of a circular disk of brass or copper, called the "mother," which had on one side a graduated circle and an alidade (a sighting device consisting of a movable strip of metal fastened to the "mother" in the center and having a sighting hole on each end). With the alidade the altitude of the sun or a star could be measured to an accuracy of about one degree.[32] An instrument of this kind had never before been at the disposal of Latin astronomers, who were now able to measure certain celestial phenomena and express them with numerical precision rather than in vague, qualitative terms. As a consequence, astronomy once more became linked with mathematics, as it had not been during the early Middle Ages. On the opposite side of the "mother" from the alidade (and fitting into its hollowed-out body) was a tympan, a thin plate of brass on which a stereographic projection of the principal circles of the heav-

Fig. 17   A Persian astrolabe, dating from A.D. 1428, now in the
                     Museum of Decorative Art, Copenhagen. The photo-
                     graph on the left shows the "mother" and the alidade.

enly sphere was engraved. In front of the tympan was the "rete," a
similar stereographic projection of the ecliptic and a number of the
brighter fixed stars, engraved on a circular disk from which most of
the metal was cut away to show the circles of the tympan behind it.
This device made it possible to use the astrolabe as a computer, by

The photograph on the right shows the tympan and
rete on the other side of the astrolabe.

which a great number of problems in spherical astronomy could be
solved simply by turning the rete and the tympan relative to one another. The most important application was determination of the exact
time of day or night from an observation of the altitude of the sun or
one of the stars mapped on the rete.

The first report of the astrolabe's use for serious observational purposes in the West is from 18 October 1092, when Walcher, prior of the Abbey of Malvern, used it to determine the time of a lunar eclipse that he observed in Italy. This was in itself an important step, but it soon led to even more momentous consequences. Walcher realized that it would be possible to keep track of the motion of the moon, so important for the ecclesiastical calendar, by calculating a table of lunations from well-known parameters fixed to a definite epoch by the eclipse observation. In 1108 he published such a table for the 76-year period 1036–1112. This table and a later version dating from 1120 seem at present to be the earliest tables constructed by a Latin astronomer.[33] As such, they not only inaugurated a hitherto unknown genre of astronomical literature in the West, but also had a catalytic effect upon subsequent developments.

The fact was that in order to exploit the astrolabe to the full, it was necessary to have a rather comprehensive set of astronomical tables, among which a table of the position of the sun for each day of the year was the most important. Such a set of tables was first provided by Adelard of Bath (fl. 1116–42), who in 1126 translated the *Astronomical Tables of al-Khwarizmi*.[34] This presented Western astronomers with a Latin version of a complete Arabic *zij* (a set of astronomical tables accompanied by precepts regarding their use), containing a wealth of previously unknown material, such as chronological tables relating to the Persian, Egyptian, Islamic, and Julian calendars, tables of the mean motions of the sun, moon, and planets with their proper "equations" or *prosthaphairesis*-angles, tables of conjunctions of the sun and moon, eclipses, a trigonometric table of sines, and much more. The introductory chapters of the *Astronomical Tables* gave brief rules or *canones* for the correct use of the tables.

At about the same time Masha'allah's standard work on the astrolabe, *De compositione et utilitate astrolabii*,[35] was made available in a Latin translation by John of Seville, while the fundamental geometrical principles of the instrument were recovered through Hermann the Dalmatian's Latin version of an Arabic translation of Ptolemy's *Planisphere*[36] (the Greek original of which is lost), in which the method of stereographic projection is explained.

The new tables were a new and powerful tool in the hands of Latin astronomers, but they were also a challenge. The challenge arose from the fact that Adelard had happened to pick a rather unsuitable *zij* for translation, al-Khwarizmi's tables being based upon the epoch of the *hegira* and the meridian of Arim, a mythological mountain in the middle of the world, actually the ancient Indian capital of Ujjain. In

1141 a certain Raymond tried to remedy some of these defects by
calculating a set of tables for Marseilles,[37] while at approximately the
same time Robert of Chester adapted Adelard's version of al-Khwarizmi's tables to the meridian of London. In 1149–50 he published another set of London tables based upon different sources.[38] Before the
end of the century we find tables for Hereford, Pisa, and several other
localities. This large body of material has not yet been sufficiently investigated, and it is too early to say anything definite about it, except
that it is based on at least two sources besides al-Khwarizmi. One is
the famous collection usually called the *Toledan Tables*, of which several Latin versions exist, based on the lost *zij* of the eleventh-century
Cordoban astronomer and instrument-maker az-Zarqali.[39] The other
source is the great (and still extant) *zij* of al-Battani, who lived in the
ninth century as a younger contemporary of al-Khwarizmi.[40] The latter was utilized by Robert of Chester for his second set of London
tables, and in the translation of Plato of Tivoli (entitled *De motu
stellarum*) it introduced Western astronomers to the finest fruits of
Islamic astronomical research.[41]

Having become familiar with the astrolabe and various sets of astronomical tables at their disposal, Latin astronomers were still at the
serious disadvantage of having little knowledge of the astronomical
theories upon which the tables were based. A little could be extracted
from the *canones* to the tables, but this was both insufficient and be-
wildering, since the various sets of tables were based upon different
theoretical systems. Al-Khwarizmi's tables, for example, had their
theoretical background in the Hindu astronomy of the great *Sidd-
hantas*, with models and parameters of planetary motions different in
many ways from those of Ptolemy's *Almagest*.[42] By contrast, al-
Battani's tables transmitted a much purer Ptolemaic tradition, while
the Toledan tables reflected a mixture of Greek and Indian methods.
Under the circumstances, it was imperative that the West acquire a
manual or exposition of theoretical astronomy in general and planetary theory in particular, and several translators attempted to meet
the need. In 1137 John of Seville produced a Latin version of a theoretical primer by the ninth-century Baghdad astronomer al-Farghani,
which (under the title *Rudimenta astronomica* or *Liber differentiarum*) provided a nonmathematical exposition of the fundamentals of
Ptolemaic astronomy.[43] Later in the century, Ptolemy's *Almagest* was
itself translated into Latin, first directly from the Greek about 1160,
and later from the Arabic by Gerard of Cremona about 1175.[44] This
was the most significant event in the history of astronomical translation.

The Latin *Almagest* provided by far the most reliable and comprehensive exposition of what Ptolemy and his predecessors had achieved in mathematical astronomy. Beyond that, it also revealed the methods by which their results had been achieved and, thus, prepared the way for a renewed understanding of astronomical research. It showed how celestial phenomena could be predicted by means of geometrical models constructed to simulate the behavior of the celestial bodies. It also revealed how the numerical parameters of such models could be derived from suitable observations made by well-designed instruments, the construction and use of which were clearly described. Finally, it did away with the mysteries of existing Latin tables by showing how Ptolemy himself had been able to reduce his numerical procedures to tabular form. It would thus seem that astronomy, after a delay of a millenium, was about to make new gains. Nor was motivation for a further development of mathematical astronomy lacking, for in the wake of the many translations from the Arabic there was a burst of astrological curiosity, which could be satisfied only by numerical calculations requiring considerable theoretical insight.

But despite this promising situation, theoretical astronomy made but slow progress, and the first century after the translation of the *Almagest* was in many respects a period of confusion. Neither the prevailing intellectual climate nor the educational system was prepared to receive the enormous influx of new knowledge provided by the translators. The old monastic and cathedral schools still taught the four disciplines of the quadrivium in very rudimentary fashion, and such schools (often manned by only a single master) were quite incapable of assimilating a complex and sophisticated work such as the *Almagest*. The only solution was to provide for more specialized studies within a more elaborate system of higher education. The response to this challenge was the emergence of the universities, with their various schools or faculties, in each of which a team of masters could make a united attack on the mountain of knowledge now placed before them.

University Astronomy

We are poorly informed regarding the scientific curriculum of the faculties of arts in European universities during the first half of the thirteenth century, but it appears from later developments that considerable effort must have been devoted to the study of Euclidean geometry, plane and spherical trigonometry, and algebraic procedures based upon both the decimal and the sexagesimal system and expressed in the new Hindu (or "Arabic") notation. The outcome of these mathematical labors began to appear about the middle of the thirteenth century, when the teaching of astronomy assumed the shape that it would retain during the remainder of the Middle Ages. From about this time on, we have a growing collection of manuals or textbooks that illustrate the changing preoccupations and developing abilities of university professors of astronomy. Often these treatises were written in the same hand, or at least bound together in a single codex, thus identifying them as parts of a more or less standardized *corpus astronomicum*.[45]

The kernel of this collection comprised three rather brief works by an Englishman named Johannes de Sacrobosco or John of Holywood, of whom very little is known except that he taught at the University of Paris from about 1230 to 1255. The first of the three works was a treatise called *Algorismus vulgaris*, which described simple arithmetical operations, including the extraction of square and cubic roots in the decimal system and the Hindu notation.[46] The second was the *Tractatus de sphaera*, a brief and nonmathematical exposition of spherical astronomy, with a few sections on cosmology and a very sketchy final chapter on the motion of the sun and the planets.[47] The last part of the triad was a *Compotus* or manual of time-reckoning as a branch of astronomy, to which the other two treatises formed an introduction.[48] It was natural, therefore, to add to this collection Grosseteste's *Calendar*, a standard work in the latter half of the thirteenth century.[49] In addition, even the earliest specimens of this small collection of writings usually contained a treatise by a certain Robert the Englishman on the so-called old quadrant—a small and inexpensive instrument for measuring altitudes, which could also be used as a sundial.[50]

Such was the scope of astronomical knowledge imparted to university students about the time when astronomy was first firmly established in the curriculum. It is necessary to stress the elementary character of these first manuals. They were clearly intended for beginners, and it would be a mistake to view them as evidence of the astronomical knowledge of more advanced scholars. This advanced knowledge appears, for example, in the impressive array of Greek and Arabic astronomical authorities quoted by Roger Bacon in his *Opus maius*, written in the early 1260s, or Albert the Great's *Speculum astronomiae*, a sort of critical bibliography of astronomical (and, therefore, licit) and astrological (and, therefore, illicit) books.[51] In short, the gap between elementary and advanced scientific studies was already clearly apparent, with the former limping far behind the latter. Since the medieval university had no higher faculty of science (apart from medicine), the only way of bridging this gap was to extend the teaching of astronomy within the faculty of arts beyond the level established by Sacrobosco's manuals. The subsequent development of the *corpus astronomicum* reveals how this was attempted.

Sacrobosco's *Sphaera* dealt with planetary theory only in its unsatisfactory final chapter. That this was considered a defect is evident from the fact that already the first versions of the *corpus* were provided with an anonymous treatise entitled *Theorica planetarum*,[52] which was destined to play a remarkable role in the history of medieval astronomy. It is still extant in more than two hundred manuscripts—a testimony to its enormous popularity in comparison with that of Ptolemy's *Almagest*, of which fewer than forty Latin manuscripts survive. The popularity of the *Theorica* as an elementary textbook was well deserved. It is divided into eight brief chapters, the first of which presents the theory of the sun of the Greek astronomer Hipparchus, according to which the sun moves uniformly around a deferent circle eccentric to the earth (that is, the center of the universe) at the rate of 59 minutes 8 seconds a day from west to east. This accounts for the sun's motion through the zodiac, which, in turn, is carried from east to west by the daily rotation of the stellar sphere. Chapter 2 contains Ptolemy's final theory of the moon, illustrated in fig. 18. The moon, *L*, moves uniformly from east to west around an epicycle, which, in turn, moves from west to east around an eccentric deferent

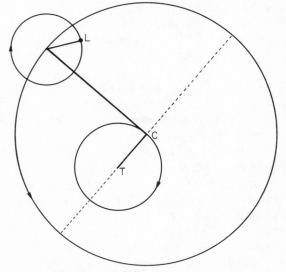

Fig. 18   The theory of the moon.

circle; meanwhile, the center of the deferent circle, C, revolves around the earth, T, from east to west. The motion of the epicycle around the deferent is not uniform with respect to the center of the deferent, C, but with respect to the earth, T, which is the "equant point" of the lunar theory.

Chapter 4 presents the theory of the superior planets—Mars. Jupiter, and Saturn. The planet, P (fig. 19), moves uniformly around its

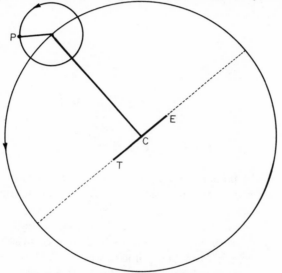

Fig. 19   The theory of the superior planets.

epicycle, while the center of the epicycle is carried from west to east around the eccentric deferent. The motion of the epicycle around the deferent is uniform with respect to the equant point, E, so situated that TCE is a straight line and TC equals CE. This is an accurate rendition of Ptolemy's final theory. The theory of the inferior planets, Mercury and Venus, is presented in the fifth chapter. The theory of Venus is essentially the same as that of the superior planets. The theory of Mercury, however, is considerably more complex. Mercury moves on an epicycle, the center of which moves from west to east on an eccentric deferent with center at C (fig. 20). Meanwhile C moves in the opposite sense about a small circle with center at Z. The motion of the epicycle about the deferent is uniform with respect to the equant point, E, on the circumference of the small circle and midway between Z and the center of the earth, T. The text takes notice of the fact that the motion of the sun (designated by the direction from the earth to the mean sun) is one of the components of which the mo-

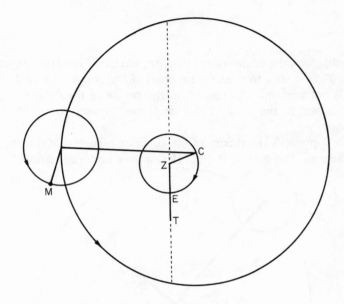

Fig. 20   The theory of Mercury.

tions of all the planets are composed; this appears as a fact which the author does not comment upon. It found no explanation until Copernicus introduced his heliocentric system.

Chapter 6 is concerned with the theory of retrograde motion and marred by an error in the determination of the stationary points of the planets, which from this point onward would be repeated again and again in medieval astronomy. The seventh chapter is a rather confused exposition of problems related to motion in latitude, eclipses, construction of tables, and other matters in which the author was clearly less at home. There is a final chapter on the astrologically important aspects of the planets—that is, their angular distances from the sun or each other—but otherwise the text is completely free of astrology. The exposition is mathematical throughout: planetary kinematics is explained by geometrical models, represented by carefully drawn figures, ultimately derived from the *Almagest* through channels that have not yet been identified. The many variables and other concepts of planetary theory are precisely defined and the fundamental relations between them explained.[53]

The historical importance of the *Theorica planetarum* follows from three of its principal characteristics. First, it definitely settled the question of the vocabulary of theoretical astronomy. Earlier expositions had reflected the uncertainties of the first translators struggling to create suitable Latin equivalents for technical terms in Arabic or Greek, often with curious arabisms as the result. By contrast, only two or three Arabic terms were retained (and Latinized) in the *Theorica*, which also established a successful one-to-one correspondence between words and concepts, thus doing away with the ambiguities of earlier texts.[54] Second, the geometrical models used in the *Theorica* were clearly derived from Ptolemy. The fact that they were presented to virtually all medieval students of astronomy during more than two centuries meant that alternative astronomical theories stood a very poor chance. Finally, the *Theorica* showed how theoretical astronomy could be taught without any reference to cosmology and with no attempt to give actual dimensions or locations in space of the geometrical devices employed.

It is apparent that the *Theorica* was soon felt to be inadequate on a number of important questions, for new treatises designed to remedy these inadequacies were soon added to the *corpus astronomicum*. The lack of numerical parameters was remedied by the inclusion of the *Toledan Tables* with their *canones*, which enabled students to calculate planetary positions according to theories conveyed in the *Theorica*. An obscure point was the precession of the equinoxes, on which the *Theorica* had sketched three different theories without choosing among them. The addition of Thabit ibn Qurra's *De motu octavae sphaerae*[55] meant that the problem was solved in terms of the theory of trepidation, according to which the sphere of the fixed stars performs an oscillatory motion relative to an empty ninth sphere on which the circles of the ecliptic are projected; this causes the equinoxes to move back and forth about a mean position, so that the rate of precession becomes a function of time, contrary to the theory of Ptolemy and al-Battani.[56] Finally, the gap between planetary theory and cosmology was bridged by two other treatises of Thabit, entitled *De hiis quae indigent antequam legatur Almagesti* and *De quantitatibus stellarum et planetarum et proportio terrae*. Here the absolute distances of the heavenly spheres were calculated on the principle that there must be no overlapping spheres and no empty space between them, as explained by Ptolemy in his *Planetary Hypotheses*.[57] A fourth brief treatise by Thabit, *De recta imaginatione sphaerae*, supplemented Sacrobosco's *Tractatus de sphaera* by providing definitions of a number of spherical astronomical concepts.

With these thirteenth-century additions, the second stage of the *corpus astronomicum* was complete. During the following century, several new developments occurred. First, the now obsolete calendar of Grosseteste was replaced by a more up-to-date work by Peter Nightingale (Petrus Philomena de Dacia), which was in general use until the middle of the fifteenth century.[58] Similarly, the *Toledan Tables* were often replaced by the *Alfonsine Tables*,[59] issued in Toledo in 1272 but preserved only in a Latin version edited in Paris about 1327 by John of Saxony. These tables were based on Ptolemaic parameters and were therefore believed to be more accurate than the *Toledan Tables*; however, modern research has shown that this is not true.[60] In addition, the *corpus* was enlarged by the addition of a number of treatises on instruments for observation or computation, of which more will be said below. Finally, its computistical contents were reinforced by the addition of a number of calendric tables and sometimes by the substitution of a commentary by Peter Nightingale for the old *Compotus* of Sacrobosco.[61]

Mathematical versus Physical Astronomy

The *corpus astronomicum* defined an astronomical curriculum that initiated university students into all of the principal aspects of astronomy: they were taught spherical astronomy and planetary theory; they were provided with calendars and tables, which enabled them to calculate the positions of the heavenly bodies and to predict particular phenomena, such as conjunctions and eclipses; and they were taught how to construct instruments for both observation and computation. The gap between elementary education and advanced studies had been successfully bridged. Nevertheless, this remarkable achievement met with strong opposition from another quarter within the faculty of arts. For while the translators recovered the Greek and Arabic traditions in mathematical astronomy, they also provided Latin versions of all of Aristotle's works, supplemented by commentaries, among which those of the twelfth-century Cordoban philosopher Averroes (or Ibn Rushd) stand out.

Aristotle had, of course, been totally unaware of the highly sophisticated mathematical methods by which Hellenistic astronomers would attempt to save the phenomena of the heavens— besides which, Aristotle clearly preferred a causal account over a mathematical one. Thus, in his *De caelo* medieval scholars encountered a completely different analysis of the heavens, based upon the fundamental principles of Aristotelian physics. In the middle of the world, Aristotle explained, was the spherical, immobile earth surrounded by the elementary spheres of water, air, and fire. Surrounding these sublunary spheres was a system of tightly fitting celestial or supralunary spheres composed of a fifth element or ether and rotating about various axes through their common center—each with a constant angular velocity. Taken together, the motions of the spheres were meant to account for the observed motions of the heavenly bodies. The fact that this conception of the universe was revived at the same time as mathematical astronomy was once again coming into its own presented thirteenth-century scholars with a problem.

On the one hand, it was clear that the universe was a material entity, the behavior of which must therefore be deducible from the laws of physics. On the other hand, it was equally clear that mathematical astronomy made use of geometrical devices that violated those laws of physics: there were deferents and epicycles with centers outside the center of the world, and there were circular motions with variable angular velocities—a clear violation of the uniform rotations that Aristotle had assigned to celestial matter. Averroes had already stressed the incompatibility of the two schemes, and his contemporary al-Bitruji had attempted to harmonize them by eliminating epicycles and eccentrics and allowing the planet to move about on its rotating sphere (in effect, placing epicycles on the surfaces of the spheres themselves).[62] Al-Bitruji's work was translated into Latin by Michael Scot in 1217 as *De motibus celorum*, but his theory was never developed to the point of tables and calculations; among its few adherents was the fourteenth-century German astronomer Henry of Langenstein, who in 1364 completed a treatise, *De reprobatione eccentricorum et epiciclorum*, in which he attacked the idea of epicycles and eccentrics.[63] The situation thus tended toward a stalemate, with strong arguments on both sides. The physicists could appeal to the generally accepted laws of Aristotelian physics as the only possible basis for a true system of astronomy. Their mathematically oriented opponents could reply that although their constructions might raise troublesome physical questions, they alone were able to make accurate predictions of celestial phenomena.

On the philosophical plane, this dilemma led to long discussions of the metaphysical status of mathematical theories in science. In the realm of astronomical theory, it led to a series of attempts to find a compromise between the Aristotelian and Ptolemaic systems, by embedding as much as possible of Ptolemy's planetary kinematics in the machinery of material spheres. This scheme was based on the suggestion of Ptolemy himself in his *Planetary Hypotheses*, and it was presented to Western astronomers in the two small cosmological treatises by Thabit ibn Qurra mentioned above. For a fuller discussion, see chapter 8 of this volume.[64] Here we shall leave the subject with the remark that in the long run the mathematical astronomers had the "trump card," since late medieval society was more interested in exact predictions of heavenly phenomena than in Aristotelian speculations on the physical nature of the universe, primarily because of the increasing interest in astrology. Ironically, it was astrology (now regarded as a superstition) which saved the delicate flower of mathematical astronomy from the hot winds of Aristotelian natural philosophy.

Instruments and Observations

Until about the middle of the thirteenth century the only instruments available to astronomers were sundials, quadrants, and astrolabes—unless we consider the water clock an astronomical instrument. The latter was used in monasteries before 1200, as we can gather from Jocelyn's account of how the fire in the church of Bury St. Edmund's was extinguished by water fetched from the clock.[65] But even with this scanty equipment it was possible to undertake fairly extensive programs of observation. In 1274 an unidentified Danish astronomer at Roskilde measured the meridian altitude of the sun day by day with an astrolabe; the results were used for determining the length of daylight by trigonometric calculation with a *kardaga sinuum*, a kind of nomogram that substituted for a table of sines.[66] The final results were then entered into the calendar of the cathedral chapter. Such a table of daylight became a standard feature of the calendars of Peter Nightingale and William of St. Cloud late in the thirteenth and early in the fourteenth centuries.

Toward the end of the thirteenth century the arsenal of astronomical equipment began to grow. A Jewish astronomer, Jacob ben Machir ibn Tibbon (ca. 1236–1305), often called Profatius Judaeus, who worked at Montpellier, invented the so-called new quadrant, which in a very ingenious way combined some of the features of the old quadrant with those of the astrolabe (see fig. 21). It was used for determining time from observations of the sun, and it became known to Latin astronomers through a *Tractatus quadrantis* by Peter Nightingale,[67] which often found its way into the *corpus astronomicum*.

The torquetum (that is, Turkish instrument) was also introduced before the end of the century.[68] It contained several brass plates with engraved and divided circles and an alidade for sighting toward a star (see fig. 22). Adjusted to the latitude of the place of observation, it could be used not only for calculating the time of observation, but also for determining the pairs of coordinates of the star in the horizontal, equatorial, and ecliptic systems. The modern astro-compass is based on similar principles.

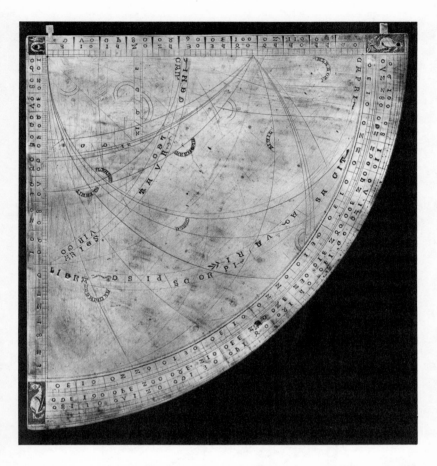

Fig. 21   The new quadrant of Profatius Judaeus. A fourteenth-
century exemplar owned by Merton College, Oxford.
For a description of the Merton instrument, see R. T.
Gunther, *Early Science in Oxford*, 2 (Oxford, 1923):
165–69.

It is possible that the *torquetum* was destined not so much for observational as for teaching purposes, since it made it possible to demonstrate the basic concepts of spherical astronomy. A simpler device
for the same purpose was the armillary sphere, or *sphaera materialis*,[69]
which was introduced in the fourteenth century in the form of a small
hand instrument with two sets of brass rings representing the fixed
and movable circles of the heavenly sphere. Constructed on a larger
scale and with graduated circles, it could be used for measurements,

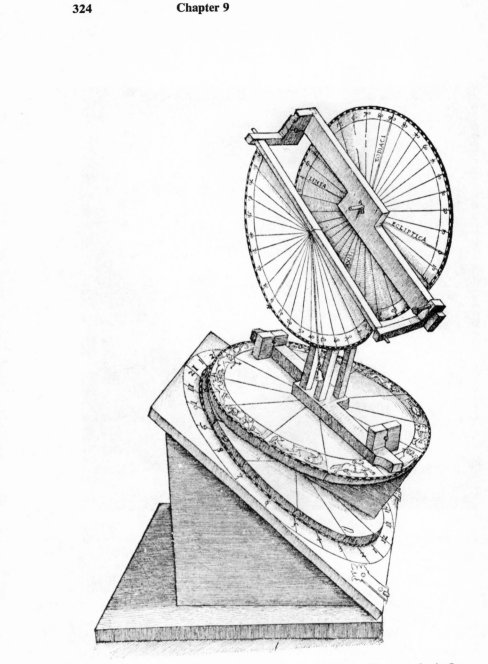

Fig. 22   A drawing of a torquetum, from Peter Apian's *Intro-*
                    ductio geographia (Ingolstadt, 1533).

just as Ptolemy's *astrolabon*,[70] from which it was ultimately derived
(see fig. 23). A purely observational instrument was the Jacob's staff
or cross-staff (see fig. 24), invented by a Jewish astronomer from

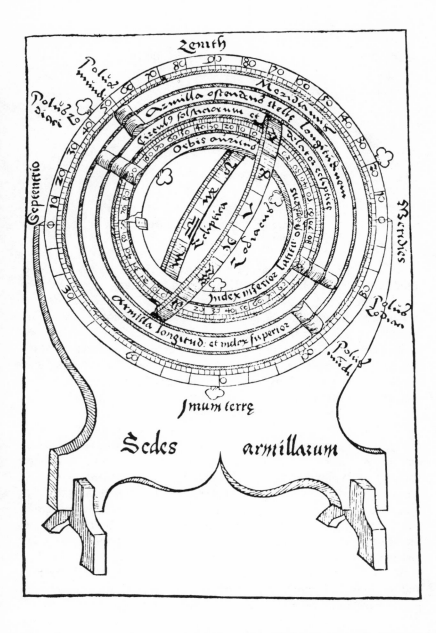

Fig. 23   An armillary sphere, from Regiomontanus, *Scripta . . . de torqueto, astrolabio armillari, regula magna Ptolemaica, baculoque astronomico . . .* (Nuremberg, 1544), fol. 20v.

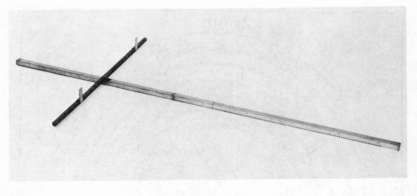

Fig. 24   A sixteenth-century Jacob's staff, now in the British Museum.

Provence, Levi ben Gerson (1288–1344).[71] It consisted of a rod fitted with a movable cross-staff; to measure the angle between two objects, for example the sun and the horizon, one end of the rod was held to the eye, and the cross-staff was adjusted until its two ends touched the sun and the horizon, after which the altitude of the sun could be read from a scale engraved on the rod. In the next century the cross-staff was taken to sea and became one of the most popular and useful navigational instruments in the age of the great discoveries; used in connection with a table of solar declinations, it enabled one to determine the latitude of the ship.

The fourteenth and fifteenth centuries saw several other types of measuring instruments of great ingenuity, but none was widely used. One of the most interesting was the *rectangulus* invented by Abbot Richard of Wallingford (ca. 1292–1335) and described in his *De arte componendi rectangulum* and *Ars operandi cum rectangulo*.[72] It was a kind of skeleton *torquetum* with graduated rods instead of graduated circles. Several other instruments from Richard's hand point to the great monastery of St. Albans as one of the principal centers for instrument-making during this period.

All of the instruments described thus far were small and portable and, therefore, required no permanent mounting or support. This explains why there were no Western observatories during the Middle Ages—if, by observatory, we mean a substantial structure for housing or supporting an instrument that was left in place between observations. The largest medieval astronomical instrument on record dates from 1318, when the French astronomer Johannes de Muris determined the latitude of Evreux, the obliquity of the ecliptic, and the time of the vernal equinox with an instrument built for this specific purpose.[73] It was probably of wood and had a radius of fifteen feet and a graduated arc of fifteen degrees, fastened in the plane of the meridian to an immovable stone. Unfortunately, no more is known of this large instrument, which seems to have been unequaled anywhere in the Latin world. The absence of large instruments and observatories reveals that medieval astronomers were not generally interested in precision measurements or long-term programs of observation. However, the many hundreds of observational records scattered throughout extant manuscripts have not yet been collected and analyzed, and, therefore, firm general conclusions cannot yet be drawn. But it does seem that most of them are brief notices of spectacular phenomena, such as eclipses and comets, and observations of this kind had little impact on theoretical astronomy. It must be noted, however, that the equinox observations of Johannes de Muris mentioned above (and a few others of the same kind) are exceptions. These were essential for checking and correcting the tables used by the astronomer, for they led to the determination of the actual amount of "precession" —that is, the motion of the fixed stars in longitude relative to the equinox—since the epoch of the tables.

One of the most important astronomical instruments of modern times is the mechanical clock, which made its first appearance in Europe during the last decades of the thirteenth century. Its highly developed form when it appeared on the European scene suggests a long evolutionary development, the history of which is not yet completely clear; water clocks, astrolabes with gear trains, and chinese mechanical timepieces may all be among its ancestors.[74] The central feature of the mechanism was the escapement, which in medieval clocks was always of the verge and foliot type. A horizontal rod with weights attached to each end oscillated on a vertical axis or verge provided with two pallets, which engaged a vertical crown wheel in such a way that the wheel advanced one tooth for each swing of the rod. The motion could be regulated by adjusting the weights on the rod; however, the period of oscillation was influenced not only by the moment of inertia of the rod and weights, but also by the friction of the entire mechanism. Thus, the entire clock took part in the process of timekeeping—with a resulting lack of accuracy and reliability. Medieval clocks had to be adjusted frequently by reference to sundials; consequently, the art of gnomonics (construction of sundials) spread rapidly in the wake of the clock.

Owing to these defects, the mechanical clock was useless for precise astronomical measurements. During the fourteenth century it came

into general use in public places, and in the fifteenth century it began
to make its way into private homes. Moreover, from an early stage
astronomers became interested in the new invention as a means of
simulating astronomical phenomena. Sophisticated trains of clockwork
made it possible not only to show the hour and minute, but also to
illustrate the motion of the sun, moon, and planets. Such highly complicated astronomical showpieces were already in evidence in the first
half of the fourteenth century, when Richard of Wallingford built and
described the famous clock of St. Albans.[75] Even more famous were
the various astronomical clocks built by Jacopo de'Dondi and his son
Giovanni (1318–89) and described by the latter in his *Tractatus
astrarii*.[76]

Recent research into late-medieval astronomy has revealed many
efforts by medieval astronomers to construct instruments by which
the calculation of planetary positions (primarily longitudes) could be
performed mechanically. The stimulus behind such efforts, of course,
was the length and difficulty of the procedures by which such astronomical functions as the "equation of center" (a correction owing to
the eccentricity of the deferent circle) and the "equation of argument"
or "*prosthaphairesis* angle" (the angle under which an observer at the
center of the earth would see the radius of a planet's epicycle) were
calculated. The first Western astronomer to show how to find these
equations mechanically was Campanus of Novara, whose *Theorica
planetarum* described for each planet a device to achieve this, later
called an *equatorium*. The basic idea was very simple. The deferent
was drawn on a round disk of parchment and divided into degrees.
To this another disk representing the epicycle was attached. Both disks
could be turned around their centers according to the mean motions
in longitude and anomaly found in tables; the longitude of the planet
could then be read off the fixed ecliptic circle drawn upon the ground
plate of the instrument by means of a thread stretched from the center
(representing the earth) to the point on the epicyclic disk representing the planet.

Although it was easy to understand and apply Campanus's instruments, they were undeniably cumbersome to construct, since a complete set comprised six or seven different devices with a total of about
twenty graduated circles. It was, therefore, a great step forward when
Peter Nightingale issued his *Tractatus de semissis* in 1293, describing
an equatorium of a much more practical type.[77] A single instrument
served for all planets, and only eight graduated half-circles were required. Its high degree of sophistication appears from the fact that the

deferent circle is common to all the planets and not represented upon
the instrument except as the locus of a pin on a rotating ruler. Essentially the same scheme was used a century later in the equatorium
usually attributed to Geoffrey Chaucer and described by him in a Middle English treatise.[78] About 1340 the French astronomer Johannes
de Lineriis made an equatorium on approximately the same principles.[79]

The End of the Middle Ages

Despite an abundance of source material, we know very little about
astronomy at the end of the Middle Ages. However, even a superficial examination of the manuscripts of the period reveals that astronomy increasingly catered to the practical needs of a changing society.
Astronomical literature was enriched by many treatises of a practical
sort, and the old *corpus astronomicum* lost some of its popularity. The
many treatises on calendar reform provoked by the popes and by the
councils of Constance and Basel bear witness to the willingness of
astronomers to serve the church by repairing the more and more obvious defects of the Julian calendar.[80] The secular enterprises of discovery and colonization also made new demands on astronomical
navigation and cartography, the latter subject being greatly advanced
by the translation of Ptolemy's *Cosmographia* early in the fifteenth
century by Jacopo Angelo da Scarperia. Finally, astrology became
part of everyday life and gave rise to vast amounts of publication of
the most varied kind. Thus, the huge *Libri tres anaglypharum*, written
about 1456 by Nicolaus de Dacia, gave a comprehensive survey in
three books of planetary theory, astrology, and astrological medicine;[81] and the many yearly calendars with astrological predictions
became one of the principal sources of income for the first generation
of printers.

Hand in hand with this growing interest in the practical applications
of astronomy went the geographical spread of astronomical knowledge
and teaching. Until the middle of the fourteenth century, the Paris
school of astronomers was predominant north of the Pyrenees. Thereafter, its importance began to decline while other schools took over.
The achievement of the scholars at Merton College, Oxford, on
mathematics and motion is well known; it seems that they had an
equally important impact on late medieval astronomy—although the
many treatises and tables produced by them and their English disciples
from the middle of the fourteenth to the middle of the fifteenth cen-

tury remain largely unexamined. At the same time, new centers of astronomical activity emerged with the spread of universities into eastern Europe, where the universities of Krakow, Prague, and Vienna were soon able to compete with earlier centers in the production of new tables, textbooks, and commentaries.

Best known is the astronomical tradition at Vienna, where Henry of Langenstein (ca. 1325–97) headed a long line of distinguished teachers. His successor was John of Gmunden (ca. 1380–1442),[82] the teacher of Georg Peurbach (1423–61), who, in turn, taught Johannes Müller, known as Regiomontanus (1436–76).[83] Peurbach and Regiomontanus worked together during the 1450s, making a serious attempt to raise the level of astronomical teaching. The old *Theorica planetarum* was discarded and replaced by a *Theoricae novae planetarum*, which was an outgrowth of a course of lectures given by Peurbach in 1454.[84] It was much longer than its predecessor, but not (as many historians have claimed) an attempt to return to the true Ptolemaic astronomy of the *Almagest*. The form and style of the old *Theorica* were preserved and some of its most obvious errors corrected, while the theory of trepidation was developed in various ways and the machinery of spheres was made an essential part of the exposition. Nevertheless, it has an important place in the history of astronomy, for Copernicus was, in all probability, initiated into the mysteries of planetary theory through a commentary on Peurbach's *Theoricae novae planetarum* written by the Cracovian astronomer Albert of Brudzewo.[85]

Peurbach's book was the last attempt to improve medieval astronomy on the basis of traditional sources and presuppositions. A new approach was adopted by the Vienna astronomers after Andrea Silvio Piccolomini introduced the new humanistic study of Greek into the university. In the humanistic tradition, Peurbach and Regiomontanus were converted to the idea of reforming astronomy by reverting to its classical sources. Their resolution to produce a new translation of the *Almagest* directly from the Greek was frustrated by the untimely death of Peurbach in 1461 and of Regiomontanus in 1476 and resulted only in an incomplete Latin paraphrase.[86] However, in the meantime, Regiomontanus took the decisive step of establishing his own private astronomical institution in Nuremberg, where he could devote all of his attention to the task of placing the study of astronomy on a new footing.

The Nuremberg institution owes its importance to Regiomontanus's new understanding of the connection between observation and theory.

Astronomy, he believed, was to be reformed through an improved knowledge of the celestial phenomena themselves, beginning with the most important element in planetary theory, the sun. Thus, he had two instruments of very large dimensions built, one of them a cross-staff five or six cubits long, the other a *triquetrum* or parallactic ruler of the Ptolemaic type for measuring zenith distances (see fig. 25).[87] Observations were to be carried out at regular intervals according to a long-term program. After the death of Regiomontanus, his observational program was continued by his assistant, Bernard Walther (1436–1504). To the astronomical equipment of the Nuremberg institution Bernard added an armillary sphere and (for the first time in the history of astronomy) a mechanical clock—though the latter instrument proved to be insufficiently accurate for astronomical purposes, and times still had to be determined from stellar altitudes. The many hundreds of Nuremberg observations were destined to be at the empirical foundations of sixteenth-century astronomy,[88] and they were employed by both Copernicus and Tycho Brahe in their attempts to reform and revitalize the science of astronomy.

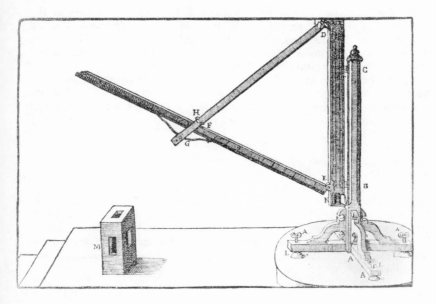

Fig. 25   A parallactic ruler of the Ptolemaic type, from Tycho Brahe's *Astronomiae instauratae mechanica* (Wandesburg, 1598).

Notes

1. Boethius, *The Consolation of Philosophy*, trans. V. E. Watts (London, 1969), 1.2 and 1.5.
2. Translated in Ernest Brehaut, *An Encyclopedist of the Dark Ages: Isidore of Seville* (New York, 1912), p. 154.
3. *The Works of Geoffrey Chaucer*, ed. F. N. Robinson, 2d ed. (Boston, 1957), p. 46. Spelling has been modernized.
4. The significance of the *qibla*-problem for the development of astronomy can be seen in al-Biruni, *The Determination of the Coordinates of Cities*, trans. J. Ali (Beirut, 1966). See also the recent paper by David A. King, "Al-Khalili's Qibla Table," *Journal of Near Eastern Studies* 34 (1975):81–122.
5. Franz Cumont, *Astrology and Religion among the Greeks and Romans* (New York, 1912).
6. John of Salisbury, *Policraticus*, 1.12 and 2.18–19.
7. Hastings Rashdall, *The Universities of Europe in the Middle Ages*, ed. F. M. Powicke and A. B. Emden (London, 1936), 1:244.
8. Petrarch's letter to Boccaccio, 7 Sept. 1363, translated in M. Bishop, ed. and trans., *Letters from Petrarch* (Bloomington, Ind., 1966), pp. 231 ff.
9. G. W. Coopland, *Nicole Oresme and the Astrologers* (Liverpool, 1952).
10. On the history of the liberal arts see J. E. Wise, *The Nature of the Liberal Arts* (Milwaukee, 1947); William H. Stahl, *Martianus Capella and the Seven Liberal Arts*, vol. 1 (New York, 1971); and J. Koch, ed., *Artes liberales von der antiken Bildung zur Wissenschaft des Mittelalters* (Leiden, 1959). See also chaps. 4 and 14 of this volume.
11. In the preface to his *Traictié de l'espere*, Nicole Oresme wrote: "La figure et la disposicion du monde, le nombre et ordre des elemens et les mouvemens des corps du ciel appartiennent à savoir à tout homme qui est de franche condicion et de noble engin" (Paris, Bibliothèque Nationale, MS Franç. 1350, fol. 1ra).
12. See Galileo's "Letter to the Grand Duchess Christina," in *Discoveries and Opinions of Galileo*, trans. Stillman Drake (Garden City, N.Y., 1957), pp. 175–216.
13. For editions of these works, see Pliny (the Elder), *Natural History*, ed. and trans. H. Rackham, 10 vols. (London, 1938–62); Macrobius, *Commentary on the Dream of Scipio*, trans. William H. Stahl (New York, 1952); and Martianus Capella, *De nuptiis philologiae et Mercurii*, ed. F. Eyssenhardt (Leipzig, 1866). Some of the astronomical chapters from Macrobius are translated in *A Source Book in Medieval Science*, ed. Edward Grant (Cambridge, Mass., 1974), pp. 27–31.
14. These are published in J.–P. Migne, ed., *Patrologiae cursus completus, series latina*, vols. 14 (Paris 1845) and 34 (Paris, 1841), respectively.
15. Plato, *Timaeus a Calcidio translatus commentarioque instructus*, ed. J. H. Waszink and P. J. Jensen (London, 1962). There is, in addition, a reference to a translation of Ptolemy's *Almagest* by Boethius in a letter from King Theodoric to Boethius, preserved in Cassiodorus, *Variarum libri xii*, in Migne, *Patrologia latina*, vol. 69 (Paris, 1848), col. 539.

16. The astronomical chapters are translated in Grant, *Source Book*, pp. 11–16, 25–27.

17. *De civitate Dei*, 5. 1–5.

18. Isidore of Seville, *Traité de la nature*, trans. J. Fontaine (Bordeaux, 1960).

19. The best analysis is that of Pierre Duhem, *Le système du monde*, vol. 2 (Paris, 1914), pp. 393–504.

20. See Bede, *De natura rerum*, in Migne, *Patrologia latina*, vol. 90 (Paris, 1850), cols. 187–273.

21. Edited in Migne, *Patrologia latina*, vol. 122 (Paris, 1853), cols. 439–1022. See also Erika von Erhardt-Siebold and Rudolf von Erhardt, *The Astronomy of Johannes Scotus Erigena* (Baltimore, 1940).

22. See O. Neugebauer, "On the Allegedly Heliocentric Theory of Venus by Heraclides Ponticus," *American Journal of Philology* 93 (1972): 600–601.

23. *Bedae Opera de temporibus*, ed. C. W. Jones (Cambridge, Mass., 1943).

24. Medieval *compotus* is described in W. E. van Wijk, *Origine et développement de la computistique médiévale*, Conférence, Palais de la Découverte, no. 29 (Paris, 1954).

25. Printed as a spurious work of Bede in Migne, *Patrologia latina*, vol. 90 (Paris, 1850), cols. 881–910.

26. See figures 14 and 15 in Ernst Zinner, *Entstehung und Ausbreitung der Coppernicanischen Lehre* (Erlangen, 1943), pp. 55–56, reproduced from Madrid, Biblioteca Nacional, MS 3707, fols. 65v–66r, written at Metz about A.D. 820.

27. Harriet Pratt Lattin, "The Eleventh Century MS Munich 14436," *Isis* 38 (1948): 205–24; H. Gray Funkhouser, "A Note on a Tenth-Century Graph," *Osiris* 1 (1936): 260–62.

28. On sundials, see Derek J. Price, "Precision Instruments: To 1500," in *A History of Technology*, ed. Charles Singer, E. J. Holmyard, A. R. Hall, and Trevor I. Williams, vol. 3 (London, 1958), pp. 582–619.

29. A. van de Vyver, "Les plus anciennes traductions latines médiévales de traitès d'astronomie et d'astrologie," *Osiris* 1 (1936): 658–91. For a list of translated treatises and manuscripts and editions, see Francis J. Carmody, *Arabic Astronomical and Astrological Sciences in Latin Translation* (Berkeley, 1956).

30. On Gerbert's career, see above, chap. 2.

31. On these treatises, see José M.ª Millás Vallicrosa, "Translations of Oriental Scientific Works (to the End of the Thirteenth Century)," in *The Evolution of Science*, ed. Guy S. Métraux and François Crouzet (New York, 1963), pp. 139–46. For the texts, see *Gerberti Opera mathematica*, ed. N. Bubnov (Berlin, 1899); and J. Drecker, "Hermannus Contractus über das Astrolab," *Isis* 16 (1931): 200–219.

32. On the astrolabe, see Willy Hartner, "The Principle and Use of the Astrolabe," in Hartner, *Oriens-Occidens* (Hildesheim, 1968), pp. 287–318; Henri Michel, *Traité de l'astrolabe* (Paris, 1947); and John D. North, "The Astrolabe," *Scientific American* 230, no. 1 (1974): 96–106.

33. On Walcher, see Charles Homer Haskins, *Studies in the History of Mediaeval Science*, 2d ed. (Cambridge, Mass., 1927), pp. 113–17.

34. For an English translation, see Otto Neugebauer, *The Astronomical Tables of al-Khwarizmi* (Copenhagen, 1962). On arabic astronomical tables in general, see E. S. Kennedy, "A Survey of Islamic Astronomical Tables," *Transactions of the American Philosophical Society* 46 (1956): 123–75.

35. Edited with a fascimile reprint of an Oxford manuscript in R. T. Gunther, *Early Science in Oxford*, vol. 5 (Oxford, 1929), pp. 137–92.

36. For a German translation, see J. Drecker, "Das Planisphaerium des Claudius Ptolemaeus," *Isis* 9 (1927): 258–78.

37. Haskins, *Studies*, pp. 96–98.

38. Ibid., pp. 120–23.

39. See G. J. Toomer, "A Survey of the Toledan Tables," *Osiris* 15 (1968): 5–174.

40. The Arabic text of al-Battani's *zij*, with a new Latin translation, is given in C. A. Nallino, *Albatenii Opus astronomicum*, vols. 1–3 (Milan, 1899–1907).

41. This version was printed as *Albategnius de motu stellarum* (Nuremberg, 1537).

42. On Indian planetary theory, see Otto Neugebauer, "The Transmission of Planetary Theories in Ancient and Medieval Astronomy," *Scripta mathematica* 22 (1956): 165–92; and S. N. Sen, "Astronomy," in D. M. Bose, *A Concise History of Science in India* (Calcutta, 1971), pp. 58–135.

43. The most recent edition is by Francis J. Carmody, ed., *Alfragani Differentie in quibusdam collectis scientie astrorum* (Berkeley, 1943). For earlier editions, see A. I. Sabra, "Al-Farghani," *Dictionary of Scientific Biography*, 4: 544.

44. On the various translations of the *Almagest*, see Haskins, *Studies*, pp. 157–65. Gerard's version from the Arabic became the standard one.

45. A more detailed analysis of the development of the *corpus astronomicum* is found in Olaf Pedersen, "The Corpus Astronomicum and the Traditions of Medieval Latin Astronomy," in *Colloquia Copernicana*, vol. 3 (Warsaw, 1975), pp. 57–96.

46. Edited, with a commentary by Peter Nightingale, in Maximilian Curtze, *Petri Philomeni de Dacia in Algorismum vulgarem Johannis de Sacrobosco commentarius* (Copenhagen, 1897). Extensive extracts are translated into English in Grant, *Source Book*, pp. 94–101.

47. The Latin text and an English translation appear in Lynn Thorndike, *The Sphere of Sacrobosco and Its Commentators* (Chicago, 1949). Chapters 1, 2, and 4 also appear in Grant, *Source Book*, pp. 442–51.

48. There are editions of Wittenberg, 1545; Antwerp, 1547; and Paris, 1550.

49. See A. Lindhagen, "Die Neumondtafel des Robertus Grosseteste," *Arkiv för Matematik, Astronomi och Fysik* 11, no. 2 (1916): 1–41.

50. Edited by Paul Tannery, "Le traité du quadrant de maître Robert Anglès," *Notices et extraits des manuscrits de la Bibliothèque Nationale et autres bibliothèques*, 35 (1896): 561–640; reprinted in Tannery, *Mémoires scientifiques*, vol. 5 (Toulouse and Paris, 1922), pp. 118–97.

51. *The Opus Majus of Roger Bacon*, ed. John H. Bridges (London, 1900), 1: 376–404; Albert the Great, *Opera omnia*, ed Borgnet, vol. 10 (Paris, 1891), pp. 629–50.

52. A critical edition of this treatise is in progress. An English translation appears in Grant, *Source Book*, pp. 451–65.

53. For a full and detailed analysis of Ptolemy's planetary theories, see Olaf Pedersen, *A Survey of the Almagest* (Odense, 1974).

54. On the terminology of medieval planetary theory, see Olaf Pedersen, "A Fifteenth Century Glossary of Astronomical Terms," in *Classica et mediaevalia Francisco Blatt septuagenario dedicata*, ed. O. S. Due, H. Friis Johansen, and B. Dalsgaard Larsen (Classica et mediaevalia, dissertationes, vol. 9) (Copenhagen, 1973), pp. 584–94.

55. Edited in Francis J. Carmody, ed., *The Astronomical Works of Thabit B. Qurra* (Berkeley, 1960).

56. Bernard R. Goldstein, "On the Theory of Trepidation," *Centaurus* 10 (1965): 232–47; Willy Hartner, "Trepidation and Planetary Theories," in *Oriente e occidente nel medioevo: Filosofia e scienze*, Accademia nazionale dei Lincei, Atti dei Convegni, vol. 13 (Rome, 1971), pp. 609–32.

57. See Bernard R. Goldstein, ed., *The Arabic Version of Ptolemy's Planetary Hypotheses*, Transactions of the American Philosophical Society, vol. 57, pt. 4 (Philadelphia, 1967); and Willy Hartner, "Mediaeval Views on Cosmic Dimensions and Ptolemy's Kitab al-Manshurat," in Hartner, *Oriens-Occidens* (Hildesheim, 1968), pp. 319–48.

58. Olaf Pedersen, "The Life and Work of Peter Nightingale," in *Vistas in Astronomy*, vol. 9, *New Aspects in the History and Philosophy of Astronomy*, ed. Arthur Beer (Oxford, 1967), pp. 3–10; Pedersen, *Peter Nightingale: A Problem of Identity with a Survey of the Manuscripts*, Cahiers de l'Institut du Moyen-Âge, no. 19 (Copenhagen, 1977).

59. A portion of the Alfonsine Tables is translated into English in Grant, *Source Book*, pp. 465–87.

60. Owen Gingerich, "The Mercury Theory from Antiquity to Kepler," in *Actes du XIIᵉ Congrès international d'histoire des sciences, Paris 1968*, vol. 3A (Paris, 1971), pp. 57–64.

61. At present Peter Nightingale's *Compotus* is known only from two manuscripts: London, British Museum, MS Harley 3647, fols. 2v–16v; and Venice, Bibl. Nazionale Marciana, MS Lat. VIII.18 (=3573).

62. See Bernard R. Goldstein, ed., *Al-Bitruji: On the Principles of Astronomy*, 2 vols. (New Haven, 1971); Al-Bitruji, *De motibus celorum*, ed. Francis J. Carmody (Berkeley, 1952).

63. See Claudia Kren, "Homocentric Astronomy in the Latin West: The *De reprobatione ecentricorum et epiciclorum* of Henry of Hesse," *Isis* 59 (1968): 269–81.

64. Pp. 280–84. Two of the more important treatises are the *Theorica planetarum* of Taddeo da Parma (Florence, Bibl. Medicea Laurenziana, MS Plut. 29 cod. 7, fols. 105r–152v); and the *Theorica planetarum* of Campanus of Novara (*Campanus of Novara and Medieval Planetary Theory: "Theorica planetarum,"* ed. Francis S. Benjamin, Jr., and G. J. Toomer [Madison, Wis., 1971]).

65. Jocelinus de Brakelonda, *Chronicle*, ed. and trans. H. E. Butler (London, 1949), p. 107. On astronomical instruments in general, see Ernst Zinner, *Deutsche und niederländische astronomische Instrumente des 11.–18. Jahrhunderts* (Munich, 1956); and Francis Maddison, "Early

Astronomical and Mathematical Instruments: A Brief Survey of Sources and Modern Studies," *History of Science* 2 (1963): 17–50.

66. See A. Otto, *Liber Daticus Roskildensis* (Copenhagen, 1933); Pedersen, "Life and Work of Peter Nightingale."

67. Unpublished, but known in fifteen manuscripts.

68. Torquetum is best known through Regiomontanus's *Scripta de torqueto*, edited by Johannes Schöner and published in *Scripta clarissimi mathematici M. Ioannis Regiomontani* (Nuremberg, 1544). See also J. Hartmann, "Die astronomischen Instrumente des Kardinals Nikolaus Cusanus," *Abhandlungen der königlichen Gesellschaft der Wissenschaften zu Göttingen, mathematisch-physikalische Klasse*, 10 (1919): 1–56.

69. See Price, "Precision Instruments," pp. 612–14.

70. Ptolemy, *Almagest*, 5.1.

71. A Latin version of Levi ben Gerson's treatise on the subject is published in Maximilian Curtze, "Die Abhandlungen des Levi ben Gerson über Trigonometrie und den Jacobstab," *Bibliotheca Mathematica*, ser. 2, 12 (1898): 97–112.

72. Edited by H. Salter in R. T. Gunther, *Early Science in Oxford*, vol. 2 (Oxford, 1923), pp. 337–70. A critical edition of all of Richard of Wallingford's scientific works has recently been published by J. D. North; see n. 75.

73. Duhem, *Le système du monde*, vol. 4 (Paris, 1916), pp. 31–32.

74. See Derek J. Price, "On the Origin of Clockwork, Perpetual Motion Devices and the Compass," *U. S. National Museum Bulletin*, no. 218 (Washington, 1959), pp. 81–112; Joseph Needham, Wang Ling, and Derek J. Price, *Heavenly Clockwork: The Great Astronomical Clocks of Medieval China* (Cambridge, 1959); Ernest L. Edwardes, *Weight-Driven Chamber Clocks of the Middle Ages and the Renaissance* (Altrincham, 1965).

75. John D. North, "Monasticism and the First Mechanical Clock," in Julius T. Fraser and N. Laurence, *The Study of Time*, vol. 2, sec. 11 (New York, 1975). Richard of Wallingford's *Tractatus horologii astronomici* has been edited and translated in J. D. North, *Richard of Wallingford*, vol. 1 (Oxford, 1976), pp. 441–526.

76. See the *Tractatus astrarii*, trans. Antonio Barzon, Enrico Morpurgo, Armando Petrucci, and Giuseppe Francescato (Vatican City, 1960); Silvio A. Bedini and Francis R. Maddison, *Mechanical Universe: The Astrarium of Giovanni de'Dondi*, Transactions of the American Philosophical Society, vol. 56, pt. 5 (Philadelphia, 1966).

77. A modern reconstruction is described in Pedersen, "Life and Work of Peter Nightingale."

78. Derek J. Price and R. M. Wilson, *The Equatorie of the Planetis* (Cambridge, 1955).

79. Ibid., pp. 188–96.

80. F. Kaltenbrunner, *Die Vorgeschichte der gregorianischen Kalenderreform* (Vienna, 1876).

81. Unpublished; I have used London, British Museum, MS Sloane 1680, fols. 48r–130r (A.D. 1476).

82. John Mundy, "John of Gmunden," *Isis* 34 (1943): 196–205.

83. Most of what we know of Peurbach and Regiomontanus has been collected by Ernst Zinner, *Leben und Wirken des Johannes Müller von Königsberg genannt Regiomontanus*, 2d ed. (Osnabrück, 1968).

84. First printed at Nuremberg, about 1473.

85. Albertus de Brudzewo, *Commentariolum super theoricas novas planetarum Georgii Purbachii*, ed. L. A. Birkenmajer (Krakow, 1900).

86. *Epytoma Joannis de Monte regio in Almagestum Ptolemei* (Venice, 1496).

87. On Regiomontanus's instruments, see *Scripta . . . Ioannis Regiomontani*.

88. Ibid.

David C.
Lindberg

The Science of Optics

Something very like the discipline that we know today as optics existed during the Middle Ages as the science of *perspectiva*. This science was concerned with such matters as the nature and propagation of light and color, the eye and vision, the properties of mirrors and refracting surfaces, image-formation by reflection and refraction, and meteorological phenomena involving light. Moreover, it took a broad view of these topics, refusing to confine itself to mathematical description or causal analysis, but insisting on a unified approach that investigated the mathematics, the physics, the physiology, and even (to a limited extent) the psychology and epistemology of the visual process.

Such an inclusive and broadly based discipline had not always existed, nor did it ever attract large numbers of adherents. It came into existence largely through the efforts of Ibn al-Haytham (known to the Latin world as Alhazen or Alhacen) and appeared in the West in the thirteenth century under the aegis of Roger Bacon; but even then the majority of scholars working in any area of what we now call optics continued to regard the mathematics, physics (sometimes accompanied by psychology or epistemology), and physiology of light and vision as distinct enterprises and preferred to practice one or another of them in isolation from the rest. In short, the old disciplinary boundary lines continued to exercise a strong influence on scholarship, and for most of the Middle Ages the perspectivists (the practitioners of *perspectiva*) struggled on behalf of a losing cause. In the end, however, Johannes Kepler took up the tradition of *perspectiva*, and through him it became the foundation for the modern science of optics.

Our first task, then, must be to consider the shape of the optical enterprise, its aims and criteria, and the various traditions that it comprised. As a vehicle for this investigation I propose to examine the development of medieval theories of vision. Not only will visual theory admirably illustrate changing conceptions of what optics was about, but in itself it was one of the central themes of medieval optics and, hence, deserving of close attention for its own sake. When we have completed this analysis of medieval visual theory, we will turn to the other major strand of medieval optics—theories of the nature and propagation of light and color.

Theories of Vision in Antiquity

Because medieval theories of vision developed out of ancient antecedents, we must begin with a brief examination of visual theory in antiquity.[1] The common premise of all ancient theories of vision was that there must be some form of contact between the object of vision and the visual organ, for only thus could an object stimulate or influence the visual power and be perceived. Now in general there appeared to be three ways in which contact could be established. The object could send its image or ray through the intervening space to the eye; the eye could send forth a ray or power to the object; or contact could be established through a medium (usually air) that intervened between the object and the eye.

The first of these alternatives was developed by the atomists, who argued that thin films of atoms depart from visible objects in all directions, maintaining a fixed configuration as they proceed, and enter the eye of an observer. Epicurus explained:

> For particles are continually streaming off from the surface of bodies, though no diminution of the bodies is observed, because other particles take their place. And those given off for a long time retain the position and arrangement which their atoms had when they formed part of the solid bodies. . . . We must also consider that it is by the entrance of something coming from external objects that we see their shapes and think of them. For external things would not stamp on us their own nature of colour and form through the medium of the air which is between them and us, or by means of rays of light or currents of any sort going from us to them, so well as by the entrance into our eyes or minds, to

whichever their size is suitable, of certain films coming from the things themselves, these films or outlines being of the same colour and shape as the external things themselves.[2]

These thin films (*eidola* in Greek, *simulacra* in Latin) were compared by Lucretius to the skin of a snake or cicada.[3] They were regarded as coherent assemblies or convoys of atoms, capable of communicating to an observer all of the visible qualities of the objects from which they issued; to receive a series of such images was to gain a visual impression of the object itself.[4]

If the atomistic theory can be called an "intromission" theory of vision, because radiation is sent to the observer, then the obvious alternative is an "extramission" theory, in which radiation is sent out from the observer's eye to "feel" the visible object. This theory, which was proposed by Euclid and further developed by the mathematician and astronomer Ptolemy, maintained that radiation issues from the observer's eye in the form of a cone and proceeds in straight lines unless reflected or refracted. If it falls on an opaque object the object is perceived, and the perception is (in some unexplained manner) returned or communicated to the sense organ.[5]

But there were obvious difficulties in both the atomistic intromission theory and the Euclidean extramission theory. Against the former it could be objected that the *eidola* of a large object would be unable to shrink sufficiently to enter the observer's eye or that *eidola* would be unable to pass through one another (when the lines of sight of two observers cross) without interference. Against the latter, one could note the absurdity of supposing that a physical ray can issue from the eye to something as remote as the fixed stars (and this in an imperceptible instant). Objections such as these led Aristotle to propose, as a third alternative, that the visible object sends its visible qualities through the intervening air (or other transparent medium) to the observer's eye. Colored bodies produce qualitative changes in the transparent medium, and these changes are instantaneously propagated to the transparent humors of the observer's eye. Thus, a green object in some sense colors the observer's eye green, and this acquisition of color constitutes the act of seeing. The eye does not *receive* the visible object, as in the atomistic theory, but *becomes* the visible object.[6]

Aristotle's theory of vision might be called (in the absence of a better term) "mediumistic," for contact between object and observer is established through the medium. An alternative mediumistic theory was defended by the physician Galen, who argued that visual spirit

descending from the brain through the optic nerve to the eye emerges
from the eye for a short distance and transforms the surrounding air,
which thus becomes an extension of the optic nerve and an instrument
of the soul. The air itself becomes percipient, perceives the object with
which it is in contact, and returns its perceptions through the transformed air to the eye and optic nerve and, ultimately, to the soul. The
crucial difference between the Galenic and Aristotelian theories is that
whereas Aristotle made the medium an instrument of the visible object and assigned the observer a passive role in vision, Galen made
the medium an instrument of the eye and soul and ascribed activity
to the observer.[7]

But to classify ancient theories of vision in terms of the direction
of radiation or the role of the medium in vision is to overlook fundamental aspects of ancient optics—and also to make the debate among
the various theories seem trivial and those who debated it for a thousand years look foolish. There is another scheme of classification,
based on the aims and criteria of visual theory, which is far more basic
and, therefore, more significant. The Euclidean theory was not merely
an extramission theory, but, more fundamentally, a mathematical
theory of vision. Euclid's purpose was to offer a geometrical explanation of the perception of space, to develop a mathematical theory
of perspective in which the visual cone accounts for the localization
of objects in the visual field and the apparent size and shape of objects
as a function of their distance from the observer and their orientation
with respect to the line of sight.[8] Although physical content unavoid-
ably crept in (Euclid apparently regarded the rays as physically real
entities), the theory was not designed with physical plausibility in
mind; it was intended as a mathematical theory of vision, and it was
to be judged by mathematical, rather than physical, criteria.[9]

By contrast, the intromission theory, whether in its atomistic or
Aristotelian form, was intended as a causal or physical account of
vision; its purpose was to explain in physical (rather than mathematical) terms how the visible qualities of objects are communicated
to the organ of sight. It was without mathematical pretensions and
was not to be judged by mathematical criteria.[10] Finally, the Galenic
theory, though containing a small amount of physical and mathematical content, was concerned principally with the anatomy of the eye
and physiology of sight; it was devised by a physician, and it was
meant, above all, to satisfy anatomical and physiological criteria and
to fulfill medical needs.

The position I wish to defend, then, is that in antiquity and well
into the Middle Ages, these three kinds of theory—mathematical,

physical, and physiological—defined the principal battle lines within visual theory. And, therefore, the debate between the intromission theory and the extramission theory was not merely a debate over the direction of radiation, for it was thoroughly intertwined with basic questions about the aims and criteria of optical theory. The intromission theory was by definition a physical or causal theory, defended by physicists or natural philosophers, and its failure on mathematical grounds was (in the view of its practitioners) of little consequence. The extramission theory, by contrast, was a mathematical theory in its Euclidean form, and, thus, immune to criticism on physical grounds, or a physiological theory in its Galenic form, designed with anatomical and physiological ends in view. The question facing visual theory, therefore, was not primarily "In which direction does the radiation proceed?" but "What criteria must a theory of vision satisfy?" And answering this latter question clearly required an investigation of the very foundations of optics.

Visual Theory in Medieval Islam

Much of the Greek achievement in optics was translated into Arabic in the course of the ninth century A.D., and almost immediately the various Greek optical traditions were reproduced on Islamic soil. The mathematical theory of Euclid and Ptolemy, built around the idea of visual rays issuing forth in conical form, was developed and defended by al-Kindi (d. ca. 866), an influential natural philosopher associated with the Abbasid court at Baghdad. Al-Kindi expressed his conclusions in an influential book (entitled *On Vision*), which circulated widely in Islam and later in Latin translation in the West.[11] Al-Kindi's exact contemporary Hunain ibn Ishaq, also associated with the court at Baghdad and himself one of the most important translators, adopted the Galenic theory of vision and disseminated it to a wider public in his *Ten Treatises on the Eye* and *Book of Questions on the Eye*.[12] Finally, the intromission theory of Aristotle found several supporters in the tenth century (including al-Razi and al-Farabi) and in the eleventh century received a full and elaborate defense by the Persian physician and philosopher Avicenna (980–1037).[13] In these newly founded Islamic traditions Greek arguments were in many cases refined and articulated, but the location and shape of the battle lines remained unchanged.

A new theory of vision having broader aims made its first appearance in the eleventh century. It was formulated by Alhazen (Ibn al-Haytham), the great mathematician, astronomer, and natural philosopher who was born about 965 in Basra (near the Persian Gulf) and later emigrated to Egypt, where he died about 1039. Alhazen's achievement was to break out of the limitations of the Aristotelian, Galenic, and Euclidean theories (each with its narrow view of the aims of optical theory) and to formulate a new intromission theory of vision that would simultaneously satisfy mathematical, physical, and physiological criteria. To reveal the difficulty of this task, let me call attention to a few of the obstacles.

Before Alhazen there were, as I have indicated, two versions of the intromission theory—Aristotle's mediumistic version and the atomistic theory of *eidola*—but neither of them provided an adequate account of the communication of visible qualities to the eye. According to the Aristotelian theory, colored bodies produce qualitative changes in all parts of the transparent medium to which they have rectilinear access. But then each part of the eye should be affected by the color of every object (and, indeed, every part of every object) in the visual field, and the result should be complete mixing and total confusion. How then is the observer able to perceive individual shapes and to discern one object off to the right, another straight ahead, and yet a third over on the left? Or to rephrase the objection, if the eye "becomes the visible object," which object (or which part of which object) in the visual field shall it become?[14] The theory fails on mathematical grounds, for it cannot explain those features of visual perception that the Euclidean theory accounted for with its visual cone.

The atomistic theory was equally defective. *Eidola* must apparently radiate from every object in such a way as to enter the pupils of a multitude of observers located in different places all at the same time —quite an impossible feat in the view of the theory's critics. Moreover, if the object is large, its *eidola* must shrink in order to enter the observer's eye, and, indeed, they must shrink exactly according to the laws of perspective if they are to account for the facts of visual perception. And once they have so shrunk, how can one know the true size of the object from which they emanated? This theory, too, fails on mathematical (not to speak of physical) grounds. Some of the objections against it were summarized by Hunain ibn Ishaq:

> All people acknowledge and agree that we see only by the hole which is in the pupil. Now if this hole had to wait until something coming from the [visible] object reached it, or a power . . . , or a form, an outline or a quality, as some people maintain, we should not know, in looking at an object, either its extent or its volume. . . . Its entering into the eyes is something which reason does not comprehend and of which nobody has ever heard, for according

to this hypothesis a complete form or outline of the viewed object would necessarily reach and enter into the eye of the beholder at the same moment. Supposing then that a great many people looked at it, say, for example, ten thousand persons, it would have to return to the eye of every one of them, and its form and outline would have to enter completely into them. But this is far from probable and must therefore be ranked among the untenable hypotheses.[15]

Hunain's arguments against the intromission theory were buttressed by those of his contemporary al-Kindi. Al-Kindi had a variety of objections to the intromission theory, but one in particular that reveals the magnitude of the difficulties facing Alhazen. Al-Kindi's position was that the perception of shape could be adequately explained only by the extramission theory. According to any of the intromission theories,[16] in his view, a circle situated edgewise before the eye would send its form or image (representing the entire circle) to the eye.[17] It is difficult to know exactly how al-Kindi conceived this transmission of forms or images in the intromission theory, but presumably they were thought to pass as coherent units through the space (or medium) between the object and the observer, maintaining the same edgewise orientation as the object from which they issued. (Figure 26

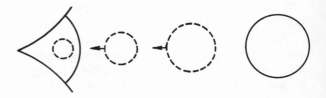

Fig. 26   Al-Kindi's interpretation of the radiation of forms according to the intromission theory.

is a rough attempt to sketch al-Kindi's view of the transmission of forms in the intromission theory.) Once inside the eye, the forms would be surrounded by the visual power and would therefore be perceived in their full circularity. However, all of this is contradicted by simple observation: in truth, a circle situated edgewise before the eye is perceived only as a line, and the intromission theory (since it makes predictions that conflict with the facts of observation) must be false. Al-Kindi's position, then, is that the intromission theory is proved false because it is incompatible with the laws of perspective.[18]

Of course, neither Aristotle nor the atomists had defended a conception of the radiation of forms or the perception of shape quite like the one al-Kindi attributed to them. However, al-Kindi was not far off on the atomists, whose theory could easily enough receive his interpretation. Aristotle had altogether ignored the problem of perceiving shapes; therefore, although he would undoubtedly have denied al-Kindi's interpretation of his theory, he had no alternative to offer. What is important about al-Kindi's argument, then, is that it reveals the inability of the intromission theory, in any of the forms in which it had thus far been articulated, to account for the perception of shape. If shapes were to be perceived by a process of intromission, it seemed necessary that there be a coherent process of radiation, by which a form or image bearing the shape of the object passed as a single unified entity to the observer's eye; but such an intromission theory was vulnerable to the objections raised against it by al-Kindi and Hunain. It is apparent that a new kind of intromission theory was required—one in which radiation into the eye would no longer be conceived as the transmission of coherent forms.

It is ironic that this new kind of intromission theory would be erected by Alhazen on a principle first stated by al-Kindi (but in a different context, not directly related to vision). Al-Kindi argued that radiation does not issue from the surface of a luminous body as a whole; rather, each point on the surface radiates light in all directions independently of other points (see fig. 27).[19] This may seem like a trivial and self-evident claim, and although al-Kindi was the first to state it clearly and explicitly, it is perhaps implicit in earlier writings. But what was surely not trivial or self-evident was the position taken by Alhazen, namely, that upon this punctiform analysis of the visible object one could build a successful intromission theory of vision. What would surely have been taken for granted by all of Alhazen's predecessors or contemporaries is that in any intromission theory a coherent visual impression can result only from a coherent process of radiation.

The obstacles facing anybody who would maintain the contrary are easily revealed. If, in fact, every point of the visible object radiates in all directions, then every point in the eye should be affected by light and color from virtually every point in the visual field, and the outcome should be total confusion. Figure 28 reveals the mixing, within the eye, of rays issuing from the endpoints of the visible object.[20] To explain vision as we know it, there must be a one-to-one correspondence between points in the visual field and points in the eye—so that every point in the eye is stimulated by one point in the visual field, and the pattern of the visual field is reproduced in the eye. But on al-Kindi's theory of incoherent punctiform radiation, which Alhazen

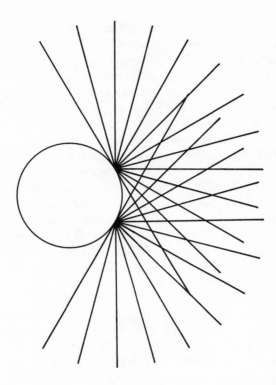

Fig. 27   Incoherent radiation from two points of a luminous
                        body.

hoped to use as the basis of his intromission theory of vision, a one-
to-one correspondence seems beyond the realm of possibility.

Alhazen undertook to solve this problem in his largest optical
work, the *Kitab al-Manazir* (translated into Latin as *De aspectibus*
or *Perspectiva*). He recognized there that every point in the eye receives a ray from every point in the visual field. But only one ray
falling on each point of the surface of the eye is incident perpen-
dicularly; or to turn the geometry around, from each point in the
visual field there falls only one ray perpendicular to the convex surface of the eye. All other rays, falling obliquely on the surface of the
eye, are refracted; and as a result of refraction they are weakened to
the point where they are incapable of stimulating the visual power.[21]

But what is of crucial importance is that these perpendicular rays
constitute a pyramid or cone, with the object or visual field as base
and the center of the eye as apex. Because the rays are rectilinear
and converge toward a single apex, they maintain a fixed arrange-

ment and fall on the crystalline humor (or lens) of the eye in precisely the same order as the points in the visual field from which they
originated (see fig. 29). They stimulate the visual power residing in
the crystalline lens, thus producing visual perception.[22] To be sure,
difficulties remain. To mention only the most serious, is it really justi-
fiable to ignore all refracted rays? Refraction may well be held to
weaken rays, but why should only slightly refracted (and therefore
only slightly weakened?) radiation be totally incapable of stimulating
the visual power? Alhazen did not make a convincing case at this
point, and it was to remain one of the principal weaknesses of the
theory until resolved by Johannes Kepler early in the seventeenth
century.[23]

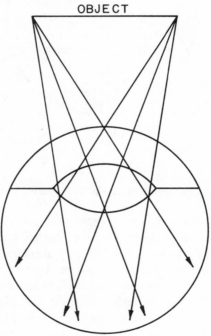

OBJECT

Fig. 28   Mixing, within the eye, of rays from the endpoints of a
visible object.

Nevertheless, Alhazen made a most important contribution to visual theory. He formulated an intromission theory which, despite certain difficulties, proved capable of explaining the principal facts of
visual perception: it explained physical contact between the object
and observer through the intromitted rays, and through its visual

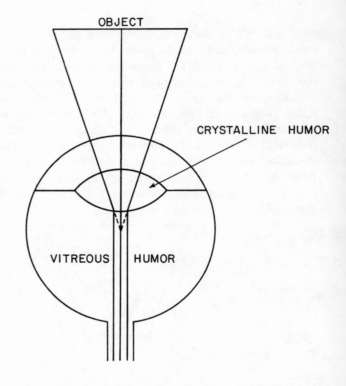

Fig. 29   The geometry of sight according to Alhazen.

cone explained the perception of shape and accounted for the laws of perspective. It thus achieved a level of success not attained by any previous intromission theory, and largely through Alhazen's influence (aided by the considerable weight of Aristotle's authority), the intromissionist character of vision would never again be seriously doubted.[24] Alhazen also established punctiform analysis of the visible object and the need for a one-to-one correspondence between points in the visual field and points in the eye as permanent and essential elements of the intromission theory—thus providing the basic framework of visual theory that has prevailed until the present.

But to say no more would be to miss the main point of Alhazen's achievement. He did not merely establish the intromission theory of vision beyond dispute and contribute many of its essential characteristics, but did so in such a way as to obliterate the old battle lines and fundamentally alter the aims and scope of optical theory. Like all intromission theories, Alhazen's gave an adequate *physical* or *causal* account of the communication of the visible qualities of the object to the observer: the "forms" that emanate from each point of the visual field communicate to the observer, apparently through some kind of modification of the medium, the pattern of light and color. But, in addition, Alhazen successfully incorporated into his intromission theory the visual cone of the extramissionists. By restricting himself to perpendicular rays, he appropriated the Euclidean visual cone and, hence, the entire *mathematical* achievement and *raison d'être* of the extramission theory. And, finally, although I do not have space to go into the matter here, Alhazen embraced the *anatomical* and *physiological* claims of the medical tradition and integrated them into his theory. What Alhazen thus achieved was to draw together the mathematical, physical, and medical traditions into a single comprehensive theory, which satisfied the criteria of all of the traditions. He was neither Euclidean nor Aristotelian nor Galenist—or else he was all three of them. He created a new optical tradition and established the aims and criteria of optics which would prevail, though not without rivals, until Kepler and beyond. Although containing traditional materials at every point, the resulting edifice was a fresh creation.

Visual Theory in Medieval Christendom

While al-Kindi and Alhazen were reconstructing visual theory in the Islamic world, the West was still struggling to retain and assimilate remnants of the Greek achievement. The Platonic theory of vision (received largely through Chalcidius's translation, early in the fourth century, of the first half of Plato's *Timaeus*) dominated until the thirteenth century. Plato's stress on the visual fire emanating from the observer's eye (coupled, of course, with external illumination and light or fire from the observed object) was reinforced by St. Augustine, who also taught an extramission theory. However, in the twelfth and thirteenth centuries virtually the full corpus of Greek and Islamic works on optics became available in translation; and this vastly complicated the situation, for the West now had at its disposal a large group of venerable (and, therefore, authoritative) works, which taught every conceivable theory of vision. Plato, Aristotle, Euclid, Galen, Augustine, al-Kindi, Alhazen, Avicenna—all were known, and all had reached different conclusions on the subject of vision. The task facing scholars of the thirteenth century was to sort out, make sense of, and, if possible, reconcile the various elements of this optical heritage.

The key figures in the process were Albertus Magnus (the teacher of Thomas Aquinas) and Roger Bacon.[25] Albertus (d. 1280) was the first great Western expositor of the whole Aristotelian corpus (which, except for logic, was largely unknown before the translations of the twelfth and thirteenth centuries), and in the 1240s and 1250s he turned his attention to those Aristotelian books that touched upon visual theory. He wrote a long refutation of the extramission theory and in its place attempted to establish the Aristotelian doctrine that vision is caused by an alteration of the transparent medium by a visible object, and the propagation of this alteration to the watery substance of the eye. However, Albertus added several non-Aristotelian embellishments, including the Galenic emphasis on the crystalline humor as the seat of the visual power and the Euclidean visual pyramid.[26] Nevertheless, Albertus remained principally an Aristotelian, and it was chiefly he who established the Aristotelian theory as a major force in the West.

Through the influence of Albertus and others, Aristotelian philosophy in general came to dominate the arts curriculum in the medieval university. By the second half of the thirteenth century, an arts student at Paris could not receive the M.A. degree without hearing lectures on all available Aristotelian works, and by the fourteenth century large numbers of masters were turning out commentaries on Aristotle's *De anima, De sensu,* and *Meteorologica*—works in which one could hardly avoid discussing visual theory. As a result, the Aristotelian theory of vision rose to a position of dominance, for it was virtually impossible to comment on the Aristotelian texts (especially *De anima* and *De sensu*) without concluding that Aristotle's arguments had demolished the extramission theory and that his own intromission theory satisfactorily accounted for the known facts of visual perception (if one ignored mathematics and physiology, as Aristotelians usually did).

The second major tradition of visual theory in the West was founded by Roger Bacon (d. ca. 1292), who took up the study of optics early in the 1260s, about the time Albertus's interest in optics was subsiding. But whereas Albertus had been a disciple of Aristotle, Roger became the disciple of Alhazen.[27] He argued that forms (or "species," as he called them) issue from each point or small part of the visible object, enter the observer's eye, and arrange themselves on the surface of the crystalline humor in the same order as the points in the visual field from which they issued. Moreover, non-perpendicular rays were to be ignored because they are weakened through refraction and do not stimulate the visual power. Bacon followed Alhazen in many other details, but it must suffice to insist that he reproduced all of the essential aspects of Alhazen's theory of vision and also adopted Alhazen's view of the aims and scope of visual theory.

However, Bacon also fancied himself the disciple of everybody else who had written on the subject. He was deeply convinced of the unity of all knowledge and the fundamental agreement of the ancient sages, and he was therefore committed to showing that all who had addressed themselves to the subject of vision had been of one mind; he would conciliate among Aristotle, Euclid, Ptolemy, Augustine, Alhazen, Avicenna, and the rest in order to reveal the underlying unity of their thought. This was by no means an impossible task. Insofar as the mathematicians had simply devised a mathematical theory of perspective, they had treated a topic ignored by Aristotle and the natural philosophers or physicists; by the same token, the natural philosophers had focused on causal issues, and the physicians on questions of anatomy and physiology. The three major traditions might thus appear complementary, and a merger along the lines suggested by Alhazen would appear quite sufficient. However, there were some points of disagreement that could not be lightly dismissed: the mathematicians had made physical claims about the nature and direction of radiation, natural philosophers had occasionally touched upon the mathematics of the visual process, and Alhazen had addressed himself not only to the mathematics and physics of vision, but also to its anatomy and physiology. Thus, if Bacon wished to demonstrate that all were in agreement, some ingenious argumentation would be required.

One point of disagreement calling for Bacon's attention was the physical nature of the radiation responsible for sight. Here he was faced with Aristotle's theory of the qualitative transformation of the medium, Alhazen's discussion of the forms of light and color, and the Neoplatonic doctrine of the multiplication of species defended earlier in the thirteenth century by Robert Grosseteste. Bacon's procedure in this case was simply to overlook the differences—to take Grosseteste's species,[28] endow them with all of the mathematical properties of Alhazen's forms, and to claim that this was what Aristotle had meant all along. In the finished theory, each point of an object was held to produce a likeness of its light and color in the adjacent transparent medium, which, in turn, produced a further likeness in the next part of the medium, and so forth:

But a species is not body, nor is it moved as a whole from one place to another; but that which is produced [by an object] in the

first part of the air is not separated from that part, since form
cannot be separated from the matter in which it is unless it should
be mind; rather, it produces a likeness to itself in the second part
of the air, and so on. Therefore there is no change of place, but a
generation multiplied through the different parts of the medium;
nor is it body which is generated there, but a corporeal form that
does not have dimensions of itself, but is produced according to
the dimensions of the air; and it is not produced by a flow from the
luminous body, but by a drawing forth out of the potentiality
of the matter of the air.[29]

This is Bacon's doctrine of the multiplication of species. It is Aristotelian in its utilization of the matter-form dichotomy and its stress on
the transformation of the medium. It harks back to Grosseteste in its
choice of terminology and in its acknowledgement that the species of
light and color are instances of a more general radiation of force or
power throughout the universe. And it assigns to species all of the
properties, mathematical and physical, of Alhazen's forms; indeed,
Bacon was careful to employ Alhazen's terminology along with that
of Grosseteste and to claim that Alhazen's "forms" and Grosseteste's
"species" are one and the same thing.

A second point of conflict among the inherited theories, which demanded Bacon's attention, was the direction of radiation. Here was a
matter addressed by every author who had written on the subject of
vision, and it would be no trivial achievement to demonstrate that all
had really been saying the same thing. Bacon approached the problem by affirming, with Aristotle, Alhazen, and the other intromissionists, that vision is basically the result of intromitted rays. But he
then argued that although intromitted rays are necessary for vision,
and, indeed, are the principal agents of vision, they are not alone sufficient for vision. It is necessary also that visual rays or species emanate from the observer's eye to excite or ennoble the medium and the
species of the visible object, thereby rendering the latter capable of
stimulating sight. Bacon pointed out that Alhazen had not disproved
the existence of visual rays, but had only demonstrated their insuf-
ficiency as the cause of sight. Bacon was thus able to produce what
he regarded as the perfect synthesis among competing schools of
thought. Visual rays exist (as maintained by Euclid, Ptolemy, al-Kindi,
and Augustine), but they perform none of the functions denied them
by the intromissionists. Each school had presented a portion of the
truth, and Bacon was convinced that he had now restored truth in its
fullness.[30]

Bacon's conclusions were presented in his *Perspectiva* (also issued as part 5 of his *Opus maius*), a book that circulated widely and was responsible, in large part, for originating the tradition of *perspectiva* in the West.[31] Bacon influenced his fellow countryman and Franciscan brother (later Archbishop of Canterbury) John Pecham (d. 1292) and the Silesian scholar Witelo (d. after 1277). Both Witelo and Pecham wrote popular optical texts (Witelo a long one entitled *Perspectiva* and Pecham a short one entitled *Perspectiva communis*) in which the science of *perspectiva* was articulated and disseminated. The perspectivist tradition persisted, though somewhat thinly, in the fourteenth century in three treatises entitled *Questiones super perspectivam* or *Questiones perspective* written by Dominicus de Clavasio, Henry of Langenstein, and Blasius of Parma. Here isolated questions from the science of *perspectiva* were submitted to analysis through the scholastic techniques of disputation. Finally, in the fifteenth and sixteenth centuries *perspectiva* became the subject of lectures at a variety of universities, including Prague, Leipzig, Krakow, Würzburg, Alcalá, Salamanca, Paris, Oxford, and Cambridge.

I have spoken only of the Aristotelian and perspectivist traditions in the West. There were others, of course, but of considerably less significance. The Galenic theory, although it experienced little or no development during the Middle Ages, could always be found in treatises on ophthalmology or general works on anatomy; and in the sixteenth century it benefited from the general revival of anatomical studies and became a significant force. Visual theory also entered into theological treatises and was put to theological use (especially in connection with psychology and epistemology) in ways that probably justify speaking of a theological tradition in medieval visual theory. Finally, the mathematical tradition of Euclid, Ptolemy, and al-Kindi had by no means disappeared; although it is exceedingly difficult to discover a Western defender of the extramission theory in its pure Euclidean form, the works of Euclid, Ptolemy, and al-Kindi had a wide circulation and were apparently being read. However, for most people in the West the label "mathematical theory of vision" no longer signified the Euclidean theory, but the Baconian. The mathematical tradition was now being carried forward by the perspectivists. In truth, the perspectivist tradition was more than mathematics, for the perspectivists were the heirs of Alhazen and were therefore concerned not only with the mathematics of vision, but also with its physics and physiology. But *perspectiva* was the only living optical tradition containing any mathematics, and perspectivists were therefore easily and properly regarded as champions of the mathematical approach.

The struggle over visual theory in the Middle Ages thus ended indecisively. Alhazen's theory of vision, with its broad view of the aims and scope of visual theory, was available in the works of Roger Bacon and his followers. But the members of competing schools had not capitulated to the arguments of the perspectivists, and, indeed, Aristotelians far outnumbered the adherents of any other theory.[32] What is of the utmost importance for the subsequent history of optics, however, is that the perspectivist tradition persisted throughout the medieval period, experienced a revival in the sixteenth century (in the work of Francesco Maurolico, the printing of Witelo's *Perspectiva* in 1535, 1551, and again with Alhazen's *De aspectibus* in 1572, and the appearance of nine printed versions of Pecham's *Perspectiva communis*), and at the beginning of the seventeenth century was seized upon by Johannes Kepler and made the foundation of his new theory of the retinal image. Kepler's achievement was, thus, the culmination of medieval developments. Kepler worked almost entirely within the perspectivist framework, attempting to build an intromission theory of vision on the punctiform analysis of the visible object.[33] He responded to the problem the perspectivists had raised, seeking a mechanism (involving refraction) by which to deal with the superfluity of rays emanating from a given point in the visual field and to establish a one-to-one correspondence between points in the visual field and points in the observer's eye. Most significantly, Kepler's conception of the scope of visual theory—his concern with the mathematics, physics, and physiology of the visual process—was taken over from the perspectivists.

There were, of course, new elements in Kepler's thought, such as the sensitivity of the retina.[34] And it is surely not in dispute that Kepler presented a new and more satisfactory answer to the question of how one sees. He argued that if all rays entering the eye are vision-producing (that is, if nonperpendicular rays must be taken into account along with perpendicular rays), then a one-to-one correspondence can be established only if all rays issuing from a given point in the visual field are returned, through refraction in the humors of the eye, to a point of focus on the retina. The result will be an inverted image of the visual field, a *pictura*, on the surface of the retina (see fig. 30).[35] But clearly this was the answer to a question with which scholars had been struggling for nearly 600 years—a new answer to a medieval question, designed to satisfy the criteria of the medieval perspectivist tradition.

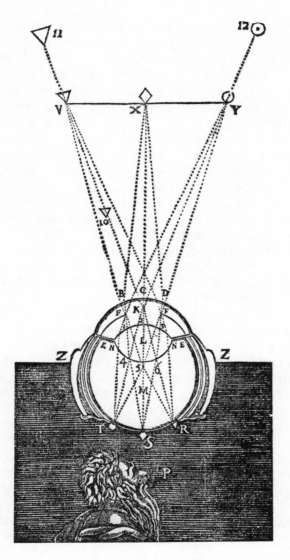

Fig. 30   Descartes' illustration of the theory of the retinal image.

The Nature and Propagation of Light and Color

If vision was one of the central themes of medieval optics, a topic that attracted equal attention was the nature and propagation of light and

color. The distinction between light and color was universal in early
optical thought, each being regarded as a property or quality of visible objects. In antiquity, light was often associated with fire or the
luminosity of fiery bodies such as the sun, while color was associated
with the quality of nonluminous objects that makes them visible. This
conception of light and color (greatly refined and extended) lies behind Aristotle's definition of light (*phos*) as a state of the transparent
medium resulting from the presence of fire or some other luminous
body—more specifically, the actualization of the transparency of the
medium, the achievement of that state in which transparency is no
longer merely potential, but actual, so that bodies separated from the
observer by the medium are visible. As for color, Aristotle defined it
as that which overlies the surface of visible objects and has the power
to set in motion (or produce qualitative changes in) a medium whose
transparency has already been actualized. Thus, in Aristotle's scheme
light is not itself visible, but signifies a state of the medium that makes
colored bodies on the other side of it visible; color, rather than light,
is the "proper object" of sight.[36]

Most Muslim scholars who discussed the matter adopted some variation of the Aristotelian theory of light and color. Avicenna, for example, accepted Aristotle's distinction between light and color and
agreed that light is a quality of the transparent, while color is a quality
of opaque bodies. However, within the category of "light" Avicenna
distinguished among (1) the brightness that one observes in fire or
the sun (*lux* in the Latin translation of Avicenna's work), the luminous quality of fiery objects by which (when a transparent medium
intervenes) they themselves are perceived; (2) the splendor (*lumen*)
shining from luminous bodies, which falls on nonluminous objects and
causes them to be visible (*lumen* might be thought of as the effect of
lux on the adjacent medium and surrounding objects); and (3) the
ray (*radius*) or radiance (*radiositas*), which Avicenna defined as "that
which appears around bodies . . . as though it were something emanating from them."[37] Avicenna added that colors exist only in potentiality
when unilluminated and unobserved: "white is not white and red is
not red unless we see them, and we do not see them unless they are
illuminated."[38]

If Avicenna's contemporary Alhazen consistently distinguished between *lux* and *lumen*, the fact was certainly obscured by the Latin
translation of his book, where *lux* and *lumen* appear to be used interchangeably. Terminology aside, however, Alhazen surely made the
same conceptual distinction. He argued that vision is produced when
the form of light (usually *lux* when employed in this phrase) and the

form of color enter the observer's eye and encounter the visual power; and it would have been perfectly obvious to any of Alhazen's Western followers that his "form of light" was identical to Avicenna's *lumen*. What is much more significant about Alhazen's theory, however, is that it makes light or *lux* an object of vision along with color, altering its status from a state of the medium required for the perception of color to a quality of luminous objects that is itself perceived.[39] Alhazen maintained that both light and color propagate their forms independently through the medium, although forms of the two kinds are propagated identically and, therefore, may become intermingled and stimulate the visual power together.[40] Alhazen also distinguished between the "essential" light of self-luminous bodies and the "accidental" light of opaque or transparent bodies that have received illumination from another source. Opaque or transparent bodies (in A. I. Sabra's words) "take possession of the light shining upon them and, having made it their own, they in turn shine as if they were self-luminous."[41]

The themes treated by Avicenna and Alhazen underwent continued development in the West. Avicenna's distinction between *lux* and *lumen* was widely employed, though (with the encouragement of Alhazen) it was widely ignored.[42] It was agreed by virtually everybody who touched upon the matter, however, that both light (usually *lux*) and color propagate their forms or likenesses through transparent media to observers. The problem that provoked the most interest and discussion was the physical nature of this propagation and the status of light and color in the medium. It was obvious that light and color (or their forms) had to pass through the medium if one were to explain visual perception; yet if the observer looked *at the medium*, rather than at the object *through the medium*, the medium did not seem luminous or colored at all—but merely transparent. Moreover, different colors could apparently occupy the same place in the medium (as when the lines of sight of two observers cross) without mixing or interference. The Spanish Muslim Averroes (d. 1198) had attempted to account for these curious facts by employing a distinction between the spiritual and the corporeal existence of light and color: in the soul, he had argued, light and color (or their forms) have a spiritual existence; in the transparent medium, an existence intermediate between the spiritual and the corporeal:

> As for those who are of the opinion that the forms of sense-objects are imprinted upon the soul in a corporeal manner, the absurdity of their view can be demonstrated by the fact that the soul can receive the forms of contraries at the same time, whereas if they

were bodies, this would be impossible. This will occur not only in the case of the soul but also in the case of media, for it is apparent that in the same space of air, the organ of sight can receive two contrary colors at the same time, as when it regards two individual things, one white and the other black. Furthermore, the fact that large bodies can be perceived by the sight through the pupil of the eye, despite its being small, . . . is proof that colors and whatever is connected with them are not conveyed to the sight materially but rather spiritually. . . . The existence of forms in media is of a kind intermediate between the spiritual and the corporeal. This is true for the reason that the existence of forms outside the soul is completely corporeal; consequently, their existence in the medium is in an intermediate stage between the spiritual and the corporeal.[43]

By "spiritual existence" Averroes meant simply existence that does not involve matter (exemplified, for example, by the immaterial celestial intelligences), so that such features of the material world as interference and mixing do not apply. How we are to conceive existence intermediate between the spiritual and the corporeal (an extremely subtle or diaphanous state of being?), Averroes does not explain.

Averroes' solution was well known and much discussed in the West.[44] Albertus Magnus, after a long and intricate discussion of what it meant for light and color (or their species) to be in the medium, concluded that in some sense they are there corporeally; as for Averroes, Albertus pointed out that he can be interpreted to have meant simply that light and color exist in the medium with extreme rarity or subtlety or that the species of color exists in the medium without the material causes that originally generated the color in its subject and without the matter of the subject.[45] Most thirteenth-century natural philosophers reached a similar conclusion. Roger Bacon argued that what is propagated is a species, a likeness or corporeal form, which is called forth successively out of the potentiality of the transparent medium; and as a corporeal form it has, of course, corporeal or material existence in the medium.[46] In the fourteenth century the problem was attacked from a wide variety of perspectives and with great subtlety. The extreme position was adopted by William of Ockham (d. ca. 1349), who dispensed altogether with species or any other intermediary between object and sense organ and embraced the idea of action at a distance. The object, he argued, can impress in the eye a quality or likeness of itself without in any way affecting the intervening medium. In defense of such an extraordinary claim, Ockham pointed out that there is no experiential evidence for the existence of species, since we have no awareness of them but posit them solely to explain our awareness of the object. However, it is unnecessary to posit species in order to explain our awareness of the object, since the object (which is disposed to act on the visual power) and the visual power (which perceives when acted upon) are sufficient by themselves to account for visual perception.[47]

To a student of modern physical optics, the medieval analysis of the physics of the propagation of light and color must surely appear futile and incomprehensible; spiritual versus corporeal existence and the *lux-lumen* distinction are not part of his idiom. But such was the nature of the medieval optical enterprise. However, the propagation of light and color could also be approached mathematically, and here the medieval and the modern theory more nearly coincide. The foundation of the mathematical approach, then as now, was the rectilinear propagation of light and color, assumed in antiquity by Euclid and perhaps demonstrated by Ptolemy.[48] During the Middle Ages the rectilinearity of light was accepted without serious question, though Alhazen (in a portion of his book never available in Latin) also supplied an experimental demonstration of the fact.[49] If light (or its form or species) is propagated in straight lines, then, of course, its path can be represented by a linear ray, and on this foundation an elaborate system of ray geometry can be erected. But ray geometry also requires an understanding of the principles of reflection and refraction. In antiquity, and, subsequently, in the Middle Ages, the law of reflection was fully and correctly grasped: the incident and reflected rays form equal angles with the reflecting surface, and the plane formed by the incident and reflected rays is perpendicular to the reflecting surface (or its tangent). As for refraction, it was established in antiquity and well known during the Middle Ages that a ray passing obliquely from a less dense to a more dense medium is refracted toward the perpendicular to the refracting interface, while a ray passing in the opposite direction is oppositely refracted. It was also recognized that the image of a visible point seen by reflection in a mirror or by refraction through a transparent interface will appear to be located where the rectilinear extension of the ray incident on the eye (in the intromission theory, or emerging from the eye in the extramission theory) intersects the perpendicular dropped from the visible point to the reflecting or refracting surface. For example, point B (fig. 31) submerged in water will appear to observer A to be located at L, where perpendicular BD dropped from the visible point to the refracting surface intersects the rectilinear extension of ray CA incident on the observer's eye.[50]

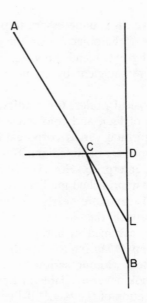

Fig. 31   A point viewed by refraction at a plane interface.

There were several attempts to supply quantitative data on refraction and even to establish a quantitative law of refraction. Ptolemy had sought such a law, and he bequeathed his numerical data to the Middle Ages.[51] This data was reproduced in the thirteenth century by Witelo, who also attempted to extrapolate Ptolemy's data to cover refraction in the reverse direction; however, it was Witelo's misfortune to misunderstand the reciprocal character of refraction (that refracted light will retrace the same path if its direction is reversed), and, consequently, he presented absurd and impossible results.[52] A quantitative law was also attempted by Robert Grosseteste early in the thirteenth century, but this amounted merely to the claim (based, undoubtedly, on principles of symmetry and simplicity) that as light passes from a less dense to a more dense medium its angle of refraction is half its angle of incidence.[53]

Employing the principle of rectilinear propagation and the laws of reflection and refraction, Alhazen and the perspectivists in the West were able to perform some remarkably sophisticated ray geometry. They could, for example, determine image locations for objects viewed in mirror surfaces or through transparent interfaces of plane or spherical figure. They analyzed in more limited fashion mirrors of conical, cylindrical, and paraboloidal form, and in some instances they were able to deal correctly with the phenomena associated with the focal

point or focal plane of a mirror. To give only a single illustration,
Witelo demonstrated that the image of *DE* (fig. 32) viewed by reflection in a concave spherical mirror by eye *B* will appear at *LN*,
where the reflected rays (*AB* and *GB*) incident on the eye intersect
DM and *EM*, the perpendiculars to the reflecting surface drawn from

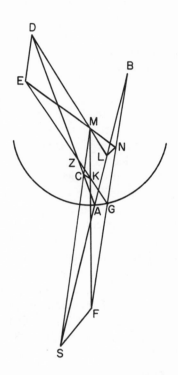

Fig. 32   Image-formation in a concave spherical mirror.

the end points of the object. If, on the other hand, the object should
be located at *CK* (between the mirror and its focal point), its image
will be at *SF*, where the extension of the reflected rays (*AB* and *GB*)
incident on the eye intersect perpendiculars *MC* and *MK*. Note that
Witelo appears to understand implicitly that image *SF* is erect and
virtual, while image *LN* is inverted and real.[54] Medieval scholars did
not develop a theory of lenses; refraction at a single interface was
analyzed with skill and sophistication, and the basic focusing properties of the burning sphere were understood, but that is as far as
the matter went. It was not the principles of optical instruments that
were being sought, but an understanding of the laws of nature, ap-

plied to the most general cases; medieval optics was not an instance of applied science, but of natural philosophy.

A final achievement of medieval geometrical optics (this in the realm of meteorology) must be mentioned, if only because it has been so celebrated. The rainbow had attracted the attention of natural philosophers in antiquity and continued to do so during the Middle Ages. Aristotle had devoted a substantial portion of his *Meteorology* to the subject, arguing that the rainbow results from the reflection of visual radiation from droplets of moisture in a cloud to the sun,[55] and some form of this reflection theory was dominant throughout the Middle Ages. However, in the thirteenth century Robert Grosseteste attempted to bring refraction into the theory, and although his own theory remains incomprehensible, it was common thereafter to employ some combination of reflection and refraction to account for the rainbow. Witelo, for example, argued that solar light is refracted by drops in a mist and then reflected to the observer's eye by the convex surface of drops deeper within the mist.[56]

However, early in the fourteenth century Theodoric of Freiberg (d. ca. 1310) presented a theory that closely resembles the modern theory, wherein solar radiation enters an individual raindrop by refraction, undergoes either one or two reflections at the rear surface of the drop (one reflection to produce the primary rainbow, two reflections to produce the secondary bow), and is then refracted once more as it emerges from the drop and proceeds to the observer's eye. Figure 33 illustrates the path of the radiation through a single drop for the primary rainbow.[57] Figure 34 illustrates the production of the primary rainbow by a collection of raindrops.[58] Theodoric claims to have performed experiments with prisms and transparent crystalline

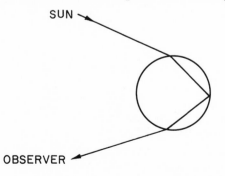

Fig. 33   Production of the primary rainbow: radiation through a single raindrop.

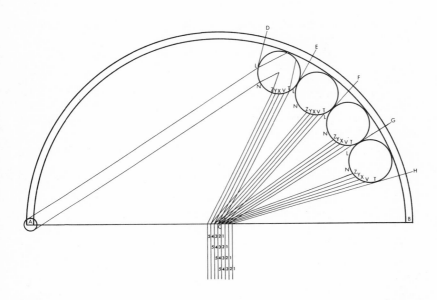

Fig. 34   Theodoric of Freiberg's theory of the primary rainbow.

spheres to confirm his theory, and it is, indeed, a remarkable achievement of the perspectivist tradition. However, to his contemporaries it
seemed no more plausible than the alternatives, and it was not widely
known or discussed during the Middle Ages. Finally, in the seventeenth century René Descartes presented virtually the same theory,
and it is through him that the theory has come down to us.[59]

Notes

1. The ideas presented in this and the next two sections are developed
and defended at much greater length in my *Theories of Vision from al-
Kindi to Kepler* (Chicago, 1976).
2. "Letter to Herodotus," in Diogenes Laertius, *Lives of Eminent Philosophers* 10. 48–49, trans. R. D. Hicks, vol. 2 (London, 1925), pp.
577–79.
3. *De rerum natura* 4. 54–61.
4. On Epicurean *eidola* see also Edward N. Lee, "The Sense of an
Object: Epicurus on Seeing and Hearing," in *Studies in Perception: Interrelations in the History of Philosophy and Science*, ed. Peter K. Machamer
and Robert G. Turnbull (Columbus, Ohio, 1978), pp. 29–30, 42–50.
5. On the Euclidean and Ptolemaic theories of vision, see Albert Lejeune, *Euclide et Ptolémée: Deux stades de l'optique géométrique grecque*

(Louvain, 1948). Euclid believed the visual cone to consist of discrete rays separated by spaces; Ptolemy considered it a continuum.

6. For Aristotle's theory, see his *De anima* 2. 7, 418a26–419a25; *De sensu* 2–3, 437a18–439b18. See also below, n. 55.

7. Aristotle's theory of vision might thus be viewed as an intromission version of the mediumistic theory, while Galen's can be viewed as an extramission version of the mediumistic theory. The fullest account of Galen's theory of vision is in Rudolph E. Siegel, *Galen on Sense Perception* (Basel, 1970); however, this work must be used with caution.

8. The location of an object within the visual field is determined by the location within the visual cone of the rays incident upon it, and the apparent size of a visible object is a function of the angle between rays issuing to its extremes.

9. Euclid's most prominent followers, Ptolemy and al-Kindi, added physical content to the Euclidean theory, but its aims remained principally mathematical.

10. For example, it made no attempt to explain how various objects or visible points are perceived at particular places within the visual field.

11. On al-Kindi's theory of vision, see Lindberg, *Theories of Vision*, chap. 2; also Lindberg, "Alkindi's Critique of Euclid's Theory of Vision," *Isis* 62 (1971):469–89.

12. Hunain translated Galen's *De usu partium* and *De placitis Hippocratis et Platonis*, the two works from which he drew most of his material on vision. An English translation of Hunain's *Ten Treatises* is available in *The Book of the Ten Treatises on the Eye Ascribed to Hunain ibn Is-hâq (809–877* A.D.), ed. and trans. Max Meyerhof (Cairo, 1928).

13. See David C. Lindberg, "The Intromission-Extramission Controversy in Islamic Visual Theory: Alkindi versus Avicenna," in Machamer and Turnbull, *Studies in Perception*; also *Theories of Vision*, chap. 3.

14. The quoted phrase is borrowed, with alterations, from Harold Cherniss, *Aristotle's Criticism of Presocratic Philosophy* (Baltimore, 1935), p. 320. This same objection had been presented in antiquity by Galen, who wrote: "Aristotle was quite correct when he said about the sudden change of bodies thus altered, that it is very nearly instantaneous, and also, with regard to this alteration, that it is the nature of bright air, when altered by colors, to transmit the alteration all the way to the organ of sight. *But Aristotle did not explain how we distinguish the position or size or distance of each perceived object*" (*De placitis Hippocratis et Platonis* 7. 7, in *Opera omnia*, ed. C. G. Kühn, vol. 7 [Leipzig, 1824], p. 638; translation by Philip De Lacy, forthcoming in the *Corpus Graecorum medicorum*).

15. *Ten Treatises*, trans. Meyerhof, p. 32.

16. Under this rubric he includes Plato's theory as well as those of Aristotle and the atomists.

17. The same claim was made by Hunain in the passage quoted above.

18. I must point out that al-Kindi's argument has been reconstructed from brief and fragmentary remarks; I believe, nevertheless, that the reconstruction is secure. See Lindberg, *Theories of Vision*, pp. 22–24, for more detail.

19. See ibid., pp. 27–30.

20. I have chosen not to indicate the refraction of radiation at the front surface of the eye and the interfaces between transparent humors, for fear of implying a level of geometrical precision that many medieval theories of vision did not, in fact, possess. Some medieval authors, such as Alhazen, had an excellent understanding of the qualitative features of refraction, and if it were clear what relative optical densities they assigned to the three humors, we could, without misrepresentation, illustrate their theories of radiation through the eye. However, this would not alter the point of the drawing—namely, that rays from all parts of the visual field mix in all (or virtually all) parts of the eye.

21. The idea that light is weakened when bent by reflection or refraction can be traced back to Aristotle; it was further defended by Alhazen, employing mechanical analogies. See Aristotle, *Meteorologica* 3. 4, 374b28–30; Lindberg, *Theories of Vision*, pp. 75–76. Although in the Latin version of Alhazen's *De aspectibus* the efficacy of perpendicular rays is explained principally in terms of their greater strength, the Arabic version reveals another explanation (which is also obscurely hinted at in the Latin translation), namely, that "there exist certain privileged directions in the lens *as a sensitive body*" (A. I. Sabra, "The Physical and the Mathematical in Ibn al-Haytham's Theory of Light and Vision," forthcoming in vol. 25 of *Boston Studies in the Philosophy of Science*). My sketch of Alhazen's theory of vision has necessarily been highly abbreviated; for additional detail see my "Alhazen's Theory of Vision and Its Reception in the West," *Isis* 58 (1967): 321–41; and my *Theories of Vision*, chap. 4.

22. At the rear surface of the crystalline humor, the rays are refracted away from their point of convergence at the center of the eye and proceed in roughly parallel fashion through the vitreous humor to the optic nerve and eventually to the junction of the two optic nerves, where vision is "completed." On this aspect of Alhazen's theory of vision, see Lindberg, *Theories of Vision*, pp. 80–84.

23. For elaboration of this point, see ibid., chaps. 4 and 8.

24. Only among Galenists and (in the West) Platonists does one find serious attempts to defend an extramission theory after Alhazen, although (as will be seen) there were intromission theories that included an extramission phase.

25. Robert Grosseteste (d. 1253), to whom much credit has traditionally been assigned for the establishment of the Western optical tradition, lived before all of the sources were available (he seems, for example, not to have known Ptolemy's *Optica* or Alhazen's *De aspectibus*) and, therefore, had only a limited influence on the ultimate shape of Western optics. However, Grosseteste is indeed of great importance for stimulating interest in optics.

26. Both ideas could have been borrowed from Alhazen (whose work Albertus knew to a limited degree) and the former from Avicenna and Averroes. Albertus's fullest exposition of visual theory is in his *Summa de creaturis*, in *Opera omnia*, ed. A. Borgnet, vol. 35 (Paris, 1896), pp. 164–228. On Albertus, see Lindberg, *Theories of Vision*, pp. 104–7.

27. These differing loyalties can be explained at least in part, in terms of general philosophical orientation, for Albertus was a firm adherent of

the Aristotelian system, while Roger (like so many Franciscans) had strong Augustinian and Neoplatonic (and, hence, mathematical) leanings. Bacon presented his visual theory in part 5 of the *Opus maius*, available in a modern edition by John H. Bridges, 3 vols. (London, 1900); and a barely acceptable English translation by Robert B. Burke, 2 vols. (Philadelphia, 1928). For a fuller discussion of Bacon's visual theory, see Lindberg, *Theories of Vision*, chap. 6.

28. The term *species* with the connotation of visual form or image was not original with Grosseteste and Bacon, but had been current in the West for centuries. It was clearly defined by Hugh of St. Victor in the twelfth century: "Species is visible form, which includes two things, [namely,] shapes and colors" (quoted by Pierre Michaud-Quantin, *Études sur le vocabulaire philosophique du Moyen Age* [Rome, 1970], p. 114).

29. *Opus maius*, pt. 5.1, dist. 9, chap. 4, ed. Bridges, 2: 71–72; translation reprinted from Edward Grant, ed., *A Source Book in Medieval Science* (Cambridge, Mass., 1974), p. 394.

30. *Opus maius*, pt. 5.1, dist. 7, chaps. 2–3, ed. Bridges, 2:49–52.

31. The perspectivist tradition is perhaps best defined by its adherence to the doctrines of Alhazen. Bacon added several new teachings, but these did not alter the basic character of the system.

32. It must be noted, however, that many Aristotelians attempted to incorporate some of the mathematics of *perspectiva* into their theories, and it is doubtful that in defending the Aristotelian theory of vision they were consciously repudiating that of Alhazen. Perhaps the best example of these syncretistic tendencies is Nicole Oresme, whose work has been examined by Stephen C. McCluskey, Jr., "Nicole Oresme on Light, Color, and the Rainbow: An Edition and Translation, with Introduction and Critical Notes, of Part of Book Three of His *Questiones super quatuor libros meteororum*" (Ph.D. dissertation, University of Wisconsin, 1974), pp. 1–22.

33. Kepler frequently cited the perspectivists and entitled his principal work (published in 1604) *Supplement to Witelo*.

34. In 1583 Felix Platter first clearly stated that the retina, rather than the crystalline humor, is the seat of the visual power. However, even this conclusion (or one very close to it) was anticipated by an anonymous medieval author in a pseudo-Galenic work entitled *De iuvamentis anhelitus*, and the whole question of retinal sensitivity had a long and intricate history prior to the seventeenth century; see Lindberg, *Theories of Vision*, chap. 8.

35. Kepler provided no figure to illustrate his final theory of vision, but the Keplerian theory of the retinal image is adequately represented in fig. 30, from Descartes' *La dioptrique*, in *Oeuvres de Descartes*, ed. Charles Adam and Paul Tannery, vol. 6 (Paris, 1902), p. 125.

36. On Aristotle's theory of light and color, see above, n. 6. For antiquity more generally, see John I. Beare, *Greek Theories of Elementary Cognition from Alcmaeon to Aristotle* (Oxford, 1906).

37. *Avicenna Latinus: Liber de anima seu sextus de naturalibus, I-II-III*, ed. S. Van Riet (Louvain and Leiden, 1972), pp. 170–72. Albertus Magnus probably hit the mark when he interpreted Avicenna's *radius* as

"the issuing forth of *lumen* along a straight line" (*Opera omnia*, ed. Borgnet, 35:184).

38. *Liber de anima*, ed. Van Riet, p. 173.

39. Avicenna's theory did the same, but Alhazen was perhaps more explicit in insisting that light and color together are the direct objects of visual perception. See *Opticae thesaurus Alhazeni arabis libri septem* (Basel, 1572), bk. 2, chap. 2, sec. 15–18, pp. 34–35.

40. Indeed, Alhazen argued that color cannot be perceived unless its form is accompanied by the form of light; see ibid., bk. 1, chap. 7, sec. 39, pp. 22–23.

41. "Physical and Mathematical." However, note that Alhazen also characterized accidental light in terms of the reflecting properties of surfaces; see ibid.

42. Roger Bacon said explicitly that he usually employed *lux* and *lumen* interchangeably; see *De multiplicatione specierum*, pt. I, chap. 1, included with the *Opus maius*, ed. Bridges, 2:409.

43. Averroes, *Epitome of Parva Naturalia*, trans. Harry Blumberg (Cambridge, Mass., 1961), pp. 15–16.

44. See Anneliese Maier, "Das Problem der 'Species sensibiles in medio' und die neue Naturphilosophie des 14. Jahrhunderts," in Maier, *Ausgehendes Mittelalter*, vol. 2 (Rome, 1967), pp. 419–51. On the related subject of the status of light and color (or their species) in a mirror, see Stephen C. McCluskey, "Images, Colors, and the Rainbow in the Fourteenth Century," *British Journal for the History of Science*, forthcoming.

45. *Opera omnia*, ed. Borgnet, 35:205–10.

46. Bacon, *De multiplicatione specierum*, ed. Bridges, 2:507–11.

47. Maier, "Das Problem," pp. 433–44. Ockham is here employing his principle of parsimony—better known as "Ockham's razor."

48. Lejeune, *Euclide et Ptolémée*, p. 38.

49. A. I. Sabra, "Ibn al-Haytham," *Dictionary of Scientific Biography*, 6:191.

50. For a convenient medieval discussion of geometrical optics, see *John Pecham and the Science of Optics: Perspectiva communis*, ed. and trans. David C. Lindberg (Madison, Wis., 1970), from which this illustration is taken.

51. See Albert Lejeune, *Recherches sur la catoptrique grecque* (Brussels, 1957), pp. 152–66.

52. A. C. Crombie, *Robert Grosseteste and the Origins of Experimental Science, 1100–1700* (Oxford, 1953), pp. 219–25. For a translation of the relevant section from Witelo's *Perspectiva*, see Grant, *Source Book*, pp. 424–26.

53. Bruce S. Eastwood, "Grosseteste's 'Quantitative' Law of Refraction," *Journal of the History of Ideas* 28 (1967):403–14; Crombie, Grosseteste, pp. 120–22.

54. An English translation of this proposition from Witelo's *Perspectiva* is given in Grant, *Source Book*, pp. 412–13. For further analysis, see Lindberg, *Pecham and Optics*, pp. 263–65 n. 123.

55. In his *Meteorology* Aristotle adopted the extramission theory of vision, presumably because in this work he was dealing with the mathe-

matics of vision. On Aristotle's theory of the rainbow, see Carl B. Boyer, *The Rainbow: From Myth to Mathematics* (New York, 1959), pp. 41–55.

56. On Grosseteste's theory and other medieval theories of the rainbow, see ibid., pp. 66–142; Crombie, *Grosseteste*, pp. 124–27, 155–62, 196–200, 226–59; Bruce S. Eastwood, "Robert Grosseteste's Theory of the Rainbow," *Archives internationales d'histoire des sciences* 19 (1966): 313–32; David C. Lindberg, "Roger Bacon's Theory of the Rainbow: Progress or Regress?" *Isis* 57 (1966):235–48.

57. A translation of the most important sections of Theodoric's account is found in Grant, *Source Book*, pp. 435–41. See also the works of Boyer and Crombie cited above, and William A. Wallace, O.P., *The Scientific Methodology of Theodoric of Freiberg* (Fribourg, 1959).

58. Copied from a fourteenth-century manuscript: Basel, Öffentliche Bibliothek, MS F.IV.30, fols. 33v–34r. The sun is at A and the observer at C. The numbered lines beneath the center of the drawing (in the vicinity of the observer) indicate how an observer fixed in one place will see different colors in different raindrops and how, as he moves, he will see different colors in the same raindrop. The path of radiation is shown within only one of the raindrops.

59. Despite the failure of historians to discover a direct link between Theodoric and Descartes, I do not think we can dismiss the possibility that Descartes had access, directly or indirectly, to Theodoric's manuscripts.

Robert P.   **The Science of Matter**
Multhauf

Ancient Theories Chemistry is peculiar among the sciences in being
of Matter a modern invention; to put it succinctly, the Greeks
did not even have a word for it.[1] We define chemistry as the study of matter through its changes.
The Greeks did have words for matter (*hyl*) and
several words for change, and they discussed both
at length. Only in the cases of Plato and Aristotle
have enough of their writings survived to enable us
to analyze their views with some confidence; and
in those writings the problems involved in understanding matter and change emerge clearly enough.

The problem of explaining the sensible world
was in its instability; things are changing into something else even while we are describing what they
are. Plato's attempt to escape this difficulty, in the
dialogue *Timaeus*, was possibly made with tongue
in cheek.[2] It was certainly done with imagination.
Plato imagines an "ideal" world which is not material and does not change, and then explains it
according to "reason," using a device which most
scientists have preferred ever since, mathematics.
The elements of nature are geometrical entities (triangles) formed out of "chaos," which he somehow
derives from empty space. Thus, Plato's explanation of nature begins with an attempt to avoid,
as far as possible, matter and change, the very
essence of chemistry! But he then goes on to apply his explanation, so far as he can, to our sensible world, equating the stablest figure (the cube)
with earth, that with the sharpest corners (the
tetrahedron) with fire, and so on. He finally calls the
whole enterprise merely an exercise in probability.

Plato's predecessors had generally concluded that there is an intermediate substance between simple matter (the *prima materia*) and the things of the sensible world which are made of it. This was the element, or elements of nature, for they finally settled on four, fire, air, water, and earth, and it is these which Plato tries to explain in terms of geometric particles. This system, however, proved to be not altogether sufficient. There was postulated another substance, ether, which was to be utilized by Aristotle as the material of the heavens, and which ultimately became a more conventional element, under the name *quintessence* (fifth essence). It was a substance which was virtually weightless and without definable qualities; hence, its existence could neither be proven nor disproven. It was in due course brought down to earth, and served through the nineteenth century to explain phenomena such as light, which seemed otherwise inexplicable. And the ether gave birth to other fluids of the same kind, some of which, notably phlogiston, played an important role in the history of chemistry. I have given these fluids the generic name *emanations*.

Plato's analysis was based on his definition of the ideal and sensible worlds in terms of existence and change. The former exists but is unchanging, while the latter changes but does not exist (by which he appears to mean that, since the sensible world is continuously changing, it cannot be said to "exist" in the full sense of the word). While Plato had in the Eleatic philosophers (for example, Parmenides) a precedent for the idea that the sensible world does not really exist, one suspects that his addiction to this peculiar definition of the sensible world was influenced by the symmetry it gave to his scheme. In any case, this was the sticking point for Aristotle, who was particularly incensed at Plato's conception of the sensible world, an entity which changes even though it does not have full existence. But Aristotle adhered to a system based on existence and change. He redefined the four possible combinations of this pair. Change without existence is impossible. "Metaphysics," treats the existent and unchanging (of which Plato's ideal world is an instance) and is approximately equal to the science of the gods, namely, theology. Mathematics deals with the nonexistent (that is, mental) and unchanging. Finally, "physics" is concerned with the sensible world, which both exists and changes; Aristotle has here narrowed the meaning of a word which already meant "nature." His physics was still not our physics, but comprehended more, including the essentials of our chemistry, which remained largely embedded in physics until it received its modern definition in the eighteenth century.

Thus, Aristotle redefined the study of nature to give it the essential form which it has had ever since. His book on physics (*Physica*) elaborated on this, but was only one of several books which dealt with overlapping subject matter. There were also books on astronomy (*On the Heavens*), on meteorology (*Meteorologica*), and on change in general (*On Coming to Be and Passing Away*—also called *On Generation and Corruption*).[3] The latter is obviously relevant to chemistry, but it reveals the difficulty of extracting that science from ancient writings. Aristotle understood change, properly, it would seem, to refer to such various phenomena as movement from one place to another, growing up, the manufacture of a chair out of a piece of wood, and chemical change. The metallurgical process, in which an ore is changed into metal, is one of Aristotle's examples of change which we can recognize as chemical. More subtle (and rather more physical, in the modern sense, than chemical) was the case of the diffusion of one substance through another to the point that one can no longer detect it. Examples are the diffusion of incense in air and the diffusion of a drop of wine through a large quantity of water. Aristotle asks whether in these cases the incense and wine actually change into air and water or whether they are still there, but does not firmly settle the matter.

This problem leads to that of the smallest possible particle of a substance, which had stymied many Greek thinkers. Most of them considered the world to be "full"; that is, they felt empty space to be a logical contradiction. How could "nothing" exist? It followed that matter must somehow be continuous, that is, not made up of independent particles, and that the apparent annihilation of something meant that it was really changed into something else. The latter conclusion led to speculation on how such a process of changing could occur. The speculation usually took the form of the question of how small a particle could be without losing its essential characteristics. It was a question which also occupied mathematicians, who were concerned with such problems as whether a line could be subdivided into an infinite, or only a finite, number of points. The profundity of such questions took the ancient scientist into philosophical fundamentals deeper than chemistry has ever reached. Indeed, it was a crucial factor in the appearance of our science of chemistry in the eighteenth century that those concerned with it tacitly agreed not to concern themselves with such questions.

Part of this tacit agreement was the acceptance, as a basis of the science, of the idea that the sensible world is made of particles which do have a limited and definite size, and which do not fill the universe completely, part of it being empty space. The Greeks' refusal to accept the possibility of empty space had much to do with their difficulty in resolving the question of the minimum possible size of particles—or even saying clearly what they meant by a particle. Yet the moderns found a suitable particle among the Greeks, in one of the numerous sects of natural philosophy which preceded Plato, the atomists. Since Democritus (5th century B.C.) the atomists have been famous; but among the Greeks they were infamous. In part this was because they attempted to explain even theology—the gods—in terms of a fortuitous mechanical conglomeration of particles. But in part their infamy may have stemmed from the oversimplified and dogmatic character of atomism. In Epicurus and Lucretius we find atomism attractively explained, and objections to it cleverly explained away; but it was clever rather than profound when considered philosophically, and the majority of Greek thinkers, whether oriented toward physics or metaphysics, found it neither attractive nor plausible.

Aristotle's most "chemical" book has generally been considered the *Meteorology*. This was a broader subject in antiquity than now, including everything that goes on between the earth and the moon. No natural changes were more striking than those which took place in meteorology, which (so Aristotle believed) included comets and other astronomical irregularities; and Aristotle made the reasonable supposition that the action of gases, vapors, and so on, which produce such spectacular phenomena in the atmosphere around us, also extend within the earth, where similar emanations roil about, producing earthquakes, volcanic activity, and generating the multifarious minerals which our planet exhibits. The generation of metals within the earth, the metallurgical operations of man, and other chemical operations were long to be explained by the "vaporous" and "smoky" vapors which Aristotle adduced to explain supposed changes within the earth.

The Origins of Alchemy

Theophrastus, Aristotle's successor as head of the Lyceum, wrote a small book on geology, one of the few subjects neglected by his predecessor. Theophrastus's *On the Earth* deals with only a few minerals, which he classifies into "precious stones," "building stones," and so on, including as one class "immature" stones, or "earths."[4] The latter include the ores of the metals, most of which are colored and which were also used as pigments, and two chemically prepared pigments, "white lead" and verdigris, which are products of the dissolution of lead and copper, respectively, by vinegar. This is about as far as the writers of classical Greece descended into the description of a chemical technology which was (as surviving artifacts show) well developed. The basic operations of metallurgy, the production of silver, lead, iron, copper, and tin (gold usually occurred in the native state), had been mastered remarkably effectively in pre-Hellenic antiquity. The efficiency of this technology is in such contrast with what seem to us the bizarre notions of the Greeks on changes in matter, that chemists have preferred to begin the history of the science with this "practical chemistry," as they call it. But chemistry is a science, while this metallurgy seems to have been purely empirical, unless one wants to accept as science vague analogies between agriculture and metallurgy. As far as theory is concerned, Aristotle's idea of vapors within the earth seems superior to the idea that the metallurgist cultivates and accelerates a natural maturation process in metals. But it was not what one would call an adequate explanation of metallurgy. The Greek philosophers, in fact, here left a void, which was to be filled by alchemy.

Alchemy probably originated in the frustration of empiricism.[5] The differences between the known metals were not then as obvious as they are now. Tin and lead, for example, were often confused, and antimony was probably known but thought to be a kind of lead. Alloys, such as bronze and brass, and especially the naturally occurring gold-silver alloy, electrum, were often thought to be independent metals. That by late antiquity the metals were generally defined as we would define them—gold, silver, copper, lead, iron, tin, and mercury—probably owed as much to a determination to limit them to the magic number seven as to any other factor. There were seven peculiar heavenly bodies (sun, moon, and the planets Mercury, Venus, Mars, Jupiter, and Saturn); hence, there should be seven metals. The idea of correspondences between one realm of nature and another has never lost its appeal for the scientist.

Concern with *all* of the metals was largely restricted to the jeweler and goldsmith, and we have from Egypt a jeweler's recipe book (written in Greek) which seems to be a prototype. This "book" is a papyrus manuscript which was purchased in Thebes in 1828 and subsequently became separated into two parts, now known as the Leyden and Stockholm papyri.[6] It is concerned with the fabrication of gold, silver, precious stones, and the purple color which was esteemed above all others. The author is generally concerned with the imitation of these precious materials. The common characteristics of the metals—

fusibility, malleability, metallic appearance—made the idea of changing one into another not unreasonable, but attempts to bring about such changes had evidently reached the point of exhaustion before the time of our Greco-Egyptian author, for he speaks not only of imitation, but sometimes of deliberately fraudulent imitation.

It has been surmised that the Leyden-Stockholm papyrus was looted from a tomb, and that it represents an ancient tradition from the last centuries before Christ, when a large middle class in Egypt created a growing demand for precious objects. We have another manuscript, related but essentially different, which reveals that others were bringing theory to bear on the problem of metal "ennoblement," theory which was in part derived from conventional Greek natural philosophy and in part from Hellenistic transmutations of natural philosophy involving astrology, magic, and theology. This author, or rather, these authors, had parted company with practical chemistry and the jewler's art and had become a kind of scientist—an alchemist.

Largely on the basis of its content of Hellenistic ideas, this latter genre of writing is supposed to have appeared in Egypt about 200 B.C. It survives in a Greek manuscript of which a number of late copies exist, notably in Venice and Paris, the oldest having been written in the tenth or eleventh century A.D., perhaps copied from a lost Byzantine original about two centuries older.[7] It contains treatises or fragments from about forty persons, totaling about 80,000 words. Alchemy emerges full of mystery in this manuscript, for hardly any of the authors can be identified or dated with any degree of confidence, and much of the text is scarcely comprehensible. Scholars have tentatively concluded that the earliest author, whose name is given as Democritus, was really a Hellenized Egyptian named Bolos of Mende, who lived in the Nile delta about 200 B.C.; and that the most important after him was another Egyptian, Zosimos, who lived about 300 A.D.

Even Bolos, our supposed earliest alchemist, seems very nearly irrelevant to the science of matter as we understand it. A comparison of his treatise with the Leyden-Stockholm papyrus leads us to think that he is attempting to advance these arts by the application of theory; but the theory seems already to have gone beyond those current among the Greeks, into the supernatural. A tendency to resort to prayer and incantation to achieve their objective, which is very pronounced among the other alchemists in this text, seems already present in Bolos. In Zosimos it is more than a tendency; and yet he still shows an awareness of mundane chemistry. His "spirits" are not altogether metaphysical or astrological. They include some products of distillation, notably, vapors of sulphur elicited by distillation processes, vapors which almost certainly include hydrogen sulphide, the "rotten-egg" gas which, as any modern student of analytical chemistry knows, joins with dissolved metals to form the multicolored sulphides which are our clues to chemical analysis.

Who was responsible for this innovation? The invention of the apparatus for distillation is attributed to Maria, who is often mentioned by Zosimos, and Agathodaimon is frequently cited as one who experimented with it. Although neither is represented by an independent treatise, this obscure pair appear to have been the pioneers of experimental alchemy. Their experimentation involved the production, by chemical means, of color changes, not only through the sulphides already mentioned, but through arsenic, the latter unknown as an element but known through two arsenic-containing pigments, realgar and orpiment (arsenic sulphide and trisulphide). It is now believed that Maria and Agathodaimon represent a more or less independent school of alchemy, perhaps an older one, stemming from Syria rather than from Egypt.

Whereas Egyptian alchemy is largely known through a single, but fairly substantial, manuscript, Syrian alchemy, although only known through fragmentary references, is known through a greater variety of sources.[8] Maria is sometimes called "the Jewess," and may have come from Asia. The Syrian city of Harran, according to an Arabic writer, had seven temples dedicated to the planets, sun, and moon, each of which had a characteristic geometric shape, color, and an image made of one of the seven metals. Their principal deities were said to be Adimun and Hermus al-Huramisah, names which are believed identical to Agathodaimon and Hermes Trismegistus. According to other Syrian legends, these worthies were the real inhabitants of the great pyramids of Egypt, thus emphasizing, whatever may be the truth of the matter, that the alchemy of Syria and that of Egypt were interconnected. Hermes, who is thought to have been a Greek adaptation of the Egyptian god Thoth, was principally reputed among the Greeks as an astrologer; but he was ultimately to be esteemed preeminent among the legendary founders of alchemy.

And, finally, Syria was the home of Apollonius of Tyana, a historical figure of the first century A.D. who is famous as a pagan sage set up briefly as a rival to Christ. An ancient biography indicates no connection between Apollonius and alchemy, but among the numerous curious treatises on the Art turned up by the industrious alchemical bibliographers of Europe in the twelfth and thirteenth centuries, none was more influential than *The Secret of Creation*, an alchemical work

by Balinus, whose name was agreed to be an Arabic corruption of
the name Apollonius of Tyana. Be that as it may, the work almost
certainly originated in Syria.

In some degree, the history of alchemy among the Arabs recapit-
ulated Greek alchemy. The *Fihrist*, a tenth-century bibliography, lists
fifty-two "philosphers who have discussed the Art." It is headed by
Hermes and Agathodaimon and includes Democritus, Balinus, and
some other familiar names, but most of the names are unfamiliar.
The same can be said of the more or less contemporary *Turba philosophorum* (as the Europeans later called it), a complex conversa-
tion among many persons which has been dubbed "a report of the
hermetic association for the advancement of alchemy."[9] In these writings the Arabs added many more shadowy personages to the considerable number already represented in the Venice-Paris manuscript. The
most famous Arabic writer from whom we actually have alchemical
writing, Jabir ibn Hayyan, is equally shadowy.[10] Not only is it still
disputed whether he was a historical personage or a pseudonym for
a group of alchemists, but his most influential writings are probably
not Arabic at all, but original Latin tracts forged under his name in
Italy in the thirteenth century. The conclusion to be drawn from all
this is that the study of Arabic alchemy is probably still in its infancy.

On the other hand, we also have tracts which describe, or at least
discuss, alchemy—tracts from Muslims who are not only historically
authentic but are among the leading thinkers of Islam, such as
al-Kindi, an Arab of the ninth century; al-Farabi, a Turk who died
about 950; and Avicenna, a Persian who died in 1037. The most important Arabic-writing alchemist was probably al-Razi (d. 923–24),
who was also one of the most famous physicians of Islam. The advances made by the Arabs in the art are largely contained in the writings of al-Razi and the Jabirian writings, and there were some real
advances. The theory that the metals are compounds of mercury and
sulphur was one (although it might better be called an influential
innovation in theory than an advance), but the study of the corrosive
properties of salts and the introduction of two new materials, sal
ammoniac (ammonium chloride) and saltpeter (potassium nitrate),
represented a focusing of attention on what was to be the most interesting group of known substances. It was ultimately to lead to the
discovery of the mineral acids.

While all learning, sacred and profane, long remained an ecclesiastical monopoly in Europe, alchemy, and science in general, was
a secular pursuit in Islam. This created a potentiality for conflict between science and religion which did not exist in Europe before

modern times. The acerbic criticisms of such as al-Razi practically assured such conflict, and a theological reaction against science (including alchemy) began in the ninth century. Ultimately, alchemy was at best an "underground" activity at the capital, Baghdad; and although it was pursued freely enough in such peripheral states as Persia and Spain, which were increasingly beyond the control of Baghdad, the period of innovation was largely over by the eleventh century. The following century saw the beginning of the transmission of the art to Latin Europe.

Alchemy in the Latin West

Our earliest evidence of a substantial contribution by Latin Europe to the study of matter relates to technology, as represented by the goldsmith's recipe book. Although Latin books of this type undoubtedly descended from earlier Greek or Arabic examples, the earliest actual example, subsequent to the Leyden-Stockholm papyrus, is a Latin manuscript of the eighth century called *Compositiones ad tinguenda*.[11] A comparison of this book with the Leyden-Stockholm papyrus shows that it is not very different in the subject matter treated. But it differs in spirit. Gone are the suggestions of metallic transmutation, real or fraudulent, replaced by mundane technical detail on materials, glues, and so forth. A few other similar treatises survive from medieval European sources, and we can see them developing into specialized literatures of several kinds, some primarily concerned with glass and ceramics, others with pigments for painting, and yet others with metallurgy of a more conventional sort. They are not concerned with gold-making, and scarcely with theory of any kind. So they have little to do with a science of matter, beyond indicating an increasing familiarity with materials and their manipulation.

A similar and more significant development occurred in medicine—more strictly speaking, in pharmacy—which Hippocrates had declared to be a branch of medicine. The basic ancient source was the *Materia medica* of Dioscorides, written in the first century A.D. [12] Like many other Greek works, it was cultivated and expanded by the Arabs and remained essentially unknown to Latin Europe until the efflorescence of translations which began in the eleventh century. In that century, Constantine of Africa, who was born a Muslim in North Africa, and died in 1087, a Christian monk at Monte Cassino, brought to Italy some awareness of Greco-Arabic pharmacy through his book *De gradibus simplicibus*. A little earlier, the same subject was being elucidated in Muslim Spain, where the Cordovan caliph possessed a

Greek manuscript of Dioscorides' *Materia medica*, which had been presented to him by the Byzantine emperor in the tenth century. This, too, gave rise to more or less independent Arabic treatises, the most important of which would be a work of Abulcasis (d. ca. 1013) called (by the Latins) *Liber servatoris*. Both this work and *De gradibus simplicibus* differed from Dioscorides in a way which was to be significant for chemistry—they gave substantially more attention to mineral remedies.

The appearance of these works in Latin more or less coincided with the translation of certain books of alchemy, most notably a book called *De aluminibus et salibus* ("On Alums and Salts"), attributed to al-Razi. Before this time we have very little evidence that alchemy, or the art of gold-making under any other name, was known at all in Europe. After this time the Europeans took to it with an enthusiasm which at least equaled that of any of their predecessors.

The European alchemists took the works of al-Razi and the Jabirian writers as a starting point, to such a degree that some of the earliest original European writings were actually attributed to those worthies. Such was the famous *Sum of Perfection*,[13] chemically the most intelligible alchemical tract produced up to its date (thirteenth century), which was certainly based on Arabic writings but probably composed in southern Italy. The Europeans took from these predecessors the practice of relying on less violent processes than those of the metallurgist, such as the property of mercury of dissolving most of the other metals (amalgamation) and the corrosive effects of such salts as sodium chloride, alum, and vitriol on others. From the Jabirian writers they received sal ammoniac (which Europe imported from Egypt), without, however, the Jabirian notion that the organic origin of the substance (it was sublimed from camel dung) indicated that the secret of success lay in the organic realm. Here they preferred al-Razi, who held that the secret lay in the exploitation of the corrosive "alums and salts."

Alchemy became a topic of respectable philosophical discourse in the writings of Roger Bacon[14] and Albertus Magnus,[15] who lived in the middle of the thirteenth century, when Europe was first becoming acquainted with it. Both speak of two kinds of alchemy, one concerned with gold-making and the other with changes in "things" as a general problem. Albertus disavows any intention of showing how some metals are transmuted into others, or how a "medical antidote" called "elisir" cures the illnesses of metals and makes manifest their occult aspects. But it is in these terms that he discusses the formation of metals in nature. Roger, in his famous evocation of "ex-

perimental science," takes examples from the "sciences" of the rainbow, the prolongation of human life, and alchemy, which he also
compares to medicine. He decides that alchemy is of two kinds, one
"speculative," which deals with the generation of all things from the
elements, and the other "operative," which is concerned with making
noble metals, colors, and many other things by art.

More than their predecessors, the Europeans occupied themselves
with the preparation of "elixirs," a term they obtained from the Arabs
(al-iksir), who had in turn received it from the Greeks (xerion). A
substance capable of inducing a chemical transformation solely by
its presence, and that in small quantity, its meaning comes very near
to that of the modern term "catalyst." Since catalytic action has been
shown to exist, it can be said that the idea of the elixir, like that of
the "improvement" of metals, was not unreasonable. But while the
impracticability of the latter would seem to have been demonstrated
by repeated failure, the failure of elixirs to produce gold was not so
evident; for this could be taken merely to signify that the effective
elixir had not yet been found. Any product of the laboratory was a
potential elixir, and the number of substances which might be tried
seemed almost infinite.

But the elixir had other connotations than those relevant to gold-
making alchemy. Some alchemists supposed elixirs to possess the
supreme medicinal property of inducing long life, or even immortality.
This, indeed, had been the principal function of the elixir in Chinese
alchemy, which was little concerned with gold-making. In Greek
alchemy the use of elixirs for any purpose seems to have been incidental. The Arabs stressed their application to gold-making, and the
Latins increasingly stressed the elixir as a medicine rather than as a
catalyst in gold-making. In this, the writings of the Catalan mystic
John of Rupescissa (fl. 1340–50) marked a turning point,[16] for although he claimed that the same elixir could perform transmutations
in metals and prolong human life, his writings were particularly influential among medicine-makers, and became the prototype of such
landmarks of the *materia medica* as the early sixteenth-century distillation books of Brunschwygk, Paracelsus's *Archidoxies*, and Andreas
Libavius's *Alchimia* (1597), the latter perhaps the first work which
can plausibly be called a textbook of chemistry.

The authors of Latin alchemical tracts continued the habit of
placing stress on the eminence and venerability of their predecessors.
For the Europeans, Hermes Trismegistus very nearly filled the latter
requirement. The former requirement, if we are to believe in the
authenticity of surviving writings, was filled by such luminaries as

Roger Bacon, Albertus Magnus, Arnald of Villanova, and Ramon Lull. Unlike the alchemical writings of Muslim savants, which vary greatly in subject matter (including some which oppose the Art), the writings attributed (in many cases wrongly) to these eminent European scholars are remarkably similar. It is a literature of "true believers," who differ mainly in that each thinks that he alone possesses the secret. Even the fifteenth-century Bernard Trevisan, the most conspicuous European critic of alchemy, concludes that he himself has, after all, discovered the secret. As for Roger Bacon and the others, although it is clear from their genuine writings that they knew of alchemy and were at least open-minded about it, most of the alchemical writing attributed to them has been judged by modern scholarship to be spurious. Indeed, while it seems likely that some of them, notably, Roger Bacon, who says that he wrote on alchemy, actually produced alchemical writings, no treatise among the numerous surviving collections attributed to them has been shown with certainty to be authentic.

There was a sharp drop in the production of alchemical writing in the fifteenth century, and its revival in the sixteenth century represented a turn toward a mystic and esoteric alchemy which had little to do with a mundane science of matter. Reviewing his history during the Middle Ages, one can hardly call the alchemist a successful scientist. At best the Art can be called (as it has been) a fossil science. The alchemist can be excused for not achieving the transmutation of metals, for this is still considered impossible by the means at his disposal. But one is astonished at the scale and duration of the effort. One is also astonished at the narrowness of the objective. Even our earliest alchemist, Bolos of Mende, was only interested in transmutations into "asem" (usually meaning electrum) and gold. His successors were generally only interested in making gold, and it is all too clear that the desirability of this objective warped the judgment of the alchemist. The glitter of gold made it impossible for him to be objective, and if disinterested research is thought to be a characteristic of the scientist, then the alchemist was no scientist.

The pretensions of the medicine-makers were not so easily exposed, and the fifteenth century saw many practitioners of the Art shift their objectives from elixirs for gold-making to elixirs for eternal life or simply superior medicines for specific cures. The shift was made easier through the fact that the elixirs were largely the same as the "medicines" the Europeans had acquired from Constantine, Abulcasis, and others. The alchemists and pharmacists of thirteen-century Europe had been working with similar materials and similar techniques toward different objectives, and it remains unknown to which group we owe some of the great discoveries. The most important of these discoveries were those of alcohol and of the mineral acids (nitric, sulphuric, and hydrochloric), which manifested themselves in the thirteenth century (although the differentiation of the three mineral acids was not fully accomplished until the seventeenth century). The key to these discoveries was distillation, for alcohol was a product of the distillation of wine, and the mineral acids were products of the distillation of alum, vitriol, saltpeter, sal ammoniac, and common salt, in various combinations. But distillation had been known to both the Greeks and Arabs. Why, then, did they not make these discoveries? The answer appears to be that their apparatus was inadequate, especially for the purpose of condensation. By about 1300, when alcohol was already fairly familiar, we first encounter a reference to the "serpente." It seems to have been a long condensing tube, sometimes spiral and sometimes carried through cooling water, a device which is still familiar to students of elementary chemistry.

The mineral acids acted, usually spontaneously and often vigorously, to decompose the majority of the substances then known. Both the alchemist and the pharmacist were overwhelmed with possibilities hitherto unsuspected. Alchemy, which had often been called the art of fire, became more the art of dissolution, and pharmacy turned increasingly to mineral remedies, many of them newly discovered salts. There was little difference between the alchemists' elixirs and the pharmacists' mineral medicines, for members of both professions were mining the same vein, the anciently known materials from which the acids had been obtained, plus some new ones, sal ammoniac and saltpeter, which had turned up in the European conquest of Eastern lore.

Alchemy also profited from the discovery of the acids, which gave it new life and made the thirteenth and fourteenth centuries the liveliest period in the history of the Art. But still no gold was made, and gold-making alchemy declined sharply in the fifteenth century. The sixteenth century is often described as the flowering time in Western alchemy, but as far as gold-making is concerned, it was an era dominated by magicians, charlatans, and avid readers of old books. The most important alchemists of this century occupied themselves with a mystic philosophy of the human spirit, rather than with gold-making. These were theorists, but their theories were extracted from the great melting pot of mysticism, Pythagoreanism, astrology, cabalism, Hermeticism, and so forth, which bubbled merrily through that century and had little to do with mundane matter. However, the majority of the denizens of the laboratory were no longer concerned with either gold-making or mysticism. Many of them had transferred their interests from gold-making to medicine-making. And just as, several centuries later, the innumerable tinkerers of the United States were to provide a reservoir of talent for a more rational engineering establishment, so the alchemists of medieval Europe were a vertiable army of potential drug-makers, and, ultimately, of chemists. This was the principal significance of medieval alchemy for the history of chemistry.

Theories of Matter in the Middle Ages

Plato and Aristotle play predominent roles in the early history of theories of matter, in part because they summarize in a relatively coherent way the considerable variety of opinions held by their predecessors. But the summation remained complex. From Plato's *Timaeus* came the idea that "things" are composed of matter and form, which Aristotle complicated by holding form to be only one of four "causes" involved in the generation of things. Moreover, form had two kinds of existence, actual and potential. Superimposed on this was the acceptance, by both Plato and Aristotle, of the four elements propounded by Empedocles—fire, air, water, and earth—as building blocks intermediate between the *prima materia* and things. As an immediate cause of the interaction of these elements, Aristotle adopted the four qualities—hot, cold, wet, and dry—which served a function much like that now given to energy. In his book *On Generation and Corruption*, he alludes twice to three "originative sources" of changes in the sensible world (including chemical changes). In the first reference, the sources are matter, the four qualities, and the four elements. But he says the latter are "less" originative, and when he returns later on to originative sources he omits the elements altogether, mentioning form (which he equates to the four qualities), matter, and a third source which he regards as "truly revolutionary," the north-south movement of the sun with the seasons.

The reasoning behind these distinctions reflects an effort to extricate natural philosophy from a web of contradictions in which it had become enmeshed through a consistant equation of existence with materialism. In general, everything which could be conceived to exist was thought to be somehow material—love, hate, even thought itself. Socrates seems to have been the first to break this habit, with his conception of "the Good," as something which really exists without the taint of materiality. The Good was the prototype of the forms on which Plato based his analysis of nature, and Aristotle, to whom metaphysics and mathematics were immaterial entities, carried the escape from universal materialism a step further. To him the material cause was only one of four causes which went into the description of things.

Thus, Plato and Aristotle rescued such entities as mathematics from matter. But they did not rescue matter—that is, the sensible world—from the intrusion of more or less immaterial entities, such as Plato's "space" and the "diaphanous" material with which Aristotle associated light. Aristotle placed the "ether" in the heavens, but it did not stay there; and this intangible substance was joined, in the writings of his successors, by other emanations, such as the Stoic pneuma and the "influences" of astrologers, an ample armamentarium for the alchemists, as we have seen.

Among the Christians these emanations were made use of for other purposes, as they were also by the Muslims; for both of them diverged from the Greeks in being unwilling to accept the distinct separation of heaven and earth into different realms, which was a characteristic of the philosophy of both Plato and Aristotle. It is curious that the Greek gods, so close to the world that they were virtually members of the human community, occupied, according to the theories of Greek cosmologists, an abode completely outside of "physics"; while the God of the Christians and Muslims, utterly remote though He was, was nonetheless fitted into the same system of physics. These monotheistic religions demanded a direct, and at least quasi-physical, connection to God.

This problem preoccupied virtually all of the leading philosophers of Christianity and Islam. St. Augustine posited a spiritual as well as a terrestrial substance—the angels, among other things, being made of the former. As Europe faded into the "Dark Ages," the further pursuit of this was left for some centuries to the Muslim scholars who emerged in the eighth century at the new permanent capital at Baghdad. Al-Kindi filled the space between God and the world with emanations, which he called "intelligence of incorporeal substance" residing between God on the one hand and the triad of soul, form, and matter on the other. Following the Koran, Muslim authors usually spoke of five "states of being," of which the terrestrial domain is only one. Al-Kindi also mentioned matter, form, place, motion, and time as "eternal substances," and al-Razi spoke of God, soul, matter, space, and time. Al-Farabi envisioned a universe of spheres made of emanations, and Avicenna regarded form itself as an emanation.

The Muslim philosophers were not indifferent to the reconciliation of this with the Greek philosophy of nature. The attempted reconciliation took the form of an increasingly complex web of subtle distinctions, especially on the materiality of entities which were defined as real without being corporeal—the "incorporeal substances." One had to inquire into the nature of their materiality. Some held it to be a different kind of matter from the ordinary; others said it was the same kind of matter, but "without quantity." Such cogitations were interrupted in the tenth century by a theological reaction against the mostly secular natural philosophers of Islam, which greatly curtailed their activities, and the further pursuit of this question was passed on to the Christians, who continued to complicate it. It was this aspect of Scholastic philosophy which led to its becoming an object of ridicule by Galileo and other reformers of the seventeenth century.

Still, the most important philosophers of Islam were yet to come, namely, Avicenna (980–1037) and Averroes (1121–98), who flourished in Persia and Spain, respectively, regions increasingly independent of the ancient political and theological capital of Baghdad. They were particularly concerned with a rather different problem, the recovery of the true Aristotle from the crowd of his interpreters. Islam had inherited the remains of the library and museum at Alexandria, a school at Jundishapur (Persia) which housed a number of Hellenistic philosophers, and other fragments of the classical age. In their large commentaries on what in their time purported to be the works of Aristotle, Avicenna and Averroes found it necessary to ponder the questions of matter and change. And they introduced some originality into the question. Avicenna proposed that in the change of one "thing" into another, the four qualities undergo a kind of qualitative diminution—they are "remitted," as he put it—and at the moment when change occurs the new substance is endowed with a new form, a "substantial form." Averroes, who practiced natural philosophy in Spain after it had declined in the heartland of Islam, took this up and pointed out that contradictions emerge if we suppose that a compound contains its own form, plus those of its ingredients, plus the forms of the elements; and he attempted to avoid these contradictions by supposing that not only the four qualities but also the "substantial forms" undergo remission. In this condition the question was transmitted to Latin Europe, where the idea of substantial forms was to become so prominent among Scholastic (that is, Aristotelian) philosophers that it was to be a principal target of the anti-Aristotelians of the seventeenth century.

The true Aristotle came to Europe with an Arabic passport, and the history of the science of matter in late medieval Europe was much a continuation of its history in Islam. The first two eminent personages to concern themselves extensively with Aristotle's works, Roger Bacon and Albertus Magnus, immediately found themselves embroiled in the question of spiritual and corporeal substance. Roger held spiritual matter not to be subject to quantity, movement, or change; the celestial bodies to be subject to quantity and movement, but not change; and sensible (terrestrial) matter to be subject to all three. Albertus acknowledged that matter cannot be the same in incorruptible heavenly bodies and in generable and corruptible things, but condemned the idea of different *kinds* of matter when it seems to contradict Genesis. But he also admitted the possibility of "two truths" on such questions. To the theologian, following Genesis, matter is one. To the philosopher, matter is the subject of change, and, consequently, different sorts of change require different sorts of matter. Thomas Aquinas, in whose time the opinions of Avicenna and Averroes had been added to those of Aristotle, adopted a compromise between the two Muslims in explaining changes in matter through a remission of the four qualities and their endowment with a new form. St. Bonaventure (1221–74), on the other hand, went on to argue that matter is not necessarily either corporeal or spiritual. It becomes one or the other according to the form it receives. Moreover, every being assumes as many forms as it has different properties; each thing has a multiplicity of forms.

All possible answers were in due course given to the question of the relation of matter and form, including that of Nicholas of Autrecourt (d. ca. 1350), who responded to the failure of doctrines of forms to solve the problems of matter by denying that we are even certain of the reality of substances. Nicholas opted for Epicurean atomism, and regarded bodies as mere effects of temporary conjunctions of atoms. Theological authority finally found it necessary to issue formal condemnation of some of these opinions. This condemnation did not correspond to the theological reaction which had occurred in Islam, for the controversialists were not only mostly clerics, but some of them were among the most saintly. Nor did it still the controversies, although it seems to have urged them in new directions.

One such direction was quantification, the road which was to lead to Galileo, although its consequence was much longer in being felt in the science of matter than in physics. The Scholastics found another avenue than atomism to the consideration of particles, in the idea of "natural minima," the smallest particles which could be conceived of as endowed with form. Such cogitations were associated with a new leader of the philosophical community, William of Ockham (d. ca. 1349), who found in natural minima an approach to quantification through what he called "numerical distinction."[17] Although little interested in mundane nature, Ockham set a fashion which led to such enterprises as the attempt of Giovanni Marliani (d. 1483) to establish "numerical distinction" in the degrees of the contraries hot and cold.

Marliani did not invent the thermometer. Nor did others with similar ideas bring them to a successful conclusion. Ramon Lull, in several works composed between about 1280 and 1308, held bodies to be composed of the four elements in "determined proportions," which are transmitted to their descendants unchanged.[18] This is true of all nature, although plants and animals, having a "generation of the second degree," are more corruptible than minerals and metals. Ramon does not appear to have attempted to measure these "determined proportions," but about the same time, Arnald of Villanova, at Montpellier, came across a work of al-Kindi, *Quia primos*, in which an attempt was made to calculate Galen's vague four intensive degrees of the qualities of medicines. Arnald did attempt, in his *Aphorismi de gradibus*, to measure the qualitative intensities of compound medicines, an inconclusive effort, but one which began a long series of such studies at Montpellier.[19]

In the mid-fifteenth century appeared what should have been the crowning work of this genre, the *Idiota de staticis experimentis*, from the pen of Cardinal Nicholas of Cusa.[20] This incidental piece from one of the best-known philosophers and churchmen of the time does not appear to have attracted much attention. It advocates the use of the balance and the comparison of weights for the solution of a wide range of problems, ranging from mechanics to medicine, and including those of chemistry. He advocates differences in weights as a guide to the evaluation of natural waters, the condition of blood and urine in sickness and health, the evaluation of the efficacy of drugs, and the identification of metals and alloys. It would be difficult to find a more specific prescription for what scientists were actually doing two centuries later.

Conclusion

The end of the Middle Ages, once neatly fixed at 1500, has been moved about, generally toward 1400, or even 1300. The Italian Renaissance occupied the last quarter of the millenium 500–1500. A technological revolution which may have had more to do than any other factor with the birth of the modern world occupied the last half of the Middle Ages. Not the least momentous event in this revolution was the introduction of modern firearms, through which the Europeans gained a decisive ascendancy over the rest of the world. For gunpowder and a superior metallurgy they were ultimately indebted to the Orient, where neither had been applied as the Europeans were to apply them.

The crucial ingredient of gunpowder was saltpeter, a chemical material long known in India, and apparently in China. Saltpeter was also crucial to the discovery of the mineral acids, a European innovation which was for the first time to expand significantly the number of materials known, and which was to give man for the first time a significant repertoire of spontaneously active materials. Chemistry, even though it still lacked its name, became perhaps the most active area of technological innovation. By 1500 the ancient goldsmith's recipe book had given birth to separate categories of treatises devoted to making pigments, ceramics, and glass, and for metallurgy. The *materia medica* had undergone a similar expansion, and mineral remedies had doubled or tripled in number; and they were no longer incidental medicaments, but had become those which seemed most promising. The spectacular developments which are only revealed in the sixteenth century in the writings of Vannuccio Biringuccio (ca. 1480–ca. 1539), Georg Agricola (1490–1555), Girolamo Cardano (1501–76), the celebrated Paracelsus (ca. 1493–1541), and others were largely made before 1500.

And what of the theorists of the nature of matter, the contemporaries of the renovators of astronomy, physics, anatomy, and physiology? No Copernicus, Galileo, Vesalius, or Harvey appeared to undo the ancient dogmas at a stroke. One could speculate that Paracelsus might have filled the role had his thought been more disciplined, but this seems doubtful. Copernicus and the other scientific luminaries of post-medieval Europe worked in branches of natural science which were tolerably well defined, and on questions which were contemporaneously recognized as crucial. But the science of matter was still undefined, and the crucial questions unrecognized. What were they: material vs. immaterial substance? essences? elements? change? mixture? Even today it is hardly possible to say which of these is more central to the science of chemistry than others. The failure of chemistry to materialize as a science in antiquity, in the Middle Ages, or any time prior to the eighteenth century, is perhaps sufficiently explained by the difficulty of the subject.

But if the students of matter failed, during the epoch of the Scientific Revolution, to produce a Copernicus, it was not for lack of effort. And the barriers to innovation were no more impervious in the science of matter than in other natural sciences. On the question of elements, new systems proliferated, as we see in Paracelsus, Cardano, Telesius, Pomponazzi, Campanella, and others of the sixteenth and seventeenth centuries. And they were indebted to the revelations, in medieval translations, that ancient natural philosophy had not been as monolithic as had been supposed. It was these translations which enabled Conrad Gesner (d. 1565) to note that there had been no less than eight systems of elements between the pre-socratic philosophers Thales and Empedocles!

That Paracelsus and the others discussed questions of matter more in the context of ordinary terrestrial materials owed much to the medieval Scholastics, who, notwithstanding their preoccupation with spiritual substances, differentiated more clearly what we would call chemical changes than had the ancients. In part this was a consequence of the wide range of their continuous disputation over matter. In part it stemmed from the success of their effort to abolish the ancient distinction between heaven and earth, which put them under the obligation of making a distinction between celestial and terrestrial substances. In part it was indebted to their awareness of alchemy, which at least produced writings which were more clearly concerned with chemical change than anything produced in antiquity.

The rise of an interest in "numerical distinction" among students of matter in the later Middle Ages seems, in retrospect, to suggest that they were in possession of the key to success, but could not find the keyhole. Long after Galileo and Kepler had shown the potency to science of numerical distinction, the chemists, as they were finally called, continued to wander in a wilderness of undirected experimentation and undisciplined argument. Indeed, by the seventeenth century they seem to have lost sight of numerical distinction. These circumstances reinforce the suggestion that the science of matter was simply too complicated, its problems too intricate, to be susceptible to the neat and succinct solutions we find Copernicus, Vesalius, and the others applying to particular problems in other sciences. The theorists of matter were, in the seventeenth century, still dealing with too many questions, and the central problem really appears to have been in reducing the number of questions which had to be asked.

Under the circumstances, Robert Boyle's rejection of all systems of elements, and Descartes' rejection not only of the traditional explanations of matter, but of the traditional questions as well, may have been salutory. But that the mechanical explanations of chemical phenomena adduced by these two and others were superior to those of the despised Scholastics is open to doubt. Lavoisier, who finally put the science of chemistry together in 1789, gained far more inspiration from numerical distinction than from the corpuscular speculations of the luminaries of the previous century. Among the numerous explanations suggested for Lavoisier's unprecedented reliance on the balance and the equivalency of weights, a reading of the medieval advocates of numerical distinction has not been included. But one can speculate that had he read some of these treatises, he might have reached the same conclusion.

Notes

1. Material for this essay is principally from Robert P. Multhauf, *The Origins of Chemistry* (London, 1966), and Etienne Gilson, *History of Christian Philosophy in the Middle Ages* (New York, 1955). Detailed references will be found in both of these books.

2. Translations and commentaries will be found in F. M. Cornford, *Plato's Cosmology* (London, 1937; also New York, 1957), and in Plato, *Timée-Critias*, ed. and trans. Albert Rivaud (Paris, 1925).

3. Among the numerous editions and translations of Aristotle's works, I recommend *De generatione et corruptione*, English trans. by H. H. Joachim (Oxford, 1922); *De la génération et de la corruption*, ed. and trans. J. Tricot, 2d ed. (Paris, 1951), and *Les météorologiques*, trans. with notes by J. Tricot (Paris, 1941).

4. Theophrastus, *On Stones*, ed. and trans. E. R. Caley and J. F. C. Richards (Columbus, 1956).

5. On alchemy see Frank S. Taylor, *The Alchemists* (New York, 1949), and E. J. Holmyard, *Alchemy* (Harmondsworth, 1957).

6. The Leyden papyrus X, which contains most of the relevant material, has been translated into English by E. R. Caley, in *Journal of Chemical Education* 3 (1926): 1149–66. Caley has also translated the Stockholm papyrus in the same journal, vol. 4 (1927), pp. 979–1002.

7. This collection has been published (in Greek and French) in M. P. Berthelot and C. E. Rouelle, *Collection des anciens alchemistes grecs*, 3 vols. (Paris, 1887–88).

8. H. E. Stapleton has been the most important expositor of alchemy in Syria and adjacent regions. Other references will be found in his last exposition, "The Antiquity of Alchemy," *Ambix* 5 (1953): 1–43.

9. John Ferguson, *Bibliotheca Chemica* (Glasgow, 1906), 2: 479.

10. The most important authority on the Jabirian alchemical treatises is Paul Kraus, *Jabir ibn Hayyan*, 2 vols. (Cairo, 1942–43).

11. Translated into English in *A Classical Technology*, trans. John Burnham (Boston, 1920).

12. Translated into English by John Goodyear (1655), in *The Greek Herbal of Dioscorides*, ed. G. T. Gunther (New York, 1934).

13. Translated into English by Richard Russell (1678), in E. J. Holmyard, ed., *The Works of Geber* (London, 1928).

14. Roger's most important works have been published in the English translation of his *Opus majus* by Robert B. Burke, 2 vols. (Philadelphia, 1928), and in *Opera quaedam hactenus inedita*, ed. J. S. Brewer, vol. 1, *Opus minus, Opus tertium, Compendium philosophiae* (London, 1859).

15. Albertus's most relevant book has been translated into English as *Book of Minerals*, trans. Dorothy Wyckoff (Oxford, 1967). Albertus's works, *Opera omnia*, ed. Auguste Bourgnet (Paris, 1890–99), run to thirty-eight volumes.

16. John of Rupescissa's most important book has been published in English as *The Book of Quinte Essence* (London, 1856).

17. S. C. Tornay, *Ockham* (La Salle, Ill., 1938), p. 35 (citing Ockham's *Summulae in libros physicorum*, 1. 18). Ockham represents a "mathematical-descriptive tendency" outlined in Curtis Wilson, *William Heytesbury: Medieval Logic and the Rise of Mathematical Physics* (Madison, Wis., 1956), pp. 18–21.

18. See Armand Llinarès, "Les conceptions physiques de Raymond Lull," *Etudes Philosophiques* 22 (1967): 439–44.

19. See Michael McVaugh, "Arnald of Villanova and Bradwardine's Law," *Isis* 58 (1967): 56–64.

20. Translated into English by Henry Viets, in *Annals of Medical History* 4 (1922): 115–35.

12

Charles H.   **Medicine**
Talbot

The Beginnings   If it is acknowledged that the Middle Ages was a
kind of tenuous extension of the late classical pe-
riod, it follows that many of the traditions, ideas,
and attitudes of medieval medicine must have had
their roots in much earlier centuries. The Middle
Ages did not create anything *ex nihilo sui et sub-
jecti*, but took over the remnants and scraps that
had survived destruction and neglect at the hands
of successive hordes of barbarians. In the sphere of
medicine, this fact becomes immediately apparent
when the texts on which the medieval physician
relied are examined. To the treatises which the
doctor used both for instruction and practice are
attached, not the name of some obscure, untutored
healer of the "Dark Ages," but the resounding
names of Hippocrates, Galen, Dioscorides, and
others from the classical age of medicine. Though
many of these works are of dubious authenticity,
their attribution to famous figures of the past showed at least that both scribe and reader considered
them to belong to the mainstream of a cherished
tradition.

Among these texts were some that described the
ideal qualities which a physician should possess
and that detailed the studies which he should have
completed before embarking on a medical career.
These studies were what we now call the liberal
arts. At the time when Hippocrates was writing,
the strictly medical part of a student's education
was carried out, not in a school, but in the house
and under the supervision of an accredited physician, so that his training was more of an apprenticeship than an academic curriculum. He learned med-

391

icine not so much from books as from experience and practice.[1] For
that reason, medicine was considered to be an art. Indeed, according
to Hippocrates it was *The Art*.

By Galen's day, though the apprentice system had not fallen completely into abeyance, a much greater emphasis was laid on learning
medicine from books, a development which Galen viewed with some
aversion. This was the period when the multiplication of elementary
manuals was occasioned by the scarcity of educational centers where
medicine could be studied and by the general tendency to call anyone who undertook the treatment of the sick a physician, no matter
how slight his acquaintance with medical literature might be.[2] Indeed, if we are to believe Galen, many of them could not even read.
The net result was that emphasis was laid more on practice than on
theory. Though Galen himself stressed the importance of medical
theory, medicine remained for him and his contemporaries an art,
not an art like architecture, music, or drama, but a conjectural art
like archery, in which practice, rather than reasoning, is paramount.[3]

Small wonder, then, that the study of medicine concentrated almost
entirely on the field of diagnosis and therapeutics. It concerned itself
little with the causes of disease and only superficially with its symptoms. What it sought was the cure of disease and the easiest means
by which this could be accomplished. The proliferation of books on
remedies during the Hellenistic period is an indication of this, and
the later compilations like the pseudo-Dioscorides, pseudo-Apuleius,
Scribonius Largus, Serenus Sammonicus, and Marcellus of Bordeaux
are a proof.

Even where educational facilities for the study of medicine were
provided, as at Alexandria, and at its later counterpart, Ravenna, the
idea that medicine was a craft on a par with the work of masons and
shipbuilders was perpetuated.[4] Though the basic course of study consisted of several texts of Hippocrates and Galen and a certain amount
of theory was taught, the real emphasis lay on practice. This can be
seen quite clearly in the choice of books selected by the teachers for
commentary. The four texts of Galen which formed the syllabus were
the *De sectis*, the *Ars medica*, the *De pulsibus*, and the *Ad Glauconem
de medendi methodo*. The first was supposed to describe the main
schools of medicine, so that the student could know which were good
and which were bad; the second gave an account of the heart; the
third described the pulse; and the fourth dealt with the cure of fevers.
So, in all, they comprised a short history of medicine, a compendium
of general pathology, a summary of diagnosis through the pulse, and,
finally, a method of curing fevers.[5]

By the ninth century, when the first pre-Salernitan texts on medicine emerged, it was only to be expected that the practical treatises
should predominate. They dealt with blood-letting, uroscopy, diet,
various kinds of remedies, gynecology, and surgery. Notable among
these down-to-earth texts were manuals like the *Lecciones Heliodori*
and the *Sapientia artis medicinae*, the direct descendants of the Hellenistic manuals, containing a mythological history of medicine, a biography of Hippocrates, ethical rules of conduct, an enumeration of
parts of the body, the names of the chief diseases, a list of surgical
instruments, and the phases of the moon when blood letting was safe.
Apart from these there were short writings of Cassius Felix, Vindician,
and Theodore Priscian, besides the translations from the larger works
of Oribasius, Alexander of Tralles, and Paul of Aegina.[6] But the main
link in the transmission of ancient medical thought to the Middle Ages
was the translation, made by Caelius Aurelianus in the fifth century,
of the work of Soranus of Ephesus on acute and chronic diseases.[7]
Aurelius's text was far too long for it to be copied in its entirety, and
as a result it suffered various changes and abridgments at the hands
of people who wished to reduce it to manageable proportions. At an
early stage, the section dealing with acute diseases was detached and
issued as an independent work under the name of Aurelius. The remaining portion on chronic diseases was reorganized in the seventh
century by a Christian physician and circulated under the name of
Aesculapius. From these two texts other smaller manuals were compiled, some practical, some theoretical. The practical compendia gave
advice on how to interrogate a patient, how to take his pulse and to
differentiate between the various kinds of pulse (see fig. 35); or they
described wounds and fractures, listed the kinds of sponges and lints
to be employed, and gave instructions on the manner of carrying out
operations. The theoretical compendia, couched in the form of question and answer, dealt with the divisions of medicine, and so on.[8]
    That this kind of literature was not addressed to a professional
class may be surmised from its incorporation into the popular encyclopedias of the times. Both the *Etymologiae* of Isidore of Seville and
the *De universo* of Rhabanus Maurus of Fulda have long sections on
medicine, which derive from the sources just mentioned.[9] These encyclopedias, written for monks and clerics, show that a knowledge
of medicine was considered to be a necessary element in their education. Though for the most part this would be pure book-knowledge,
it ensured that medical texts studied in one monastery or cathedral
school would pass almost automatically to another monastery or school
in an adjoining region, and in this way a rapid diffusion of medical

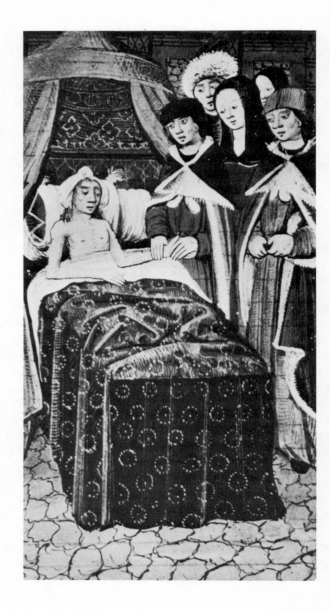

Fig. 35   A physician taking a patient's pulse. Glasgow, University Library, MS Hunter 9, fol. 76r (15th c.). On this illustration, see Loren MacKinney, *Medical Illustrations from Medieval Manuscripts* (London, 1965), p. 16.

texts was assured. The books on medicine, for instance, brought from southern Italy to the north of England in the seventh and eighth centuries and carefully copied there soon found their way into the German monasteries established by Boniface on his missionary journeys.[10]

Monasteries, however, were not merely repositories of medical literature; they were also centers of medical practice. In the large communities, provision had to be made for those monks who fell sick. Not only had buildings to be constructed with special kitchens and dining halls, with bathrooms and toilets, with separate rooms for doctors and infirmary staff, and with wards isolating the dangerously diseased from the mildly sick and convalescent, but gardens also had to be laid out where herbs could be grown as medicaments.[11] In such circumstances a knowledge and experience of the medicinal efficacy of plants was essential. Since many of the herbs listed in the available medical texts were not native to the more northerly countries of Europe, nor even to some Mediterranean climates, substitutes had to be found, and while the basic remedies, like rue, mint, rosemary, and fennel were common, others were unobtainable.[12] Moreover, though some of the herbals gave the precise amount of the drug to be administered, the dosage in many cases was either vague, totally unspecified, or grotesquely misconstrued by copyists.[13] Experiment was therefore necessary, not only because the properties of many plants were not known, but also because even when the plants were familiar, their safe dosage could not otherwise be ascertained. Hence, more and more emphasis was laid upon the practical aspects of medicine. An expert physician like Heribrand, who expounded the *Aphorisms* of Hippocrates to his pupils at Chartres, was renowned, not so much for his knowledge of texts, as for the fact that "he was an expert in pharmacology, pharmacy, and botany."[14]

This essentially practical attitude to medicine became more pronounced as time went on. A visible proof lies in the dwindling space allotted in medical manuscripts to the description of the symptoms of disease and to the increasing accumulation of recipes which fill their pages.

Up to the middle of the eleventh century, then, medical ideas remained more or less static. Though the multiplication of medical texts had increased considerably from the eighth and ninth centuries onward, the basic teaching derived from the translations and adaptations of Greek texts had lain undisturbed. The Carolingian renaissance of the ninth century, which in many other fields had given rise to striking developments, had not fundamentally affected the course of medical enlightenment. When we consider that Alcuin, Lupus of Ferrières, and Gerbert of Aurillac scoured Europe in their search for classical literary texts, it seems strange that medical texts should not have aroused the same passionate inquiry. Why, for instance, did the work of Celsus, *De medicina*, of which two copies existed in Italy, remain unused and uncopied from the ninth century onward? And why did the complete and unadulterated text of Caelius Aurelianus's translation of Soranus lie neglected in the library of Lorsch? Perhaps the explanation lies in the reverence paid to the liberal arts, in contrast to the lack of esteem accorded to a mere craft.

The Awakening

In the south of Italy at the monastery of Monte Cassino a converted Muslim named Constantine the African (d. 1085) devoted himself to translating some of the medical works produced by the Arabs.[15] He began, in the tradition of his times, to make Latin versions of the practical treatises, such as Isaac Israeli's book on urines, which he followed up with a translation of the same author's work on fevers and diets. Later he turned his attention to the theory of medicine and produced versions of the *Isagoge* of Johannitius, the *Aphorisms* and *Prognostics* of Hippocrates with the *Regimen in Acute Diseases*, the last two of which were accompanied by the commentaries of Galen. Finally he made available the *Pantegni* of Haly Abbas, the *Viaticum* of al-Djezzar and the *Megategni* of Galen, three compendia which covered the whole field of medicine, both theoretical and practical. He began with the smaller tracts because, apart from their practical outlook, they could be carried about, frequently consulted, and committed to memory, but his later efforts were concentrated on providing a complete conspectus of medicine.

It might be expected that the impact of this new material on the masters at the school of Salerno would have been immediate and revolutionary. But this was not so. The medical school at Salerno,[16] which had enjoyed a high reputation since at least the tenth century, had not produced a writer of distinction during the whole period of its existence, and, though the names of Gariopontus and Petrocellus, were vaguely associated with it, their writings, though orderly, contained essentially the same material as the Aesculapius-Aurelius complex that had been known since the ninth century.[17] Even the writers who emerged after the publication of Constantine's translations were imbued with this old-fashioned outlook, as the titles of their books proclaim: *Practica* Platearii, *Practica* Bartholomaei, *Practica* Petrocelli, *Curae* Johannis Afflatii, and so on.[18] In all these works the early tradition of presenting a brief description of a disease followed by a list of remedies was continued without an iota of medical theory. Indeed, in some cases neither the description of symptoms nor the application of remedies was as detailed or as clear as it had been in previous times. Even in the textbook compiled for the use of students from the writings of the Salernitan masters and entitled *De aegritudinum curatione*, there was little sign of theoretical teaching, whereas there were signs of regression in the incorporation of elements from folk-lore, magic, and superstition. It was, however, in this book that the first signs of Constantine's influence began to manifest themselves, and gradually, as time wore on, the stirrings of a new life in medical teaching began to appear. Perhaps the most interesting element here was a revival of the study of anatomy, a subject that had been neglected by Christians and Arabs alike. Though the Arabic compendia contained large sections on the theoretical part of anatomy, they had never mentioned actual anatomical demonstrations. Now at Salerno the structure of the body began to be taught from the skeletons of men found dead or drowned, and the inner organs and nerves by the dissection of pigs.[19] This was a great step forward that was to lead to the development of another branch of medicine, the practice of surgery.

The first book on surgery to emerge from Salerno was the so-called Bamberg surgery, a compilation made from Constantine's translation of Haly Abbas, material from a treatise on phlebotomy attributed to Hippocrates, part of Oribasius's *Euporistes*, and ideas taken from Vindician. This treatise held the field until the latter part of the twelfth century, when it was superseded by the methodical work of Roger Frugardi, which attained instant popularity not only in Italy but also in the rest of Europe.[20] In another field, that of pharmacology, the Salernitans also made progress. The *Circa instans* of Platearius, the *Antidotarium* of Nicholas, and several other works of a like nature, with the commentaries on them, showed that not only was the new material from Arabic works being absorbed, but it was also being expanded.

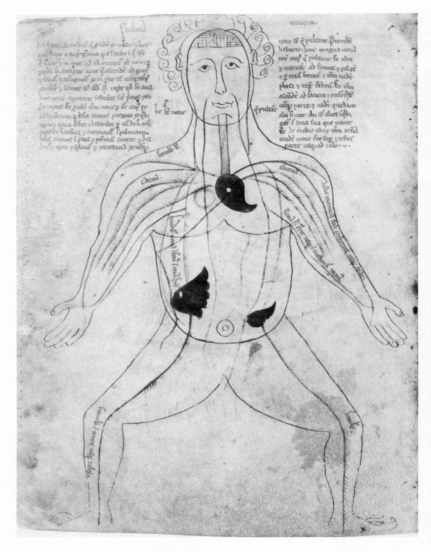

Fig. 36A   A pseudo-Galenic diagram of the nerves (A) and bones
                     (B) of the human body. Cambridge, Gonville and
                     Caius College Library, MS 190/223, fols. 2v–3r (11th
                     or 12th c.).

Still in the sphere of the practical, the Salernitans emphasized the
value of diet. Though there had been many tracts on this subject in
earlier times, and the ideas in them had been greatly amplified by

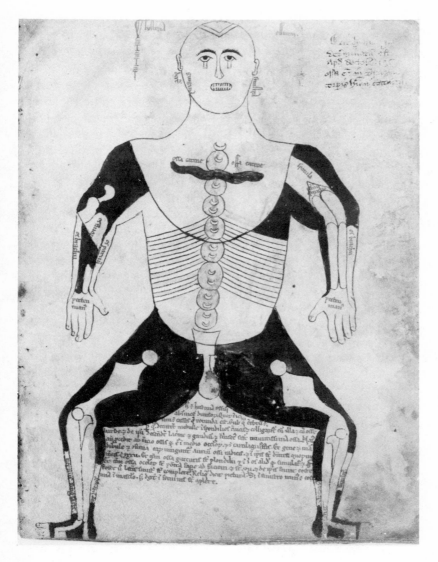

Fig. 36B

the translations of Isaac Israeli's books, the *Regimen Salernitanum* couched in a few hundred verses, surpassed them all in popularity.[21] Its precepts, easily committed to memory, commended themselves to all classes of society, high and low, and found a ready acceptance in every country of the West. No less popular was the compendium on gynecology attributed to Trotula,[22] a prose tract later reduced to verse and translated, at one period or another, into all the European languages.

Beneath these outcrops of patently practical medicine lay a solid substratum of theory which can be discerned here and there in the lectures given to students. The uses made of Hippocrates' *Aphorisms*, the assumption of a basic knowledge of the humors and complexion, the references to diagnosis by urine and pulse, and much else point to an accepted curriculum of medical knowledge, which was later to be known as the *Articella*.[23] Whether this course of medical study originated at Salerno or was already an established syllabus adopted from a school of liberal arts has not yet been satisfactorily determined, but its appearance fully fledged in the south Italian school about the middle of the twelfth century is an undisputed fact. This undoubtedly attracted to it students from other intellectual centers, where medicine was dealt with in a perfunctory fashion. John of Salisbury, enmeshed in his veneration for the liberal arts, treated such students as sordid seekers after gain, unfit to study philosophy and incapable of comprehending theology, the reason being that at this period, when Hugh of St. Victor was elaborating the hierarchy of the arts and sciences,[24] medicine was relegated to a lowly place among the mechanical crafts, midway between navigation and theatricals.

In the meantime, Constantine's translations of medical works from the Arabic had been overtaken by two new developments, the establishment of a group of translators at Toledo and the appearance in Italy of several scholars concerned with Greek manuscripts. The group at Toledo worked mostly on Arabic texts containing works of dialectic, geometry, astrology, philosophy, and medicine, while the scholars in Italy restricted themselves to the writings of Aristotle and Galen. Their combined output was immense, Gerard of Cremona (at Toledo) accounting for no less than seventy hitherto unavailable texts.[25] Among those which were destined to have a profound effect on the future direction of medicine were the works on physics by Aristotle and the medical compilations of Avicenna, Rhazes, Abulcasis, and al-Kindi. Already by the end of the twelfth century Aristotle's ideas had penetrated the school of Salerno, the evidence appearing in the writings of Urso of Calabria; but the full impact was not to be felt until the founding of the universities.

The universities were, in substance, a trade guild organized by both masters and students to protect themselves against exploitation by townspeople in the matter of rents, the prices of food and clothing, and the taxes imposed on ordinary citizens by the municipalities, and to secure autonomy from the church hierarchy. But they also had the higher aim of establishing certain standards of teaching and devising courses of study, which candidates were obliged to follow if they were to be admitted to their fellowship.[26] In Bologna, where this development first took place among the teachers and students of law, a change took place which hardened the attitude of the students toward their teachers, and within a short time they were able to dictate which teachers they would have, and the kind of conduct and lectures expected from them. In Paris and other university centers where the professors were clerics and did not rely absolutely on fees from their students, the atmosphere was somewhat different, but the fundamental organization remained much the same. In all of them the study of medicine was dependent upon prior study in the faculty of arts.

Medicine was not considered an independent study but a science subordinate to psychology. Since the basic text for the study of psychology was the *De anima* of Aristotle, and the prime commentary on this text was that of Avicenna, it will be obvious that the medical student was trained initially in the philosophical ideas of these two writers and that he gained all his ideas about the senses, sight, taste, touch, and hearing from philosophers and not from physicians. When the student eventually came to study medical texts exclusively, his approach to them was already molded in a philosophical rather than a physical or medical cast, and his method of dealing with them relied more on dialectic and argument than on observation and practical experiment. The common curriculum laid down for medical students was the *Articella*, a small group of writings dealing with a general outline of medical principles, treatises on the pulse and urines, the *Aphorisms* of Hippocrates, the *Prognostics* and the *Regimen in Acute Diseases* accompanied by a commentary of Galen, and Galen's *Ars parva*. In essence, this course was the same as that laid down by the Alexandrian iatrosophists in the eighth century and considered sufficient to enable a man to practice medicine, though falling short of the complete training that would produce a well-rounded physician. The general impression it gave, therefore, was that medicine was predominantly a practical art.

The incorporation into the medical curriculum of Avicenna's *Canon* was to put an end to this.[27] From the outset, Avicenna made it clear that philosophy was on surer grounds than medicine because it dealt, not with the fallacious knowledge provided by the senses, but with immutable principles which could not be contradicted. He therefore made it his business to show that medicine was a science, not an empirical art, and throughout the whole of the opening chapters of the *Canon* he was at pains to point out that where philosophy was at loggerheads with medicine, the findings of philosophy must always be preferred. The views of Aristotle, consequently, must take precedence over those of Galen when they do not coincide, and the natural philosophers who follow his teachings must be exalted above the level of the mere practitioners who rely on experience. Here we meet the earlier Arabic distinction between the *physicus* and the *medicus*.

The acceptance by the universities of the writings of Aristotle on natural philosophy together with the Arabic commentaries on them gave rise to a great intellectual ferment and brought with it problems of a philosophical and theological nature, such as those on the immortality of the soul, the eternity of the world, the unity of the intellect, and much else, which culminated in the temporary banishment from the University of Paris of all Aristotle's physical and metaphysical works. Whether or not this ban, which was not lifted until the end of the first quarter of the thirteenth century, had the effect of deterring authors from putting their thoughts to parchment until they felt free to use the new material without fear of censure, is a moot point. But the fact remains that it was not until the 1230s and 1240s that medical writings permeated with Aristotelian and Arabic modes of thought began to appear.[28] But when they did, they displayed, in contrast to medical works of the previous century, a complete transformation. The writings of Gilbertus Anglicus (ca. 1230), for instance, placed side by side with those of the Salernitans (whom he knew and quoted), are startling in their contrast—a compact mass of arguments and syllogisms, of philosophical analysis and meticulous distinctions, and, above all, a formidable confrontation of the "moderns," backed up by Aristotle, Avicenna, and Averroes, with the "ancients" of a generation before.[29] It is true that much of the old tradition was also still in evidence—the tendency to employ folk remedies, the acceptance of the practice of surgery as part of the physician's competence, and an emphasis on the value of experience. But the total impression left after these elements have been eliminated is one of medicine being subjected to a hard and wide-ranging set of theories.

No one knows whether Gilbertus was a product of Montpellier or Paris or both, but it is interesting to note that William of Saliceto (1210?–76/80), a contemporary (more or less) in northern Italy showed no signs of this brittle, scholastic attitude. His *Summa conservationis et curationis* is written with a certain elegance and ease that stands in sharp contrast to the crabbed and spiky syllogisms of Gilbertus.[30] Yet Aristotle and Avicenna were not unknown at Bologna and Verona, where Saliceto was teaching. Perhaps it was his commitment to surgery, rather than to theoretical medicine which saved him from too philosophical an approach. The school of Salerno, which had suffered an eclipse after the founding of the university at Naples, still shed its aura on the northern cities, and its practical outlook directed medical men toward surgery, rather than to the study of natural philosophy. Certain it is that Hugh of Lucca, Bruno Longoburgo, Theodoric of Cervia, Roland of Parma, and Lanfranc, all surgeons, are the men who come to mind when medicine in Italy at this period is mentioned. Yet the scholastic approach to medicine was to dominate here also, and Taddeo Alderotti was its champion.

Alderotti (1223?–95) was a Florentine by origin, but he had studied grammar and the liberal arts at Bologna and progressed toward medicine by way of logic. In his teaching he laid great emphasis on the dialectical approach and applied all the refinements of philosophy and logic to the elucidation of medical texts. Thereby he set a precedent that was to be followed by all the eminent teachers in the chairs of medicine throughout Italy. He also set a precedent of a different kind when he demanded exorbitant fees for his extracurricular services, for in spite of his stark theoretical treatment of medicine in the schools, he was an expert clinician in dealing with cases of sickness. His demand for a hundred ducats a day as his salary from Pope Honorius IV and his eventual recompense of ten thousand ducats highlight the esteem in which his knowledge and his practical competence were held. His written consultations provide some insight into the ingenuity he employed in applying his theoretical principles to concrete cases of sickness, and they also show that he had a fair understanding of chemistry.[31] The Bolognese were not slow to recognize in him a man of outstanding talents, and as a reward they granted him exemption from paying taxes, allowed him, though not a citizen, to buy property, and allowed his students the same rights and privileges as those enjoyed by students of law.

This last privilege was a signal honor, which placed medicine on an equal footing with the most highly regarded of the sciences. At Paris, medicine had succeeded in freeing itself from the tutelage of the arts only after a long struggle, and it was not until about 1272 that it finally attained the status of an independent faculty. The proposition put forward by Avicenna and contradicted by Averroes that medicine was a science had evidently led to much wrangling there between the medical men and the pure philosophers.[32] We hear an echo of it in the writings of Arnald of Villanova at Montpellier, who took to task those lecturers who propped up their colossal volumes of Avicenna on a lectern and insisted on speaking in "universals."[33]

But we get a better idea of the acrimony it aroused among the contestants from Peter of Abano, who spent about ten years in Paris at the end of the thirteenth century. He stoutly believed that medicine was not merely a science, but, because it had existed long before the discovery of any other speculative science, the first of the sciences.[34] The objections raised by the philosophers and theologians were that medicine was reducible to a form of divination, that it was a servile craft which plied for money, and, finally, that it dealt solely with the mortal and corruptible body, a subject less noble than the soul with its spiritual faculties. These arguments were supported by attacks on the character of medical men: the doctor was an inexhaustible sea of envy; he was the mouthpiece of detraction; he was driven on by unflagging ambition; he was an unrepentent upholder of his own ignorance, but a prating contradictor of other scholars' truths; finally, he was an inexcusable neglector of the sick. As if this litany of mischiefs was insufficient, it was further contended that medicine lay under the influence of the four planets, Scorpio, Mars, Taurus, and Venus, the first two of which impelled the physician to evil and cruelty, and the last two to licentiousness and debauchery. The medical man was, therefore, by nature a creature of loose morals and should not aspire to the company of philosophers or place his profession among the sciences. But aspire he did. And in order to show that medicine and philosophy are sisters, and that whereas philosophy may contemplate the whole of nature, medicine deals with man, the noblest creature in nature, Peter compiled his *Conciliator*, a record of two hundred and ten disputations on the subject held in Paris.

Disputations were an essential part of university training no matter what the faculty might be. Since information was imparted to students by way of commentary on an approved text, difficult points in the exposition gave rise to questions, and these questions were discussed usually at the end of a lecture. Discussions were often continued outside the lecture room under the supervision of masters, and where the masters disagreed on a solution it would give rise to a disputation which, as often as not, took place in public. Students from their earliest days at the university became inured to these intellectual wrestling matches and were called upon, as the final stage before graduating, to show their prowess in these mental gymnastics. Among the professors themselves these disputations afforded an opportunity for displaying their dialectical skill and for proving to the authorities, who had engaged their services, that their money was well spent. Viewing the subjects of these wrangles with a dispassionate eye and at a distance of some centuries, it is hard to see what all the fuss was about. Whether one should take a large meal at lunch or dinner seems an innocuous question to pose, but the teacher of medicine at Siena, Jacopo de Prato (fl. 1347), did not think so, and after putting various contestants to rout, he published his disputation on it. Other questions, like that on the graduation of medicines, were more intractable and not so easily disposed of, and all the subtle writings of Roger Bacon, Arnald of Villanova, Tommaso del Garbo, Gentile of Foligno, and many others could not deliver them from the final resort to experiment. Disputants might disagree about the degrees of hot or cold in a compound containing scammony, but their theorizing was aimless until the medicine was tried out, as Bernard of Gordon counsels, "first on birds, then on dumb animals, then [on patients] in hospitals, and then on the Friars Minor."[35] A similar dispute concerned the effect of medicine on the body. Did the medicine act by itself, or was the potency of the medicine educed into act by the body? Gentile of Foligno (d. 1348), reviewing previous solutions to this question, asserted that the answer of Taddeo Alderotti was to be reprobated, that of Gentile de Cingulo was attacked by all, that of William of Brescia, Pancio de Lucca, and Dino del Garbo was derisory, that of Torrigiano was not acceptable, while that of his former colleague, Master Martin, was the object of reproach.[36] Evidently in his own estimation Gentile was the only person capable of giving the correct answer. The impression remains that if in such disputations the professors were ardently seeking the truth, they were also enjoying themselves in tearing each other to shreds. After a long, bitter, and blistering argument on some such question, the remarks of Ugo Benzi (1376–1439) sum up the whole position: "And consider how much useless labor doctors Francesco de Zanellis and Giovanni of Santa Sophia have expended on this matter in their long disputations at Siena. . . ."[37]

In the meantime, medicine of a very practical nature was being pursued in the hospitals. These had originally started out as hospices where the sick, the old, and the indigent could be hospitably received and nursed back to health. All along the pilgrim routes that stretched through France and Spain to Compostella, along the rough tracks through the high alpine passes to the plains of Lombardy, and on the highways reaching to the ports where people embarked for the Holy Land, hospices of various kinds were erected. Hospitality, shelter, rather than medical care, was their main objective. But with the increasing economic prosperity of the towns, the establishment of religious orders whose vocation was to tend the sick, and the easier availability of trained medical staff, the founding of hospitals in large urban centers became a possibility. The buildings, modeled on the huge cathedral-like infirmaries of the great Benedictine and Cistercian monasteries sprang up all over Europe.[38] Gradually the idea took root that hospitals should not indiscriminately admit sufferers from every kind of disease, but there should be various establishments for maternity cases, for exposed children, for incurable diseases and the dying. No longer were the hospitals restricted to giving nursing care; they were now staffed by competent physicians and surgeons, engaged by the municipalities at a fixed salary and under contract to give their services freely to rich and poor alike. The municipalities also vied with one another in engaging well-qualified physicians to deal with the sick in their homes, the surgeons, particularly, being employed as a kind of police force to report all cases of wounding or homicide. Positions of such responsibility were not filled by practitioners of inferior rank, but were eagerly competed for by people of high standing, and in most university centers one of the obligations of the teaching professors of medicine was that they should be prepared at all times to give their services to ailing citizens, to the rich for a reasonable fee, to the poor freely.

Apropos of the engagement by cities and municipalities of physicians and surgeons, the insistence that in time of war both physicians and surgeons should accompany the combatants to battle is worthy of note. Only sickness or some physical disability exempted them from this obligation. In the case of the surgeon, a wagon provided with medicaments was at his disposal, and he received a fee for each wounded soldier he attended, but there seems to be some doubt as to whether he treated only the more important personages on the field of battle. The records that survive appear to show that those whom the surgeon brought back in his wagon to be tended at a hospital or in their own homes belonged to the upper classes. Furthermore, the handful of medical men attached to an army would be quite inadequate to deal with the hundreds, and sometimes thousands, of casualties. Documents do not disclose the fate of physicians captured in time of war, but surgeons were treated as ordinary prisoners, and if they wished to regain their freedom, they were obliged to pay their own ransom, the municipality taking no responsibility at all in this matter.[39]

Although there seems to be no explicit mention in any contract that the surgeon was to be an executor of justice, the inference to be drawn from certain phrases in municipal documents is that surgeons were often called upon to inflict dismemberment on criminals when the authorities called for it. There is explicit mention, on the other hand, of surgeons having to tend the wounds of those upon whom such a sentence had been passed. In the case of false-moneyers, who had their hands chopped off; or of adulterers, who had their noses cut off; or of homosexuals, who were hideously disfigured, the surgeon had to visit the prison each day and tend their wounds. This was, perhaps, part of their responsibility as police, and the reason why in certain regions they received the appellation of *carnifex*, butcher.

In most contracts between municipalities and their medical staff it was stipulated that there should be no collusion with the apothecaries. This was intended to eliminate criminal dealing in poisons and to ensure that commercial considerations did not interfere with honest prescribing. Strict rules for the selling of drugs had been part of the legislation of Frederick II in the first half of the thirteenth century, and in the course of time these rules had been adopted, amplified, and issued by most large towns. To prevent mistakes in the compounding of drugs, apothecaries were obliged to be men of education, to know at least Latin and their native tongue, and to possess copies of the approved *antidotaria*. Each year their premises were to be examined by members of the local college of physicians so that their drugs could be tested for efficacy, and anything that in the college's judgment was useless, falsified, or adulterated could be thrown out. This was intended to counteract the common practice among apothecaries of substituting useless products for costly drugs which bore the same outward appearance. In spite of all legislation some doctors succeeded in owning apothecary shops and, thus, supplemented their already considerable incomes.[40]

During the whole of the Middle Ages the credulity of the public made the growth of charlatanism inevitable. Most of the quacks engaged in the peddling of secret remedies, and though for the most part their herbs were harmless and only cured those whose faith was boundless, they were hounded relentlessly from the cities both by the guilds of apothecaries and by the municipalities. The itinerant surgeons were in another category. Some of them were undoubtedly men of experience and great skill, who by long practice surpassed the achievements even of university-trained personnel. They usually devoted themselves to one specialty, either hernia, cutting for stone, or couching for cataract. But there were others who passed from town to town, who operated, took their fees, and then departed before the full effect of the operation could be judged. In this way many patients lost either their eyesight or their lives. In most cities regulations forbade any practice of medicine by strangers, and if such were caught, they were fined and thrust into prison. It was in the countryside, therefore, that they usually operated.

The position of the midwife in medicine had been accepted from earliest times and never impugned. Obstetrics and gynecology had been accepted, particularly when clerics were the sole purveyors of medical treatment, as essentially a woman's sphere. But after the transference of medicine to lay hands, women took an even greater part in the profession of healing. During the thirteenth century the mention of women physicians, surgeons, and apothecaries became more frequent; for the most part they were taught their craft by fathers who intended to hand over their clientele when they retired or died. At Paris, Vienna, and several other places the medical faculty instituted law cases against them and inflicted heavy penalties for breaching the regulations against unqualified practitioners, that is, against those who had not attended a university course. But in Italy and particularly in the kingdom of Naples a more tolerant attitude was adopted, and women were not only allowed but actively encouraged, after they had passed an examination by the royal physicians and surgeons, to undertake the cure of surgical and gynecological cases. No less a person than Ugo Benzi encouraged midwives to take a more scientific interest in women's ailments and gave them lessons on the anatomy of the womb, while Michael Savonarola wrote for them a complete treatise in the vernacular on most aspects of gynecology.[41]

The Later Middle Ages

At the beginning of the fourteenth century the medical world was loud with the clamor of new commentators on Hippocrates, Galen, Avicenna, and other texts. At Montpellier, Bernard Gordon and Arnald of Villanova were propounding, in their different ways, what might be termed an empirical type of medicine. In this they were closely followed at Oxford by John Gaddesden, the dutiful but not uncritical author of the *Rosa medicinae*. At Paris, John of St. Amand (d. ca. 1300) was expatiating on the *Antidotarium Nicolai*, while in another quarter of the city Henri de Mondeville (d. ca. 1316), a scholastic to his fingertips, was illustrating with colored drawings his views on anatomy and surgery. Peter of Abano at Padua, though captivating his students by his audacious forays into the secrets of astrology, was stirring up antagonism among his colleagues by advocating beliefs that went counter to orthodox theology, while, at Bologna, Taddeo Alderotti, was carefully paving the way for a school of expositors, who would carry his methods into all the universities of Italy.

To commentators on the *Ars parva* of Galen it must have been a source of extreme exasperation to reach its final paragraphs and find how few of the texts described there could be obtained. Of the hundred and fifty or more titles listed by Gentile of Foligno, medieval scholars possessed no more than twenty, and these were by no means sufficient to provide an overall understanding of Galen's teaching. To fill this void, Niccolò da Reggio, a physician at the court of the king of Naples, made translations of about fifty new works, which had been sent at the request of Robert of Sicily by the Emperor Andronicus III from Constantinople.[42] Between 1308 and 1348 Niccolò pored over these Greek manuscripts and made elegant Latin versions, some of which were sent to Guy de Chauliac (d. 1368), papal surgeon at Avignon.[43]

Since the only new anatomical work translated by Niccolò was concerned with the eyes, no additional information had accrued to the corpus already formed earlier. Consequently, when Mondino de'Luzzi (d. 1326), addressed himself to the task of writing a completely novel treatise on human anatomy, he was restricted to the use of Galen's *De usu partium* and other less informative treatises. Before his day, the teaching of anatomy had been based on the dissection of animals, but when toward the end of the thirteenth century he had occasion to perform autopsies on two pregnant women, he grasped the opportunity to make a full examination of their bodies and to give a more exact description of the internal organs.[44] Until that time dissection on humans had been viewed askance by civil as well as by ecclesiastical authorities, and medical science was further hampered by too wide an interpretation of Boniface VIII's Bull forbidding the boiling of crusaders' and others' bodies in order that their bones could be interred in their native land. The fact that Mondino was dissecting and teaching anatomy at the time the Bull was promulgated points to some relaxation of attitudes, and his pioneering efforts were soon followed by others. Though at a later stage municipal authorities gave permission for the bodies of dead criminals to be dissected by duly accredited surgeons for the purpose of public instruction, the permission was hedged about with so many conditions that complete freedom to dissect was never enjoyed. It was forbidden, for instance, to dissect a dead criminal who had been a citizen of the place where the anatomization was to take place; it was not allowed if the relatives of the dead person objected. On the other hand, the anatomists did not want old or diseased bodies, but only those that were perfectly formed and in a good state of health when death supervened.

Since a dissection could take as long as ten days to perform, the season of the year also had to be taken into consideration: the winter months were the most suitable, but the summer, when the body putrefied quickly, made it impossible. Expense was another limiting factor. Dissections were carried out manually by surgeons, while the anatomical explanations were given normally by the chief professor of medicine. All demanded a fee for their work. Besides these payments, there were the costs of the feasts which usually followed each session of an anatomization. Many teachers and scholars refused to dip into their own pockets for what, in many cases, proved to be a mere social gathering, and so, even when a corpse was at their disposal, the opportunity of dissecting it was often passed over. In some medical faculties, where permission for dissection was readily available, no anatomizations took place for more than forty years.

It was from one of Mondino's pupils, Bertuccio, that Guy de Chauliac (1298–1368) gained his knowledge of anatomy, but it was seemingly from books that he learned his surgery. His voluminous compilation, based on the writings of his predecessors and heavily biased in favor of the Arabs, bears little evidence of actual surgical practice. He may have used the knife when embalming the bodies of popes, but he was careful to avoid it on living patients. It is surprising, therefore, to see with what disdain he treated Henri de Mondeville, who had acquired his surgical skill on the field of battle, and how high-handedly he spurned the notion that healing by first intention was safer, quicker, and more rational. Yet, ironically, his book became the classic manual for years to come.[45]

The true Galenic attitude to surgery was that it could not be taught from books, but that it should be learned by experience and from an early age. Normally, surgical knowledge was handed down from father to son. We have evidence from Bertapaglia himself that his son assisted him in surgical operations at the age of eleven, and it was evident from the statutes of surgeons' guilds that the young apprentice's training began by preparing the instruments, bandages, and ointments that his master used. Since the bleeding of patients was one of the basic skills to be acquired, the apprentice first had to learn the position of the chief veins, which were cut for different ailments. The aspirant then learned how to make his own knives and prepare a good cutting edge, and later, under the supervision of an accredited barber-surgeon, he performed simple operations until he acquired enough skill to work on his own. The final step in his career was his admission to a surgeon's guild, where his competence in the theoretical field was tested by surgeons and physicians. Though the teaching of surgery was not allowed at universities like Paris and Oxford because it was a manual craft, courses in surgery were part of the curriculum at some Italian universities. But judging by Dino del Garbo's commentary on Avicenna's section on surgery, it was an arid and purely speculative affair, which never mentioned a surgeon's needle, much less a knife or a bandage. Progress in surgical techniques came from men like Jan Yperman, the Flemish pupil of Lanfranc, or from John Arderne, who perfected the operation for *fistula in ano*. Since knights in armor spent much of their time on horseback, they suffered intensely from hemorrhoids, anal inflammations, and fistulae, and because permanent injury often resulted from use of the knife, no one dared to treat them except by cautery. Arderne (1307–78), whose book knowledge equaled that of any university professor, although he had never attended the schools, developed an ingenious technique to deal with these afflictions and had an enormous clientele among the nobles of Europe. Since a personal technique was a trade secret and an economic asset, many surgical inventions never found their way into textbooks, such as that of Chauliac; however, when Arderne retired from active service and had nothing to lose, he described his operations in detail, illustrating every stage in his technique with drawings.[46]

In the matter of producing books the surgeons were at a disadvantage in comparison with the physicians teaching at universities. The surgeon, if he wrote a book, would have one or at most two pupils to whom his work would be addressed, and it could remain uncopied and unpublished for years. Thomas Morstede's *Booke of Fayre Surgerye*, written in his own hand in 1446, has come to light only in recent years.[47] The professors of medicine, on the other hand, had shoals of scholars who, if they did not make notes at the actual lectures, were able to buy copies of them at the university stationers. One of the regulations insisted upon at the leading universities was that every teacher should hand to the bedel and make available to the stationer a fair copy of his lectures and disputations, so that any student could make a copy for himself or have one written out by the professional scribes provided for that purpose. Many of the texts we now read we owe to this process. "Written by me," one says, "at Padua, where a war was being waged against the Duke."[48] "Copied by me," says another, "at Milan, while I was suffering from quartan fever."[49] But the composing of books by professors did not always run smoothly. Dino del Garbo tells us in his *Dilucidarium* that after teaching at Bologna for two years, he had to leave because the city was laid under interdict and he was obliged to migrate to Siena where,

Fig. 37   Medieval operations for cataract (above) and nasal
polyps (below). Oxford, Bodleian Library, MS Ash-
mole 1462 (12th c.), fol. 10r. For commentary on this
and the next figure, see MacKinney, *Medical Illustrations*, pp. 70, 79.

though there was no university, some schools were open. Two years
later he returned to Bologna, continued his writing, and eventually
accepted an invitation from Padua to reform the Studium there. "Here
I began my book again and made some progress . . . but finally I was
compelled by the deplorable state of the city to leave and return to
Florence, where I picked up the threads of my work once more and
finished it in 1319,"[50] that is, about fifteen years after he had started
it.

These brief phrases alert us to one of the disturbing elements in
medieval university life, particularly in Italy. In the Middle Ages,
when Florence was antagonistic to Pisa, when Padua under Venetian
rule was opposed to Milan, when Ferrara and Bologna were on different political sides, the stability of the universities was always at
risk. This involved both teachers and students. The prestige or decline of a teaching center could depend on civic disturbances, on economic factors, or on political alliances, and the conditions that favored
having many illustrious professors and a large influx of students could
easily be undermined. Students unwilling to have their course of
studies interrupted usually followed their teachers, but later on this
means of escape was closed to them by the legislation of the various
states, which forbade their subjects to study anywhere but in the state
university.

Another cause of disruption was the sporadic outbreak of plague.
Though medieval people had some idea of contagion, they seemed at
a loss to explain infection. The fact that from early times the free
movement of lepers had been restricted shows that they understood
how this disease could be spread by contact. The strict regulations
which banished the leper from the towns and considered him as legally dead, which forced him to carry a clapper and warn wayfarers
of his approach, and to use a stick to indicate the goods he wished
to buy were carried out with brutal severity. But the idea of contagion
appears to have been confined solely to leprosy, so that when plague
struck, and particularly the Black Death, the only explanation offered by doctors was that it was caused by the conjunction of malefic

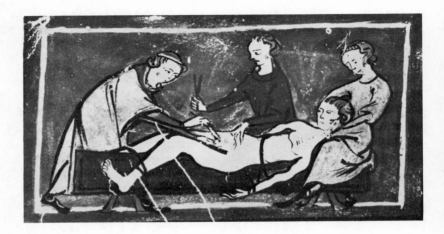

Fig. 38   Operation for scrotal hernia, from Montpellier, Bib-
                           liothèque Interuniversitaire, Section Médecine, MS 89
                           bis, fol. 23r (14th c.).

planets. This belief in astrology had taken firm root through the diffusion of the *Secreta secretorum* of pseudo-Aristotle and had been
confirmed by the advocacy of Roger Bacon and Peter of Abano. In
some universities it was taught in the medical faculty as a serious
course. According to the practitioners of astrological medicine, conjunction of Saturn and Jupiter had poisoned the upper atmosphere,
and the envenomed vapors had entered through the pores into the
bodies of men. This had upset the balance of the humors and resulted
in the buboes or swellings that appeared either in the groin or under
the armpit. The only treatment physicians could advise was to avoid
the rank stench of cemeteries, to burn aromatic herbs, to abstain from
hot baths and heating foods that opened the pores, and to cleanse the
system by copious drafts of purgatives. Only in Italy were any sensible
measures pursued. At Perugia, where Gentile of Foligno daily attended the sick, he persuaded his patients to answer a questionnaire,
and from their replies he drew up a plan that could be put into action
should any future crisis arise. These plans were adopted later in other
cities, some of them being that travelers should not be allowed to
enter their gates unless they could produce a clean bill of health.[51]

The hygenic state of most northern cities left much to be desired.
In spite of the numerous books written by physicians on personal
cleanliness, little was written by them about the dirt in the towns.

Owing to the rapid economic growth during the thirteenth and fourteenth centuries, many cities had become overcrowded. The most elementary arrangements for drainage and sanitation had been neglected,
and the narrow streets became clogged with filth. Artisans like black-
smiths, tanners, brewers, and dyers threw all their refuse into the
highway; butchers and poulterers left their offal under the stalls in
the marketplace; while private citizens poured their slops from win-
dows onto passers-by and deposited dung from their stables at street
corners. Few streets were paved, and in winter or after heavy rain the
narrow passages between houses became a muddy morass, where
pedestrians and riders jostled each other for room, often leading to
quarrels and fighting. These conditions were not improved by the
pigs, hens, sheep, and other animals that were allowed to wander
about uncontrolled. Quite early the Italian cities had enacted statutes
to deal with these problems, but it was not until the fourteenth century that they were generally enforced. Then regulations were made
to prevent garbage being placed in the streets, taxes were imposed for
paving the roads and squares, scavengers were appointed to remove
rubbish at stated intervals, and guards were instituted to police the
towns at night. Stricter laws were enacted to ensure the quality of
meat, fish, bread, and other commodities displayed for sale in the
shops, and grave penalties were inflicted on those who polluted the
streams or fountains. For the municipal authorities it was an uphill
task, but one manfully pursued.[52]

In contrast to the paucity of texts on public health, the treatises on
personal hygiene were legion. Two of the earliest examples came from
the pen of Bartholomew of Salerno (fl. 1150), written for King Louis
VII of France and for Peter the Venerable, abbot of Cluny. After the
introduction of Avicenna's *Canon* into the medical curriculum of the
universities, the theme became more popular, and from the fourteenth
century onward treatises on the preservation of health increased enormously. The writers of such treatises were usually private physicians
attached to a court, but many were also drawn up by eminent university professors. In spite of the lapse of time between the first Salernitan texts and the later productions, no improvement is discernible; in
fact, most of them repeat Avicenna word for word. Linked with these
were essays on the prolongation of life. For the most part, these were
the continuation of arguments first raised in the thirteenth century on
the question of the *humidum radicale*, the fundamental moisture from
which the body was formed. The problem was whether this *humidum
radicale*, which was assumed to be dried up by the natural heat of
the body, could be restored, or whether the process by which it was

dried up could be arrested. If the process could be stopped, then there was no doubt that life could be prolonged beyond the three score years and ten assigned by the Scriptures to mankind's existence. From being a mere subject of discussion in the schools, where it was long debated, and where it provided little else than an opportunity for intellectual acrobatics, it became the concern of princes, popes, and others who held the reins of power, as familiar a topic as the question of preserving one's health.[53]

Linked with this was the *Consilia* literature, initiated as a genre of medical expertise by Taddeo Alderotti and continued with aplomb by Gentile of Foligno, Antonio Cermisone (d. 1441), Bartolomeo Montagnana (d. 1460) (see fig. 39), Ugo Benzi, Ferrari da Grado (d. 1472), and Baverio de Baveriis (ca. 1480). These *Consilia* provide little that is new in the field of medical theory or practice, being for the most part simple applications of the ideas in Johannitius's *Isagoge*, but they are interesting from a social point of view, insofar as they recount in some detail the ailments and diseases of the high and mighty.[54]

While the followers of Taddeo Alderotti, that is, Dino del Garbo (d. 1327), Tommaso del Garbo (d. 1370), William of Brescia (d. 1336), and Torrigiano, though disagreeing among themselves on certain points of interpretation, continued his tradition of theoretical analysis of medical texts, a more practical approach was favored by the writers of Montpellier, like Jean de Tornamire (1330–95), Jean de Piscis (d. 1443), Valescus of Taranta (fl. 1400), and the school of Paris.[55] In England no writer had emerged to succeed John of Gaddesden. But about the middle of the fourteenth century, when Guy de Chauliac was sniffing petulantly at the *Rosa medicinae*, Simon Bredon, the compiler of a huge volume on diagnosis from feces and urine, showed a decided interest in mathematics, astronomy, and judicial astrology.[56] This was not surprising in a fellow of Merton College, surrounded as he was by colleagues whose main concern was mathematical science. But in Padua, where no such tradition existed, Giovanni de'Dondi was a new phenomenon in the medical field, versed not only in astronomy but also in the more intricate workings of practical mechanics. His planetarium, the fruit of sixteen years labor, was one of the wonders of the age.

The infiltration of the study of mechanics and mathematical physics into the medical school of Padua about this time can be traced to the influence of the Oxford school.[57] The acceptance of the philosophical novelties of William of Ockham and the interest shown in the scientific discoveries of the Merton teachers brought about a distinct change

Fig. 39   Bartolomeo Montagnana (d. 1460), a Paduan physician, inspecting a flask of urine. Munich, Bayerische
Staatsbibliothek, MS Lat. 25, fol. 1v (A.D. 1434). On
this figure, see MacKinney, *Medical Illustrations*, p. 13.

in the attitude of Italian professors. It was the custom to recruit from
the professors of philosophy and, particularly, of logic, those men
who were to fill the higher echelons of the faculty of medicine, a
tradition that persisted in Italy well into the eighteenth century. Ugo
Benzi, for instance, who at the end of the fourteenth century filled
the chair of logic at Piacenza and then at Bologna (where he was
hailed as "prince of philosophers"), began to teach theoretical medicine at Siena after only six months tuition by Marsilio da Sancta
Sophia. The philosophical approach to medical questions, already
embedded in the tradition of Taddeo Alderotti, was further reinforced
by the absorption of ideas from the writings on logic by such men as
Dumbleton, Swineshead, and Heytesbury and the physical theories

adumbrated in England and developed at the University of Paris by John of Saxony and Nicole Oresme. The new doctrines were avidly accepted in Italy and not only disseminated in the teaching of philosophy but also applied to the theoretical interpretation of medicine. As a result, the new sets of disputations and questions that dripped from the pens of medical writers at this period have a different flavor from those of earlier days. They were no longer concerned with the problem of whether it was good to take a bath after medicine, whether medicine should be given at the first onset of an illness, or whether incantations and engraved gems (as recommended by Peter of Abano) could ward off disease. The new questions had a distinct bias in another direction, made manifest by the importation into medicine of terminology and theories taken from the science of mechanics. One of the earliest propagandists of these ideas was Jacopo of Forlì (d. 1413), who had taught logic for twenty-seven years before being called to the chair of medicine, and whose questions became, by decree of the University of Padua, the norm for succeeding professors of theoretical medicine.[58] That this new trend was viewed with caution by some teachers is apparent from a marginal note inserted about this time into a text of Galen's *De complexionibus*: "Note, that it is better for a physician not to reason too much, but to put his trust in experience. If he wishes to reason, let him defer it until his intellectual powers are fully developed, because a small error in dialectic can lead to great blunders in medicine."[59]

A contemporary of Jacopo at Padua was Marsilio of Sancta Sophia (d. 1405), whose treatise on fevers was later printed together with works on the same theme by Ricardus Anglicus and Marsilio's nephew, Galeazzo. The Sancta Sophia family gave several eminent men to medicine, the founder of the dynasty, Giovanni, having been a pupil of Peter of Abano. Galeazzo, whose approach to medicine was less tinged with mechanics than that of Jacopo of Forlì, was called from Padua in 1398 to be the first professor of medicine at the newly founded University of Vienna, and under his supervision the first dissection ever to be made in that city was performed. Until that time most students of medicine from the northern and central parts of Europe had traveled either to Paris or to the Italian universities for their education, and there formed one of the strongest "nations"; and even after their own universities were founded, those close links persisted. The latter claim is best illustrated by the texts that formed the basis of their medical education, for besides the usual books of Hippocrates, Galen, and Avicenna, the favorite authors were Jacques Despars of Paris (d. 1458), Jacopo of Forlì, Gentile of Foligno,

Mondino de'Luzzi, Saladin di Ascoli, Michael Savonarola, Antonio Guainerio (d. 1445), and other Italians. At Prague, on the other hand, founded in 1348 with eight professors from the University of Paris, the writers from Montpellier appear to have played the foremost role, the result being that medical compositions issuing from Prague, like those of Albicus, are more concerned with sick people than with theories of sickness.

In spite of the fact that most physicians' libraries in Italy of the fifteenth century contain the works on logic by Swineshead and Heytesbury and the medical writings of Jacopo of Forlì, the books on medicine most popular at that time and chosen by printers toward the end of the century were concerned with *Practica*. This is all the more surprising in the case of Jacopo's pupils, Antonio Guainerio, later professor at Pavia, and Michael Savonarola, professor at Ferrara. Guainerio wrote a commentary on the *Almansor* of Rhazes, printed in 1497–98, which was dubbed by its editors as "a work of which practitioners are in great need" and which was devoid of all argumentative tendencies.[60] Michael Savonarola, on the other hand, commented on Avicenna and introduced his work with a plea to students to stop wasting their time by wrangling at street corners on specious theories and give more time to practice, which would bring them favor with the people and give them a more lucrative occupation. Two of his contemporaries at Padua, Cermisone and Montagnana, also concerned themselves more with attention to the sick than with debatable questions, leaving behind them hundreds of *consilia*, which were excellent examples of diagnosis, prognosis, and treatment. Giovanni Arculano, teaching at Padua from 1427 onward, was not afraid to instance the questions of Jacopo of Forlì and Ugo Benzi (probably having in mind the latter's disputation *De malitia complexionis diversae*) as a complete waste of time, and made good his exhortations by including in his book original observations on certain pathological states and illustrating surgical techniques with new instruments.[61] Other professors at Pavia followed the same path. Gianmatteo Ferrari da Gradi wrote, apart from his *consilia*, a *practica*, and his pupil Marco Gatenaria followed suit, both of them writing commentaries on the ninth book of the *Almansor*.[62] But they had as a colleague Giovanni Marliani, who, from being a mere warden of the scholars in 1433 at a stipend of twelve florins, rose to be ordinary professor of medicine in 1467 at a salary of 500 florins. His only contribution to practical medicine appears to have been a pamphlet on plague— his most important writings, even while teaching medicine, bearing on mathematics and physics.[63] Giangaleazzo Maria Sforza, to whom he acted as physician, called him "another Aristotle in philosophy, another Hippocrates in medicine, and another Ptolemy in astronomy," though it is doubtful whether his opponents in the various debates in which he became involved would have subscribed to such a eulogy.

A surprising element in many of these writings was the medical treatment derived from folklore. There had been an undercurrent of folk medicine from the earliest times, noticeable even in certain works of Galen, and not absent from the *materia medica* of Avicenna. But its reemergence at a time when the Renaissance, in some places, was in full spate, showed that even the most brilliant intellects could be swayed by popular belief. This phenomenon manifested itself mainly in cases concerned with impotence and sterility, which seldom responded to the therapeutic means available at the time. It was humiliating for an eminent professor or practitioner to confess that he was powerless in face of such indispositions, and the easiest way out of such an impasse was to attribute the defect to preternatural agencies—the evil eye, incantations, spells, or other works of the devil. Recourse, therefore, was had to age-old folk remedies, most of them irrational and some of them obscene. Even Michael Savonarola, who admitted that he had never actually encountered such a case but had learned from hearsay, filled two columns of his book with bizarre, though reputable, counteragents. And whereas in the twelfth century and earlier the incubus had been regarded as the result of overeating or drinking or of some other bodily surfeit, the doctors now began to accept the idea that the incubus was a devil, who moved by night and attacked people in the darkness. In cases of difficult childbirth, Galeazzo di Sancta Sophia recommended tying a piece of coral to the right thigh or placing a magnet over the navel to ensure easy delivery.[64] Others, in teaching ways of counteracting poisons, not merely the bites of rabid dogs or stings of venomous snakes, but the poisons administered by enemies, suggested the wearing of rings adorned with precious stones, such as sapphires and emeralds, a safeguard derived from Pliny and repeated by Rhazes. When academics countenanced such beliefs it is small wonder that the illiterate populace clung to them.[65]

Though the writings of the physicians in the fifteenth century far outdistanced in number those of the surgeons, surgery was not a neglected art. Based as it was on experience, ingenuity, and bold action, surgery had no need of long, discursive arguments. Though it was taught in the Italian universities, in contrast to Paris and similar institutions, academic surgery made no spectacular progress and seemed unable to loose itself from the apron strings of the past. Pietro d'Argellata of Bologna (d. 1423), who was hailed in his own day as an outstanding surgeon, was little more than a mouthpiece for those who had preceded him, his compilation in six books borrowing heavily not only from Guy de Chauliac, but also from Lanfranc, Mondeville, Saliceto, and a host of others. Though he worked in the local hospital, accompanied by assistants, he does not appear to have tried out any new techniques apart from modifications in suturing wounds, but he did perform a cesarian operation and saved the unborn child.[66] It was left to the so-called empirics to show where progress could be made. In Catania, Sicily, a family of surgeons, Branca by name, had rediscovered the secrets of skin grafting—rediscovered, because the operation had been described centuries before by Celsus, had been known to Byzantine surgeons, but had been passed over in silence both by the Arabs and the Salernitan school. This operation, a veritable salvation for those who had been mutilated in nose or lips or ears as a punishment for crime, had remained a family secret for several generations until it was taught, about 1460, to a German surgeon, a man unschooled, as were his teachers, in academic medicine. Heinrich von Pfolspeundt (or Pfalspaint) was one of those Teutonic knights derided by Guy de Chauliac as "curing all wounds with incantations, potions, oil, wool, and cabbage leaves." There is, indeed, a great deal of this in Pfolspeundt's book, because he had learned his trade, not at a desk or from books, but in the hard school of war. He had healed between three and four thousand wounded comrades in the battles against Poland, and whereas, before he introduced his methods of extracting missiles, seventy or eighty out of every hundred died, the mortality rate under his hands sank to less than ten.[67] It is often said that he was the first to deal with gunshot wounds, but there is nothing in his text to support this claim. He was, however, the first to publicize Branca's technique of making new noses and to give a detailed description, not in Latin, the language of the learned, but in his mother tongue, that all might benefit thereby. Rather different were two of his countrymen, Hieronymus Brunschwig and Hans von Gersdorff, both surgeons of Strassburg. Brunschwig (1450–1512) had read, on his own admission, three thousand texts and quoted some of them to support his ideas, but his observations, based on experience, raised his book above the level of a compilation.[68] Gersdorff (ca. 1455–1529), who had taken part in the wars against Charles of Burgundy, was also well-read. Though he owed much to Guy de Chauliac, his vast experience at the hospital of St. Antony in Strassburg, where he performed between one and two hundred amputations, lent authority to his book, which was further enhanced by remarkable illustrations showing the new instruments employed in the extraction of shot and arrows and in the mending of broken limbs.[69] Unlike his forerunners he used no anesthetics, frightened, perhaps, by Chauliac's story of the bishop who died through an overdose.

The use of vernacular by these surgeons was not an indication of ignorance of Latin. Not only in Germany, but in other countries also, the native tongue had become a common means of expression, coinciding, perhaps, with growing nationalist feeling. In Italy, France, and England, more and more medical works appeared in the vernacular, works concerned not merely with regimens of health, but with more specialized subjects, like gynecology. At Paris, even the university authorities bowed to this convention and allowed lectures on surgery to be given in French to surgeons of the short gown, though it must be admitted that in this case the decision was based on the barbers' unfamiliarity with Latin.[70]

It was about this time that a new star appeared above the medical horizon. His name was Antonio Benivieni (1440–1502), a Florentine physician. During his lifetime he attracted no attention, but after his death, when his case notes were being examined by his brother, it was discovered that he had systematically carried out autopsies on his patients in order to discover the causes of their illnesses and subsequent deaths. This was something completely different from the ritual dissections carried out in the medical schools, where the cutting up of corpses had no other end than to corroborate and illustrate the anatomical teachings of Galen. There had been isolated instances during the thirteenth century when autopsies had been performed to discover the causes of unusual deaths, but this was a far cry from the methodical pursuit of Benivieni, who successfully persuaded parents and relatives of his dead patients to allow him to probe into the hidden depths of disease. Though he attempted objectively to assess the meaning of what he found, his deductions were invariably wrong, clouded as his judgment was by the medical theories he had learned in the schools. His pioneer effort, however, has earned for him the title of "founder of pathology," and his book, printed in the years when syphilis was becoming endemic, undoubtedly stimulated physicians all over Italy to embark on wholesale autopsies in the search for the source of that scourge.[71] A contemporary of Benivieni was Alessandro Achillini (1463–1512) at Bologna, a man who came late to medicine after a lifetime of teaching philosophy. He wrote a book on anatomy, in which he corrected Galen on several important points, the tone of which sounded like the instructions given to a prosecutor at a dissecting class: "cut this, and you will see," "open this and you will find," and so on.[72] It heralded a new attitude, a critical approach to ancient writers, observable also in Alessandro Benedetti (1460–1525) and Gabriele Zerbi, that was to find full expression in the writings of Andreas Vesalius.[73]

Conclusion

The great weakness that seems to underlie all medieval medicine stems not so much from its reliance on the ideas inherited from the past as on the discordant interpretations placed upon them. There were too many authorities all at variance with one another, and the conflicts they aroused among their followers distracted minds from the real task of creating a synthesis that would be acceptable in itself and credible to all. It was not merely a question of medicine versus philosophy, of Galen versus Aristotle, or even of Avicenna versus Averroes, but of what kind of medicine and of what kind of philosophy. When a contestant took up first one cudgel and then another, choosing according to his own subjective attitude, it was not possible to conclude the affair with any degree of satisfaction. And so, at the end of several centuries of internecine warfare, the field was in complete disarray. Small wonder that Paracelsus, intent on establishing an entirely new approach to medicine, should have symbolized his plea for progress by throwing his volumes of Galen and Avicenna into the flames. Only out of such ashes could a phoenix arise. Development, therefore, in medieval medicine was minimal in comparison with the enormous amount of energy and intellectual labor expended on it, and few are the instances in which it can truly be said that a great step forward is clearly discernible. The whole stretch of time between the ninth and the end of the fifteenth century may, consequently, be divided into three periods, a period of conservation, a period of exploration, and a final period of elaboration. The opening of the gates to a flood of new developments was not to take place until much later.

Notes

1. Henry E. Sigerist, "On Hippocrates," *Bulletin of the History of Medicine* 2 (1934): 203–5.
2. For a fuller and more balanced account see I. E. Drabkin, "On Medical Education in Greece and Rome," *Bulletin of the History of Medicine* 15 (1944): 333–50; Owsei Temkin, "Greek Medicine as Science and Craft," *Isis* 44 (1953): 213–25; F. Kudlein, "Medical Education in

Classical Antiquity," in *The History of Medical Education*, ed. Charles
D. O'Malley (Los Angeles, 1970), pp. 3–37.

3. "Unde [artes] conjecturales quoque dicuntur: ex quarum numero
medicina fuerit, quemadmodum rhetorica, gubernatoria et sagittandi ars"
(*Galeno ascripto introductio seu medicus*, in *Claudii Galeni opera omnia*,
ed. C. G. Kühn, vol. 14 [Leipzig, 1827], pp. 685–86). By "conjectural"
the writer evidently meant that the outcome of the action was a hit-or-
miss affair and could not be predicted with certainty.

4. Augusto Beccaria, "Galeno nei commenti della scuola di Ravenna,"
Italia medioevale e umanistica 14 (1971): 11.

5. Owsei Temkin, "Studies on Late Alexandrian Medicine," *Bulletin of
the History of Medicine* 3 (1935): 420ff; Augusto Beccaria, "Sulle tracce
di un antico canone latino di Ippocrate e di Galeno, III," *Italia medioevale e umanistica* 14 (1971): 1–23.

6. Rudolf Laux, "Ars medicinae: Ein frühmittelalterliches Kompen-
dium der Medizin," *Kyklos* 3 (1930): 417–34, with accompanying bibliography; Augusto Beccaria, *I codici di medicina del periodo presalernitano* (Rome, 1956), passim.

7. *Caelius Aurelianus on Acute Diseases and on Chronic Diseases*, ed.
and trans. I. E. Drabkin (Chicago, 1950).

8. Valentin Rose, *Anecdota Graeca et Graecolatina*, vol. 2 (Berlin,
1870), pp. 183–280. Examples of this literature, illustrated with photographs, can be found in Henry E. Sigerist, "Early Medical Texts in Manuscripts at Vendôme," *Bulletin of the History of Medicine* 14 (1943):
68–113.

9. William D. Sharpe, "Isidore of Seville: The Medical Writings,"
Transactions of the American Philosophical Society, n.s., vol. 54, pt. 2
(Philadelphia, 1964); Rhabanus Maurus, in J.-P. Migne, ed., *Patrologiae
cursus completus, series Latina*, vol. 111 (Paris, 1852), bk. 18, chap. 5.

10. C. H. Talbot, *The Anglo-Saxon Missionaries in Germany* (London, 1954), pp. 62–149.

11. See the plan of the projected infirmary at the Abbey of St. Gall in
Loren C. MacKinney, *Early Medieval Medicine with Special Reference
to France and Chartres* (Baltimore, 1937), p. 215, plate 1.

12. Walahfrid Strabo, *Hortulus*, trans. Raef Payne, commentary by
Wilfrid Blunt (Pittsburgh, 1966); MacKinney, *Early Medieval Medicine*;
and MacKinney, *Bishop Fulbert and Education at the School of Chartres*
(Notre Dame, Ind., 1957).

13. In A.D. 983 the Emperor Otto II died through an overdose of aloes.
He was given four drachms, approximately the dose recommended in the
nineteenth century for a horse. Jonathan Pereira, *The Elements of Materia
Medica and Therapeutics* (Philadelphia, 1852–54), 2: 202.

14. *Richeri Historiarum liber*, 4. 50, Scriptores Rerum Germanicarum
in usum scholarum (Hannover, 1877), p. 153.

15. Rudolf Creutz, "Der Arzt Constantinus von Monte Cassino. Sein
Leben, sein Werk und seine Bedeutung für mittelalterliche medizinische
Wissenschaft," *Studien und Mitteilungen zur Geschichte des Benediktiner-
Ordens* 47 (1929): 1–44; Creutz, "Die Ehrenrettung Konstantins von
Afrika," ibid., 49 (1931): 35–44; see also Michael McVaugh, "Con-
stantine the African," *Dictionary of Scientific Biography*, 3: 393–95.

16. Paul Oskar Kristeller, "The School of Salerno: Its Development and Its Contribution to the History of Learning," *Bulletin of the History of Medicine* 17 (1945): 138–94.

17. See the analysis of the *Passionarius* of Gariopontus by Rose, *Anecdota Graeca et Graecolatina*, 2: 180; C. H. Talbot, "Some Notes on Anglo-Saxon Medicine," *Medical History* 9 (1965): 156–69.

18. All these can be found in S. De Renzi, *Collectio Salernitana*, 5 vols. (Naples, 1852–56).

19. Copho, "De anatomia porci," *Collectio Salernitana*, 2: 388–401; George Washington Corner, *Anatomical Texts of the Earlier Middle Ages: A Study in the Transmission of Culture* (Washington, D.C., 1927). This contains the *Anatomia porci*, the second Salernitan anatomical demonstration, the anatomy of Master Nicolas, and the anatomy of Ricardus Anglicus.

20. For this and later surgical texts see Karl Sudhoff, "Beiträge zur Geschichte der Chirurgie im Mittelalter," *Studien zur Geschichte der Medizin*, pts. 11 and 12 (Leipzig, 1918).

21. Sir John Harington, *The School of Salernum* (London, 1922).

22. *The Diseases of Women by Trotula of Salerno*, trans. Elizabeth Mason-Hohe (Hollywood, Calif., 1940).

23. See below.

24. *Didascalicon, a Medieval Guide to the Arts*, trans. Jerome Taylor (New York, 1961).

25. Karl Sudhoff, "Die kurze 'Vita' und das Verzeichnis der Arbeiten Gerhards von Cremona, von seinen Schülern und Studiengenossen kurz nachdem Tode des Meisters (1187) zu Toledo verabfasst," *Archiv für Geschichte der Medizin* 8 (1915): 73–82.

26. The reader is referred to chapter 4 of the present volume.

27. The idea that medicine was akin to philosophy had a long tradition. It was expressed in the sentence, used by Isidore and taken over from the Iatrosophists, in which it is maintained that "philosophy is the medicine of the soul, while medicine is the philosophy of the body," and also in the words, attributed to Aristotle, to the effect that philosophy and medicine are sisters. But little attention had been paid to these sentiments in the West.

28. It should be noted that Averroes, styled by medieval writers as "the Commentator," followed Aristotle in considering medicine to be a mechanical art. But the condemnation of his theological and philosophical views brought with it the neglect, for a considerable time, of his other ideas.

29. *Compendium medicinae* (Lyon, 1510), fols. 57v, 313, and elsewhere.

30. Printed at Piacenza in 1476 and at Venice in 1489.

31. *I "Consilia,"* ed. Giuseppe Michele Nardi (Torino, 1937). The introduction to this volume should be supplemented by Francesco Puccinotti, *Storia della Medicina*, vol. 2, pt. 2 (Livorno, 1859), pp. 289–340.

32. *Conciliator differentiarum philosophorum et medicorum* (Venice, 1483), bk. 1, diff. 1–5.

33. *De considerationibus operis medicinae*, chap. 4: ". . . videtur quod reputant se contentos si videre aut legere possint, et in magnis cathedris

sarcinam voluminis [Avicennae] ostentare"; *Breviarium practicae*, bk. 4, chap. 10: "Et medici Montispessulani . . . magis aspiciunt ad curationes particulares et didascola et vera experimenta quam semper universalibus latrare." Both passages, much abbreviated here for reasons of space, should be read in their entirety.

34. *Conciliator*, bk. 1, diff. 3.

35. Quoted by Michael R. McVaugh, "Quantified Medical Theory and Practice at Fourteenth-Century Montpellier," *Bulletin of the History of Medicine* 43 (1969): 403.

36. *Questiones et tractatus extravagantes* (Venice, 1520), fol. 65r-v.

37. *Expositio super libros Tegni Galieni* (Venice, 1498), bk. 2, fol. 14v.

38. John D. Thompson and Grace Goldin, *The Hospital: A Social and Architectural History* (New Haven, 1975), pp. 1–50.

39. Ugo Stefanutti, *Documentazioni cronologiche per la storia della medicina, chirurgia e farmacia in Venezia dal 1258 al 1332* (Venezia, 1961), passim. Though this deals with only one city, the regulations reflect those common to many other places both in Italy and elsewhere.

40. Stefanutti, pp. 37–39; H. Denifle and A. Chatelain, *Chartularium universitatis Parisiensis* (Paris, 1891), 4: 406–7.

41. Savonarola, *Il trattato ginecologico-pediatrico in volgare "Ad mulieres ferrarienses de regimine pregnantium et noviter natorum usque ad septennium,"* ed. L. Belloni (Milan, 1952).

42. Lynn Thorndike, "Translations of Works of Galen from the Greek by Niccolo da Reggio (c. 1308–45)," *Byzantina Metabyzantina*, 1 (1946): 213–35; Robert Weiss, "The Translators from the Greek of the Angevin Court of Naples," *Rinascimento* 1 (1950): 211–26.

43. Guy de Chauliac, *Inventorium seu collectarium in parte chirurgicali medicine* (Venice, 1513), chap. 1: "In hoc tempore in Calabria magister Nicolaus de Regio . . . requirente Rege Roberto multos libros Galeni translatavit et eos in curia nobis transmisit."

44. Mundinus, *Anatomia* (Padua, 1484).

45. The Latin text was printed many times in the sixteenth century and was translated into French, Italian, Spanish, and English. The English medieval version has recently been edited in part by Björn Wallner (Lund, 1964) and completely by Margaret S. Ogden for the Early English Text Society (London, 1971).

46. *De arte phisicali et de cirurgia*, trans. Sir D'Arcy Power (London, 1922); *Treatises of Fistulae in ano . . . by John Arderne*, ed. D'Arcy Power (London, 1910).

47. First described by C. H. Talbot, *Medicine in Medieval England* (London, 1967), pp. 193–96, with an extract.

48. Oxford, Bodleian Library, MS Canon. Misc. 177, fol. 61.

49. University of Pavia, MS 314, colophon.

50. *Dynus super quarta primi* (Venice, 1514), fol. 166.

51. Anthony Weymouth, *Through the Leper Squint: A Study of Leprosy from Pre-Christian Times to the Present Day* (London, 1938); Karl Sudhoff, *Archiv für Geschichte der Medizin*, vol. 4 (1911) and succeeding volumes until 14 (1923) has printed a series of plague tracts covering the period after the Black Death; Arnold Klebs, *Die ersten gedruckten Pest-*

schriften: Geschichtliche und bibliographische Untersuchung (Munich, 1926); Philip Ziegler, *The Black Death* (London, 1969).

52. Lynn Thorndike, "Medieval Sanitation, Public Baths and Street Cleaning," *Speculum* 3 (1928): 192–203.

53. The following must suffice as examples: *De regenda sanitate consilium di Oderico da Genova*, Scientia Veterum, vol. 25 (Genoa, 1961); *Il 'de conservatione sanitatis' di Maestro Benedetto da Norcia*, Scientia Veterum, vol. 32 (Genoa, 1962); *Il libro di Arnaldo di Villanova sul modo di conservare la gioventù e ritardare la vecchiaia*, Scientia Veterum, vol. 38 (Genoa, 1963); S. A. Tissot, *Della salute dei letterati*, Scientia Veterum, vol. 42 (Genoa, 1963).

54. Dean Lockwood, *Ugo Benzi: Medieval Philosopher and Physician, 1376–1439* (Chicago, 1951), pp. 86–146, 238–348.

55. Joannes de Tornamira, *Clarificatorium super nono Almansoris* (Venice, 1507); for Joannes de Piscis see C. H. Talbot, "The Accumulationes receptarum of Joannes Piscis," *Bulletin of the History of Medicine* 34 (1960): 123–36; Valescus de Taranta, *Philonium* (Venice, 1521).

56. C. H. Talbot, "Simon Bredon (c. 1300–1372), Physician, Mathematician and Astronomer," *British Journal for the History of Science* 1 (1962–63): 19–30.

57. For a fuller discussion of mathematical physics at Oxford, and its penetration into other disciplines, see chap. 7 of the present volume.

58. Girolamo Tiraboschi, *Storia della letteratura Italiana*, vol. 5, pt. 1 (Florence, 1807), p. 261.

59. From a manuscript in the Wellcome Library, London.

60. His *Practica* was printed seven times between 1481 (Pavia) and 1534 (Lyons).

61. *Practica* (Venice, 1493), fols. 2r, 66r, 150v.

62. *De curis egritudinum particularium noni Almansoris practica uberrima* (Lyons, 1506).

63. According to the colophon at the end of this tract (Vatican, MS Barberini Lat. 186, fol. 52r), it was published in the vernacular at Venice in 1478 and later at Rome in 1482, but no trace of these printed versions has been found.

64. *Opus medicinae practicae* (Haganae, 1533), chap. 88.

65. Dan McKenzie, *The Infancy of Medicine: An Enquiry into the Influence of Folklore upon the Evolution of Scientific Medicine* (London, 1927); C. J. S. Thompson, *The Hand of Destiny: The Folk-Lore and Superstitions of Everyday Life* (London, 1932).

66. *De Chirurgia* (Venice, 1499), bk. 5, tract. 9, chap. 7.

67. *Buch der Bundth-Ertznei*, ed. H. Haeser and A. Middeldorpf (Berlin, 1868), preface.

68. *Das Buch der Cirurgia des Hieronymus Brunschwig* (Strassburg, 1497).

69. *Feldbuch der Wund-arznei* (Straussburg, 1517), passim.

70. H. Denifle et A. Chatelain, *Chartularium Universitatis Parisiensis*, 4: 594.

71. *De abditis nonnullis ac mirandis morborum et sanationum causis*, printed together with M. Gatinaria, *Super nono Almansoris* (Venice,

1516); there is also a translation by Charles Singer (Springfield, Ill., 1954).

72. *Annotationes Anatomicae* (Bologna, 1520), passim.

73. For all of these see L. R. Lind, *Studies in Pre-Vesalian Anatomy: Biography, Translations, Documents* (Philadelphia, 1975).

13

Jerry
Stannard

Natural History

Natural history in the Middle Ages was a complex amalgam of fact and fancy, in which reports concerning fabulous and exotic animals, plants, and minerals were indistinguishable from everyday experiences concerning indigenous species and domesticated varieties. Because of the guidelines established by theology and philosophy, the wide range of data derived from practical experience and the crafts rarely conflicted with the dicta derived from the Scriptures or excerpted from Patristic writings—the hexaemeral literature being a case in point.[1]

The methods, vocabulary, and conceptual framework employed by medieval writers who touched upon the world of nature were shaped by a plan loftier than the empirical study of animals, plants, and minerals. As a result, medieval natural history might be compared to a scrapbook: the beliefs and claims regarding the creatures of nature are like so many clippings, each page, so to speak, representing the random notices concerning a single species with little attempt at verification or scientific accuracy. For these reasons, then, it is preferable to speak of "natural history" than of "biology" and to organize selected materials in a manner that accords with medieval modes of thought and experience.

The complex relations between the world of nature and the world of texts is everywhere apparent in writings on and allusions to natural history throughout the Middle Ages. As a consequence, the enormous debt owed to classical antiquity must always be taken into consideration.[2] Many medieval reports concerning the fauna, flora, and mineralia have their origin in Greco-Roman texts. It is from such texts (including excerpts, epitomes, paraphrases, and the like) that much of the detailed information (and misinformation) was transmitted, later to be amplified or modified depending upon circumstances;[3] for example, lexica, poems, sermons, and so forth, imposed different requirements upon the use of ancient and possible pagan material.[4]

Thus, the chronological boundaries of medieval natural history are as vague and poorly defined as is its subject matter. On the one hand, they stretch back, though indirectly, to Greco-Roman writers—especially Aristotle, Dioscorides, and Pliny. On the other hand, they extend forward through the seventeenth century. At that time, developments in the nascent biological sciences quickened the rate at which some of the older claims were relegated to folk belief, where isolated examples may still be detected. It is, therefore, a delicate task to distinguish, date, and trace the origins of the various intellectual currents which came together in the medieval accounts.

A case in point is provided by the high medieval bestiaries or animal books, as they are popularly but misleadingly sometimes called. All of them, in one fashion or another, were descended from the anonymous *Physiologus*, originally written in Greek, and its early Latin versions. With the passage of time, the number of animals was increased, their names and the order of presentation changed, various plants and mineral substances added, and the illustrations and their extra-biological interpretations varied in accordance with changes in taste. These variations are readily understandable when it is borne in mind that the ultimate purpose served by the bestiary was didactic and that the assorted creatures described therein were merely the vehicles for arresting the reader's attention. But despite all the variations, the beliefs concerning the animals described but not invariably illustrated were endlessly embellished, with little regard for factual details based on observation. As such, those beliefs thoroughly permeated medieval literature and art, sacred and profane alike. Armed with such texts, surrounded by legends and fables in which the traits of imaginary creatures were as credible as those of real ones, and reinforced daily by their symbolic associations, the independent study of animals for their own sakes was rarely undertaken. In the strict sense, then, there was no scientific study of animals such as we associate today with zoology.

The situation is somewhat different with respect to plants because the descriptions of the several hundred herbs, shrubs, trees, and their products found in herbals were designed to serve the eminently practical purpose of medical therapy. As such, fabulous and imaginary plants, though often cited in medieval texts of various kinds, found little space in herbals. Owing to the very old and widespread belief in the virtues or curative properties (*virtutes*) attributed to many plants and to the training required in collecting the correct species for a specified purpose, the compilers of herbals, many of whom are now unknown, wisely restricted themselves, for the most part, to practical information. In this respect, they differed considerably from the bestiarists, some of whom seem to have had inflated ideas of their literary abilities.[5] The fund of useful and practical information found in herbals was based on many centuries of experience with local plants and the accumulated wisdom associated with the uses, internal and external, of those plants. As a consequence of the practical needs which the herbal was designed to meet, the descriptions of many of the species indigenous to Western Europe were based on or supplemented by personal knowledge of the plants in question.[6] In this respect, the herbal may be regarded as a forerunner of the local floras which emerged in the first half of the sixteenth century. Although these post-incunabula are commonly considered by bibliographers as monuments of Renaissance typography, they retain much of the conceptual vocabulary of the late Middle Ages and, thus, provide a good example of the continuity of scientific ideas.[7]

Like the bestiaries, lapidaries, or books of stones, seemingly describe natural objects. In many instances, the precious and semi-precious minerals and gemstones described therein can be identified, just as can some of the picturesque beasts of the bestiaries. But, to continue the analogy, the descriptions of the gems and jewels figured in the lapidaries are based as much on imagination, fables, classical mythology, and misunderstandings of earlier reports, ultimately Greco-Roman, as are those of bestiaries. In addition to recounting some of the physical properties of minerals (usually restricted to color and hue), their medicinal, magical, and apotropaic virtues are given a place as prominent as is the absence of any experimental data. The compilation of lapidaries was, with a few conspicuous exceptions, as close as medieval writers on natural history came to what later developed as geology and its several subdisciplines, such as crystallography and paleontology.[8]

The aforementioned literary genres—bestiaries, herbals, and lapidaries—provide a basis for understanding the principal contributions of medieval natural history and, as such, furnish a convenient point of departure for examining in greater detail its form and content. But as popular as they were—and of that, the surviving manuscripts and incunabula amply testify—they are not the only sources available to the modern student.[9] As will be noted in the following pages, a wide range of collateral material may be consulted profitably, most of which is easily accessible in the major libraries. This includes leech-books; books of antidotes and recipes; encyclopedias; lexica; hunting, fishing, and agricultural manuals; books of secrets and other instances of what is known on the Continent as *Fachliteratur* or *Fachprosa*; travel accounts; plus, of course, the indefinably large class of *belles-lettres* and theology.

In the absence of established or testable general principles such as apply today in the biological sciences and allied disciplines (for example, evolutionary mechanisms, the physicochemical transfer of genetic material, demographic studies, computer simulation, and so forth), the following presentation is arranged according to the traditions of natural history. By this means, the separate or compartmentalized study of animals, plants, and minerals—the so-called three Kingdoms of Nature—will be seen most clearly; by the same token, the lack of fundamental interconnections of a demonstrable scientific nature will be evident.

Thus, we shall examine *seriatim* animals, plants, and minerals with their medieval subdivisions whenever appropriate.

Animals

With respect to the study of animals, the bestiary is perhaps the most typical and accessible class of sources.[10] Bestiaries, along with encyclopedias, provide the best means of ascertaining both what was known and what was believed about animals in the Middle Ages.[11] To a limited extent, herbals and books of medical recipes are also useful, insofar as they describe various invertebrates (especially mollusks and insects), portions of which were popularly believed to possess therapeutic value. Other, more specialized, tracts will be mentioned below.

Despite the fact that the general purpose of bestiaries, encyclopedias, and herbals differed, the methods employed in their compilation were much the same, namely, excerpting previous texts and only rarely introducing material gained through personal observation. Accordingly, consistency and agreement in matters of detail are not to be expected. This is much more the case with bestiaries than with herbals, for while the order of presentation in the latter frequently is alphabetical and the sources are more limited, in the case of bestiaries the order sometimes follows the *Physiologus*, but at other times seems to have been determined by the compiler according to criteria now unknown.[12] Furthermore, since the names of animals were not standardized, greater freedom was shown in the order of presentation in the vernacular bestiaries. Finally, one particular class of bestiaries can be identified readily by the addition of a short etymological excursus on the name of the animal in question. These etymologies, usually as fanciful as they were charming, ultimately stem from Isidore of Seville's *Origines sive Etymologiae*, one of the earliest of the encyclopedias and, subsequently, a common source for medieval writers.[13]

Thus, because it is difficult to correlate the accounts of the same object in different texts, the technique adopted by the medieval encyclopedists provides a convenient starting point.[14] They divided the animal kingdom into *quadrupedia* (essentially mammals), *aves* (birds and other flying creatures), *pisces* (fish, aquatic mammals, and assorted marine creatures), *serpentes* (reptiles, including amphibia), and *vermes* (literally, worms, but including insect larvae, spiders, and almost anything that was not readily accommodated elsewhere). If to these classes are added the fabulous animals and composite monsters, we have a fairly complete roster of the kinds of animals described or referred to in the Middle Ages.

Quadrupedia

Following the example of the *Physiologus*, which probably originated in Alexandria, perhaps as early as the second century A.D.,[15] a number of African and Asian species were included in the bestiaries and encyclopedias, but only rarely in the herbals. Because little, if any, accurate information regarding such exotic beasts was available to the compilers, the descriptions in classical texts, which supplemented the *Physiologus,* were slavishly copied or misunderstood or both. As a consequence, the compilers were at liberty to alter the accounts in whatever manner they chose. Usually, however, it was either to sharpen the moral or allegorical significance or to embellish the original account by exaggerating some characteristic of the animal in question, for example, size, ferocity, or speed.

Despite these fanciful additions, some of which quickly became a part of general European animal lore,[16] most of the larger mammals can be identified. With respect to extra-European species, their Greek names or synonyms (often in a debased form and ultimately deriving largely from Aristotle by way of Pliny, Aelian, Solinus, and Isidore), plus crude but recognizable illustrations, enable us to recognize the lion, tiger, panther, elephant, camel, onager, and hyena.[17] The iden-

Fig. 40   An elephant, from a thirteenth-century English bestiary. Oxford, Bodleian Library, MS Bodley 764, fol. 12r.

tification of the elephant is a case in point. Owing to the convergence of its name in Greek and Latin, then later in several of the vernacular languages, references to its large size and thick skin, the value placed upon its tusks, and some fanciful but still elephantine illustrations, there is little doubt about the animal intended by writers, most of whom probably never saw a live specimen.[18] Each of these animals became the subject of fables and anecdotes, many variations of which appear in the descriptions.

Two other kinds of information, plus the illustrations, frequently, but not invariably, appear as part of the description. One, an etymology of the animal's name, has been mentioned above. A second kind of information is characteristic of bestiaries but less frequent in encyclopedias and altogether absent in herbals, namely, the moral, allegorical, or mystical significance of the animal. Oftentimes set off by rubricated initials or by a marginal device, the *significatio* was the symbolic analogue of the more prosaic, but not necessarily more accurate, description. Following a somewhat fanciful, yet naturalistic, description of the eagle, it is written: "Because of sins which take their origin from the mother, man is like the eagle here, but he is renewed thus: he soars above the clouds, and feels the sun's fires, despising the world and its pomp."[19] In a similar fashion, the onager signified or symbolized the devil, while the panther usually signified Christ in the bestiaries but not in the encyclopedias.[20] Whether or not the *significationes* were believed literally, the fact remains that they provided the preacher with entertaining illustrations for his sermons and, thus, reached a large audience.[21] The great emphasis on the *significatio* lends credence to the opinion that moral edification, rather than zoological observation, was, indeed, the basic motive for the compilation of bestiaries.

Considerable space is devoted in medieval writings to two other classes of quadrupeds—indigenous species and domesticated varieties. Among the former of those two classes, the wolf, bear, hedgehog, and fox attracted much natural and much more unnatural history in the form of fables, anecdotes, proverbs, and superstitions. The wolf and bear, for example, were feared by all, not only because of the physical harm which they were capable of inflicting, but because of another kind of harm as well. Throughout the whole of Europe, including the British Isles, the wolf was the subject of a widespread superstition concerning werewolves.[22] In order to secure protection from the wolf, "vampire," and other beasts into which humans could be, and were, it was commonly believed, transformed, much energy was devoted to compiling prayers and charms and to manufacturing amulets, periapts, and other protective devices.[23] Not uncommonly, the latter were made from or incorporated a portion of the animal against which it was regarded as effective. This latter, a good instance of sympathetic magic, is a reminder that medieval as well as post-medieval natural history supplied as much to folklore as it received.[24]

Despite the practical experience that must have been gained by hunting wolves, boars, stags, and bears—the latter for the purpose of bear-baiting—accounts of these animals in bestiaries and encyclopedias are not appreciably more accurate or realistic than those concerning more exotic species. But by consulting hunting manuals, and comparing them with allusions to the chase in literary texts and with the abundant iconographic evidence, a balanced account can be reconstructed.[25] The notices concerning the stag (*cervus*) provide a good example of the limitations of the bookish accounts found in bestiaries. Except for some miscellaneous remarks which indicate otherwise, one would little guess that stag hunting involved considerable preparation and that many observational data were ready at hand by merely consulting the hunters and others engaged in the chase.[26] The notices concerning the stag indicate, rather, how deeply ingrained were the fables emanating from the *Physiologus*. It may be noted, however, that the popularity of the *Physiologus* and the bestiaries did not interfere with the attempts made by miniaturists and artists to portray what they saw.

The iconographical evidence pertaining to medieval natural history is much too large to be discussed here. But it must be taken into account as an additional source of information and as a corrective to some one-sided accounts which emphasize the unworldliness and otherworldliness of the Middle Ages.[27] By consulting the iconographical data, not only can the influence of the bestiaries be visually traced,[28] but the popularity of certain themes can be better understood within the medieval context—for example, the enduring appeal of the fox or the frequency with which the hare appears in fables, proverbs, and recipes.

Domesticated animals, the other class of quadrupeds to be discussed, include the ox, cow, horse, donkey, sheep, goat, and, of course, the dog. Pigs, which foraged for themselves and were only partially domesticated, bore little resemblance to the short-legged, heavily-built hogs raised today. Horses, oxen, and donkeys were the basic source of energy for transportation, hauling, and plowing. As such, they were a normal part of the medieval scene, and the average citizen could be expected to know something of their habits. Little

culuſ aureuſ in naribʒ ſuiſ· mulier pulchra
& fatua.Item ſuf carnaliũ cogitatōnes for
dide· ex quibʒ opa praua tielur decocta ꝑtdut
in yſcuia. Qui comedunt carnem ſuillam· &
iuſ ꝑfanũ in uaſiſ cor· id eſt ĩ cordibʒ cor.

per a ferutate uocatur· ablata· f luiū·
& ſubrogata· p.Unde & apud greos
ſiagros id eſt feruſ dr: Omne enim q

Fig. 41   A wild boar, from Oxford, Bodleian Library, MS Bod-
                ley 764, fol. 38v (13th c.).

of this common knowledge appears in the bestiaries and encyclopedias. But by virtue of their importance in the economy, various specialized tracts were circulating, though on a plane quite distinct from that of the bestiaries. The former included agricultural tracts and those pertaining to veterinary medicine.[29] The military uses of horses, finally, provided another rich source of firsthand information. But when transmuted by chivalric romances, there was as little resemblance to campaigns as the bestiaries bore to plowing.

Aves

Medieval knowledge of and beliefs about birds cover a broad spectrum, the polar opposites of which may be represented by the empirically based *De arte venandi cum avibus*,[30] written by the Emperor Frederick II, and the Anglo-Saxon allegorical poem *Phoenix*, long attributed to Cynewulf.[31] If to these be added the loose collection of popular tales, exemplified by Chaucer's lines on Chanticleer and Pertelote,[32] a basis is furnished for understanding three of the major attitudes toward birds in the Middle Ages—practical, symbolical, and anecdotal.

There is very little doubt that a good portion of the information which went into the various tracts on hawking and falconry was based on a close examination of several species of accipiters.[33] Details regarding their diet and longevity, mating habits and change of plumage, diseases and their remedies were based on many years of experience. The different patterns of moulting, for example, are described thus: "Raptores, that are in constant need of their flight feathers to aid in capturing their prey, have a regular form of moult, so that they never entirely lose their flying ability. Harmless birds that are not in such urgent need of wing power to gain a living (i.e., those whose provender does not fly away from them) moult in less orderly fashion; but as they require flying powers to secure shelter and to avoid dangers, the moult is not entirely without plan. Waterfowl (i.e., swimmers), on the other hand, make a complete and unusual moult influenced by the fact that they do not escape dangers nor obtain their sustenance by flight. By living in the water they attain both objectives."[34] If the authors of the treatises on birds were acquainted with the bestiary tradition, they wisely ignored the heroics of the eagle and such like; in the same fashion, there is no evidence that the bestiarists knew or cared to know anything about falconry.

The heavily allegorical accounts of the birds of prey in the bestiaries represent, so to speak, another world. While some of the fables

Vsio appellatus quod muribz mfest
sit. Hunc uulgus catum a captura
uocant. Alii dicunt quod captat id ẽ
uidet. Nam tam acute cernit ut fulgoxe lumi
nis noctis tenebras supet. Unde a gɾeco uenit
catus id est mgeniosus. a΄ποτογκαϬεϬται.

Vs pusillum
aiumal grecu̅
illi nomen est
quicquid ii ex eo cfitur
latmum fit. Alii dicunt
mures quod ex humoxe terre nascantur. Nam
mus terra. unde & humus. His in plenilunio

have their peculiar charm and may even possess some slight literary value, they were based not on real birds but on literary types in which birds are reminders of our moral duties. The turtle dove, for example, was praised in many bestiaries, and other genres as well, for her quiet, retiring nature, her shyness, constancy, and fidelity. As such, a model was provided for mortal man to remain constant in belief and steadfast in the face of adversity.[35] The pelican, caladrius, even the ostrich are further examples of such types about which stories were constructed.[36]

When we come to domestic fowl, we see still a different world. The cock and hen, and usually nearby the wily fox, the doves in the loft, ducks in the pond, and geese in the barnyard were part of the real world, only fragments of which appear in the bestiaries.[37] But their presence is presupposed in other texts, even though the storyteller's art sometimes may supply them with a set of all-too-human attributes.[38]

Pisces

In general, fish did not attract the attention that mammals or birds did, nor, but for different reasons, reptiles. After all, fish could not be tamed nor easily domesticated, and their life histories were virtually unknown. Smoked, salted, or fresh, they formed a dietary staple for many. But there was little romance or poetry connected with that; only hard, and sometimes dangerous, physical labor.[39]

This does not mean that fish, in the broad sense, were ignored by the bestiarist, encyclopedist, physician, or cook.[40] Grouping together different literary genres, approximately fifty different species of fish are mentioned (for example, bream, carp, cod, flounder, herring, mackerel, mullet, perch, pike, salmon, tench, trout, turbot, and wrasse) along with some twenty-five aquatic creatures which today can be assigned to other phyla (for example, crab, dolphin, and sea urchin) or dismissed as imaginary (for example, the hydrus).[41]

Despite the fact that Aristotle had differentiated between whales and dolphins, on the one hand, and bony and cartilaginous fish, on the other, the whale was commonly regarded as a fish throughout the Middle Ages. Since the sight of a living whale was restricted to fishermen, among whom few *literati* normally would be included, reports of its great size became exaggerated in direct proportion to the distance from the coastline. Such reports, moreover, fitted nicely with the biblical account of Jonah and his misadventures. As a consequence, the whale became a popular subject, not only in bestiaries but in ecclestiastical art and sculpture as well.[42]

Even more fabulous than the whale was the *serra*. Although the characteristics attributed to it are purely imaginary, it is worth noting as a typical example of a rather common phenomenon in the history of zoology. An ordinary animal, albeit an unusual one and usually uncommon, undergoes a double transformation: first, one or more of its own characteristics becomes exaggerated, and second, properties of another animal are grafted onto it. In the present case, the underlying animal seems to be a sawfish (*Pristis*), to which was later added a somewhat vague report concerning flying fish (family *Exocoetidae*), many genera of which are able to glide above the surface of the water by virtue of their winglike pectoral fins. The union of these two different fishes produced a strange creature: "There is a beast in the sea which we call a *serra* and it has enormous fins. When this monster sees a ship sailing on the sea, it erects its wings and tries to outfly the ship, up to about two hundred yards. Then it cannot keep up the effort; so it folds up its fins and draws them in, after which, bored by being out of water, it dives back into the ocean," and "*Serra* is called thus because he has a serrated cock's comb and, swimming under the vessels, he saws them up."[43]

As noted above, medieval writers often included, along with fish, a variety of other aquatic, usually marine, organisms. Mussels and clams, for example, were prized as delicacies, while the oyster was valued even more for its pearl, though there is little indication that its formation was understood.[44] Mollusks, moreover, were used in medicine for numerous complaints. It may be noted that in the absence of a taxonomic system, terrestrial mollusks, such as the common garden snail and slug, were considered *vermes*, while the squid and octopus, also mollusks, were considered *pisces* or marine creatures.[45]

Serpentes

The numerous chapters on reptiles in bestiaries and their widespread occurrence in various art forms cannot be explained on ecological or toxicological grounds alone. The repugnance felt toward serpents and, on the other hand, their fascination for poets and theologians rest on the same grounds. Both attitudes go back to a literal acceptance of the creation story in which the serpent plays a role no less important than the other two mythological figures. But the story of Adam and

Eve in the garden of paradise, which was, incidentally, the reputed habitat of several species of plants and animals, was not the only subject with which serpent lore was connected.[46]

The viper (or adder, *Vipera berus*), for example, appears in a variety of contexts. The strong language of Matthew 3:7 ("brood of vipers") naturally became a text with endless possibilities for allegory. And because only viper's flesh, so it was believed, was acceptable in the preparation of the famous antidote *theriaca andromachi*, much effort and not a little cunning were involved in preparing it, sometimes fraudulently, for therapeutic use.[47]

The asp and the crocodile are further examples of reptiles with a rich tradition in medieval literature and art.[48] The latter, especially, is a good example of how the accounts of exotic creatures become exaggerated and merge into the fabulous, resulting, in the case of reptiles, in a strange brood, for example, the basilisk, dragon, griffon, and amphisbaena.[49]

Vermes

Used as a catch-all taxonomic term, *vermes* served naturalists until the time of Linnaeus. In medieval texts, it was not only a taxonomic term, but a term which connoted vague yet pronounced value judgments, hence, the origin of our word "vermin." Insofar as the entities covered by this class word can be sorted out, insects, both mature and larval forms, various arachnids, and the common earthworm are among the most prominent.[50]

Because of their abundance and probably also because of their annoying habits, mosquitoes, houseflies, lice, and fleas were objects to be eliminated rather than to be studied. The scorpion, however, because of its larger size and its painful, poisonous sting was an object of dread and, as such, received some notice.[51] For much the same reason, spiders, too, were specifically commented upon. Their webs, moreover, lent themselves to poetic fancy but were also used for medicinal purposes.[52]

Among the insects, the bee and the ant received the most favorable notice. The former, one of the very few domesticated insects, was well known because of its economic importance; honey and wax, along with vinegar, were three of the widely used natural products, and neither household nor physician nor apothecary could do without them.[53] Ants, like bees, because of clearly observable social habits, lent themselves to moralization and became established in literature as well as in folklore.[54]

For butterflies and moths the case is somewhat different. Neither vicious nor harmful (the voracious larval forms, or caterpillars, were not associated with the mature adults), their colorful patterning attracted the attention of miniaturists. A wide range of lepidopterous species are thus depicted in illuminated manuscripts along with a few of the more conspicuous coleoptera.

Monsters and Fabulous Creatures

No account of the animal life recognized in the Middle Ages would be complete without at least a brief mention of monsters and the like. This category included imaginary animals, mythological creatures, and composite forms, some of which were semi-human, for example, sirens, satyrs, and assorted giants.[55] The extent to which the existence of these creatures was seriously believed in depends partially on the entity in question. Some of the imaginary zoophytes, for example, were taken seriously through the eighteenth century. But whether believed or not, monsters and their ilk played no small role in art and literature and occur not only in bestiaries and encyclopedias, but also in lexica, travel accounts, and theological writings.

Some of these creatures, for example, the phoenix and siren, owed their existence to the survival of classical mythology.[56] Others were the products of Christian bibliolatry, in which, for example, the dragon, like the devil (or the angels), plays an almost necessary role in the cosmic drama.[57] Finally, there are some entities about whose status modern scholars are undecided. Some, like the *leucocrota* and *manticora*, are wholly imaginary.[58] Others, however, may have resulted from the fusion of several real animals, for example, the *eale* and the *autolops* (both of which seem to have been based on an African antelope) or the unicorn, called *monoceros* in some bestiaries.[59] The latter is perhaps the best known of the fabulous creatures which helped to make the medieval forest—usually their favorite haunt—as exciting as it was dangerous.

Plants

It is even more the case with plants than with animals that the medieval records represent two distinct levels: (1) an empirico-practical level in which the descriptions and uses are based upon a knowledge of the living plant and (2) a learned-scholastic level in which the discussions of and references to plants are based on little, if any, empirical data.

Herbals, with their descriptions (sometimes accompanied by illustrations) of plants, may appear at first glance to be the exact counterpart of bestiaries.[60] Herbals resemble bestiaries in the following respects: (1) they contain an enumeration of the virtues of the "species" (usually medicinal, in herbals, rather than allegorical, as in bestiaries); (2) there is a separate section devoted to each "species"; (3) the order of the sections is determined by nontaxonomic criteria; and (4) the basis of many of the descriptions, especially of exotica, is to be found in Greco-Roman writings.

A closer comparison, however, of herbals and bestiaries (and lapidaries, which stand closer to the latter) will reveal some fundamental differences both with respect to method and with respect to content. First, because the cultivated, indigenous, and naturalized species were often locally abundant and/or widely distributed, many of them were well known and commonly used for alimentary and other domestic purposes, for example, cereal grasses, fruit trees, herbs, and legumes. Second, the habitat of such plants, often described as "known to everyone," is often carefully and accurately specified in order to facilitate their collection in the fresh state. Third, except for the exotica and the magical plants (which latter seldom occur in herbals), the fabulous aspects are absent. In short, herbals were designed as practical manuals and were, to the extent that they were used, continually subjected to testing and refinement. Bestiaries and lapidaries, on the other hand, were designed for moral edification, and additions or alterations were the results of reading other texts, rather than a closer reading of nature.

Whatever might be the limitations of medieval herbals with respect to taxonomic niceties, they provide us with the fullest information about the largest numbers of plants known and used. On matters of detail and for various specialized problems, they must be supplemented by other sources. Especially useful are leechbooks, recipe books, encyclopedias, and lexica, many texts of each of these genres supplying details not found in herbals.[61]

A typical chapter in an herbal may contain the following information:[62] name and synonyms (sometimes accompanied by an etymology); description of the plant, including habitat and other practical information which subserved its therapeutic uses, for example, phenological data, especially the proper time to collect and the part or portion to be used;[63] the virtues of the plant in question; and instructions regarding preparation, administration, dosage, and storage. For example, coltsfoot (*Tussilago farfara L.*), a European plant now sporadically naturalized in the United States, is described as follows:

It is called *ungula caballina* in Greek and Latin and *phatanum* in Arabic. The authorities state that it has broad leaves like those of the waterlily and that they are green on the upper surface but white on the underside. Pliny, in place of describing it, merely states that it possesses great virtue. It is very good for running sores when the leaves are laid thereon. Also it is good for children who suffer from a loose cough, for it expells much phlegm and moisture. The leaves laid upon a burn caused by fire or gunpowder draw out the heat. The juice of the plant has all the virtues of the leaves. Its juice, mixed with that of fumitory, when smeared upon the spots which the sun has burned, causes the redness of the skin to vanish and makes the skin white and fair.[64]

The format employed here was first enunciated in Dioscorides' *De materia medica*.[65] It was followed, with only minor modifications, throughout the Middle Ages; it endured to the early eighteenth century, and one can still find vestiges of it today in popular writings and in folk medicine.

The single most important modification—and one that had far-reaching effects, insofar as it delayed the evolution of the herbal as a botanical work—was the rearrangement of its contents. Originally *De materia medica*, divided into five books, was nonalphabetical, and some effort toward natural grouping was evident; for example, five consecutive chapters, each treating a species of *Allium*, are followed by an equal number of chapters on various *Cruciferae*. But medieval herbals tended to follow the plan of the so-called *Dioscorides alphabeticus*, which, though not the earliest Latin translation, became the most popular; in the case of vernacular herbals, such as the *Hortus sanitatis germanice* (1485), the arrangement follows that of its Latin prototype.

The alphabetical arrangement, it must be noted, well served the herbal's therapeutic purpose by facilitating ready reference in time of need. Taxonomic considerations were, thus, foreign to the practical purpose served by an herbal, though in the text itself, two or more taxa, usually of similar appearance, might be distinguished on the practical grounds that one taxon was more beneficial or, alternatively, was inert or toxic. The encyclopedias, on the other hand, with their different, that is, nonmedical, purpose divided the plant kingdom into a few simple but large classes, for example, herbs and trees, either of which might be further subdivided into aromatic (that is, exotic) and common (that is, indigenous) species. Within these classes, the individual sections were similar to those of herbals, and an alphabetical arrangement frequently was retained. Despite their different aims,

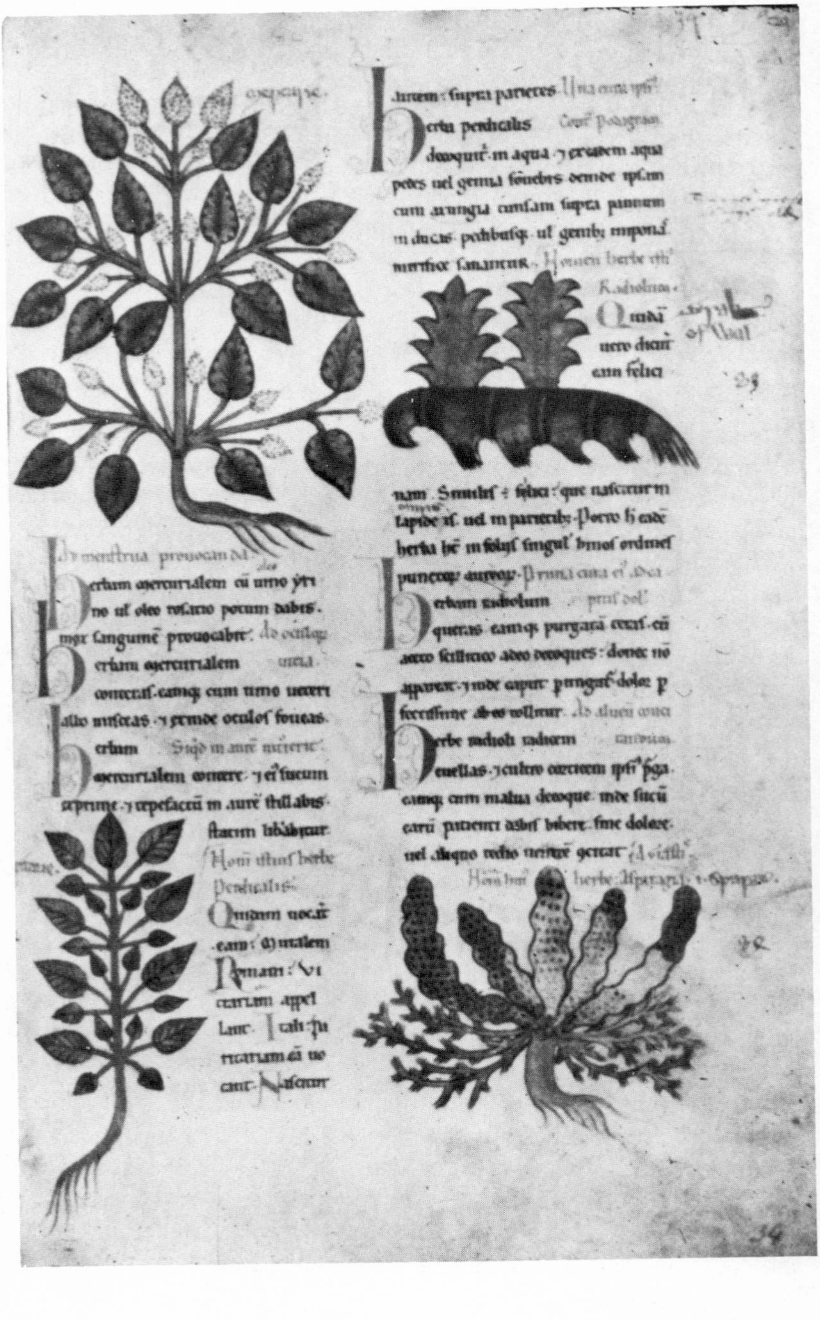

Fig. 43   A page from a medieval herbal, showing dog's mercury
(*Mercurialis perrenis* L.), pellitory or pellitory of the
wall (*Parietaria officinalis* L.), polypody (*Polypodium
vulgare* L.), and asparagus or sparrow grass (*Asparagus officinalis* L.). Oxford, Bodleian Library, MS Ashmole 1462, fol. 34r (early 13th c.).

the practically oriented herbals and the scholastically motivated encyclopedias supply an abundance of information, some details of which follow.

In the main, most of the larger and more conspicuous flowering plants indigenous to or naturalized in Western Europe were believed to possess one or more virtues; consequently, their descriptions outnumber those of any other class of plants in medieval writings. On the basis of the accumulated experience of many centuries, it was known that certain plants led to marked and predictable physiological responses on the part of the user: for example, as diuretics, purges, emetics, sternutatories (substances that provoke sneezing), vermifuges, and so forth, a number of common plants were routinely used,[66] many of which remained official until the early nineteenth century.[67] Many further species were used whose rationale is not so obvious, for example, as galactagogues (which promote the secretion of milk), sudorifics (which induce sweating), aphrodisiacs, and rubefacients (which cause redness of the skin), or which were used, often in compounds, because of a pronounced taste or odor. The analgesic and narcotic effects attributed to many plants in the Middle Ages require close examination when one considers the crude form in which plant material was taken and the size and frequency of dosage.[68] Finally, many plants were alleged to be specific for a particular complaint, but the concept of disease specificity, like that of pathogenic microorganisms, rested on no experimental evidence.[69] Nevertheless, because of the necessity of self-medication, local plants were widely used in times of illness.[70] Their recognition and collection, preparation and storage, was part of the local tradition and contributed its share to natural history. Whether the motive was medical, alimentary, or pecuniary, a knowledge of the local flora was often the beginning of a medical or scientific career. Herborizations and the formation of herbaria, a part of the normal training for physicians in the Renaissance, were outgrowths of established medieval custom.[71]

In the case of cryptogams (for example, mosses, ferns, and fungi), the medieval records are comparatively scanty.[72] Although a few of the cryptogams which were mentioned by medieval writers can be identified, principally the ferns and horsetails, they played on the whole a small role. Since their life cycle was unknown, a bit of mystery was associated with them. This was especially the case with the fungi which, because of their gross morphology and the highly poisonous species, were treated in a fashion analogous to the *vermes* noted above.

Exotic plants, like animals, were virtually unknown in their native habitat, though a few reports from travelers to the Holy Land indicate some recognition of an alien flora.[73] But, in the form of spices, for both medicinal and culinary purposes, the dried seeds, bark, leaves, and roots formed a major part of the Levantine imports.[74] For example, pepper, ginger, cloves, cinnamon, and nutmeg were available in most of the larger cities;[75] another dozen or so plant products are occasionally encountered, a figure that was quickly increased upon the discovery of the New World. In the same fashion, various gums and resins, some of which were used for liturgical purposes, others for various economic purposes, were transshipped to Europe. Some of these substances, for example, frankincense, myrrh, and dragon's blood (a resin from the species *Dracaena*), were sufficiently valuable as to have created myths of their own.[76] Not surprisingly, fabulous accounts of the origin of such substances found a ready place in certain genres of medieval writings.

As noted above, descriptions of imaginary and fabulous plants did not often appear in herbals.[77] But, as is the case with nearly all medieval writings, there were exceptions; for example, the barnacle goose tree and the mandrake both appear in herbals, while the *arbor vitae*, the *arbor scientiae*, the *peridexion*, and the like, found an even wider audience, aided perhaps by their iconography.[78]

But the near absence of fabulous plants and trees from herbals should not be taken to mean that the natural history of the vegetable kingdom was devoid of unnatural elements. Along with the medieval romances and hagiographical texts, where one, like Brendan's companions, expected the miraculous,[79] many plants were invested with extraordinary properties, usually therapeutic, but capable, at the hands of enchanters, necromancers, and the like, of being used for nefarious purposes.[80] Many of the plants which fall under this heading were real plants, for example, betony, peony, rue, sage, and verbena.[81] But when these plants were subjected to rituals or incorporated into artifacts (amulets, talismans, and so forth), their virtues were greatly augmented. Such plants, which might be called "magiferous" plants, dot the pages of herbals, leechbooks, and books of medical recipes, while memories of their extraordinary properties still survive in present-day folk medicine.

Some mention must be made, finally, of the nonmedicinal, non-magical uses of plants and their products. While there exist an abundance of iconographical evidence and literary references to the economic uses of plant material, occasionally supplemented by physical evidence, only a few texts specifically devoted to such uses have been published.[82] Among these documents might be mentioned agricultural and horticultural texts which exhibit a range of knowledge based on practical experience with only a minimum of theorizing.[83]

In addition to the separate tracts noted above, there exist many data regarding the diverse uses of plant material in the household and for sundry artistic and technological purposes. Together, they indicate a close familiarity with the physical properties of woods, fibers, and other portions of plants which were necessary for the maintenance of daily life and its amenities.

Minerals

For present purposes, it will be convenient to consider the lapidary as the mineralogical analogue of the bestiary and, though to a lesser extent, of the herbal.[84] Though somewhat simplified, this procedure serves to bring out several similarities of lapidaries to the other two genres: (1) The order of presentation varies from one lapidary to another and is not determined by taxonomic criteria. (2) Major emphasis is placed upon the medicinal and magical virtues of the stones. (3) Rocks and minerals native to Western Europe are subordinated to precious and semi-precious gems of exotic origin. (4) A Greco-Roman basis underlies the accounts of many of the gems.[85] In three further respects, lapidaries show a closer similarity to bestiaries than to herbals. (5) Material of a fabulous nature constitutes an essential part of the description. (6) The stones are often interpreted allegorically. And (7) biblical allusions are common.

As the foregoing summary makes clear, lapidaries are to modern mineralogy as bestiaries are to modern zoology.[86] The compilers of lapidaries (and the corresponding books in encyclopedias and entries in glossaries devoted to minerals) had little knowledge of minerals beyond a cursory inspection of the cut, polished, and mounted gem-stones which were widely used in ecclesiastical art. Moreover, they exhibited little interest in the allied technological processes, for example, mining, ore-separation, and the like. With the exception of

Albert the Great's *De mineralibus*, which demonstrates his curiosity about and some firsthand information on techniques, the lapidaries must be supplemented by "Books of Secrets."[87] The latter, which have been too long neglected by historians of science, contain a wealth of information of a most practical kind.[88] It can only be noted here that their authors demonstrate a thorough familiarity with many complicated techniques pertaining to the preparation and use of mineral substances.

Because, in the descriptions of the stones, all but the most superficial of physical properties were lacking, lapidaries became pious catalogs of miracles. The variations in nomenclature, orthography, and order of presentation, moreover, make it hazardous to estimate with any degree of precision the number of different mineral, metallic, and other substances (for example, amber, *karabe*, and so forth) described therein. Suffice it to say, an interest in their miraculous powers and, consequently, in ownership did not appreciably further experimental investigations.

Nonetheless, like the bestiaries, lapidaries served the minimal function of acquainting their readers with the existence and names of substances whose exotic origin and wonderful properties stimulated the curious. While belief in their virtues may have arrested temporarily the development of the natural sciences, their descriptions, howsoever inadequate, consolidated and completed a system which remained intact for centuries to come. That system unified on *a priori* grounds the three kingdoms of nature. And by establishing the fundamental relations between those kingdoms, it made them amenable to human understanding and, hence, capable of being communicated. The invention of printing with movable type in the middle of the fifteenth century was the culmination of that process and, at the same time, the precursor of another kind of natural history.

Conclusion

In balance, it appears that the two most significant features of medieval natural history are, first, the fund of practical, reliable, firsthand information that was available, ready to be used by scholars and clerks, and second, the brute fact that, save on rare occasions, the weight of tradition prevented its most effective use. If not hostile to the collection of empirical data from nature's realm, the prevailing tradition was indifferent to the creatures of nature except as they fulfilled a superhuman plan. Insofar as data were collected, that activity subserved either a practical immediacy, as in the case of *materia medica*, where the virtues of the object or substance to be collected were determined in advance; or a form of scholarship, as in the case of encyclopedias, where the voice of authority assured assent. Whatever discrepancies there were between writ and fact, they were, for the most part, silently ignored. The tradition which made this possible, however, became increasingly top-heavy with each succeeding century. Slowly and piecemeal it was replaced, initially by the recognition that all living matter shared certain properties, and later by the still-incomplete reduction of those properties to physical laws—as invariable and as mysterious as ever, but testable and quantifiable.

Notes

1. As examples, see Basil, *Hexaemeron*, in J.-P. Migne, *Patrologia cursus completus, series Graeca*, vol. 29 (Paris, 1886), cols. 3–208; Ambrose, *Hexaemeron*, in J.-P. Migne, *Patrologiae cursus completus, series Latina*, vol. 14 (Paris 1845), cols. 134–288; Bede, *Hexaemeron*, in ibid., vol. 91 (Paris 1850), cols. 9–190. Of related interest are J. Levie, "Les sources de la septième et de la huitième Homélie de Saint Basile sur l'Hexaméron," *Musée Belge* 23 (1920): 113–49; L. Fonck, "Hieronymi scientia naturalis exemplis illustratur," *Biblica* 1 (1920): 481–99; and P. Plass, *De Basilii et Ambrosii excerptis ad historiam animalium pertinentibus* (Marburg, 1905), published dissertation.

2. Especially valuable are Max Manitius, *Geschichte der lateinischen Literatur des Mittelalters*, 3 vols. (Munich, 1911–32); R. R. Bolgar, *The Classical Heritage and its Beneficiaries* (New York, 1964); and Paul Lehmann, *Pseudoantike Literatur des Mittelalters* (Leipzig, 1927).

3. See, for example, Karl Rück, "Die Naturalis historia des Plinius im Mittelalter," *Sitzungsberichte der bayerischen Akademie der Wissenschaften*, Phil.-hist. Classe, 1898, 1:203–318, and, for the utilization of such material for specialized purposes, Jerry Stannard, "Greco-Roman Materia Medica in Medieval Germany," *Bulletin of the History of Medicine* 46 (1972): 455–68.

4. As an example of the pagan background of what later became an acceptable part of Latin Christianity, see Francesco Sbordone, "La fenice nel culto di Helios," *Rivista Indo-Greco-Italica* 19 (1935): 1–46.

5. This is particularly true of the so-called *bestiares d'amour*; see *Li Bestiaires d'Amours di Maistre Richart de Fornival e li Response du Bestiaire*, ed. Cesare Segre (Milan, 1957); *Le Bestiaire d'Amour Rimé*, ed. Arvid Thordstein (Lund, [1941]).

6. Jerry Stannard, "Botanical Data and Late Medieval Rezeptliteratur," in *Fachprosa-Studien: Beiträge zur mittelalterlichen Wissenschafts- und Geistesgeschichte*, ed. Gundolf Keil (Berlin, 1978).

7. For example, see Gerhard Eis, "Ein deutscher Herbarius von 1531," *Deutsche Volksforschung in Böhmen und Mähren* 2 (1943): 97–112;

and Erwin Stresemann, "Das Tierbuch des Lazarus Rotin," *Atlantis* 12 (1940): 397–400.

8. For a good overview, see Karl Mieleitner, "Geschichte der Mineralogie im Altertum und im Mittelalter," *Fortschritte der Mineralogie* 7 (1922): 427–80.

9. Lynn Thorndike and Pearl Kibre, *A Catalogue of Incipits of Mediaeval Scientific Writings in Latin*, rev. ed. (Cambridge, Mass., 1963); Arnold C. Klebs, *Incunabula Scientifica et Medica* (Bruges, 1938).

10. Carl Appel, ed., "Aiso son las naturas d'alcus auzels e d'alcunas bestias," in *Provenzalische Chrestomathie*, 4th ed. (Leipzig, 1912), pp. 201–4; P. T. Eden, ed., *Theobaldi Physiologus* (Leiden, 1972); E. B. Ham, ed., "The Cambrai Bestiary," *Modern Philology* 36 (1939): 255–37; C. Hippeau, *Le Bestiaire Divin de Guillaume Clerc de Normandie* (Paris and Caen, 1852–77; facsimile reprint Geneva, 1970); M. R. James, ed., *A Peterborough Psalter and Bestiary of the Fourteenth Century* (London, 1921); Alfons Mayer, ed., "Der waldensische Physiologus," *Romanische Forschungen* 5 (1890): 392–418; E. Walberg, ed., *Le Bestiaire de Philippe de Thaün* (Lund, 1900). In addition, see Brunetto Latini, *Li livres dou tresor*, ed. Francis J. Carmody (Berkeley, 1948), bk. I, chaps. 130–99.

11. Much useful information will be found in Florence McCulloch, *Medieval Latin and French Bestiaries* (Chapel Hill, N.C., 1962); Max Friedrich Mann, "Zur Bibliographie des Physiologus," *Anglia Beiblatt* 10 (1899–1900): 274–87.

12. See the concordance in Max Friedrich Mann, "Der Bestiare Divin des Guillaume le Clerc," *Französische Studien* 6, no. 2 (1888): 105–6.

13. McCulloch, *Bestiaries*, pp. 28–30. On Isidore's natural history, see A. Schmekel, *Isidorus von Sevilla, sein System und seine Quellen* (Berlin, 1914).

14. The sections devoted to animals in available published medieval encyclopedias include: Isidore of Seville, *Etymologiarum sive originum libri XX*, ed. W. M. Lindsay (Oxford, 1911), bk. 12; Rabanus Maurus, *De universo*, in Migne, *Patrologia Latina*, vol. 111 (Paris, 1852), cols. 9–614, bk. 8; Hildegard of Bingen, *Physica*, in Migne, *Patrologia Latina*, vol. 197 (Paris, 1855), cols. 1117–1346, bks. 5–8; Thomas of Cantimpré, *Liber de natura rerum*, ed. H. Boese (Berlin, 1973), bks. 4–9; Konrad of Megenburg, *Das Buch der Natur*, ed. Franz Pfeiffer (Stuttgart, 1861; facsimile reprint Hildesheim, 1962), bk. 3; Gershon Ben Shlomoh, *The Gate of Heaven*, trans. F. S. Bodenheimer (Jerusalem, 1953), bks. 4–7.

15. Ben E. Perry, "Physiologus," *Pauly-Wissowa, Real-Encyclopädie der classischen Altertumswissenschaft* 20 (1950): 1074–1129. For alternative interpretations, see Max Wellmann, "Der Physiologos," *Philologus*, supp. 22, no. 1 (1930): 1–116, and Friedrich Lauchert, *Geschichte der Physiologus* (Strassburg, 1889).

16. As specimens, see Carlo Frati, "Ricerche sul Fiore di virtù," *Studj di filologia romanza* 6 (1893): 247–448; Friedrich Lauchert, "Der Einfluss des Physiologus auf den Euphuismus," *Englische Studien*, 14 (1890): 188–210; Milton Garver, "Some Supplementary Italian Bestiary Chapters," *Romanic Review* 11 (1920): 308–27; G. Polivka, "Zur Geschichte des

Physiologus in den slavischen Literaturen," *Archiv für slavische Philologie* 14 (1892): 374–404.

17. See Alexandra Konstantinowa, *Ein englisches Bestiar des zwölften Jahrhunderts* (Berlin, 1929); Samuel Ives and Hellmut Lehmann-Haupt, *An English Thirteenth-Century Bestiary* (New York, 1942); Adolph Goldschmidt, "Frühmittelalterliche illustrierte Enzyklopädien," *Vorträge der Bibliothek Warburg, 1923/1924* (Leipzig, 1926), pp. 215–26.

18. See, *Hortus sanitatis germanice* (Mainz, 1485), chaps. 172 *Ebur* and 371 *Spodium* (both of which have the identical woodcut). For other passages pertaining to the elephant, see Paul Meyer, "Le Bestiaire de Gervaise," *Romania* 1 (1872): 431, under "elephanz"; Carl Appel, "Aiso son las naturas," p. 204, under *"Del orifan."*

19. *Theobaldi Physiologus*, ed. Eden, p. 31. For a more elaborate moralization, see Gustav Heider, "Physiologus. Nach einer Handschrift des XI. Jahrhunderts," *Archiv für Kunde österreichischer Geschichtsquellen* 5 (1850): 574.

20. For the onager, see Francis J. Carmody, *Physiologus latinus versio Y* (Berkeley, 1941), p. 121; Verner Dahlerup, "Physiologus i to Islandske Bearbejdelser," *Aarbögger for Nordisk Oldkyndighed og Historie*, ser. 2, vol. 4 (1889), p. 266; Gustav Heider, "Physiologus," p. 561; Philippe de Thaun, *Bestiaire*, ed. T. Wright (London, 1841), p. 107. For the panther, see Carmody, *Physiologus*, p. 124; Dahlerup, "Physiologus," p. 271; Heider, "Physiologus," p. 554; Paul Meyer, "Le Bestiaire de Gervaise," *Romania* 1 (1872): 428; "The Bestiary [British Museum MS Arundel 292]," in *Selections from Early Middle English*, ed. Joseph Hall, vol. 1 (Oxford, 1920), p. 195.

21. For examples of sermon-exempla, some of which may have an observational basis, others of which were drawn from bestiaries and herbals, see G. R. Owst, *Literature and Pulpit in Medieval England*, 2d ed. (Oxford, 1961), pp. 149–209. See also Adolf Tobler, "Lateinische Beispielsammlung mit Bildern," *Zeitschrift für romanische Philologie* 12 (1888): 57–88, and Lynn Thorndike, "The Properties of Things of Nature Adapted to Sermons," *Medievalia et Humanistica* 12 (1958): 78–83.

22. An explicit statement that humans are transformed does not appear in the bestiaries, but such a belief underlies the statement that a man will become dumb if seen by a wolf; see Hugh of St. Victor, *De bestiis et aliis rebus*, 2. 20, in Migne, *Patologia Latina*, vol. 177 (Paris, 1879), col. 67. The same belief is repeated by Thomas of Cantimpré, who adds an explanation based on the humors; see his *Liber de natura rerum*, p. 143.

23. A typical example, insofar as the material combines Greco-Roman, folk-pagan, and Christian elements, is represented by *Lacnunga*, in *Anglo-Saxon Magic and Medicine*, ed. J. H. G. Grattan and Charles Singer (Oxford, 1952), pp. 96–205. See also Marie Brie, "Der germanische, insbesondere der englische Zauberspruch," *Mitteilungen der schlesischen Gesellschaft für Volkskunde* 8 (1906): 1–36; Heinrich Schneegans, "Sizilianische Gebete, Beschwörungen und Rezepte in griechischer Umschrift," *Zeitschrift für romanische Philologie* 32 (1908): 571–94; and Willy Braekman, *Middelnederlandse Bezweringsformulieren en Toverdranken* (Ghent, 1964).

24. See Godfrid Storms, *Anglo-Saxon Magic* (The Hague, 1948), for an edition of A-S charms and their analysis, and especially pp. 85–86, 154–58 for the incorporation of animal substances into medicines and amulets.

25. See H. Werth, "Altfranzösische Jagdlehrbücher nebst Handschriftenbibliographie der abendländischen Jagdliteratur überhaupt," *Zeitschrift für romanische Philologie* 12 (1888): 146–91, 381–415; 13 (1889): 1–34; Kurt Lindner, *Die Jagd im frühen Mittelalter* (Berlin, 1937); and Gunnar Tilander, *Cynegetica*, 8 vols. (Lund and Uppsala, 1953–61).

26. Marcelle Thiébaux, "The Mediaeval Chase," *Speculum* 42 (1967): 260–74.

27. For animals, see F. Klingender, *Animals in Art and Thought to the End of the Middle Ages* (Cambridge, Mass., 1971). For plants, see Lottlisa Behling, *Die Pflanzenwelt der mittelalterlichen Kathedralen* (Cologne, 1964); and *Die Pflanze in der mittelalterlichen Tafelmalerei*, 2d ed. (Cologne, 1967).

28. In addition to the references to the many studies by G. C. Druce cited in subsequent notes, see his "The Medieval Bestiaries: Their Influence on Ecclestiastical Art," *Journal of the British Archaeological Association*, n.s., 25 (1919): 41–82; 26 (1920): 35–79.

29. Much useful information, including extracts and bibliography, is found in Léon Moulé, "Glossaire vétérinaire médiéval," *Janus* 18 (1913): 265–72, 363–79, 439–53, 507–35. For recent studies in the history of medieval veterinary medicine, see Gerhard Eis, *Mittelalterliche Fachliteratur*, 2d ed. (Stuttgart, 1967), pp. 29–34.

30. Fridericus II, *De arte venandi cum avibus*, ed. Carl Willemsen (Leipzig, 1942); also, translation by Willemsen, *Über die Kunst mit Vögeln zu jagen* (Frankfurt, 1964). English version by C. A. Wood and F. M. Fyfe, trans., *The Art of Falconry* (Stanford, 1943). (A facsimile reprint of the Vatican MS. Pal. lat. 1071 was announced in 1969 but I have not seen it.) In general, see Gunnar Tilander, "Etude sur les traductions en vieux français du traité de fauconnerie de l'empereur Fréderic II," *Zeitschrift für romanische Philologie* 46 (1926): 211–90.

31. N. F. Blake, ed., *The Phoenix* (Manchester, 1964).

32. See Pauline Aiken, "Vincent de Beauvais and Dame Pertelote's Knowledge of Medicine," *Speculum* 10 (1935): 281–87; and Patrick Gallacher, "Food, Laxatives, and Catharsis in Chaucer's Nun's Priest's Tale," *Speculum* 51 (1976): 49–68.

33. Paul Meyer, "Traités du fauconnerie," *Romania* 15 (1886): 277–83; Hartmut Kleineidam, "Li volucraires," *Zeitschrift für romanische Philologie* 86 (1970): 1–21; C. H. Haskins, "Some early treatises on falconry," *Romanic Review* 13 (1922): 18–27.

34. *The Art of Falconry, being the De Arte Venandi cum Avibus of Frederick II of Hohenstaufen*, ed. and trans. Casey A. Wood and F. Marjorie Fyfe (Stanford, 1943), p. 99.

35. For representative passages, see F. J. Carmody, "Physiologus latinus versio Y," *University of California Publications in Classical Philology* vol. 12, pt. 7 (1941): 131; Meyer, "Traités de fauconnerie," p. 440; Joseph Hall, ed., *Selections from Early Middle English* (Oxford, 1920),

1: 194 (an edition of a bestiary from British Museum, MS Arundel 292, [13th c.]).

36. See E. G. Stanley, ed., *The Owl and the Nightingale* (Edinburgh, 1960); and the Old Irish *Fiachairecht andso sis*, ed. R. Best, in *Eriú*, 8 (1916): 120–26. On the caladrius, see G. C. Druce, "The Caladrius and its Legend," *Archaeological Journal* 69 (1912): 381–416.

37. For domestic fowl, see W. H. Riddell, "The Domestic Goose," *Antiquity* 17 (1943): 148–55; and Christian Hünemörder, *Phasianus: Studien zur Kulturgeschichte des Fasans* (Bonn, 1970), published dissertation. For the ways in which reports based upon observation of the habits of common birds enter into popular literature, see James Hardy, "Popular History of the Cuckoo," *Folk-Lore Record* 2 (1878): 47–92.

38. See especially D. S. Brewer, ed., *The Parlement of Foulys* (Edinburgh, 1962). For the rich fable literature, see Léopold Hervieux, *Les Fabulistes Latins depuis le siècle d'Auguste jusqu'à la fin du moyen âge*, 5 vols. (Paris, 1884–99); for example, compare Avianus' *Fabula* 15, "De Grue et Pavone" (3: 272) with the later, reworked fable, "Quomodo Grus et Pavo litigabant" (3: 335). Finally, for the use of bird-imagery to illustrate human behavior, see J. E. Wülfing, "Das Bild und die bildliche Verneinung im Laud-Troy-Book," *Anglia* 27 (1904): 535–80 ("Die Vögel," pp. 568–73).

39. Gerhard Hoffmeister, "Fischer- und Tauchertexte vom Bodensee," in *Fachliteratur des Mittelalters. Festschrift für Gerhard Eis*, ed. Gundolf Keil et al. (Stuttgart, 1968), pp. 261–75 (edition of text, plus bibliography). In general, see Franz Joseph Mone, "Ueber die Flussfischerei und den Vogelfang vom 14.-16. Jahrhundert," *Zeitschrift für die Geschichte des Oberrheins* 4 (1853): 67–97; and Christian Hünemörder, "Die Geschichte der Fischbücher von Aristoteles bis zum Ende des 17. Jahrhunderts," *Deutsches Schiffahrtsarchiv* 1 (1975): 185–200.

40. For the alimentary uses of animal and plant substances, see Anton Birlinger, "Ein alemannisches Büchlein von guter Speise," *Sitzungsberichte der bayerische Akademie der Wissenschaften*, 1865, 2: 171–206; Luis Faraudo de Saint Germain, "El Libre de Sent Soví," *Boletín de la Academia Buenas Letras de Barcelona* 24 (1951–52): 1–77; Anita Feyl, "Das Kochbuch des Eberhard von Landshut," *Ostbairische Grenzmarken* 5 (1961): 352–66; and Jerry Stannard, "The Botanical-Medical Background of Baptista Fiera's Coena de herbarum virtutibus," in *Civilità dell'Umanesimo: Atti del VI, VII, VIII Congresso del Centro di Studi Umanistici "A. Poliziano"* (Florence, 1972), pp. 327–44.

41. For descriptions of many of the more common species, see Juliana Berners, *A Treatyse of Fysshynge wyth an Angle. Originally printed by Wynkyn de Worde, 1496*, ed. "Piscator" (Edinburgh, 1885). For fabulous fishes and "marine monsters," see Thomas of Cantimpré, *Liber de natura rerum*, ed. Boese (see n. 14), pp. 231–55; Hugh of St. Victor, *De bestiis*, ed. Migne (see n. 22), 3. 55 (cols. 105–11); and C. Speroni, "More on the Sea-Monsters," *Italica* 35 (1958): 21–24.

42. Cornelia Coulter, "The Great Fish in Ancient and Medieval Story," *Transactions of the American Philological Society* 57 (1926). For details, see V.-H. Debidour, *Le Bestiaire sculpté du moyen âge en France* (Paris, 1961), p. 386, under "baleine."

43. T. H. White, *The Bestiary, A Book of Beasts being a Translation from a Latin Bestiary of the Twelfth Century* (New York, 1960), pp. 199, 201. For a detailed study, with specimen texts, see G. C. Druce, "The Legend of the Serra or Saw-Fish," *Proceedings of the Society of Antiquaries of London*, ser. 2, 31 (1919): 20–35.

44. Gerhard Eis, "Austerschalen," in *Studien zur altdeutschen Fachprosa* (Heidelberg, 1951), pp. 11–28.

45. See Mia Gerhardt, "Knowledge in Decline: Ancient and Medieval Information on Ink-fishes and Their Habits," *Vivarium* 4 (1966): 144–75. Snails and slugs were frequently used in medieval medicine and pharmacy; for example, both Marcellus of Bordeaux and Benedictus Crispus recommend their collection and use; see Jerry Stannard, "Marcellus of Bordeaux and the Beginnings of Medieval Materia Medica," *Pharmacy in History* 15 (1973): 47–53; and Stannard, "Benedictus Crispus, an Eighth Century Medical Poet," *Journal of the History of Medicine* 21 (1966): 24–46. For a typical medieval account, see Ursula Schmitz, ed., *Hans Minners Thesaurus Medicaminum* (Würzburg, 1974), p. 139, under "limacia."

46. Out of an abundant literature, see H. Pogatscher, "Von Schlangenhörnen und Schlangenzungen," *Römische Quartalschrift für christliche Altertumskunde* 12 (1898): 162–215; Oswald Zingerle, "Das Paradiesgarten der altdeutschen Genesis," *Sitzungsberichte der Akademie der Wissenschaften in Wien*, Phil.-hist. Classe, 112 (1886): 785–805; Gustav Münzel, "Das Frankfurter Paradiesgärtlein," *Das Münster* 9 (1956): 14–22. For the reception and acceptance of the Adam-Eve-serpent motif into folk belief, see Barbara Renz, "Schlange und Baum als Sexualsymbole in der Völkerkunde," *Archiv für Sexualforschung* 1 (1915–16): 341–44.

47. For the early history of *theriaca* and its preparation, see Jerry Stannard and Peter Dilg, "Observations on *De theriacis et mithridateis commentariolus*," in *Joachim Camerarius und seine Zeit*, ed. F. Baron (Munich, 1977), pp. 152–86.

48. See Florence McCulloch, "The Metamorphoses of the Asp," *Studies in Philology* 56 (1959): 7–13; G. C. Druce, "The Symbolism of the Crocodile in the Middle Ages," *Archaeological Journal* 66 (1909): 311–38.

49. See Anna Bisi, *Il grifone* (Rome, 1965); F. Wield, *Drachen in Beowulf und andere Drachen* (Vienna, 1962); G. C. Druce, "The Amphisbaena and Its Connexions in Ecclestiastical Art and Architecture," *Archaeological Journal* 67 (1910): 285–317.

50. See J. Cortelyou, *Die altenglischen Namen der Insekten, Spinnen- und Krustentiere* (Heidelberg, 1906).

51. S. Eitrem, "Der Skorpion in Mythologie und Religionsgeschichte," *Symbolae Osloenses* 7 (1928): 53–81.

52. Richard Riegler, "Spinnenmythus und Spinnenaberglaube," *Schweizerisches Archiv für Volkskunde* 26 (1925): 55–69, 123–42. For the collection of spider webs, see Jerry Stannard, "Hans von Gersdorff and some Anonymous Strassburg Apothecaries," *Pharmacy in History* 13 (1971): 55–65.

53. Thomas of Cantimpré's *Bonum universale de apibus* (*Liber apum*) or *Bienboeck* (see Klebs, *Incunabula* p. 320), the first monograph on

bees, was not accessible to me. Much relevant information is found in
L. Armbruster, "Die Biene im Wissen um 1200," *Archiv für Bienenkunde*
22 (1941): 49–144. For a sample of moralizing based on the bees' habits,
see *Moralia Ricardi Heremite De natura apis*, in K. Sisam, *Fourteenth-
Century Verse and Prose* (Oxford, 1955), pp. 41–42.

54. Mia Gerhardt, "The Ant-Lion," *Vivarium* 3 (1965): 1–23; G. C.
Druce, "An account of the Μυρμηκολέων or Ant-Lion," *Antiquaries Journal* 3 (1923): 347–64.

55. Jean Bayet, "Le symbolisme du cerf et du centaure à la porte rouge
de Notre-Dame de Paris," *Revue archéologique* 44 (1954): 21–68.

56. See Edmond Faral, "La queue de poisson des sirènes," *Romania*
74 (1953): 433–506; F. Sbordone (see n. 4); G. C. Druce, "Some Ab-
normal and Composite Human Forms in English Church Architecture,"
Archaeological Journal 72 (1915): 135–86.

57. Richard Newald, "Die Teuffelliteratur und die Antike," *Bayerische
Blätter für das Gymnasialschulwesen* 63 (1927): 340–47.

58. On the *leucocrota* and *manticora*, see McCulloch, *Bestiaries* (see
n. 11), pp. 136, 142; Brunetto Latini, *Li livres dou tresor* (see n. 10), pp.
167, 168; and Thomas of Cantimpré, *Liber de natura rerum*, pp. 146,
150.

59. On the eale or yale, see G. C. Druce, "Notes on the History of the
Heraldic Jall or Yale," *Archaeological Journal* 68 (1911): 173–99. For
the abundant literature on the unicorn, see especially Odell Shepard, *The
Lore of the Unicorn* (London, 1930).

60. In addition to the books on plants and trees in the encyclopedias
mentioned above, n. 14, see Gösta Brodin, *Agnus Castus* (Uppsala,
1950); Otto Bessler, "Das deutsche Hortus-Manuskript des Henricus
Breyell," *Nova Acta Leopoldina* 15 (1952): 190–266; Heinrich Ebel,
Der Herbarius communis des Hermannus de Sancto Portu (Würzburg,
1940); Henrik Harpestraeng, *Liber Herbarum*, ed. Poul Hauberg (Copenhagen, 1936); Erhard Landgraf, "Ein frühmittelalterlicher Botanicus,"
Kyklos 1 (1928): 114–46; Macer Floridus, *De viribus herbarum*, ed.
Ludwig Choulant (Leipzig, 1832); *The Herbal of Rufinus*, ed. Lynn
Thorndike (Chicago, 1949); Heinrich Werneck, "Kräuterbuch des Jo-
hannes Hartlieb," *Ostbairische Grenzmarken* 2 (1958): 71–124.

61. Jerry Stannard, "Mittelalterliche Botanik: Themen und Quellen"
(a lecture delivered at the Frankfurter Abend, 2 May 1974), summarized
in Wolfgang-Hagen Hein, "Themen und Quellen der mittelalterlichen
Botanik," *Deutsche Apotheker-Zeitung* 114 (1974): 735 ff.

62. Jerry Stannard, "Medieval Herbals and their Development," *Clio
Medica* 9 (1974): 23–33.

63. Ernest Wickersheimer, "Nouveaux textes médiévaux sur le temps
de cueillette des simples," *Archives internationales d'histoire des sciences*
29 (1950): 342–55.

64. *Hortus sanitatis germanice* (Mainz, 1485), chap. 420: *Ungula
caballina - Brantlattich.*

65. Jerry Stannard, "Dioscorides and Renaissance Materia Medica,"
Analecta Medico-Historica 1 (1966): 1–21.

66. Willem Daems, ed., *Boec van medicinen in Dietsche* (Leiden,
1967); Margaret S. Ogden, ed. *The Liber de diversis medicinis* (London,

1938); Henry E. Sigerist, "A 15th Century Text on the Medicinal Uses of Madder," *Bulletin of the History of Medicine* 4 (1936): 57–60; Jerry Stannard, "Squill in Ancient and Medieval Materia Medica, with Special Reference to Its Employment for Dropsy," *Bulletin of the New York Academy of Medicine* 50 (1974): 684–713.

67. Antonius Imbesi, *Index plantarum quae in omnium populorum pharmacopoeis sunt adhuc receptae* (Messina, 1964).

68. See Theodor Husemann, "Die Schlafschwämme und andere Methoden der allgemeinen und örtlichen Anästhesie im Mittelalter," *Deutsche Zeitschrift für Chirurgie* 42 (1896): 517–96; 54 (1900): 503–50; W. Daems, "Spongia somnifera," *Beiträge zur Geschichte der Pharmazie* 22, no. 4 (1970): 25–26; and Gustav Klein, "Historisches zum Gebrauche des Bilsenkrautextraktes als Narkoticum," *Münchener medizinische Wochenschrift* 22 (28 May 1907): 1098–99.

69. See Georg Sticker, "Entwicklungsgeschichte der spezifischen Therapie," *Janus* 33 (1929): 131–90, 213–34, 245–70.

70. Ernest Wickersheimer, "Faits cliniques observés à Strasbourg et à Haslach en 1362 et suivis de formules de remèdes," *Bulletin de la Société française d'histoire de la médecine* 33 (1939): 69–92.

71. Peter Dilg, *Das Botanologicon des Euricus Cordus* (Marburg, 1969); Konrad Böhner, "Die Herbationen des Nürnberger Aerztekollegiums," *Süddeutsche Apotheker-Zeitung* 66 (1926): 16–19, 25–27. For records of early herbaria, see Jerry Stannard, "Botanical Nomenclature in Gersdorffs Feldtbüch der Wundartzney," in *Science, Medicine and Society in the Renaissance: Essays to Honor Walter Pagel*, ed. Allen G. Debus (New York, 1972), 1: 87–103.

72. No good study exists for the medieval knowledge of cryptogams, but useful historical material will be found in A. H. R. Buller, "The Fungus Lore of the Greeks and Romans," *Transactions of the British Mycological Society* 5 (1914–16): 21–66; Franz Ferk, "Volkstümliches aus dem Reiche der Schwämme," *Mitteilungen des Naturwissenschaftlicher Verein für Steiermark* 47 (1910): 18–52. The first scientific study of cryptogams was that of Clusius (1525–1609); see E. Rose, "Le petit traité des champignons comestibles et pernicieux de la Hongrie décrits au XVIe siècle par Charles de l'Escluse d'Arras," *Bulletin de la Société Mycologique de France* 15 (1899): 280–304; 16 (1900): 26–53; and Jerry Stannard, "Classici und rustici in Clusius' Stirpium pannonicarum historia," in *Festschrift anlässlich der 400 jährigen Wiederkehr der Wissenschaftlichen Tätigkeit von Carolus Clusius* (Eisenstadt, 1973), pp. 253–69.

73. See Reinhold Röhricht and Heinrich Meisner, "Ein niederrheinscher Bericht über den Orient," *Zeitschrift für deutsche Philologie* 19 (1886): 1–86, especially pp. 69 ff., which describe *seriatim* stones, animals, birds, and plants purportedly seen between ca. 1338 and 1348.

74. Francesco Balducci Pegolotti, *La pratica della mercatura*, ed. Allan Evans (Cambridge, Mass., 1936), especially pp. 293–97 ("Nomi di Spezierie").

75. See F. A. Flückiger, "Die Frankfurter Liste," *Archiv der Pharmazie* 201 (1872): 433–64, 508–26; Flückiger, "Das Nördlinger Register," *Archiv der Pharmazie* 211 (1877): 97–115; and Wolfgang-Hagen Hein,

"Die Preisverzeichnisse des Grazer Codex 311," *Pharmazeutische Zeitung* 118 (1973): 1146–48, 1510–12; 119 (1974): 500–2.

76. Jerry Stannard, "Vegetable Gums and Resins in Medieval Recipe Literature," forthcoming in the proceedings of the International Congress for the History of Pharmacy, Bremen 1975, edited by Wolfgang-Hagen Hein.

77. On fabulous plants, see Henry Lee, *The Vegetable Lamb of Tartary* (London, 1887); Edward Heron-Allen, *Barnacles in Nature and Myth* (London, 1927); Moshé Lazar, "La légende de l'arbre de paradis ou bois de la croix," *Zeitschrift für romanische Philologie* 76 (1960): 34–63; Jerry Stannard, "The Plant Called Moly," *Osiris* 14 (1962): 254–307; Philipp Strauch, "Palma contemplationis," *Beiträge zur Geschichte der deutschen Sprache und Literatur* 48 (1924): 335–75.

78. See M. Laurent, "Le phénix, les serpents et les aromates dans une miniature du XIIe siècle," *Antiquité classique* 4 (1935): 375–401; Julius von Schlosser, "Tacuinum sanitatis in medicina. Ein veronesisches Bilderbuch und die höfische Kunst des XIV. Jahrhunderts," *Jahrbuch der Kunsthistorischen Sammlungen in Wien* 16 (1895): 144–214; J. Corblet, "Etude iconographique sur l'arbre de Jessé," *Revue de l'art chrétien* 4 (1860): 49–61, 113–25, 169–81. For illustrations of fabulous plants in herbals, see T. G. Leporace et al., *Un inedito erbario farmaceutico medioevale* (Florence, 1952); and Elena Toesca, "Un erbolario del '300," *La Bibliofilia* 39 (1937): 341–53.

79. On Brendan's voyages and the reports of his miraculous natural history, see *Navigatio Sancti Brendani Abbatis*, ed. Carl Selmer (Notre Dame, Ind., 1959).

80. Thirty such practices are condemned in the anonymous *Indiculus superstitionum et paganiarum* (*Monumenta Germaniae Historica, Legum sectio II; Capitularia Regum Francorum*, ed. Alfredus Boretius, vol. 1 [Hanover, 1883], pp. 222–23). For other examples, see Peter Browe, "Die Eucharistie als Zaubermittel im Mittelalter," *Archiv für Kulturgeschichte* 20 (1930): 134–54; Joseph Klapper, "Das Gebet im Zauberglauben des Mittelalters," *Mitteilungen der schlesischen Gesellschaft für Volkskunde* 9 (1907): 5–41; Jan Gessler, "Een gebed en een recept uit de 10e eeuw," *Oostvlaamsche Zanten* 12 (1937): 1–12.

81. These plants are studied by Jerry Stannard, "Magiferous Plants and Magic in Medieval Medical Botany," *Maryland Historian*, in press.

82. W. Crecelius, "Recepte für Bereitung von Kräuterbier," *Jahrbuch des Vereins für Niederdeutsche Sprachforschung* 4 (1878): 89–90; Willem Daems, "Der Misteltraktat des Wiener Kod. 3811," *Sudhoffs Archiv* 49 (1965): 90–93; Adalbert Jeitteles, "Färbemittel und andere Recepte," *Germania* 29 (1884): 338–40; Joachim Telle, "Altdeutsche Eichentraktate," *Centaurus* 13 (1968): 37–61; Josef Werlin, "Weinrezepte aus einer südtiroler Sammelhandschrift," *Archiv für Kulturgeschichte* 45 (1963): 243–52; J. Werlin, "Drei deutsche Wacholder-Traktate des Spätmittelalters," *Sudhoffs Archiv* 49 (1965): 250–54; Hans Wiswe, "Mittelalterliche Rezepte zur Färberei," *Jahrbuch des Vereins für niederdeutsche Sprachforschung* 81 (1958): 49–58.

83. Of an enormous literature, see Alicia Amherst, "A Fifteenth-Century Treatise on Gardening by Mayster Ion Gardener," *Archaeologia* 54

(1894): 157–72; Roswitha Ankenbrand, *Das Pelzbuch des Gottfried von Franken* (Heidelberg, 1970), published dissertation; [Charlemagne,] *Capitulare de villis*, in *Monumenta Germaniae Historica, Legum sectio II: Capitularia Regum Francorum*, ed. Boretius, 1: 82–91; *The Book of Husbandry by Master Fitzherbert*, ed. W. Skeat (London, 1882); Dorothea Oschinsky, ed., *Walter of Henley and other Treatises* . . . (Oxford, 1971); Palladius, *On Husbondrie*, ed. B. Lodge (London, 1883); Peter of Crescenzi, *Opera di agricoltura* (Venice, 1553); Walafridus Strabus, *Hortulus*, trans. Raef Payne, commentary by Wilfrid Blunt (Pittsburgh, 1966).

84. Published lapidaries include Marbod of Rennes, *Lapidarius*, in Migne, *Patrologia Latina*, vol. 171 (Paris, 1854), cols. 1737–70; Giulio Bertoni, "Il lapidario francese estense," *Zeitschrift für romanische Philologie* 32 (1908): 686–97; Paul Meyer, "Les plus anciens lapidaires français," *Romania* 38 (1909): 44–70, 254–85, 481–552; Joan Evans and Mary Serjeantson, *English Mediaeval Lapidaries* (London, 1960); Friedrich Wilhelm, *Denkmäler deutscher Prosa des 11. und 12. Jahrhunderts* (Munich, 1914) ("Das Prüler Steinbuch de XII. lapidibus," pp. 37–39). For useful discussions of lapidaries, see John Riddle, "Lithotherapy in the Middle Ages," *Pharmacy in History* 12 (1970): 39–50; R. Besser, "Über Remy Belleaus Steingedicht," *Zeitschrift für neufranzösische Sprache und Literatur*, 8 (1886): 185–250.

85. For example, Valentin Rose, "Aristoteles De lapidibus und Arnoldus Saxo," *Zeitschrift für deutsches Alterthum* 18 (1875): 321–455.

86. Although outdated, the editor's introduction to Guillaume's *Bestiaire Divin* is one of the few attempts to deal with bestiaries, herbals, and lapidaries; see C. Hippeau, *Le Bestiaire Divin*, pp. 3–72. See also F. H. Garrison, "Herbals and Bestiaries," *Bulletin of the New York Academy of Medicine* 7 (1931): 891–904; and Christian Hünemörder, "Botanisches und Zoologisches bei Alanus ab Insulis," in *Festgabe für Kurt Lindner* (Berlin, 1971), pp. 125–31.

87. Albertus Magnus, *Book of Minerals*, trans. Dorothy Wyckoff (Oxford, 1967).

88. Willy Braekman, *Medische en technische Middelnederlandse Recepten* (Ghent, 1975); H. G. T. Frencken, *t'Bouck va Wondre, 1513* (Roermond, 1934); Piero Giacosa, "Un ricettario del secolo XI esistente nell'Archivio Capitolare d'Ivrea," *Atti della Reale Accademia di Torino*, ser. 2, 37 (1886): 643–63; Hjalmar Hedfors, *Compositiones ad tingenda musiva* (Uppsala, 1932); D. V. Thompson, "The De clarea of the So-Called Anonymous Bernensis," *Technical Studies in the Field of Fine Arts* 1 (1932): 8–19, 69–81; Hans Vermeer, "Technischnaturwissenschaftliche Rezepte," *Sudhoffs Archiv* 45 (1961): 110–26.

14

James A.
Weisheipl,
O.P.

The Nature, Scope, and Classification of the Sciences

The modern word *science* has come to mean different things to different people, particularly in modern times. In order to arrive at a precise meaning it is important to understand how it was used in Greek antiquity and in the Middle Ages before the Scientific Revolution of the seventeenth century. A brief historical survey will show how the ancients used the word *science*, how scientific knowledge differed from other kinds of knowledge, and how various kinds of science can be classified and studied. Provisionally, *science* can be taken to mean a given area of study, worthy of human investigation and knowledge. It is clear from the very start that there are many areas of human learning, all deserving of investigation for the betterment of mankind, as well as for the betterment of the individual. In order to understand the history of science we must see how the word was used in Greek antiquity and how it came to be understood in the Latin West.

Science in Greek Antiquity

By the beginning of the classical period, Greek education consisted first in a study of the "liberal arts," meaning by that music, gymnastics, poetry, grammar, rhetoric, and dialectics (the art of reasoning and disputing).[1] These were thought necessary for the well-rounded Greek mind in civilized society. They were called "arts" (*tekne*), as distinct from philosophy, because in each case something was produced or constructed at least mentally. They were taught as productive skills, or practical techniques, necessary for the cultivated life of the free man, the citizen in a free society. In early Greek thought "philosophy" (which literally

461

means "the love of wisdom") was taken to mean a kind of true, objective, unsophisticated knowledge, vastly different from the liberal arts. This wisdom (*sophia*) was the attempt of the human mind to explain in a rational way some of the natural processes in nature, the peculiar properties of number, or the moral values of conduct. These were recognizably different areas of human knowledge worthy of the learned man. Such speculative knowledge, whether it be of natural processes (physics), numbers (mathematics), or moral values (ethics) was called "scientific" (*episteme*), or objectively demonstrable knowledge, to describe a human study essentially different from the arts, whether those arts be "liberal" or "technical." The implication that philosophical (or "scientific") studies should be pursued only after the liberal arts had been mastered was endemic to the Greek concept of scientific knowledge.[2] Thus there were at least three important areas of objective truth worthy of pursuit by the mature man after he had acquired the arts, namely, physics, mathematics, and moral values. But these parts of philosophy or "scientific" knowledge could not be studied without the necessary art of logic, which was a liberal art, the art of right reasoning.

The earliest Greek philosophers before Socrates were intensely interested in discovering the ultimate source or root of all physical things observable in nature.[3] For Thales of Miletos, one of the seven "wise men of Greece," the ultimate principle of all natural beings was some kind of moisture or water. For Anaximander it was some "limitlessness" distinct from the obvious four elements: earth, water, fire and air. For Anaximines, it was some kind of life-giving air or breath (spirit); and for Xenophanes, it was the tension between earth and water which produced a kind of Super One, from whom all things come. All of these early philosophers tried to find a physical explanation in some given thing that could be identified as one. But it was Heraclitus, an irritable Ionian from the city of Ephesus, who provoked the response that was to dominate the formation of "scientific" explanations in Greek thought. For him, "all things are in flux," as fire that must ever be driven on by its own contraries. "All things come into being by conflict of opposites, and the sum of all things flows like a stream." In other words, there is no stability anywhere in the universe, and all true wisdom lies in the recognition of constant change. Heraclitus had no sympathy for the flourishing school of Pythagoreans, with their abstract and immobile numbers and figures; nor had he sympathy for the poet-educator Hesiod.

Heraclitus's absolute denial of stability in nature seems to have brought an immediate response from Parmenides of Elea about the year 500 B.C. In a relatively long poem, preserved by Simplicius and later writers, Parmenides distinguished between two types of human knowledge: the way of popular opinion and the way of objective and absolute truth. The way of popular opinion followed by most people is no more than the deception offered by the senses. Among these deceptions are the plurality of things in the universe and the continual mutability of things in nature. The truth of the matter, according to Parmenides, was that "all is one and absolutely motionless." That is to say, he denied the multiplicity and mutability apparent to the senses. The basic principle of Parmenides' philosophy was that only "being" can exist; whatever is "not being" cannot exist.[4] By "being" Parmenides clearly meant a physically existing body; every sort of vacuity would be "nothing" and, hence, incapable of existence.[5] From this principle it logically follows that change and plurality are impossible. In every change something "new" supposedly comes into existence. Whence comes this new thing? It cannot come from nothing, for nothing can come from nothing (*ex nihilo, nihil fit*); it cannot come from being, for then it would already have been (*quod est non fit*). Similarly, all plurality implies separation and distinction, so that one is somehow different from the other at least in place; but this would require the existence of a "nothing" to distinguish them, and a "nothingness" cannot exist. Therefore, the way of truth and reason requires us to say that all is one and immutable, even though the senses tell us otherwise. If all is one and changeless, then the search pursued by the ancient physicists is futile. There is nothing to be accounted for. The logical dilemma of Parmenides revolutionized Greek thought in the sense that henceforth some rational account had to be given to explain the possibility of both change and plurality.

The "most systematic and consistent theory," Aristotle tells us, was advanced by the atomists Leucippus and Democritus.[6] In order to justify the obvious multiplicity and mobility in the world, Leucippus of Miletos (fl. 430 B.C.) made one change in the Eleatic principle of Parmenides and his followers: he insisted on the reality of "non-being," or "nothingness," which could be conceived as the void. Leucippus maintained that "*what is* is no more real than that *which is not*," and that "both are alike causes of the things that come into being."[7] In other words, by postulating the real existence of Being and Nothingness, one could account for plurality and movement in the world. In the last analysis, however, the Greek atomists could not explain how atoms themselves could be generated or corrupted, that is, changed; these atoms had to be immutable, as Parmenides had argued. Apparent generation and corruption of "substantial unities" could be explained by the rearrangement of atomic particles, and local movement of those particles in the void could justify sense experience that at least local motion is possible, even though substantial change in the atoms themselves, that is, their generation and corruption, is untenable. Consequently, the world of sense experience, the "way of opinion," tells us that there are substantial entities which are generated and corrupted. But reason, "the way of truth," knows that there are only immutable atoms in a real void.

Again the bifurcation of knowledge remained: the way of opinion and popular belief on the one hand, and scientific knowledge of the truth (the void and the local motion of atoms) on the other. The assumption is that evidence of the senses does not deserve the title of true science, but only of "opinion," for it is a world of ever-changing shadows, as Heraclitus and Plato noted.

Plato's philosophy of nature owes much to Heraclitus, Pythagoras, and Parmenides. Like Heraclitus, Plato viewed the sensible world as ever-changing and not susceptible to scientific inquiry. Like Pythagoras, Plato postulated the separate existence (that is, separate from material things) of geometrical figures and numbers, which never change and can be studied. And like Parmenides, Plato held that only the immutable is real and eternal; these immutable and eternal Forms or Ideas exist apart from the world of change, and true "wisdom" consists in contemplating them until one logically reaches the highest Idea of the One and the Good, namely the supreme God.

The grades of knowledge, for Plato, correspond directly to the grades of being: the more abstract the knowledge, the more immaterial and perfect the being. This view is partially represented by the well-known allegory of the line (*Republic* 6. 509D–511E). The gradation of being and speculative knowledge can be represented as is shown in chart 1.

In other words, Plato considered the study of nature to be no more than "a likely story" to explain how things might have come about, as described in his *Timaeus*. Since the world of sense experience is ever-changing, as Heraclitus had said, the study of nature does not merit the name of "science." Higher than the world of sense experience and physics is the domain of mathematics, including geometry, arithmetic, astronomy, and harmonics. These were the areas studied by the Pythagoreans. Since the world of mathematics is made up of immutable, eternal truths, it deserves the name and character of "science." For Plato, the world of mathematical reality consists of subsistent numbers and figures, existing apart from the world of sen-

Visible world — **Opinion (*doxa*)**

1. The shadows of physical bodies, such as shadows cast by objects in light; to these correspond imaginative *conjectures* in interpreting the shadows as to what the real object might be.

2. The physical, sensible, and visible objects, which in reality are but "shadows" and reflections of a subsisting Idea beyond the physical, for example, animals, plants, and inanimate bodies. To this grade corresponds *belief*; this is the domain of PHYSICS, or the study of nature, to which all of his predecessors had been devoted.

Intelligible world — **Science (*episteme*)**

3. Intelligible objects to which the mind must turn as to the model of physical objects; but to study these intelligible Forms, the mind must utilize certain "hypotheses" in order to reach a true conclusion. Corresponding to this grade is discursive reason, which is the domain of MATHEMATICS, or disciplinary science (geometry, astronomy, arithmetic, and harmonics).

4. Intelligible objects apprehended without the aid of any sensible object, the mind passing only from one Idea to another Idea. Corresponding to this grade of true and pure being is pure contemplation, or understanding (*noesis*) and perfect "science" which is DIALECTICS (later called "first philosophy" or metaphysics); this perfect science uses hypotheses not as first principles, but as points of departure into a world which is above hypotheses in order to rise beyond them to the first, unique principle of the whole.

Chart 1

sation, intermediate between the world of pure "Ideas" and the world of material objects perceived by sensation. But even higher than the world of mathematicals is the domain of subsistent forms and ideas, separate from all matter. Among the ideas that are immaterial and subsistent in their own right are the ideas of pure man, fire, water, hair, mud, and dirt, existing apart from the physical, changing world.[8] The contemplation of these forms is "wisdom"; and the contemplation of the highest form, namely the Good-in-itself, is "theology." Plato, of course, made no distinction between dialectics, the highest wisdom, and moral philosophy, since for Socrates and Plato, knowledge of the transcendental good necessarily produces virtuous actions. Only after Aristotle was moral philosophy generally distinguished from the speculative sciences, because for Aristotle mere knowledge of virtue does not necessarily produce virtuous actions.

It is clear that Plato's tripartite division of speculative knowledge is a hierarchical one: physics at the bottom, mathematics (geometry, astronomy, arithmetic, harmonics) in the middle, and dialectics (which was later called first philosophy or metaphysics) at the top. The main point is that for Plato, mathematical being is more fully real than physical, sensible being; it is conducive to opening the doors of "dialectics," which is about all separated forms or ideas, the highest of which is the One and Good-in-itself (God). The crux of Platonic philosophy is that things exist in reality in the same way as they are conceived by the mind. Since mathematical being is conceived without sensible matter, so they must exist without sensible matter.

A problem arises, however, concerning the forms of material things, such as tables, humanity, animality, and the like; these are conceived in a universal manner by the mind without individual matter. But the very nature and definition of these forms involve "sensible" matter. For this reason Aristotle argued that they cannot subsist in themselves without matter; the nature of man involves flesh, blood, and bones—not *this* blood and *these* bones—but flesh, blood, and bones that require sensible matter even in their definition. It was precisely over the existence of separate forms for each species that Aristotle broke with Plato. The only "separated substances" that Aristotle would admit to exist were those demonstrated in natural philosophy as movers of the heavens.

For Aristotle, the tripartite division given by Plato does not correspond to a hierarchy of scientific knowledge or to a hierarchy of being. Instead of Plato's hierarchical grades of abstraction, Aristotle showed that mathematics alone is really "abstract"; the mathematical sciences are more removed from reality than physics, and, hence, must be judged by natural philosophy, not the other way around. For this reason, mathematics can be taught to young men as a preliminary to the sciences of being: physics, ethics, and metaphysics or "first philosophy." Moreover, whereas Plato had denied the scientific status of physics, Aristotle spent most of his effort rehabilitating the "science of nature," which he described in general terms in the *Physics*, and continuing his researches into all of its branches including biology and psychology. Further, Aristotle gave a new foundation to "first philosophy." Rejecting subsistent immaterial ideas, Aristotle established his metaphysics on the proved existence of "separated substances" (such as the celestial movers), which are the cause of motion and sensible being studied in physics. Finally, he distinguished the practical science of ethics or "moral philosophy" from the contemplation of the supreme Good attained in metaphysics. Thus, for the first time a difference is recognized between speculative sciences and practical sciences.

Aristotle clearly distinguished three types of academic discipline: the "arts," the practical sciences of moral philosophy, and the three branches of speculative philosophy (physics, mathematics, and metaphysics). However, in the order of training or acquiring learning, Aristotle would have youths trained first in the liberal arts (gymnastics, grammar, statuary, music, logic, rhetoric, and poetry) and then in the "disciplinary" sciences of mathematics (arithmetic and geometry) before beginning the extensive science of nature, to which could be added "the more physical parts of mathematics," namely, optics, harmonics, and astronomy. Only after a considerable part of natural science had been mastered, especially psychology, should one study the moral sciences, to play his role in the political life of the city and to prepare himself for the supreme science of metaphysics, the most important part of which is called theology. Thus, for Aristotle, the acquisition of learning and wisdom is a lifelong process, where true wisdom or metaphysics is not expected of a man until he is fifty years old and where a life of virtue is expected to be practiced in a social and political context.

With these qualifications in mind, we can construct a division of the sciences drawn from the incidental statements scattered throughout Aristotle's *Metaphysics* and *Nicomachean Ethics* (see chart 2).

There are two important points to be noted regarding Aristotle's classification of the speculative sciences: (1) despite its superior position in the schema, mathematics is not to be considered a higher or nobler science than physics, "the science of nature," but only as a preparation for natural science and metaphysics, even though it is truly "abstract"; (2) the sciences which Aristotle calls "the more physical parts of mathematics," namely, optics, astronomy, harmonics, and mechanics, occupy a unique position in the schema of the sciences, for they are *intermediate* between pure mathematics and physics and later became known as *scientiae mediae*, or intermediate sciences; these intermediate sciences use mathematical principles and formulas to study certain areas susceptible of mathematical consideration, such as celestial motions, radiant lines (optics), tonal ratios (harmonics), and mechanical problems of every kind, and every other kind of knowledge involving the application of mathematics to nature.

It was Aristotle who most clearly presented the precise nature of truly "scientific" knowledge. This he did in his many writings on logic, particularly his two books on demonstrative knowledge, the *Posterior Analytics*. The most perfect kind of demonstration is a proof through the proper, immediate, and commensurate *cause* of an

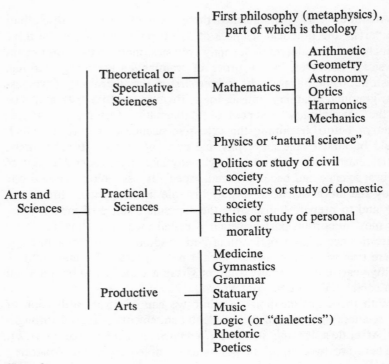

Arts and Sciences

- Theoretical or Speculative Sciences
  - First philosophy (metaphysics), part of which is theology
  - Mathematics
    - Arithmetic
    - Geometry
    - Astronomy
    - Optics
    - Harmonics
    - Mechanics
  - Physics or "natural science"
- Practical Sciences
  - Politics or study of civil society
  - Economics or study of domestic society
  - Ethics or study of personal morality
- Productive Arts
  - Medicine
  - Gymnastics
  - Grammar
  - Statuary
  - Music
  - Logic (or "dialectics")
  - Rhetoric
  - Poetics

Chart 2

effect or fact; this is called *propter quid* demonstration. It is a knowledge and proof of a fact through a true cause or reason for the fact. These are hard to discover. In the absence of such perfect demonstration, the scientist can still "demonstrate" the fact of a thing or proposition through a proof known as *quia*, either through some remote cause of an effect (*per causam remotam*) or through a proper effect (*a posteriori*). These are different kinds of logical demonstrations which beget true "scientific" or epistemic knowledge. Of course, no human science is made up solely of *propter quid* demonstrations. Such demonstrations are often hard to come by or even impossible, and a human science must often be content with *quia* demonstrations, statements of fact, conjectures, workable hypotheses, and even the considered opinions of other learned men. Thus, each "science" is a huge area of learning that consists of proofs or "demonstrations" of what is clearly and certainly known to be the case. But investigation of these areas often involves possible theories, hypotheses, opinions, clarification, and simple classification.

The Early Middle Ages

Soon after Aristotle, several important philosophical schools emerged quite independently of Plato and Aristotle. The most important of these for the Roman world was the Greek philosophy of the Stoics, who placed the highest goal in life in their stoic ethics. They divided philosophy into three branches or areas of study: (1) logic, (2) physics, which included whatever bits of metaphysics they would admit, and (3) ethics, as the highest "science" of human life. While Stoic philosophy played an important role in molding the Roman mind and way of life, it played an insignificant role in the development of medieval science, despite the fact that St. Augustine adopted the Stoic division of philosophy, thinking it to be Plato's.[9]

Rome inherited a long-standing tradition in educating youths in the arts, particularly in grammar and rhetoric. Grammar was taught not only at the elementary level, but also at an advanced level. Eminent grammarians flourished in Roman society; they laid the foundations of the Latin language. But more important than the grammarians were the rhetoricians, because rhetoric was considered indispensable for the cultivated Roman who wished to fulfill his role in the life of Roman society. Under the Empire, Roman education copied much that was Greek, but with some interesting differences. The study and practice of Roman Law, even at Byzantium in the East, was in the Latin language. Philosophy and medicine, on the other hand, never became integral parts of Roman education, not even at its highest levels; these had to be studied in the Greek language. Even before the reign of Augustus Caesar, boys of the Roman aristocracy learned the liberal arts in both Latin and Greek. Cicero was an outstanding example of the educated Roman who knew Greek philosophy and played a significant part in the conduct of Roman society.

The earliest Latin classification and exposition of the liberal arts seems to have been incorporated in the now lost work of Terence Varro (116–27 B.C.) entitled *Disciplinarum libri IX*. Varro's compendium of the disciplinary or "encyclical" studies embraced successively: (1) grammar (fragments of which are still extant), (2) logic, or "dialectics," (3) rhetoric, (4) geometry, (5) arithmetic, (6) astronomy, (7) music, (8) medicine, and (9) architecture. In later classifications of the "arts," architecture and medicine were considered parts of the mechanical or "servile" arts. The remaining parts of Varro's classification are the well-known trivium (grammar, dialectics, and rhetoric) and quadrivium (geometry, arithmetic, astronomy, and music). Early in 387, St. Augustine started to write an encyclopedia of the seven "liberal arts" and philosophy in his *Disciplinarum libri*, but only the grammar, six books on music, and the beginning of "the other five disciplines, namely dialectics, rhetoric, geometry, arithmetic, and philosophy" seem to have been actually completed by him.[10]

The "Dark Ages" that followed the fall of the Roman Empire and Latin civilization left little for posterity except the seven liberal arts and "theology," which was basically the study of the Bible. More than anyone else, Manlius Severinus Boethius deserves to be remembered for the elementary textbooks he provided for later centuries that groped their way to a new civilization we know as "medieval." The new civilization was gothic, but its roots were Roman. Boethius (ca. 475–524) has been called "the last Roman and the first Scholastic,"[11] because he preserved the ideal of the classical Roman tradition when the Roman world was crumbling all about him; and he established the foundations of medieval Latin scholasticism both in theology and in philosophy. Boethius translated many of Aristotle's logical works into Latin, providing the only real link between Aristotle and the Dark Ages. Boethius himself was a convinced Platonist, but he realized the importance of Aristotle and wished to translate and "harmonize" all the works of Plato and Aristotle.[12] The Roman world had an abundance of works on Latin grammar and rhetoric in the writings of Donatus, Priscian, Cicero, and others. But it provided nothing for logic, philosophy, and the quadrivium (a term first used by Boethius to signify the four mathematical sciences). Boethius, therefore, composed elementary adaptations from the Greek for music, arithmetic, and geometry. It would seem that Boethius also translated Ptolemy's "astronomy" and Archimedes' "mechanics," but nothing is known of these translations or summaries today.[13] The summaries and translations of Boethius plus the Roman works that survived the barbarian invasions constituted the foundation of a liberal education for the next six hundred years in Western Europe. To these must be added the important study of the Bible throughout these centuries.

Although Boethius discussed the nature and division of "philosophy" in his two commentaries on Porphyry's *Isagoge*, the *locus classicus* for the division commonly accepted in medieval schools is found in his short theological treatise *De trinitate*. For Boethius, "philosophy," which is the love of wisdom, is divided into speculative and practical. Practical philosophy is divided into (1) ethics of the indi-

vidual, (2) ethics of the family, and (3) politics, which is the highest
science in the practical order. Speculative sciences are also divided
into three kinds or "branches": (1) physics, which deals with material bodies in motion, both in reality and in our consideration of
them; (2) mathematics, which leaves out of consideration the afore-
said material bodies in motion, but considers only their changeless,
measurable and numerable aspects; and (3) metaphysics or "natural
theology," which leaves aside material bodies in motion and concen-
trates on truly changeless and immaterial things, such as "being" and
the Divine Substance, which is without matter and motion. Boethius
introduces his treatise *De trinitate* with the division of philosophy to
show that the subject matter of his treatise is without matter and without motion, in other words, a treatise of pure theology. This Boethian
division of philosophy is the old tripartite division given by Plato and
Aristotle, but with one important difference. For Boethius, the "object" of each science is a "form" of some kind, namely, a form in
matter and motion (physics), a form without matter and motion in
consideration by the mind but not in reality (mathematics), and a
form without matter and motion in reality and in our consideration
(theology). According to Boethius, the forms "abstracted" seem to
constitute a hierarchy ascending from physics, through mathematics,
to metaphysics and God. The schema and description shown in chart
3 represents the way he was read during the later Middle Ages:

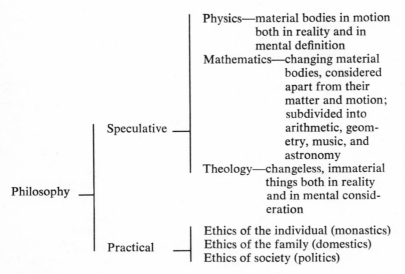

Chart 3

Marcus Aurelius Cassiodorus, a junior contemporary of Boethius, wrote a highly influential *Institutiones* for his monks at Vivarium about the year 544–45. This is a manual of divine and secular literature, divided into two books. The first book is a compendium of sacred scripture, exegesis, hagiography, and religious discipline. The second book is a summary of the seven liberal arts: grammar, rhetoric, dialectics (logic), arithmetic, music, geometry, and astronomy. At the beginning of his summary of dialectics, Cassiodorus discusses the definition and division of "philosophy," a procedure which was frequently followed throughout the Middle Ages. His division of philosophy is identical with that of Boethius, but for each part Cassiodorus gives a brief definition, indicating its nature and scope within the classification. *Natural* philosophy discusses the nature of each material thing which is produced naturally; *doctrinal* philosophy (mathematics) is the science which considers abstract quantity, that is, quantity which has been mentally "separated" from matter and from other accidents; and *divine* philosophy, which considers the ineffable nature of God and spiritual creatures. The rest of the second book discusses the seven liberal arts. This second book of the *Institutiones* was often copied separately and expanded in later centuries.

Early in the seventh century, Isidore of Seville composed an encyclopedic work called the *Etymologiae* in twenty books, which served as a common reference work throughout the Middle Ages. He gives a summary of the seven liberal arts in the first three books: 1, grammar; 2, rhetoric and dialectics; 3, arithmetic, geometry, music, and astronomy. Following Cassiodorus, Isidore discusses the definition and division of "philosophy" at the beginning of his compendium of dialectics, but he gives two schemes for subdividing "philosophy." The first is the familiar Stoic division, which St. Augustine attributed to Plato, namely, logic, physics, and ethics. To this Stoic division he gratuitously added his own subdivision (see chart 4).

Side by side with the Stoic division and his own subdivision of philosophy, Isidore simply repeated the familiar tripartite division of speculative philosophy into physics, mathematics, and metaphysics (theology) and of practical philosophy into ethics of the individual, ethics of the family, and politics. The point is that at this time nobody knew what to make of "philosophy" or "science" strictly so called.

During the early Middle Ages, that is, up to the introduction of the Aristotelian corpus into the Latin West in the twelfth century, Schoolmen preserved the division of Boethius even though they had little or no idea of what was meant by physics, metaphysics (natural

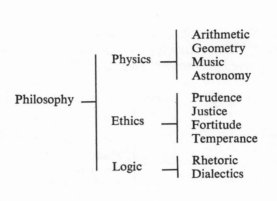

Chart 4

theology), or ethics. Some, following Isidore, thought that physics meant the quadrivium; some, like Cassiodorus following Boethius, thought that metaphysics meant Christian theology based on the Bible; some, like Isidore, thought that ethics meant the four cardinal virtues. It is a tribute to medieval scholarship that the medieval schoolmen continued to promulgate Boethius's division of philosophy without having the Aristotelian books that would have given meaning and content to the schema of classification. Throughout the early Middle Ages, scholars—such as Alcuin of York, Rabanus Maurus, Scotus Eriugena, and the entire school of Chartres—gave one or the other of the two classical divisions of "philosophy," and sometimes combined the two.[14] In practice, no damage was done, even though no advance was made in philosophy. Education simply comprised the seven liberal arts, which continued to be the foundation of human knowledge, and the study of the Christian faith found in the Bible and in the writings of the Fathers of the Church.

In the twelfth century a more thorough synthesis of the two ancient classifications was presented in the various *Didascalia*, or general introductions to the *artes*. These summary treatises follow the general pattern of the traditional *Disciplinarum libri*, discussing the nature and classification of learning, and briefly explaining the nature of each art. The best known of these is the *Didascalicon* of Hugh of St. Victor (1096–1141). In this remarkable treatise seven mechanical arts are introduced as parts of "philosophy" in order to balance the seven liberal arts; all seven liberal arts, including grammar, find a place in this classification; and it is a successful combination of the Boethian and Stoic divisions of "science." Hugh of St. Victor says, "Philosophy is divided into theoretical, practical, mechanical, and logical; these four branches embrace all scientific knowledge."[15] Except for the mechanical arts, the basic division of scientific knowledge

is that of the Stoics. In this case "physics" is taken to be equivalent
to "theoretical" and coextensive with Boethius's tripartite classification of speculative "philosophy" (see chart 5).

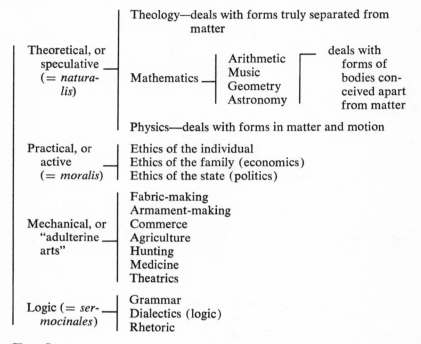

Chart 5

Hugh of St. Victor's classification of scientific knowledge was
taken over by Clarenbaud of Arras around the middle of the twelfth
century, and it enjoyed continuous circulation from the time it was
written. But in practice it did not affect the actual teaching of arts
in the schools of Chartres, Paris, or elsewhere. Although men like
Peter Abelard fully recognized the scientific character of logic, it was
taught mainly as an art, in keeping with Boethius, who had said
that logic "is not so much a science as an instrument [or tool] of
science."[16]

Science in the High Middle Ages

The period between roughly 1170 and 1270 was the high watermark
for treatises on the nature and division of the sciences. In the late
twelfth century, works of Arabic authors who had discussed the

classification and meaning of "science" were translated into Latin in Spain. Dominic Gundissalinus not only translated al-Farabi's important *De scientiis* and *De ortu scientiarum*, but also compiled a lengthy work of his own on the division of the sciences drawn from Arabic and early Latin writers.[17] Later Gerard of Cremona produced a new version of al-Farabi's *De scientiis*, and Michael Scot composed a treatise *De divisione philosophiae*, fragments of which have been preserved in the encyclopedic *Speculum doctrinale* of Vincent of Beauvais (fl. 1244–64). The impressive encyclopedia of Vincent presents at least ten definitions and divisions of the sciences, drawn from earlier and contemporary sources.[18]

The work that contributed most to a more profound understanding of the nature and scope of the sciences was Aristotle's *Posterior Analytics*, which was translated three times in the second half of the twelfth century. But, as John of Salisbury noted, it was so difficult to understand that there was scarcely a master willing to teach it.[19] The earliest Latin scholastic to comment upon it and make it his own was Robert Grosseteste, between 1200 and 1209. Albertus Magnus and Thomas Aquinas also commented upon it at a later date. The influence of Aristotle's *Posterior Analytics* on the minds of these two giants is most conspicuous in their grasp of the nature of science in general, and of physical science in particular.

The essence of scientific knowledge was clearly revealed to the scholastics by the translation of the actual texts of Aristotle and the incorporation of them into the university curriculum. Until it could be seen what Aristotle meant by physics, ethics, and metaphysics, it was useless to speculate on the classification given by Boethius in his *De trinitate*. The problems encountered in incorporating the Aristotelian *libri naturales*, or "books of natural science," into the faculty of arts are well known. The condemnations of Aristotle's *libri naturales* in 1210 and 1215, as well as succeeding proscriptions, eventually fell into disuse; and by 1255 the faculty of arts at the University of Paris required that almost all of these books of Aristotle's natural philosophy be read in the schools. Despite ecclesiastical condemnations of Aristotle as dangerous to the Christian faith, the medieval university adopted Aristotle and his writings as its very own. In fact, the *libri naturales* were thought to be the only books of their kind that could illumine the mind about man and the physical world in which we live. Despite numerous corrections that had to be made to the views of Aristotle in the Middle Ages through further observation and experimentation, nothing was written to replace the Aristotelian books on physical theory (*Physics*), physical astronomy (*De caelo*), primitive chemistry (*De generatione et corruptione*), meteorology (*Metheora*), psychology (*De anima*), and biology (*De animalibus*). Even in the seventeenth century, when criticism of Aristotle was at its height, nothing comparable was written to replace his books in the schools of Europe.

When the books of the new Aristotelian learning were incorporated into the medieval universities, it was simply a matter of adding physics, ethics, and first philosophy (metaphysics) to an existing structure of the seven liberal arts. There was no need for a new mathematics, for Euclid, Ptolemy, and Boethius were already taught as parts of the quadrivium; similarly, logic, grammar, and rhetoric were already in the university as the trivium. The new Aristotelian learning was simply tacked on to the liberal arts program as "the three philosophies": natural philosophy, moral philosophy, and first philosophy (metaphysics). With the addition of these new "sciences," the faculty of arts became known as the faculty of arts and sciences.

With the introduction of the new disciplines, the temptation became increasingly great to see a hierarchical gradation of the sciences after the manner of Plato. For the Platonists, mathematics supplied the principles for understanding the ever-changing images in nature, and the door to an understanding of metaphysics. The three parts of speculative philosophy enunciated by Boethius were seen as "three degrees of abstraction," whereby the mind ascended from the transience of nature to the stability of mathematics and, finally, to the immateriality of metaphysical being. In late scholastic philosophy, the doctrine of "the three degrees of formal abstraction" is fully enunciated: the *first degree* is formed by abstraction from "individual" matter, leaving the subject of consideration with only "sensible" matter, the realm of the natural sciences; the *second degree* is formed by abstraction from "sensible" matter, leaving only "intelligible" matter, the realm of the mathematical sciences; the *third degree* is formed by abstraction from *all* matter, leaving only pure being as such, the object of metaphysics.

At the very beginning of his paraphrase of the *Metaphysics*, composed probably between 1265 and 1270, Albertus Magnus directed his attack on "the error of Plato, who said that natural things are based on mathematical things, and mathematical things on divine things, just as the third cause is dependent on the second, and the second on the first; and so [Plato] said that the principles of natural things are mathematical, which is completely false."[20] The basis of this error, Albert explains, is that Plato had seen a certain ascending order from natural bodies to mathematical, to divine being, but he had misunderstood the explanation of this order. Perceiving that all changeable beings in nature are continuous, and that all continuous beings are simple, Plato had thought that the principles of natural science are mathematical, and that the principles of mathematical being are metaphysical, or "divine." "And this is the error which we have rejected in [our commentary on] the Books of the *Physics*, and which we shall again reject in the following books of this science [of metaphysics]."[21] The metaphysical error that Plato made was frequently pointed out by Thomas Aquinas and by Albertus Magnus: Plato equivocated on the word *one*. The "one" which is convertible with "being" so that we can say "a being" is "one" is not the same kind of "oneness" which is the principle of number, that is, the "one" we use in counting.[22] Because of this error Plato could identify the "one" in mathematics with the "One" in metaphysics (God) and make all knowledge and all reality dependent upon the "One" and "Good" *secundum se*, which is contemplated in the divine science of metaphysics.

From the vehemence of Albert's criticisms, it is clear that the "error of Plato" that he attacks was promulgated by "the friends of Plato" in Albert's own day.[23] These Platonists could easily be Grosseteste, pseudo-Grosseteste, Roger Bacon, and possibly Robert Kilwardby, for all of whom the *key* to natural science is mathematics. For Albertus Magnus, mathematical being, such as number and figure, is a mere abstraction of the mind from the quantified constitution of physical bodies, and not an intrinsic constituent of natural species. Since such mathematical abstractions are made *from* natural bodies, they are thereby equally applicable to them to the extent to which those bodies are quantified. This is particularly true in the aforementioned *scientiae mediae*, which use mathematical principles to investigate natural phenomena. That is to say, the "intermediate sciences," such as astronomy or optics, deal only with quantified aspects of natural phenomena and not the whole phenomenon.

A *scientia media* for the medieval scholastics was intermediate between physics and pure mathematics in such a way that mathematical principles are used to "demonstrate" certain properties of a subject that is truly quantified in some determinable respect. Astronomy, for example, was a true "science" dependent on geometry, because geometrical principles are applied to the quantitative aspects of celestial motions, such as speed, size, distance, and proportionality. The purpose of astronomy, it was thought, was to explain rationally the movements of the heavens in such a way that all phenomena seen in the heavens could be "preserved" or "saved," predicted, and explained. Thus, the role of astronomy was to "save the phenomena" in the heavens through mathematical principles. Simplicius, the sixth-century Aristotelian commentator, explains the origin and nature of astronomy as follows: "Eudoxus of Cnidos was the first Greek to concern himself with hypotheses of this sort, Plato having, as Sosignes says, set it as a problem to all serious students of this subject to find what are the uniform and ordered movements by the assumption of which the phenomena in relation to the movements of the planets can be saved (*sozein tà phainómena*)."[24] As Simplicius explains it, the purpose of astronomy is "to save the appearances, or the phenomena."[25] In order to render such phenomena open to mathematical treatment, all nonquantitative aspects must be disregarded, such as the nature of celestial bodies and the causes of celestial motion, as well as their purpose. Just as mathematics itself must "abstract" from all nonquantitative aspects of natural bodies, so too its applicability is limited to the quantitative aspects of natural phenomena. Since potency, act, form, substance, causality, and even motion itself are nonquantified factors in natural philosophy, they are not open to mathematical treatment. They are considerations and aspects that necessarily elude mathematical treatment. However, since all physical bodies and motions of all kinds are quantified in some respect, they are open to mathematical treatment; such mathematical treatment constituted the medieval sciences of optics, harmonics, astronomy, statics, dynamics, and kinematics. Although Aristotle raised the problem of "intermediate sciences" in his *Posterior Analytics*, his description left much to be desired, and the thirteenth-century scholastics, notably Robert Grosseteste, Albertus Magnus, and Thomas Aquinas, were in a much better position historically to amplify and specify the nature of a *scientia media* between mathematics and natural philosophy. In the seventeenth-century Scientific Revolution the study of such intermediate sciences became considerably expanded as new mathematical principles came to be applied to a wider range of natural phenomena. Thus, Sir Isaac Newton could present his revolutionizing volume entitled *Principia mathematica philosophiae naturalis*, "The Mathematical Principles of Natural Philosophy."

We cannot leave the medieval period, however, without considering the most ambitious and astute consideration of the nature, scope, and classification of the then-known sciences in the thirteenth century. It was the widely known treatise entitled *De ortu scientiarum* ("On the Origin of the Sciences") by the famous Robert Kilwardby, an Englishman who was a renowned master of arts at Paris around the middle of the thirteenth century.[26] Kilwardby, who had become a Dominican of the English Province shortly before 1250, was apparently asked by his religious superiors to write a comprehensive treatise on all the arts and sciences as an aid to younger men in the Order. Its popularity is attested to by the eighteen complete and two incomplete manuscripts extant, but it was not printed until our own day. In it Kilwardby not only presented a classification of all the known sciences, but also raised fundamental metaphysical questions about the sciences discussed. The sixty-seven chapters of this highly balanced work give not only the origin, nature, scope, definition, and division of all the known sciences and arts, but also their interrelationships and the basic problems with which each science is concerned. The skeleton division here presented reminds one very much of Hugh of St. Victor's classification, but one would be greatly deceived if he assumed from this that Kilwardby's treatise is a mere recapitulation of Hugh's, for it is infinitely more sophisticated, rich, and thorough (see chart 6).

After Robert Kilwardby's *De ortu scientiarum*, philosophers ceased to write treatises on the nature, scope, and classification of the sciences. All such information was given briefly as part of an introduction to a specified course of study.

Early medieval treatises on the nature and classification of the sciences, such as those of al-Farabi, Hugh of St. Victor, Michael Scot, Gundissalinus, and Kilwardby, were special introductions to the study of the human sciences in general. Such introductions or treatises, it was well recognized, belong to the office of the metaphysician reflecting on the whole of human knowledge. For example, Roger Bacon states: "A noble part of metaphysics, since it is common to all sciences, is to show and demonstrate the origin, distinction, number, and order of all the sciences and what is characteristic of each."[27] Although the writing of such treatises belonged to the metaphysician, the intention of such introductory literature was to show beginners the scope and nobility of each science, and why it should be studied with diligence. Treatises contemporary with Kilwardby, such as those of Nicholas of Paris,[28] Arnulph of Provence,[29] and John of Dacia,[30] had the same purpose, but they are all conspicuously inferior to the masterly treatment by Kilwardby.

For some reason, scholastics of the fourteenth century were no longer interested in writing this type of literature. While the division of the sciences was well known to all students, fourteenth-century scholastics were no longer interested in discussing problems of the nature, scope, and classification of the sciences. They were more

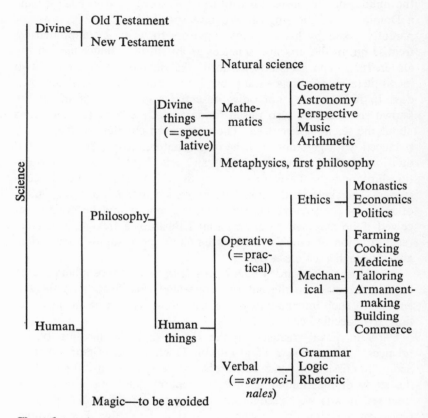

Chart 6

interested in the new logic and the new problems of physics that arose in the early fourteenth century, first at Oxford, then on the Continent. The age of recovering the "new Aristotelian learning" was past. A new age was inaugurated with new problems of its own leading to the seventeenth-century revolution in scientific knowledge.

Notes

1. On the ideal of Greek education, see Henri Marrou, *A History of Education in Antiquity* (New York, 1956), pp. 61–115.

2. For further detail, see James A. Weisheipl, "The Concept of Scientific Knowledge in Greek Philosophy," in *Mélanges à la mémoire de Charles De Koninck* (Quebec, 1968), pp. 487–507.

3. On the origin of Greek scientific knowledge, see George Sarton, *A History of Science: Ancient Science through the Golden Age of Greece*

(Cambridge, Mass., 1952), pp. 168–73; G. E. R. Lloyd, *Early Greek Science: Thales to Aristotle* (London, 1970), pp. 1–49.

4. H. Diels, *Fragmente der Vorsokratiker*, 5th ed. (Berlin, 1934–1938), 28 B frag. 6.

5. Aristotle, *De caelo* 3. 1. 289^{b}21; Diels, *Fragmente der Vorsokratiker*, 28 B frag. 6.

6. Aristotle, *De generatione et corruptione* 1. 8. 324^{b}35–325^{a}3.

7. Theophrastus, *Physicorum Opiniones* 1, frag. 8, 19–21, in H. Diels, *Doxographi Graeci* (Berlin, 1929), p. 483.

8. Plato, *Parmenides* 130 B–D; cf. Aristotle, *Metaphysics* 12. 3. 1070^{b}19–20.

9. Augustine, *De civitate Dei* 8. 4, in *Patrologiae cursus completus, series latina*, ed. J.-P. Migne, vol. 41 (Paris, 1845), col. 228.

10. Augustine, *Retractationes* 1. 6. This is the reading printed in the edition of Pius Knöll (Corpus scriptorum ecclesiasticorum latinorum, vol. 36 [Vienna, 1902], p. 28); one might have expected "astronomica" in place of "arithmetica," to give a closer parallel to *De ordine* 2. 15. On the other hand, "philosophy" should perhaps be viewed as including astronomical matters.

11. Martin Grabmann, *Die Geschichte der scholastichen Methode* (Freiburg, i. B., 1909), 1: 148–77.

12. Boethius, *Commentarii in librum Aristotelis Peri Hermenias*, editio secunda, 2. 3, ed. C. Meiser (Leipzig, 1880), vol. 2, pt. 1, pp. 79–80.

13. Cassiodorus, *Liber variorum*, 1, ep. 45, *Patrologia latina*, ed. Migne, vol. 69 (Paris, 1848), vol. 539.

14. See James A. Weisheipl, "Classification of the Sciences in Medieval Thought," *Mediaeval Studies* 27 (1965): 54–90.

15. *Didascalicon*, 2. 1, ed. C. H. Buttimer (Washington, D.C., 1939), p. 24.

16. Boethius, *Super Porphyrium*, editio secunda, 1. 3 (Corpus scriptorum ecclesiasticorum latinorum, vol. 48 [Vienna, 1906], p. 140, line 13 to p. 143, line 7).

17. Cf. Dominicus Gundissalinus, *De divisione philosophiae*, ed. Ludwig Baur (*Beiträge zur Geschichte der Philosophie des Mittelalters*, vol. 4, pts. 2–3 [Münster, 1903]); Alfarabi, *Über den Ursprung der Wissenschaften*, ed. Clemens Baeumker (ibid., vol. 19, pt. 3 [Münster, 1916]).

18. Vincent of Beauvais, *Speculum doctrinale* (Douai, 1624), vol. 1, cols. 9–21.

19. *Metalogicon*, 4. 6, ed. C. C. J. Webb (Oxford, 1929), p. 171.

20. Albert, *Metaphysica*, 1.1.1, ed. Bernhard Geyer, in *Opera omnia*, ed. Geyer, vol. 16, pt. 1 (Cologne, 1960), p. 2, lines 31–35.

21. Ibid., lines 45–47.

22. Thomas, *Summa theologiae* 1, quest. 11, ans. 1, ed. Institute of Medieval Studies (Ottawa, 1953), vol. 1, pp. 56a–57a; *In I Metaphysica*, lect. 7–8, ed. M. -R. Cathala (Turin, 1926), pp. 38–46. Cf. Plato, *Republic* 478 B; *Parmenides* 144 B; *Sophist* 237 D – 238 A; Albert, *Metaphysica*, 1.4.13, ed. Geyer, vol. 16, pt. 1, pp. 65–67.

23. See James A. Weisheipl, "Albertus Magnus and the Oxford Platonists," in *Proceedings of the American Catholic Philosophical Association*

33 (1958): 124–39; and "Classification of the Sciences in Medieval Thought," pp. 72–90.

24. Simplicius, *In II De caelo*, 12, comm. 43 (Venice, 1548), fol. 74r–v.

25. On this role, see Pierre Duhem, *To Save the Phenomena: An Essay on the Idea of Physical Theory from Plato to Galileo*, trans. E. Doland and C. Maschler (Chicago, 1969).

26. Robert Kilwardby, *De ortu scientiarum*, ed. Albert G. Judy, O.P. (London, 1976).

27. Bacon, *Communia naturalium*, in *Opera hactenus inedita*, ed. R. Steele, vol. 2 (Oxford, 1905?), p. 5; also *Communia mathematica*, in ibid., vol. 16 (Oxford, 1940), p. 1.

28. Oxford, Merton College, MS 261, fols 67ra–69vb.

29. Ibid., fols. 13ra–18va.

30. *Johannis Daci Opera*, ed. A. Otto (Corpus philosophorum Danicorum medii aevi, vol. 1, pt. 1 [Copenhagen, 1955]), pp. 3–44.

15

Bert
Hansen

Science and Magic

People have long associated magic with the Middle
Ages. An "age of belief" is assumed to foster the
occult; Merlin and witches are standard props in
everyone's picture of pre-modern Europe. Modern
scholarship, however, has reassessed and clarified
this picture in several respects. Almost single-
handedly Lynn Thorndike demonstrated that magic
in the Middle Ages was not marginal to intellectual
life, nor an activity of ignorant, credulous, or su-
perstitious people—or at least not of these alone.
Magic, as he revealed, formed an important part
of medieval thought and experience. Thorndike
also amassed tomes of documentation for his claim
that magic was, in the Middle Ages and early modern period, a source of the empirical and experimental approach to nature.[1] More recently a number of other historians have reassessed Neoplatonic
magic and Hermeticism in the Renaissance and
have perceived them to be factors in the creation
of "modern science" in the seventeenth century.[2]
These two important developments in the historiography of magic and science condition the task of
this study: to set forth the basic outlines of the
magical world-view of the Middle Ages, comparing
and contrasting magic with the contemporary natural philosophy, and briefly relating this to the interactions of magic and science in the sixteenth and
seventeenth centuries.

At the outset, two comments on my approach
are in order. I distinguish here between the ubiqui-

tous "folk magic" of the population in general and the "intellectual magic" of those who read and wrote books on the subject; my treatment is limited to the latter.[3] Secondly, as this essay examines the conceptions and philosophical tenets of medieval and Renaissance magicians, the result is a *doctrinal* history that neglects, at some risk, the social history within which the doctrinal developments must eventually be interpreted. My only justification for this omission, other than obvious limitations of space, is that essential first tasks are to ascertain a rough anatomy of magic and to suggest its connections with science. One can then try to analyze science and magic together as related ideological components in feudalism and in its transition to bourgeois society.[4]

A Characterization of Magic

By the very phrasing magic *and* science, people have often been misled into conceiving the two enterprises—even in the Middle Ages—as entirely distinct. Generalizing from his fieldwork in the South Pacific, the anthropologist Bronislaw Malinowski wrote: "Science is founded on the conviction that experience, effort, and reason are valid; magic on the belief that hope cannot fail nor desire deceive. The theories of knowledge are dictated by logic, those of magic by the association of ideas under the influence of desire."[5] This readily accepted distinction emerged gradually during the early modern period, but it was definitely not the conception of medieval and Renaissance thinkers themselves, for desire and hope were certainly part of the Aristotelian natural science, and reason and experience not alien to medieval magic.

Although reason versus hope is not a useful distinction, I believe that understanding (*episteme*) versus use (*techne*) is. At its most general, then, magic may be characterized as the utilization of "occult forces" (that is, forces either supernatural, or natural but hidden) to accomplish specific desired ends, often by means of words or symbols.

Let us observe first that magic's aim is some real accomplishment, not mere knowledge or understanding. This instrumentality constitutes a technology rather than a science. It is differentiated from other arts and technologies by the nature of the forces it employs.

Second, magic distinguishes the natural from the supernatural. This dichotomy in the Middle Ages, however, was quite different from the modern distinction between these same terms, for phenomena regarded as natural by medieval people were only those which

occur *most of the time*, in nature's *habit* or usual course. The law of
nature within the Aristotelian conceptual framework was not one of
rigid necessity, nor one of mathematical probability, but simply that
of the usual or ordinary occurrence. Thus, a rare enough event might
be considered unnatural, even though its causes did not involve divine
or preternatural interference. Monstrous births (a two-headed calf,
for example) were thus regarded as unnatural, though today we
would suppose that, although the cases are unusual and deviate from
the normal course, they are yet governed by natural laws and processes. This identity of the "normal" and the "natural" in the tradition
of Greek natural philosophy is one of its greatest differences from
modern science and is, therefore, crucial to understanding medieval
magic on its own terms.[6]

In the modern conceptualization, nature's laws are inexorable and
unbreakable. No phenomenon—no matter how remarkable—is an
"outlaw," except those we segregate entirely as purely spiritual or
supernatural. But in the thirteenth and fourteenth centuries there
were two kinds of phenomena which, though fully recognized as
within the physical world, were considered as contrary to nature
(that is, contrary to normal modes of operation): the marvelous and
the artificial. The latter category included both mechanical devices
(of the sort Roger Bacon recommended) and constraints upon nature's tendencies, as, for example, the "unnatural" or forced movement of a heavy body upward. Roger Bacon analyzed the "marvelous
power of art *and* nature," thus placing his fantasy inventions outside
the natural realm:

> Some there are that ask which of these two be of greater force
> and efficacy, Nature or Art, whereto I answer and say that al-
> though Nature be mighty and marvelous, yet Art using Nature for
> an instrument is more powerful than natural virtue, as is to be
> seen in many things. But whatever is done without the operations
> of Nature or Art is either no human work [that is, it is angelic or
> demonic] or if it be [human], it is fraudulently . . . performed
> [that is, by sleight-of-hand]. . . .
> Now will I begin to recount unto you strange things, performed
> by Art and Nature, and afterwards I will show you the causes and
> manners of things, wherein shall be nothing magical; so that you
> shall confess all magic power to be inferior to these and unworthy
> to be compared with them. And first of all by the figuration of Art
> itself. There may be made instruments of navigation without men
> to row in them, as huge ships to brook the sea only with one man
> to steer them, which shall sail far more swiftly than if they were

full of men. And chariots that shall move with an unspeakable
force, without any living creatures to stir them, . . . yea instruments
by which to fly, so that one sitting in the middle of the instrument,
and turning about [a device] by which the wings being artificially
composed may beat the air after the manner of a flying bird.

Moreover instruments may be made wherewith men may walk
in the bottom of the sea or river without bodily danger, which
Alexander the Great used to the end [that] he might behold the
secrets [that is, the unseen things] of the seas. . . . And these have
been made not only in times past but even in our days. And it is
certain that there is an instrument to fly with, which I never saw,
nor know any man that hath seen it, but I full well know by name
the learned man that invented the same.[7]

Three centuries later Francis Bacon attempted to overcome the
limitations of such a division. In the *Advancement of Learning*, published in English in 1605 and in Latin in 1623, he wrote:

The division which I will make of Natural History is founded
upon the state and condition of nature herself. For I find nature in
three different states, and subject to three different conditions of
existence. She is either free, and follows her ordinary course of
development; as in the heavens, in the animal and vegetable
creations, and in the general array of the universe; or she is driven
out of her ordinary course by the perverseness, insolence, and
frowardness of matter, and violence of impediments; as in the
case of monsters; or lastly, she is put in constraint, moulded, and
made as it were new by art and the hand of man; as in things
artificial. Let Natural History therefore be divided into the History
of Generations, of Pretergenerations, and of Arts; which last I
also call Mechanical and Experimental History. . . . And I am the
more induced to set down the History of the Arts as a species of
Natural History, because an opinion has long been prevalent, that
art is something different from nature, and things artificial different
from things natural. . . . But there is another and more subtle
error which has crept into the human mind; namely that of considering art as merely an assistant to nature, having the power
indeed to finish what nature has begun, . . . but by no means to
change, transmute, or fundamentally alter nature. . . . Whereas
men ought on the contrary to be persuaded of this; that the artificial does not differ from the natural in form or essence, but only
in the efficient [that is, the active agent]. . . . Gold is sometimes
refined in the fire and sometimes found pure in the sands, nature
having done the work for herself. So also the rainbow is made in
the sky out of a dripping cloud; it is also made here below with a

jet of water. Still therefore it is nature which governs everything; but under nature are included these three; the *course* of nature, the *wanderings* of nature, and *art*, or nature with man to help.[8]

By opposing the Aristotelian (and scholastic) differentiation between art and nature which labeled mechanical and other contrivances "unnatural," and sometimes magical or marvelous, Francis Bacon was developing a new concept of nature which was thereby more encompassing than the medieval notion. Yet he still maintained the sense that nature's laws are not fully deterministic: her law can be broken, she can be forced. At first he seems to challenge the view, held by alchemists especially, that art is "merely an assistant to nature, having the power to finish what nature has begun"; but in the end he is seen to concede the issue in part by characterizing art as "nature with man to help."

In the third place, the characterization of magic that I have proposed here distinguishes certain natural forces as occult, or hidden, and others as generally known. This appears in modern terms to be an epistemological rather than an ontological distinction, but not so then: occult forces were assumed to be of a different character from those of everyday experience. Thus, a person using the generally unknown properties of some herb was doing magic, even though the effect might be perfectly "natural" in modern terms. This characteristic of magic, namely, that any particular phenomenon can lose its magical character merely by becoming widely enough known, explains in part the demise of magic in the seventeenth century under the impact of expanding conceptions and experiences of what is natural and common.[9]

Finally, the assumed power of words or other symbols characterizes medieval Latin magic as it does folk and tribal magic of other times and places. Pentagrams and other geometrical figures, symbols for celestial bodies and metals, and unusual and foreign-sounding words were as important among the magician's paraphernalia as wands, gems, talismans, and herbs. The nonphysical nature of words and symbols, however, militated against their ready acceptance as true causes of physical events, and many attempts were made to fit this part of magic into the Aristotelian framework, which demands contact between a cause and its effect.[10] A slight digression to consider Thomas Aquinas's treatment of the power of magical words will clarify some of these issues.

Thomas first demonstrated that "the works of magicians do not result only from the influence of heavenly bodies"—thereby explicitly denying that magical effects and apparitions might be merely imagined. He then asked, "whence the magic arts derive their efficacy (a question that will present no difficulty if we consider their mode of operation)." Thomas cited the "use of certain significative words in order to produce certain definite effects," and reasoned that "words, in so far as they signify something, have no power except as derived from some intellect." Then, after showing that it cannot be the intellect of the speaker that has the power over things, he concluded that it must be the intellect of some other being, thus demonstrating the necessary existence of demons, which he then showed must be evil.[11] By such reasoning, Aquinas transformed the credulous reports of supposed magical effects into accepted facts of natural philosophy and, by the same stroke, strengthened a philosophical justification for demons, demonology, and witchcraft.[12]

Nothing about good or evil is indicated in the present characterization of magic, even though magic has seldom been considered a neutral activity. Distinctions between "white" and "black" magic were ancient and widespread; yet upon examination, they become almost meaningless. Without undue cynicism, it is possible to view white magic as whatever one is doing and black magic as what is done by one's enemies (or the enemies of one's faith, be it political or religious). Theoretically, black magic employed the devil's powers or agents, but as there was no practical way to assess whether a particular spirit was divine or diabolic, this distinction must be recognized as theoretical and polemical rather than empirical. Generally, writers opened magic books with condemnations of some *other* kind of magic. For example, Giambattista della Porta wrote: "There are two sorts of Magick: the one is infamous, and unhappie, because it hath to do with foul spirits, and consists of Inchantments and wicked Curiosity; and this is called Sorcery; an art which all learned and good men detest; neither is it able to yeeld any truth of Reason or Nature, but stands meerly upon fancies and imaginations, such as vanish presently away, and leave nothing behinde them. . . . The other Magick is natural; which all excellent wise men do admit and embrace, and worship with great applause. . . ."[13] Whatever the effect of della Porta's words on his contemporaries, we must acknowledge that their authenticity is in their polemic, not in their apparently descriptive, character.

Even if the difference between black and white magic is judged to lie in the eye of the beholder, it must be acknowledged that people did have very strong commitments to the difference. Only historians are able to take lightly a distinction over which people lost their property, reputations, and even their lives. In 1326 the Decretal "Super illius specula" of Pope John XXII declared that magicians— for their dealings with the devil or his agents—were *ipso facto* excommunicated. (Thomas Aquinas's contribution to the connection of magic and demons has already been noted above.) Even in the early seventeenth century, Shakespeare's audience (contemporaries of Francis Bacon, Campanella, and Galileo) thoroughly accepted the reality of angels and demons whom magicians could invoke. The wonders of white magic performed by Prospero in *The Tempest* were perceived as fundamentally alien to the magic of his "black" counterpart Sycorax, or of the witches in *Macbeth*. White magic could be accepted within some Christian frameworks, and, in fact, many of the early critics of magic, for example, Jean Gerson in the early fifteenth century or Meric Casaubon in the seventeenth, were conservative religious thinkers concerned primarily with magic's "black" character.

Underlying Assumptions of Medieval Magic

Let us proceed to a description of the metaphysics underlying the magical approach to nature, that is, of the assumptions about causality, forces, quality, and quantity, acknowledging throughout the resonance with the science of the period.

Medieval people lived in a world that was, above all, a world of *essences*. They perceived things as having inner "natures" or qualities characteristic of them and inherent in them. Aquinas's notion of *essentia* has been epitomized as "the inner structure of existence, [operating] as the principle of form for each kind of being."[14] An object's essence thus corresponded closely with a merging of two of Aristotle's four "causes," the formal and the final. Since essences were thought to be not simply conceptual entities, but to be real and have existence, they were regarded as discoverable and even at times transferable (as in alchemical operations and in the use of talismans).[15] Fortified by this sense of the reality of particular essences and qualities, medieval scholars had no difficulty in assuming hidden or occult qualities. (Hidden did not mean permanently unknown, but, rather, generally unrecognized or known only to various sorts of wise people: sages, *magi*, and sometimes initiates or recipients of a personal revelation.) Much of the lore of magic dealt with occult qualities, how to discover them and how to apply them.

Such an ontology of essences supported a system of qualitative explanation. That is, to know *what* some thing is, or *what virtue* (power) it has, explains it. Although in principle qualitative explanations need not exclude quantitative ones, it seems that they tend to preclude them in fact. Medieval magic, like Aristotelian natural philosophy, used numbers, if at all, only as entities that have certain qualities, known or occult, and never achieved a significant interest in counting or measuring. Certain numerals were "perfect," others lucky or unlucky; and, in addition, various geometric figures had special virtues.

Just as certain numbers were assigned absolute qualities of nobility, perfection, and so on, so natural objects were assigned status or value according to their essence. This meant that values of goodness, perfection, and so on were considered to inhere in nature. An object might be good not simply in the functional sense of being *good for* some purpose, but in an absolute sense.[16] This structure was manifest in many hierarchies of value discovered in nature: witness the chain of being in Aristotelian biology with its "higher" and "lower" forms, the "superior" (that is, supralunar) and "inferior" (that is, sublunar) regions of the cosmos, the distinction between "perfect" (sexually generated) animals and "imperfect" (spontaneously generated) ones, and so on. Similarly, pure earthy matter was considered cold and dry, not simply by comparison to warmer and moister things, but in an absolute sense. When, therefore, a plant like peppermint came to be recognized as a "hot" plant by virtue of its peculiar quality of being peppermint, it was immediately recognized as useful for countering some other entity regarded as "cold." Many magical operations were simply manipulations devised to concentrate or utilize a certain virtue known to inhere in a substance or object. In many ways, such an enterprise seems like a crude pharmacology, with the important difference that in not segregating material from symbolic forces it was based on quite a different notion of causality than modern pharmacology.

The causality of medieval magic, as its notions of being, thus flowed in the streambed of Aristotelian thought. To be sure, some turbulence arose at times as thinkers tried to understand and rationalize more ancient magical phenomena in terms of the Aristotelian notions, but always the fundamental context remained the same. All developments seem to have been revisions within the system rather than radical departures from it. (On the mathematicism of Renaissance magic as a possible exception to this generalization, see below.) For example, the tenet of Aristotle's physics that there must be material contact between the mover and the moved was challenged by various natural philosophers in the ancient world, in Islamic civilization, and in the Latin West on the basis both of observed phenomena like projectile motion (lacking an apparent continuing mover) and of the reported phenomena of magical tradition, like the effects of the evil eye. That some people, especially old women, have the ability to "fascinate" (that is, to bewitch) people merely by looking at them did not fit the theory, and many thinkers tried to rationalize it as involving some kind of contact.[17] Roger Bacon, for example, explained fascination in terms of the multiplication of species.[18] This same tension between scientific doctrine and magical phenomena probably lies behind the frequent discussions of the nature of celestial influence that was postulated both in astrology and in talismanic magic.[19]

Final cause or purpose, a notion largely abandoned by modern science, has been seen as a character distinguishing magic from science; but in the Middle Ages it was, of course, a character that expressed their unity. It is hard for modern people to realize fully the experience of living in a world infused with purpose—all the way from the lowly ant, whose thrifty home economy was established as a parable to teach one how to live; to the grand motions of the celestial bodies, established both to declare the glory of God and to give light and heat; and even to the *primum mobile*, placed above the heavens to cause all other motions in the cosmos. In such a world, the assertion that some herb has a certain color or shape (as for example the liver-colored hepatica useful for liver ailments) in order to help people know it for some particular use is both rational and meaningful. In an organically unified experience of the world, the color is considered one expression of the essence of the herb.

Such an example leads us to consider the assumed power of similarity (whether of color or some other quality). The like and the unlike appeared to have special and real powers based on the simple fact of similarity or dissimilarity. Like things shared a relationship deeper and wider than they do in our modern view, according to which one must specify by which exact parameters they are alike, and not jump from some commonalities to the grander category of the like. Two principles sum up the interactions of affinity: "like effects like" and "like affects like." William of Auvergne (bishop of Paris, 1228–49) wrote: "Some people have defined nature as an innate force in things of procreating like things out of like; and Aristotle says that every operation of nature is by similitude."[20] The former of the two principles, more commonly phrased as "like engenders like," has many examples: the continuity of plant and animal species, the good (or evil) effects of good (or evil) actions, the use of a small amount of "seed" gold by the alchemist to produce more gold, and so forth.

The second principle is the basis for all forms of sympathetic magic in the widest variety of cultures and for most of Western science before the seventeenth century. In the late thirteenth-century *Book of the Marvels of the World*, falsely attributed to Albertus Magnus, we read:

> It was known of Philosophers that all kinds of things move and incline to themselves, because an active . . . virtue is in them, which they guide and move, as well to themselves as to others, as fire moveth to fire, and water to water.
>
> Also Avicenna said, when a thing standeth long in salt, it is salt, and if any thing stand in a stinking place, it is made stinking. And if any thing standeth with a bold man, it is made bold; if it stand with a fearful man, it is made fearful. . . . And generally it is verified of them by reasons, and diverse experience, that every nature moveth to his kind. . . . And in all dispositions there is nothing which moveth not to itself, according to his whole power. And this was the root, and the second beginning of the works of secrets. And turn thou not away the eyes of thy mind.
>
> After that, this was grafted in the minds of the Philosophers and they found the disposition of natural things. For they knew surely that great cold is grafted in some, in some boldness, in some great wrath, in some great fear, in some barrenness is engendered, in some ferventness of love is engendered; either after the whole kind, as boldness and victory is natural to a Lion, or *secundum individuum*, as boldness is in an harlot, not by Man's kind, but *per individuum* [that is, not by virtue of her membership in the human species, but by her individual character]. There came of this great marvels and secrets able to be wrought.
>
> . . . And the turners of one metal into another called Alchemists know that by manifest truth, how like nature secretly entereth, and rejoiceth of his like. And every science hath now verified that in his like. And note thou this diligently, for great marvellous works shall be seen upon this.
>
> Now it is verified and put in all men's minds, that every natural kind, and that every particular or general nature, hath natural amity and enmity to some other. . . . As the Sheep doth fear the Wolf, and it knoweth not only him alive, but also dead; not only by sight, but also by taste. . . . And some have this *secundum totam speciem* [according to the whole species], and at all times, but some only *secundum individuum* [according to the individual], and at a certain time. And it is the certifying of all Philosophers,

that they which hate each other in their life, hate their parts altogether after they die. For a skin of a Sheep is consumed [by] the
skin of the Wolf; and a timbrel, tabor or drumslade made of the
skin of a Wolf causeth [that] which is made of a Sheep's skin not
to be heard, and so is it in all others. And note for this a great
secret.[21]

The phenomenal range of such interactions in medieval magic is
merely suggested by this passage from *The Book of Marvels*. Such
sympathies were observed everywhere and based on a wide variety
of connections: similar substance, similar geographical origin, similar numerical identity, similar properties (color, shape, texture, etc.),
similar names, and so on.[22] For example, the planet Venus (associated with copper, since both the goddess and the mineral ore originally came from Cyprus) was assumed to have influence over the
working of copper.

Although the power of sympathetic magic is one of the most characteristic features of magic in all cultures,[23] its universality must not
obscure the essential resonance that it has with medieval Christian
philosophy of nature based on Aristotle. The fact of an opposition,
at the end of the Middle Ages, of Renaissance Hermeticism to scholastic science has led some writers to locate the uniqueness of Hermeticism in such a factor.[24] In fact, the idea that the attraction of
like for like is a cause of motion was first introduced into natural
philosophy by Plato, and then incorporated by Aristotle into a cosmology of four elements, each with its natural place. The idea of
sympathy among parts of the body, found in the Hippocratic writings,
was transferred by the Stoics to the cosmos as a whole.[25] It did
become a principal tenet within the corpus of Hermetic writings
(assembled in the third and fourth centuries A.D.), but was by no
no means limited to this tradition, as can be seen from the examples
given above.

Magic Books and Other Sources

The literature of medieval magic forms no coherent genre; few books
before the Renaissance have "magic" in their titles.[26] Books on natural history and on natural philosophy often contain at least something on magic. William of Auvergne's *De universo* is a rich source.
The encyclopedias of Alexander Neckam, Thomas of Cantimpré,
Bartholomew the Englishman, Vincent of Beauvais, and others always report marvels and magical practices. Scholastic treatises, many

of them in the form of university exercises (*questiones, quodlibeta,* and so on), contain citations of magical phenomena and reveal their author's attitudes toward magic in general. Books on "errors" of philosophy, episcopal condemnations of philosophical doctrines, and papal proclamations form an incidental, but useful, source.[27]

Books of "experiments" or "secrets" constitute a more numerous group of sources.[28] These collections of facts and recipes were often attributed to famous scientists, especially Aristotle and Albertus Magnus, or to sages like Plato, Moses, and Hermes. Some of the medieval *Secrets* incorporated translations from ancient or Islamic sources, but many were of later production, despite their attribution to venerable authorities.

A similar but more philosophical genre is formed by books of "problems," or questions about (often unrelated) phenomena. Although the *Problemata* of pseudo-Aristotle is the prototype here, the problem-literature in the Middle Ages has its own history, more significant than that of one Aristotelian text.[29] Over many centuries the same problems reappear in various contexts. This genre expanded so much in the early modern era that eventually entire monographs were devoted to a single "question." Let us briefly review, as one example, the problem of the corpse which begins to bleed in the presence of its murderer. The following (incomplete) list of scholars who treated this problem demonstrates its continuing relevance. Most of these thinkers (including Mersenne and Descartes) accepted the verity of the phenomenon and devoted their efforts to explaining it, naturalistically or otherwise. Doctrines of sympathy, antipathy, and other kinds of action at a distance received much attention in these discussions. Note that in at least three cases the treatment amounted to a full monograph: Urso of Calabria (late twelfth century), Alexander Neckam (ca. 1200),[30] Giles of Rome (1290),[31] Peter of Abano (1310),[32] Nicole Oresme (ca. 1350), Galeotto Marzio da Narni (late fifteenth century), Cornelius Gemma (1575), Guilielmus Adolphus Scribonius (1583), Andreas Libavius (1594; 314 theses on this problem), Peter Binsfeld (1596), James Stuart (1597 [the future James I of England]), Heinrich Kornmann (1610), Marin Mersenne (1623), Francisco Torreblanca (1623), Johannes Frankenius (1624; an entire book), Father Juan Eusebio Nieremberg (1633; several chapters are relevant), Théophraste Renaudot (1640–41), Johann Amos Comenius (1643), René Descartes (1647), Lazarus Gutierrez (1653), Walter Charleton (1654), Gaspar Schott (1657–59), Gaspar Caldera de Heredia (1658), Josua Arndius (1664), J. M. Schwimmer (1672), Johann Christian Frommann

(1675), M. F. Geuder (1684), Gottfried Voigt (1685; an entire book), and Christian Friedrich Garmann (1709). Curiosity about particular peculiar phenomena continued as one of the most lively interests of the correspondents of the Royal Society of London even to the end of the seventeenth century.[33]

Magical Traditions and the New Science

Some role for magic in the emergence of the "new science" in the sixteenth and seventeenth centuries has been established by recent scholarship. The older, simpler view of science and magic as opposing forces—progress versus superstition, empiricism versus mysticism —is no longer tenable. Nonetheless, magic's place in the Scientific Revolution cannot yet be fully specified, in part because much of the more interesting current scholarship has focused on individuals (Bruno, Dee, Harvey, Paracelsus, Newton), thus leaving many questions as to the general changes unanswered. For this reason I shall not attempt a full interpretation, but, rather, shall limit myself to several propositions on the issue, utilizing the foregoing description of *medieval* magic both to clarify what is known at present and to highlight areas for further research.[34]

Magic of any sort is fundamentally a technology, not a science. As medieval magic's view of the world was fully that of scholastic natural philosophy, magic's contributions to the *ideas* of the new science were rare and particular, rather than systematic. For example, Hermetic ideas about the central importance of the sun conditioned Copernicus's thinking about heliocentricity; and Harvey's ideas on the circulation of the blood reveal the impact of the magical and alchemical tradition of seeking microcosmic analogues to the macrocosm. The *spiritus* of the alchemists and Hermetic writers as well as the magical conception that nature is animate and active are important constituents of the intellectual ambience of Newton's struggle with forces and active principles, in response to the purely mechanistic natural philosophy of Descartes.[35]

Although at the conceptual level magic only assisted and suggested, at the practical level it fostered a new direction. Aristotelian natural philosophy was not unempirical, as it has sometimes been portrayed, but it had the passive empiricism of observation, and not the active empiricism of *praxis*. In striking contrast to the natural philosopher, the magician manipulated nature.[36] Even the nonempirical mysticism of Cabala had a practical side, attempting to capture the powers of superior things.[37] Della Porta wrote that magic is "the practical part

of natural philosophy. . . . The works of Magick are nothing else but
the works of Nature, whose dutiful hand-maid Magick is."[38] In
Christopher Marlowe's *Doctor Faustus*, the great magician solilo-
quizes about his ambition for power over nature:

> These metaphysics of magicians
> And necromantic books are heavenly:
> Lines, circles, signs, letters and characters—
> Ay, these are those that Faustus most desires.
> O what a world of profit and delight,
> Of power, of honor, of omnipotence,
> Is promised to the studious artisan!
> All things that move between the quiet poles
> Shall be at my command. Emperors and kings
> Are but obeyed in their several provinces,
> Nor can they raise the wind or rend the clouds;
> But his dominion that exceeds in this
> Stretcheth as far as doth the mind of man.
> A sound magician is a mighty god:
> Here, Faustus, try thy brains to gain a deity![39]

Magic, in both the scholastic and the Hermetic traditions, was
firmly linked in people's minds with mechanical things. The devices
of Roger Bacon are perhaps the most notable example in the Middle
Ages.[40] John Dee is said to have been considered a "conjurer" primarily for having designed elaborate stage machinery enabling an
actor to fly above the stage for Aristophanes' play *Peace*.[41] In 1648
John Wilkins's book on applied mechanics, simple machines and
their uses, was entitled *Mathematical Magic*.[42]

Renaissance Neoplatonism seems to have reintroduced a place for
mathematics in magic not found in scholastic magic. To be sure,
some simple numerology (magic numbers, magic squares, and so on)
had been present, but this does not compare with the flourishing
number mysticism which the Renaissance thinkers derived from
Plato, Pythagoras, and the Cabala.[43] Giovanni Pico claimed (and
others repeated the claim after him), among the eighty-five mathematical conclusions in his 900 "theses," that "By numbers a way is
had to the searching out and understanding of every thing able to be
known."[44] The Old Testament adage that the world was made in
"number, weight, and measure" (Book of Wisdom of Solomon
11:21) became a common slogan. John Dee tried to establish applied sciences within a Neoplatonic mathematical framework.[45] Kepler's work is suffused with mathematical mysticism.

The connections between magic and science in this transitional period cannot be understood apart from attention to wider social, economic, and religious developments. Here, a purely "internal" history of science seems most inadequate. Mining, military technology, and navigation blossomed in the sixteenth century. It was no coincidence that it was John Dee, an adviser on navigation to explorers, who published the first English *Euclid* (1570), or that William Gilbert's famous *De magnete* (1600) owes much to *The Newe Attractive* (1581) by Robert Norman, an "unlearned mathematician," who was a sailor for twenty years and a maker of compasses.[46] Intellectual developments must be analyzed against the changing background of religious and social change, for the vicissitudes of the Reformation, Puritanism, university reform, a new social mobility, commercial expansion, and the dangers of atheism and radicalism posed many of the problems—both social and intellectual—for the scientists of the sixteenth and seventeenth centuries.[47]

One of the most suggestive and integrative schemata of this transition has been proposed tentatively by Frances Yates.[48] She would have us consider a chronological succession of three types of workers: the Renaissance magus, closely allied with artistic expression, giving way in turn to the Rosicrucian type of magus, more interested in mechanical and mathematical magic, more influenced by Paracelsian alchemy and medicine, and more secretive. At this second stage, philanthropic impulses, often of an irenic or utopian character, come into play. Then, from the Rosicrucian emerges the scientist, abandoning the secretive tendency and applying mechanical technology in the service of humankind.

This attention to the ways in which Renaissance magic helped reorient people's approach to nature, thereby nurturing impulses connected with the new science, must not obscure the ungrateful way science "repaid" its debt: by bankrupting magic's metaphysics. The change was not effected overnight, but in the long run the new quantitative and mechanistic approach eventually established a metaphysics which left no room for essences, animism, hope, or purpose in nature, thus making magic something "unreal," or supernatural in the modern sense.

One example may clarify this transformation. Consider the phenomena of harmony and resonance in vibrating strings.[49] In a world of essences and sympathies—the world of both natural magic and Aristotelian science—the qualitative differences between strings made of sheep gut and of wolf gut sufficed to explain why the strings would be discordant, for there would be a natural antipathy between sheep and wolves.[50] The early seventeenth-century mathematician Jean Leurechon doubted such an explanation but could only ask, when confronted with sympathetic vibration: "Is it the occult sympathies, or is it in the strings being wound up to take like notes?"[51] By the 1630s the work of Galileo, Mersenne, and others had established a mathematical relationship for the vibrating string, which came to be known as Mersenne's Law. It states that the frequency of a string in vibration is proportional to the inverse of its length multiplied by the square root of the ratio of the tension to the mass-per-unit-length. This scientific triumph epitomizes a transformation both of the terms of physical explanation and of the character of the fundamental units of nature.

When Galileo and others identified a string's pitch—a "secondary quality"—with the frequency of its vibration and could demonstrate that this frequency is determined entirely by the string's length, density, and tension, the occult virtues were banished from a new world of "primary qualities," consisting only of mass, dimension, and time.[52] As the metaphysics of this new science came to supplant its predecessor, magic lost its intellectual foundation. In the dialectic of change from a dichotomized, Aristotelian world of separate terrestrial and celestial physics, open to explanation but not to exploitation, the unified, animated macrocosm that could be operated on by Rosicrucian number-magic formed a temporary transition which then gave way to an impersonal, objective, geometrized universe in which knowledge was power.

Notes

1. In 1905 Lynn Thorndike published *The Place of Magic in the Intellectual History of Europe* (New York, 1905; reprint ed. New York, 1967). His main work, however, was *The History of Magic and Experimental Science*, covering the first through the seventeenth century, 8 vols. (New York, 1923–58). At his death in 1968 he was still contributing important studies. A bibliography of his writings from 1905 to 1952 is given in *Osiris* 11 (1954):8–22.

2. Their arguments are discussed more fully at the end of this chapter, but it is perhaps appropriate here to clarify some terms. The fifteenth century saw a revival of interest in the full corpus of Plato's writings (the Latin Middle Ages had known only part of the *Timaeus* and a few other short pieces) and in the writings attributed to Hermes, a figure of vague and hoary antiquity, associated with the Egyptian god Thoth and said to have been Moses' teacher. In fact, as was to be demonstrated in the seventeenth century, the Hermetic writings were produced in the third and fourth centuries of the Christian era and were a rather eclectic combina-

tion of Neoplatonic, Stoic, Gnostic, and Christian elements. These elements seemed—as long as the false dating was accepted—to be proofs of Hermes' foreshadowing of Platonic and Christian wisdom. Many alchemical texts of antiquity and the Middle Ages exhibit the assumptions of this tradition. The living character of Hermetic thought in the writings and debates of the men who, in the sixteenth and seventeenth centuries, made the Scientific Revolution has been demonstrated by many studies cited elsewhere in this chapter. See especially the works of P. M. Rattansi, Allen G. Debus, and Richard S. Westfall.

Despite the number of important works on this subject, there has yet been no full and unified assessment of the contributions of Hermetic thought to science. An important part of any such assessment will be the question: How much of the undeniable effects of magical thought should be attributed to Hermetic magic, and how much to medieval scholastic magic?

3. Of course these two categories are not mutually exclusive. At times the experiences and practices of folk magic were found among the phenomena treated by the scholars. For example, the long-standing and widespread belief in fascination (the evil eye) was frequently analyzed by thinkers attempting to understand it in terms of Aristotelian physics and psychology; see p. 491. And with regard to beliefs in demons and the powers of witches, scholars exerted a quite unfortunate influence on the expressions of popular belief. Aquinas's influence is treated on p. 488. For a broader analysis of the impact of theoreticians on the form of the witchcraft mania, see H. R. Trevor-Roper, *The European Witch-Craze of the Sixteenth and Seventeenth Centuries and Other Essays* (New York, 1969), especially pp. 93–105.

4. Keith Thomas, in *Religion and the Decline of Magic* (New York, 1971), offers a wealth of factual material on the daily practices of magic and astrology and on their social context in England. His conclusion, however, that magic declined in the seventeenth century because people took on a new outlook of tolerance, self-help, and optimism regarding the ills of the world (pp. 659–63) seems to beg the question by leaving the sources of this new attitude unexplained. Another interpretation of the development of science and magic in England that takes into account the social and religious context is P. M. Rattansi, "The Social Interpretation of Science in the Seventeenth Century," in *Science and Society, 1600–1900*, ed. Peter Mathias (Cambridge, 1972), chap. 1.

5. *Magic, Science, and Religion and Other Essays* (New York, 1954 [original essay, 1925]), p. 87.

6. Greek notions of the natural are examined at length in Arthur O. Lovejoy and George Boas, *Primitivism and Related Ideas in Antiquity* (Baltimore, 1935). The shifting fortunes of magic—both its intellectual status and its popularity—seem to be connected to attitudes about a supposed golden age when things, being more primitive, were better. This connection illuminates the high evaluations placed on nature and on (medieval) magic by romantics and neoromantics.

7. "Letter to William of Paris on the Secret Works of Art and Nature and on the Nullity of Magic," in *Fr. Rogeri Bacon Opera quaedam hac-*

tenus inedita, ed. J. S. Brewer (London, 1859). I am quoting from the anonymous Elizabethan translation entitled "The Admirable Force and Efficacy of Art and Nature," published with Bacon's *Mirrour of Alchimy* (London, 1597), pp. 54, 64–65. See also A. G. Molland, "Roger Bacon as Magician," *Traditio* 30 (1974): 445–60.

8. *The Works of Francis Bacon*, ed. James Spedding, Robert Leslie Ellis, and Douglas Denon Heath (New York, 1869–72), vol. 8, pp. 409–11 (bk. 2, chap. 2). The Latin text is in vol. 2, pp. 189–90. The moral and esthetic debate of art versus nature has often been presented; see, for example, Shakespeare's *The Winter's Tale*, 4. 4, and the extensive literature on it.

9. The notion that marvels are simply phenomena not fully understood has a long and varied history. Augustine's version is discussed by Claude Jenkins, "Saint Augustine and Magic," in *Science, Medicine and History: Essays on the Evolution of Scientific Thought and Medical Practice written in honour of Charles Singer*, ed. E. Ashworth Underwood (London, 1953), 1: 131–40. In the mid-thirteenth century, Thomas Aquinas wrote: "We marvel at something when, seeing an effect, we do not know its cause. And since one and the same cause is at times known to certain people and unknown to others, it happens that, of those seeing an effect, some marvel and some do not; for an astronomer does not marvel at an eclipse of the sun because he knows the cause, but a person ignorant of this science, not knowing the cause, must marvel at it" (*De veritate catholicae fidei contra gentiles seu Summa philosophica* 3, 101 [Rome, 1926]). A century later, Nicole Oresme utilized the same notion to criticize belief in magic and marvels by arguing that a natural cause can be found for *any* supposedly marvelous effect (see the introduction to my forthcoming edition of Oresme's *Quodlibeta* (*Nicole Oresme and the Marvels of Nature* [Leiden, in press]).

At the very beginning of the seventeenth century, Tommaso Campanella, a contemporary of Francis Bacon, described the influence of new technological developments on the notion of magic: "Everything done by skilled men in imitating nature or helping it by unknown art is called magical work, not only by the vulgar crowds but by all men in general. . . . It was considered magic when Archytas made a dove which could fly like a natural one. And at the time of the Emperor Ferdinand in Germany, a German constructed an artificial eagle and a fly which flew by themselves. But since the art was not understood, all this was called magic. Later, all this became ordinary science.

"The invention of gunpowder . . . and of printing and the use of the lodestone were magical. But today they are common knowledge. The reverence for making of clocks and for the other mechanical arts is lost, for their methods have become obvious to everybody." Campanella, *Del senso delle cose e della magia* (1604), 4. 5, ed. A. Bruers (Bari, 1925), pp. 241–42. This translation is revised from that of Peter Munz in Eugenio Garin, *Science and Civic Life in the Italian Renaissance* (Garden City, N.Y., 1969), p. 145.

10. These attempts are discussed further in the next section of this chapter.

11. *De veritate catholicae fidei contra gentiles seu Summa philosophica* 3. 104–6, ed. cit., pp. 204–10; for further discussion of the nature of demons, see 3. 107–10, pp. 210–19. These texts of Aquinas and other relevant ones are excerpted in *Witchcraft in Europe, 1100–1700: A Documentary History*, ed. Alan C. Kors and Edward Peters (Philadelphia, 1972), pp. 53–73.

12. See also C. E. Hopkin, *The Share of Thomas Aquinas in the Growth of the Witchcraft Delusion* (Philadelphia, 1940); and Trevor-Roper, *European Witch-Craze*, pp. 115ff., 178, 184–85.

13. John Baptista Porta, *Natural Magick* (London, 1658); reprint ed. New York, 1957), p. 1. This is an anonymous English translation of the enlarged twenty-book edition (Naples, 1589) of a work first published in four books (Naples, 1558).

14. This phrasing is Herbert Marcuse's ("The Concept of Essence" in *Negations: Essays in Critical Theory*, trans. Jeremy J. Shapiro [Boston, 1968], pp. 46–47; original publication in *Zeitschrift für Sozialforschung* 5 [1936]). Marcuse argues that the Middle Ages made essence static, "as a structural law of the created world," thereby destroying the critical, dynamic character it formerly had for Plato and Aristotle.

15. Alchemy, we recall, was founded on the notion of the reality of forms and primitive undifferentiated matter, related in such a way that the form of gold might under proper conditions be imposed on the matter of a base metal after the removal or destruction of its own original form.

16. Such absolutes are fundamental to Aristotle's philosophy. His characterization of the prime mover, for example, is based on the fact that, "only what is primary can move the primary, what is simple the simple" (*De caelo* 288b1–5).

17. A survey of the discussions of fascination from antiquity to the seventeenth century would constitute a virtual history of the vicissitudes of the concept of action at a distance. For texts from Pliny, Avicenna, al-Ghazzali, Roger Bacon, Aquinas, Giles of Rome, and Oresme on fascination see my *Nicole Oresme and the Marvels of Nature* (commentary and notes to the passage at 70v27–71r8); for broader analyses of the problem of action at a distance, see Mary Hesse, *Forces and Fields: The Concept of Action at a Distance in the History of Physics* (London, 1961), and Max Jammer, *Concepts of Force* (rev. ed., New York, 1962). On the belief itself see Frederick T. Elworthy, *The Evil Eye: An Account of this Ancient and Widespread Superstition* (London, 1895; reprint ed. New York, 1970), and the many references in Thorndike, *History of Magic and Experimental Science*.

18. "Letter to William of Paris on . . . Art and Nature," in *Opera*, ed. Brewer, pp. 528–29. On "multiplication of species" see chap. 10, above, pp. 351–52.

19. In the fourteenth century, Nicole Oresme's naturalistic critique of astrology (*Questio contra divinatores*, ed. Stefano Caroti, in *Archives d'histoire doctrinale et littéraire du moyen âge* 43 [1976]: 201–310) belabors the problem of "unknown (*ignota*)" celestial influence. For Ficino and Pico, the *spiritus mundi* fills this need (see Frances A. Yates, *Giordano Bruno and the Hermetic Tradition* [Chicago, 1964], chaps. 4 and 5).

20. "Unde et naturam diffinierunt aliqui viri, vim rebus insitam procreandi similia ex similibus, et Aristoteles omnem operationem naturae dicit per similitudinem esse" (*Liber de vitiis et peccatis* in *Opera omnia* [Paris, 1674; reprint ed. Frankfurt am Main, 1963], vol. 1, p. 270, col. 2).

21. The *Marvels* was generally appended to the *Book of Secrets*, also falsely attributed to Albertus. These tracts were very popular, being reprinted many times in Latin, English, French, Italian, and German. The passage quoted here is from *The Book of Secrets of Albertus Magnus: Of the Virtues of Herbs, Stones and Certain Beasts; Also A Book of the Marvels of the World*, a recent reprint of the first English edition (ca. 1550), edited by Michael R. Best and Frank H. Brightman (Oxford, 1973), pp. 74–76. According to Thorndike the *Book of Secrets* was an important source for della Porta's *Natural Magick* (see *History of Magic and Experimental Science*, 6: 418–21).

22. The doctrine of the power of names took on a new importance in the seventeenth-century search for a natural or "philosophic" language in which words would correspond directly to the essences of things. By some this was supposed to be the language used by Adam when he gave all things their names (Genesis 2:18–20). Such interests hark back to Raymond Lull's *ars notoria* and continue through the efforts of John Amos Comenius and John Wilkins to the combinatory art which becomes the calculus of Leibniz. The mnemonic tradition within this complex of ideas has been analyzed by Frances A. Yates in *The Art of Memory* (Chicago, 1966), especially chap. 17. On the Adamic language, see pp. 179–83 of Hans Aarsleff, "Leibniz on Locke on Language," *American Philosophical Quarterly* 1 (1964): 165–88. See also Paolo Rossi, *Aspetti della rivoluzione scientifica* (Naples, 1971), chap. 7; also Rossi's *Clavis universalis: Arti mnemoniche e logica combinatoria da Lullo a Leibniz* (Milan, 1960), especially chap. 7; and Hans Aarsleff's articles on Jacob Boehme, John Amos Comenius, and John Wilkins in *The Dictionary of Scientific Biography*.

23. On contagion, one of the most important kinds of sympathetic magic, see the brilliant historical and crosscultural analysis by Mary Douglas, *Purity and Danger: An Analysis of Concepts of Pollution and Taboo* (London, 1966). Ludwig Edelstein has argued that sympathy became a scientific principle of Greek medicine in the 4th century B.C. and only later became magical with the rise of Neoplatonic philosophy and Christianity, when sympathetic effects were explained not by physical but by psychic causes; see his essay "Greek Medicine in its Relation to Religion and Magic," *Bulletin of the Institute of the History of Medicine* 5 (1937):230–34.

24. Consider the remarks of Richard S. Westfall in his valuable recent study of "Newton and the Hermetic Tradition," in *Science, Medicine and Society in the Renaissance: Essays to Honour Walter Pagel*, ed. Allen G. Debus (New York, 1972), 2: 183–98: "I intend the phrase 'Hermetic tradition' to have a general meaning rather than a specific one. . . . I do not insist that individual ideas I call Hermetic were original or unique with that system of philosophy. . . . As I use it here, the 'Hermetic tradition' refers primarily to three closely related aspects of a conception of

nature. . . . First, nature was seen as active. . . . Second, nature was seen as animate. . . . Third, nature was seen as psychic. Projecting the mind onto nature, the Hermetic tradition understood the characteristic actions of bodies in similar terms. Bodies exercise influences on each other, sympathies and antipathies. . . . The attraction of like for like, however it was expressed, was one of the characteristic ideas of the Hermetic tradition. . . . 'Occult qualities' could refer to many things, but above all the phrase referred to the modes of explanation prevalent in the Hermetic tradition" (pp. 183–85). Westfall's analysis of the importance of *magical* modes of thought in Newton's environment seems sound (and important); I would only suggest that most of these factors were present in Western science long before the revival of Hermeticism. The writings of Yates, Rattansi, and Westfall have assured a recognition of some role for magic in the Scientific Revolution; it remains, however, to clarify the peculiar contribution of Hermetic magic.

25. See Jammer, *Concepts of Force*, pp. 31–32 on Plato and Aristotle. On pp. 43–44 Jammer shows that in antiquity *pneuma* was at times assigned the role of the material substratum necessary for the operation of sympathy.

26. For this reason it may be useful to mention a few important secondary treatments that will guide the reader to the (largely unpublished) primary sources. The most important is Thorndike's *History of Magic and Experimental Science*, 8 vols., followed by Yates's *Giordano Bruno*. Also useful are Henry Charles Lea, *Materials toward a History of Witchcraft*, 3 vols. (arranged and edited by Arthur C. Howland, New York, 1957); Wayne Shumaker, *The Occult Sciences in the Renaissance: A Study in Intellectual Patterns* (Berkeley, 1972); Thomas, *Religion and the Decline of Magic*; D. P. Walker, *Spiritual and Demonic Magic from Ficino to Campanella* (London, 1958); and Marie Boas, *The Scientific Renaissance* (New York, 1962), chap. 6.

27. A listing of all the questions of hundreds of Quodlibets reveals many questions about magical phenomena; see P. Glorieux, *La littérature quodlibétique*, 2 vols. (Kain [Belgium]/Paris, 1925–35). Lists of "errors" can be found in Giles of Rome, *Errores philosophorum* (ed. Josef Koch and trans. John O. Reidl [Milwaukee, 1944]), and in the famous condemnation by bishop Stephen Tempier at Paris in 1277 (text in the *Chartularium Universitatis Parisiensis*, ed. H. Denifle and E. Chatelain [Paris, 1889–97]), 1: 543–48; translation in *Medieval Political Philosophy: A Sourcebook*, ed. Ralph Lerner and Muhsin Mahdi [New York, 1963]).

28. See John Ferguson, *Bibliographical Notes on Histories of Inventions and Books of Secrets*, 2 vols. (London, 1959); and Thorndike, *History of Magic and Experimental Science*.

29. Brian Lawn, *The Salernitan Questions. An Introduction to the History of Medieval and Renaissance Problem Literature* (Oxford, 1963); and Nancy Siraisi, "The *Expositio Problematum Aristotelis* of Peter of Abano," *Isis* 61 (1970): 321–39.

30. Lawn, *Salernitan Questions*, pp. 162–63, 184.

31. Glorieux, *La littérature quodlibétique*, 1: 145.

32. Siraisi, *"Expositio Problematum,"* p. 339, for Peter. All the rest (except Renaudot) are cited in Thorndike, *History of Magic and Experimental Science*; for Renaudot see Lawn, *Salernitan Questions*, pp. 142–44. A brief review of the juridical use of this phenomenon as a test of an accused murderer's innocence is W. G. Aitchison Robertson, "Bier-Right," in the proceedings of *Ve Congrès International d'Histoire de la Médecine* (Geneva, 1926), pp. 192–98.

33. See Thomas Birch, *The History of the Royal Society of London for Improving of Natural Knowledge from Its First Rise*, 4 vols. (London, 1756–57; reprint ed. New York, 1968). A very useful index of proper names has recently been published by Gail Ewald Scala, "An Index to Birch, *History of the Royal Society*," *Notes and Records of the Royal Society of London* 28 (1974):263–329.

34. In addition to the various studies on Hermeticism and science cited elsewhere in this chapter, the following must be noted: Allen G. Debus, "Renaissance Chemistry and the Work of Robert Fludd," in *Alchemy and Chemistry in the Seventeenth Century*, William Andrews Clark Memorial Library Seminar (Los Angeles, 1966); Debus, "The Chemical Dream of the Renaissance," Churchill College Overseas Fellowship Lecture (Cambridge, 1968); Debus, *The English Paracelsians* (New York, 1966); Walter Pagel, *Paracelsus: An Introduction to Philosophical Medicine in the Era of the Renaissance* (Basel, 1958); P. M. Rattansi, "Newton's Alchemical Studies," in *Science, Medicine and Society in the Renaissance*, ed. Debus, 2: 167–82; P. M. Rattansi and J. E. McGuire, "Newton and the Pipes of Pan," *Notes and Records of the Royal Society of London* 21 (1966):108–43; Paolo Rossi, *Francis Bacon: From Magic to Science*, trans. Sacha Rabinovitch (London, 1968); Rossi, "Hermeticism, Rationality and the Scientific Revolution," in *Reason, Experiment and Mysticism in the Scientific Revolution*, ed. M. L. Righini Bonelli and W. Shea (New York, 1975); and Frances A. Yates, *Theatre of the World* (Chicago, 1969).

Mary Hesse attacked this historiography of magic and seventeenth-century science in "Hermeticism and Historiography: An Apology for the Internal History of Science," in *Historical and Philosophical Perspectives of Science*, ed. Roger H. Stuewer (Minneapolis, 1970), pp. 134–60, and in "Reasons and Evaluations in the History of Science," in *Changing Perspectives in the History of Science: Essays in Honour of Joseph Needham*, ed. Mikuláš Teich and Robert Young (London, 1973), pp. 127–47. That her philosophically based objections do not withstand the comparison with historical evidence is effectively shown by P. M. Rattansi in "Some Evaluations of Reason in Sixteenth- and Seventeenth-Century Natural Philosophy," in *Changing Perspectives*, ed. Teich and Young, pp. 148–66.

35. Copernicus cited Trimegistus (i.e., Hermes Trismegistus) in bk. 1, chap. 10 (fol. 9v) of the *De revolutionibus orbium coelestium* (Nuremberg, 1543); see also Thomas W. Africa, "Copernicus' Relation to Aristarchus and Pythagoras," *Isis* 52 (1961):403–9, with Edward Rosen's critique, "Was Copernicus a Pythagorean?" *Isis* 53 (1962):504–8, and Africa's reply, ibid., p. 509. See also Rosen's "Was Copernicus a Herme-

tist?" in *Historical and Philosophical Perspectives of Science*, ed. Stuewer, pp. 163–71. Rosen answers both questions in the negative, concluding that "out of Renaissance magic and astrology came, not modern science, but modern magic and astrology."

On Harvey, see Walter Pagel, "William Harvey and the Purpose of Circulation," *Isis* 42 (1951):22–38, and his *William Harvey's Biological Ideas* (Basel, 1967).

On Newton, see Westfall's "Newton and the Hermetic Tradition"; also Rattansi, "Some Evaluations," and "Newton's Alchemical Studies."

36. Recent discussions of magic and technology include Paolo Rossi, *Philosophy, Technology, and the Arts in the Early Modern Era*, trans. Salvatore Attanasio, ed. Benjamin Nelson (New York, 1970; original ed., Milan, 1962); Rossi, *Francis Bacon*, chap. 1; and Peter J. French, *John Dee: The World of an Elizabethan Magus* (London, 1972), chap. 7 and passim.

37. Pico della Mirandola's efforts to unite practical Cabala with natural magic are treated in Yates, *Giordano Bruno*, pp. 84–116.

38. *Natural Magick*, p. 2.

39. *Tragedy of Doctor Faustus* (1588), 1: 52–66.

40. Compare the descriptions given by Roger Bacon (at the beginning of this chapter) with the analysis of Campanella quoted in note 9, above.

41. French, *John Dee*, p. 24.

42. The full title reads: "Mathematical Magic: or, the Wonders that may be performed by Mechanical Geometry. In two books. [Bk. 1] Archimedes, or Mechanical Powers. [Bk. 2] Daedalus, or Mechanical Motions. Being the most easy, pleasant, useful (and yet most neglected Part) of the Mathematics. Not before treated of in this language." See, too, Alex Keller, "Mathematical Technologies and the Growth of the Idea of Technical Progress in the Sixteenth Century," in *Science, Medicine and Society in the Renaissance*, ed. Debus, 1: 11–27, which includes examples of magical devices.

43. See Allen G. Debus, "Mathematics and Nature in the Chemical Texts of the Renaissance," *Ambix* 15 (1968):1–28; Ernst Cassirer, *The Individual and the Cosmos in Renaissance Philosophy*, trans. Mario Domandi (New York, 1963; original ed. Leipzig, 1927); French, *John Dee*; and Nicholas H. Clulee, "John Dee's Mathematics and the Grading of Compound Qualities," *Ambix* 18 (1971):178–211.

44. Giovanni Pico della Mirandola, *Conclusiones sive theses DCCCC Romae anno 1486 publice disputandae, sed non admissae*, ed. Bohdan Kieszkowski (Geneva, 1973), p. 74.

45. John Dee, "Mathematical Preface," in Euclid, *The Elements of Geometry*, trans. Henry Billingsley (London, 1570), and reprinted in facsimile, New York, 1975, with Introduction by Allen G. Debus; I. R. F. Calder, *John Dee Studied as an English Neoplatonist* (Ph.D. dissertation, University of London, 1952), pp. 628–45. I am presently preparing an annotated edition of Dee's *Preface*.

46. On Norman and Gilbert, see Edgar Zilsel, "The Origins of William Gilbert's Scientific Method," *Journal of the History of Ideas* 2 (1941): 1–32; and Rossi, *Philosophy, Technology and the Arts*, pp. 4–5.

47. For England some headway has been made in this regard. See the following studies: Rattansi, "Social Interpretation of Science"; Hugh Kearney, *Scholars and Gentlemen: Universities and Society in Preindustrial Britain, 1500–1700* (Ithaca, New York, 1970); Kearney, *Science and Change, 1500–1700* (New York, 1971). For Europe, see Frances A. Yates, *The Rosicrucian Enlightenment* (London, 1972), and R. J. W. Evans, *Rudolf II and His World: A Study in Intellectual History, 1576–1612* (Oxford, 1973).

48. Frances A. Yates, "The Hermetic Tradition in Renaissance Science," in *Art, Science and History in the Renaissance* ed. Charles S. Singleton (Baltimore, 1968), pp. 255–74.

49. The details of this example are drawn from Sigalia Dostrovsky, "Early Vibration Theory: Physics and Music in the Seventeenth Century," *Archive for History of Exact Sciences* 14 (1975): 169–218; some of this material is summarized in her doctoral dissertation, *The Origins of Vibration Theory: The Scientific Revolution and the Nature of Music* (Princeton University, 1969). See also Claude V. Palisca, "Scientific Empiricism in Musical Thought," in *Seventeenth Century Science and the Arts*, ed. Hedley Howell Rhys (Princeton, 1961), pp. 91–137.

50. The idea is a commonplace; della Porta explains it in his *Natural Magick*, bk. 20, chap. 7, p. 403. Similarly, drums of wolf hide supposedly would silence ones of sheep skin (see the remarks of "Albertus" on this matter in the *Marvels*, quoted above).

51. *Mathematical Recreation. Or a Collection of sundrie problems . . . now delivered in the English tonge, with the examinations, corrections and augmentations* [by W. Oughtred] (London, 1633), p. 126. Leurechon originally published this under the pseudonym of Hendrik van Etten in Lyon, 1627. About the middle of the century we find Walter Charleton still accepting the discord of the lute strings made from sheep and wolf gut, although he rejects the phenomenon of the wolfhide drum silencing the sheepskin drum (see Thorndike, *History of Magic and Experimental Science*, 7: 461).

52. On the ascendency of primary over secondary qualities, see E. A. Burtt, *The Metaphysical Foundations of Modern Physical Science* (New York, 1924), and Alexandre Koyré, "The Significance of the Newtonian Synthesis," in his *Newtonian Studies* (Cambridge, Massachusetts, 1965), pp. 3–24.

Suggestions for
Further Reading

General

Crombie, A. C. *Augustine to Galileo*. London, 1952. Reissued as *Medieval and Early Modern Science*, 2 vols. Garden City, N.Y., 1959.

Dictionary of Scientific Biography. 15 vols. New York, 1970–.

Dijksterhuis, E. J. *The Mechanization of the World Picture*. Translated by C. Dikshoorn. Oxford, 1961.

Duhem, Pierre, *Etudes sur Léonard de Vinci*. 3 vols. Paris, 1906–13.

———. *Le système du monde*. 10 vols. Paris, 1913–59.

Grant, Edward. *Physical Science in the Middle Ages*. New York, 1971.

———, ed. *A Source Book in Medieval Science*. Cambridge, Mass., 1974.

Haskins, Charles H. *Studies in the History of Mediaeval Science*, 2d ed. Cambridge, Mass., 1927.

Lewis, C. S. *The Discarded Image*. Cambridge, 1964.

Maier, Anneliese. *An der Grenze von Scholastik und Naturwissenschaft*, 2d ed. Rome, 1952.

———. *Metaphysische Hintergründe der spätscholastischen Naturphilosophie*. Rome, 1955.

———. *Die Vorläufer Galileis im 14. Jahrhundert*, 2d ed. Rome, 1966.

———. *Zwei Grundprobleme der scholastischen Naturphilosophie*, 2d ed. Rome, 1951.

———. *Zwischen Philosophie und Mechanik*. Rome, 1958.

Murdoch, John E., and Sylla, Edith D., eds. *The Cultural Context of Medieval Learning: Pro-*

ceedings of the First International Colloquium on Philosophy, Science, and Theology in the Middle Ages—September 1973. Dordrecht, 1975.

Pedersen, Olaf, and Pihl, Mogens. *Early Physics and Astronomy: A Historical Introduction.* London, 1974.

Sarton, George. *Introduction to the History of Science.* 3 vols. Baltimore, 1927–48.

Taton, René, ed. *La science antique et médiévale.* Paris, 1957. Translated as *Ancient and Medieval Science.* New York, 1963.

Thorndike, Lynn. *A History of Magic and Experimental Science.* 8 vols. New York, 1923–58.

Weisheipl, James A., O.P. *The Development of Physical Theory in the Middle Ages.* New York, 1959.

The Ancient Background

Africa, Thomas W. *Science and the State in Greece and Rome.* New York, 1968.

Clagett, Marshall. *Greek Science in Antiquity.* London, 1957.

Cohen, Morris R., and Drabkin, I. E., eds. *A Source Book in Greek Science.* Cambridge, Mass., 1958.

Farrington, Benjamin. *Greek Science.* 2 vols. London, 1944–49. Reissued in one volume. London, 1953.

Lloyd, G. E. R. *Aristotle.* Cambridge, 1968.

———. *Early Greek Science: Thales to Aristotle.* London, 1970.

———. *Greek Science after Aristotle.* London, 1973.

Neugebauer, Otto. *The Exact Sciences in Antiquity,* 2d ed. Providence, 1957.

Sambursky, S. *The Physical World of the Greeks.* Translated by Merton Dagut. London, 1956.

Islamic Science

Arnold, Sir Thomas, and Guillaume, Alfred, eds. *The Legacy of Islam.* London, 1931.

Fakhry, Majid. *A History of Islamic Philosophy.* New York, 1970.

Nasr, Seyyed Hossein. *Science and Civilization in Islam.* Cambridge, Mass., 1968.

Peters, F. E. *Allah's Commonwealth: A History of Islam in the Near East, 600–1100 A.D.* New York, 1973.

Schacht, Joseph, and Bosworth, C. E., eds. *The Legacy of Islam,* 2d ed. Oxford, 1974.

Early Latin Science

Brehaut, Ernest. *An Encyclopedist of the Dark Ages: Isidore of Seville.* New York, 1912.

Chenu, M.-D., O.P. *Nature, Man, and Society in the Twelfth Century.* Translated by Jerome Taylor and Lester K. Little. Chicago, 1968.

Courcelle, Pierre. *Late Latin Writers and their Greek Sources.* Translated by Harry E. Wedeck. Cambridge, Mass., 1969.

Fontaine, Jacques. *Isidore de Seville et la culture classique dans l'espagne wisigothique.* 2 vols. Paris, 1959.

Haskins, Charles H. *The Renaissance of the Twelfth Century.* Cambridge, Mass., 1927.

Laistner, M. L. W. *Thought and Letters in Western Europe, A.D. 500–900,* 2d ed. London, 1957.

Riché, Pierre. *Education et culture dans l'occident barbare, VIᵉ–VIIIᵉ siècles.* Paris, 1962.

Stahl, William H. *Roman Science.* Madison, Wis., 1962.

Stahl, William H., with Johnson, Richard, and Burge, E. L. *Martianus Capella and the Seven Liberal Arts,* vol. 1, *The Quadrivium of Martianus Capella, Latin Traditions in the Mathematical Sciences 50 B.C.–A.D. 1250.* New York, 1971.

The Philosophical Setting of Medieval Science

Carré, Meyrick H. *Realists and Nominalists.* Oxford, 1946.

Copleston, Frederick, S.J. *A History of Philosophy,* vol. 2, *Medieval Philosophy, Augustine to Scotus.* Westminster, Md., 1950.

Gilson, Etienne. *History of Christian Philosophy in the Middle Ages.* New York, 1955.

————. *Reason and Revelation in the Middle Ages.* New York, 1938.

Leff, Gordon. *The Dissolution of the Medieval Outlook: An Essay on the Intellectual and Spiritual Change in the Fourteenth Century.* New York, 1976.

————. *Medieval Thought: St. Augustine to Ockham.* Baltimore, 1958.

Moody, Ernest A. *Studies in Medieval Philosophy, Science, and Logic: Collected Papers, 1933–1969.* Berkeley, 1975.

New Catholic Encyclopedia. 15 vols. New York, 1967.

Pieper, Josef. *Scholasticism: Personalities and Problems of Medieval Philosophy.* Translated by Richard and Clara Winston. New York, 1960.

van Steenberghen, Fernand. *Aristotle in the West.* Translated by Leonard Johnston. Louvain, 1955.

Vignaux, Paul. *Philosophy in the Middle Ages.* Translated by E. C. Hall. London, 1959.

Wallace, William A. *Causality and Scientific Explanation*, vol. 1, *Medieval and Early Classical Science.* Ann Arbor, 1972.

Weinberg, Julius R. *A Short History of Medieval Philosophy.* Princeton, 1964.

Medieval Universities

Baldwin, John W. *The Scholastic Culture of the Middle Ages, 1000–1300.* Lexington, Mass., 1971.

Beaujouan, Guy. "Motives and Opportunities for Science in the Medieval Universities." In *Scientific Change*, edited by A. C. Crombie. London, 1963. Pp. 219–36.

Cobban, A. B. *The Medieval Universities: Their Development and Organization.* London, 1975.

Daly, Lowrie J., S.J. *The Medieval University.* New York, 1961.

Gabriel, Astrik L. *Garlandia: Studies in the History of the Mediaeval University.* Notre Dame, Ind., 1969.

Haskins, Charles H. *The Rise of Universities.* New York, 1923.

Kibre, Pearl. *The Nations in the Mediaeval Universities.* Cambridge, Mass., 1948.

————. "The Quadrivium in the Thirteenth Century Universities (with Special Reference to Paris)." In *Arts libéraux et philosophie au moyen âge. Actes du quatrième congrès international de philosophie médiévale, Université de Montréal, 27 August–2 September 1967.* Montreal, 1969. Pp. 175–91.

————. *Scholarly Privileges in the Middle Ages.* Cambridge, Mass., 1962.

Leff, Gordon. *Paris and Oxford Universities in the Thirteenth and Fourteenth Centuries.* New York, 1968.

Rashdall, Hastings. *The Universities of Europe in the Middle Ages*, 3 vols. Edited by F. M. Powicke and A. B. Emden. Oxford, 1936.

Siraisi, Nancy G. *Arts and Sciences at Padua: The Studium at Padua before 1350.* Toronto, 1973.

Weisheipl, James A., O.P. "Curriculum of the Faculty of Arts at Oxford in the Early Fourteenth Century." *Mediaeval Studies* 26 (1964): 143–85.

————. "Developments in the Arts Curriculum at Oxford in the Early Fourteenth Century." *Mediaeval Studies* 28 (1961): 151–75.

Wieruszowski, Helene. *The Medieval University.* Princeton, 1966.

Mathematics

Boyer, Carl B. *A History of Mathematics*. New York, 1968.
Cantor, Moritz. *Vorlesungen über Geschichte der Mathematik*, vols, 1–2. Leipzig, 1880–92 and subsequent editions.
Clagett, Marshall. *Archimedes in the Middle Ages*. 3 vols. Madison-Philadelphia, 1964–.
Juschkewitsch, A. P. *Mathematik in Mittelalter*. Leipzig, 1964.
Zeuthen, H. G. *Histoire des mathématiques dans l'antiquité et le moyen âge*. Paris, 1902.

The Science of Weights

Brown, Joseph E. "The *Scientia de Ponderibus* in the Later Middle Ages." Ph.D. dissertation, University of Wisconsin, 1967.
Clagett, Marshall. "Leonardo da Vinci: Mechanics." *Dictionary of Scientific Biography*, 8: 215–34.
———. *The Science of Mechanics in the Middle Ages*. Madison, Wis., 1959.
Duhem, Pierre. *Les origines de la statique*. 2 vols. Paris, 1905–6.
Moody, Ernest A., and Clagett, Marshall. *The Medieval Science of Weights*. Madison, Wis., 1952.

The Science of Motion

Clagett, Marshall. *Nicole Oresme and the Medieval Geometry of Qualities and Motions*. Madison, Wis., 1968.
———. *The Science of Mechanics in the Middle Ages*. Madison, Wis., 1959.
Crosby, H. Lamar, Jr. *Thomas of Bradwardine: His Tractatus de Proportionibus. Its Significance for the Development of Mathematical Physics*. Madison, Wis., 1961.
Grant, Edward. *Nicole Oresme: De proportionibus proportionum and Ad pauca respicientes*. Madison, Wis., 1966.
Maier, Anneliese. See above, under *General*.
Molland, A. G. "The Geometrical Background to the 'Merton School'." *British Journal for the History of Science* 4 (1968–69): 108–25.
Moody, Ernest A. "Galileo and His Precursors." In *Galileo Reappraised*, edited by Carlo Golino. Berkeley, 1966. Pp. 23–43.
Murdoch, John E., and Sylla, Edith D. "Swineshead, Richard." *Dictionary of Scientific Biography*, 13: 184–214.
Wilson, Curtis. *William Heytesbury: Medieval Logic and the Rise of Mathematical Physics*. Madison, Wis., 1960.

Cosmology

Grant, Edward. "Medieval and Seventeenth-Century Conceptions of an Infinite Void Space Beyond the Cosmos." *Isis*, 60 (1969): 39–60.

———. "Medieval Explanations and Interpretations of the Dictum that 'Nature Abhors a Vacuum'." *Traditio* 29 (1973): 327–55.

———. *Nicole Oresme and the Kinematics of Circular Motion.* Madison, Wis., 1971.

Oresme, Nicole. *Le livre du ciel et du monde.* Edited and translated by A. D. Menut and A. J. Denomy. Madison, Wis., 1968.

Thorndike, Lynn. *The Sphere of Sacrobosco and Its Commentators.* Chicago, 1949.

Astronomy

Benjamin, Francis S., Jr., and Toomer, G. J. *Campanus of Novara and Medieval Planetary Theory: Theorica planetarum.* Madison, Wis., 1971.

Dreyer, J. L. E. *A History of Astronomy from Thales to Kepler*, 2d ed. New York, 1953.

Kuhn, Thomas S. *The Copernican Revolution: Planetary Astronomy in the Development of Western Thought.* Cambridge, Mass., 1957.

North, J. D. "The Astrolabe." *Scientific American* 230, no. 1 (January 1974): 96–106.

Pedersen, Olaf. *A Survey of the Almagest.* Odense, 1974.

Pedersen, Olaf, and Pihl, Mogens. See above, under *General.*

Optics and Meteorology

Boyer, Carl B. *The Rainbow: From Myth to Mathematics.* New York, 1959.

Crombie, A. C. *Robert Grosseteste and the Origins of Experimental Science, 1100–1700.* Oxford, 1953.

Eastwood, Bruce S. "Grosseteste's 'Quantitative' Law of Refraction: A Chapter in the History of Non-Experimental Science." *Journal of the History of Ideas* 28 (1967): 403–14.

———. "Medieval Empiricism: The Case of Grosseteste's Optics." *Speculum* 43 (1968): 306–21.

Lindberg, David C. *John Pecham and the Science of Optics.* Madison, Wis., 1970.

———. *Theories of Vision from al-Kindi to Kepler.* Chicago, 1976.

———. "The Theory of Pinhole Images from Antiquity to the Thirteenth Century." *Archive for History of Exact Sciences* 5 (1968): 154–76.

———. "The Theory of Pinhole Images in the Fourteenth Century." *Archive for History of Exact Sciences* 6 (1970): 299–325.

Rosen, Edward. "The Invention of Eyeglasses." *Journal of the History of Medicine and Allied Sciences* 11 (1956): 13–46, 183–218.

The Science of Matter

Holmyard, E. J. *Alchemy*. Harmondsworth, 1957.

McMullin, Ernan, ed. *The Concept of Matter in Greek and Medieval Philosophy*. Notre Dame, Ind., 1963.

Multhauf, Robert P. *The Origins of Chemistry*. London, 1966.

Read, John. *Prelude to Chemistry: An Outline of Alchemy*, 2d ed. London, 1939.

Taylor, F. Sherwood. *The Alchemists*. New York, 1962.

Dales, Richard C. "Marius 'On the Elements' and the Twelfth-Century Science of Matter." *Viator* 3 (1972): 191–218.

Medicine

Bonser, Wilfrid. *The Medical Background of Anglo-Saxon England*. London, 1963.

Bullough, Vern L. *The Development of Medicine as a Profession: The Contribution of the Medieval University to Modern Medicine*. Basel, 1966.

Corner, George W. *Anatomical Texts of the Earlier Middle Ages: A Study in the Transmission of Culture, with a Revised Latin Text of the Anatomia Cophonis and Translations of Four Texts*. Washington, D.C., 1927.

Demaitre, Luke. "Theory and Practice in Medical Education at the University of Montpellier in the Thirteenth and Fourteenth Centuries." *Journal of the History of Medicine and Allied Sciences* 30 (1975): 103–23.

Kibre, Pearl. "The Faculty of Medicine at Paris, Charlatanism, and Unlicensed Medical Practices in the Later Middle Ages." *Bulletin of the History of Medicine* 27 (1953): 1–20.

———. "Hippocratic Writings in the Middle Ages." *Bulletin of the History of Medicine* 18 (1945): 371–412.

Kristeller, Paul Oskar. "The School of Salerno: Its Development and Its Contribution to the History of Learning." *Bulletin of the History of Medicine* 17 (1945): 138–94.

MacKinney, Loren C. *Early Medieval Medicine, with Special Reference to France and Chartres*. Baltimore, 1937.

———. *Medical Illustrations in Medieval Manuscripts*. London, 1965.

McVaugh, Michael. "The "Humidum radicale' in Thirteenth-Century Medicine." *Traditio* 30 (1974): 259–83.

———. Introduction to Arnald of Villanova's *Aphorismi de gradibus* in *Arnaldi de Villanova Opera medica omnia*, vol. 2. Granada and Barcelona, 1975.

———. "Quantified Theory and Practice at Fourteenth-Century Montpellier." *Bulletin of the History of Medicine* 43 (1969): 397–413.

Rubin, Stanley. *Medieval English Medicine*. Newton Abbot, 1974.

Saffron, Morris H. *Maurus of Salerno, Twelfth-Century "Optimus physicus," with his Commentary on the Prognostics of Hippocrates*. Transactions of the American Philosophical Society. Vol. 62, pt. 1. Philadelphia, 1972.

Schipperges, Heinrich. *Die Assimilation der arabischen Medizin durch das lateinische Mittalalter. Sudhoffs Archiv für Geschichte der Medizin und der Naturwissenschaften*, Beiheft 8. Wiesbaden, 1964.

Seidler, Eduard. *La médecine à Paris au XIVe siècle*. Paris, 1966.

Singer, Charles, trans. *The Fasciculo di medicina, Venice 1493*. 2 vols. Florence, 1925.

Talbot, Charles H. *Medicine in Medieval England*. London, 1967.

Temkin, Owsei. *Galenism*. Ithaca, 1973.

Theodoric of Cervia. *The Surgery of Theodoric, ca. A.D. 1267*. Translated by Eldridge Campbell and James Colton. 2 vols. New York, 1955.

Ziegler, Philip. *The Black Death*. London, 1969.

Natural History

Albertus Magnus. *Book of Minerals*. Translated by Dorothy Wyckoff. Oxford, 1967.

Brodin, Gösta. *Agnus Castus*. Uppsala, 1950.

Eamon, William, "Books of Secrets and Popular Science in England." Ph.D. dissertation, University of Kansas, 1977.

Frisk, Gösta, ed. *A Middle English Translation of Macer Floridus De viribus herbarum*. Uppsala, 1949.

Larkey, Sanford V., and Pyles, T., eds. *An Herbal [1525]*. New York, 1941.

McCulloch, Florence. *Medieval Latin and French Bestiaries*. Chapel Hill, 1962.

White, T. H. *The Bestiary: A Book of Beasts, being a Translation from a Latin Bestiary of the Twelfth Century*. New York, 1960.

Classification of the Sciences

Aquinas, Saint Thomas. *The Division and Methods of the Sciences*
3d ed. Translated by Armand Maurer. Toronto, 1963.
Taylor, Jerome. *The Didascalicon of Hugh of St. Victor: A Medieval
Guide to the Arts.* New York, 1961.
Wiesheipl, James A., O.P. "Classification of the Sciences in Medieval
Thought." *Mediaeval Studies* 27 (1965): 54–90.

Magic and the Occult Arts

Cohn, Norman. *Europe's Inner Demons: An Enquiry Inspired by
the Great Witch-Hunt.* London, 1975.
Molland, A. G. "Roger Bacon as Magician." *Traditio* 30 (1974):
445–60.
Thomas, Keith. *Religion and the Decline of Magic.* New York, 1971.
Thorndike, Lynn. See above, under *General.*
Walker, D. P. *Spiritual and Demonic Magic from Ficino to Cam-
panella.* Studies of the Warburg Institute, vol. 22. London, 1958.
Wedel, Theodore O. *The Mediaeval Attitude toward Astrology, Par-
ticularly in England.* New Haven, 1920.
Yates, Frances A. *Giordano Bruno and the Hermetic Tradition.* Chi-
cago, 1964.

Notes on Contributors

Joseph E. Brown Received his Ph.D. in history of science from the University of Wisconsin in 1967. He also has a degree in theology from Kenrick Theological Seminary in St. Louis. He spent 1969–73 at the Institute for Advanced Study (Princeton) as visiting member and research assistant and is presently assistant professor of history and political science at Rensselaer Polytechnic Institute. His principal research interest is the medieval science of weights, on which he wrote his doctoral dissertation: "The 'Scientia de Ponderibus' in the Later Middle Ages."

Edward Grant Received his Ph.D. in history of science from the University of Wisconsin in 1957. He is professor of history and the history and philosophy of science at Indiana University. He has published *Nicole Oresme: De proportionibus proportionum and Ad pauca respicientes* (Madison, Wis., 1966); *Nicole Oresme and the Kinematics of Circular Motion. Tractatus de commensurabilitate vel incommensurabilitate motuum celi* (Madison, Wis., 1971); *Physical Science in the Middle Ages* (New York, 1971); and *A Source Book in Medieval Science* (Cambridge, Mass., 1974). With the focus of his current research in medieval cosmology, he is completing studies on the medieval concepts of space and place.

Bert Hansen Earned his Ph.D. in the history of science at Princeton University in 1973. He is now assistant professor of history at the State University of New York at Binghamton. His *Nicole Oresme and the Marvels of Nature: A Critical Edition of his "Quodlibeta" with English Translation and Commentary* will appear in *Studien und Texte zur Geistesgeschichte des Mittelalters* (Leiden, in press). He has also contributed articles to the *Dictionary of Scientific Biography* and reviews to *Isis* and the *American Historical Review*. His current research is on John Dee and early modern science.

Pearl Kibre

Earned her Ph.D. in history from Columbia University in 1936 under the direction of Lynn Thorndike. She is currently professor emeritus of history at Hunter College and the Graduate Center of the City University of New York, where she is still in residence. At the present time she is president of the Fellows of the Mediaeval Academy of America. Her many publications include the following books: *The Library of Pico della Mirandola* (New York, 1936); (with Lynn Thorndike) *A Catalogue of Incipits of Mediaeval Scientific Writings in Latin* (Cambridge, Mass., 1937; rev. ed. 1963); *The Nations in the Mediaeval Universities* (Cambridge, Mass., 1948); and *Scholarly Privileges in the Middle Ages: The Rights, Privileges, and Immunities of Scholars and Universities at Bologna, Padua, Paris and Oxford* (Cambridge, Mass., 1961). She is now working on "Hippocrates Latinus," of which the first two installments have been published in *Traditio* (1975 and 1976).

David C. Lindberg

Holds an M.S. in physics from Northwestern University and the Ph.D. in history and philosophy of science from Indiana University (1965). He is professor of the history of science at the University of Wisconsin–Madison. His publications have fallen principally within the domain of medieval and Renaissance optics and include *John Pecham and the Science of Optics: Perspectiva communis* (Madison, Wis., 1970); *John Pecham: Tractatus de perspectiva* (St. Bonaventure, N.Y., 1972); *A Catalogue of Medieval and Renaissance Optical Manuscripts* (Toronto, 1975); and *Theories of Vision from al-Kindi to Kepler* (Chicago, 1976).

Michael S. Mahoney

Received his Ph.D. in the history of science in 1967 from Princeton University, where he is presently associate professor. His research has been principally in the history of mathematics; his publications include *The Mathematical Career of Pierre de Fermat* (Princeton, 1973); "Another Look at Greek Geometrical Analysis," *Archive for History of Exact Sciences* 5 (1968): 318–48; and "Die Anfänge der algebraischen Denkweise im 17 Jahrhundert," *Rete* 1 (1971): 15–30.

Robert P. Multhauf

Has a Ph.D. in history from the University of California, Berkeley (1953). He is on the staff of the Smithsonian Institution, where he was director of the Museum of History and Technology from 1966 to 1969; he is also adjunct professor, George Washington University. He has been editor of *Isis* (the journal of the History of Science Society) since 1964 and is former president of the Society for the History of Technology. His publications include *The Origins of Chemistry* (London, 1966) and *A History of Common Salt* (Baltimore, forthcoming).

John E. Murdoch

Earned his Ph.D. in philosophy from the University of Wisconsin in 1957. He is professor of the history of science at Harvard University. His many publications on medieval mathematics and natural philosophy include: "The Medieval Language of Proportions: *Elements* of the Interaction with Greek Foundations and the Development of New Mathematical Techniques," in *Scientific Change*, ed. A. C. Crombie (London, 1963), pp. 237–71, 334–43; "The Medieval Euclid: Salient Aspects of the Translations of the *Elements* by Adelard of Bath and Companus of Novara, *Revue de synthèse*, ser. 3, nos. 49–52 (1968), pp. 67–94; "Mathesis in philosophiam scholasticam introducta: The Rise and Development of the Application of Mathematics in Fourteenth Century Philosophy and Theology," in *Arts libéraux et philosophie au moyen âge* (Actes du quatrième congrès international de philosophie médiévale, Montréal, 27 August – 2 September 1967) (Montreal, 1969), pp. 215–54; and "From Social into Intellectual Factors," in *The Cultural Context of Medieval Learning*, ed. John E. Murdoch and Edith D. Sylla (Dordrecht, 1975), pp. 271–339.

Olaf Pedersen

Was educated at the University of Copenhagen, where he earned degrees in theoretical physics and medieval philosophy. He later studied the history of philosophy and science in Paris and founded the Institute for the History of Exact Sciences at the University of Aarhus. His publications include *Nicole Oresme og hans naturfilosofiske system* (Copenhagen, 1956); *Early Physics and Astronomy* (London, 1974), with M. Pihl; *A Survey of the Almagest* (Odense, 1975); and several textbooks on mechanics. His current research is on the history of medieval astronomy.

Nancy G. Siraisi

Received the B.A. from Oxford University and the Ph.D. from The City University of New York in 1970. She is associate professor of history at Hunter College of the City University of New York. Her publications include *Arts and Sciences at Padua* (Toronto, 1973); "The *Expositio Problematum Aristotelis* of Peter of Abano," *Isis* 61 (1970): 321–39; "The Music of Pulse in the Writings of Italian Academic Physicians," *Speculum* 50 (1975): 689–710; and "The Libri Morales in the Faculty of Arts and Medicine at Bologna: Bartolomeo da Varignana and the Pseudo-Aristotelian *Economics*," *Manuscripta* 20 (1976): 105–18. Her current research focuses on the faculties of arts and medicine in the Italian universities of the thirteenth and fourteenth centuries.

Jerry Stannard

Received his Ph.D. in philosophy and classics from the University of Illinois in 1958. He is professor of the history of science and medicine in the department of history at the University of Kansas. He has published approximately sixty papers on various aspects of ancient and medieval botany, *materia medica*, and pharmacy, including "Greco-Roman Materia Medica in Medieval Germany," *Bulletin of the History of Medicine* 45 (1972): 455–68; "Medieval Herbals and Their Development," *Clio Medica* 9 (1974): 23–33; and "Squill in Ancient and Medieval Materia Medica, with Special Reference to Its Employment for Dropsy," *Bulletin of the New York Academy of Medicine*, ser. 2, 50 (1974): 684–713. He is presently completing a monographic study entitled "History of the Medicinal Uses of Vegetable Gums and Resins," and is planning an English translation of Dioscorides' *De materia medica*.

Brian Stock

Studied the natural sciences at the University of Toronto before proceeding to his A.B. at Harvard and his Ph.D. at the University of Cambridge in 1967. He is presently senior fellow of the Pontifical Institute of Mediaeval Studies and professor of history in the University of Toronto. Among his many publications on medieval literature and intellectual history are *Medieval Latin Lyrics* (Boston, 1971); *Myth and Science in the Twelfth Century. A Study of Bernard Silvester* (Princeton, 1972); "The Philosophical Anthropology of Johannes Scottus Eriugena," *Studi medievali*, ser. 3, 8 (1967): 1–57; "The Middle Ages as Subject and Object: Romantic Attitudes and Academic Medievalism," *New Literary History* 5 (1973–74): 527–47; and "Experience, Praxis, Work, and Planning in Bernard of Clairvaux: Observations on the *Sermones in Cantica*," in *The Cultural Context of Medieval Learning*, ed. John E. Murdoch and Edith D. Sylla (Dordrecht, 1975), pp. 219–68.

Edith D. Sylla

Received her Ph.D. in history of science from Harvard University in 1971. She is associate professor of history at North Carolina State University at Raleigh. She has published "Medieval Quantifications of Qualities: The 'Merton School'," *Archive for History of Exact Sciences* 8 (1971): 9–39; "Medieval Concepts of the Latitude of Forms: The Oxford Calculators," *Archives d'histoire doctrinale et littéraire du moyen âge* 40 (1973): 223–83; and "Autonomous and Handmaiden Science: St. Thomas Aquinas and William of Ockham on the Physics of the Eucharist," in *The Cultural Context of Medieval Learning,* ed. John E. Murdoch and Edith D. Sylla (Dordrecht, 1975), pp. 349–91. Her current research continues to focus on the intellectual tradition at Oxford University in the fourteenth century.

Charles H. Talbot
Received his Ph.D. and B.D. from the English College of Rome in 1928 and 1930. From 1947 to 1953 he was affiliated with the Warburg Institute of the University of London; he was Birkbeck Lecturer at Cambridge University in 1962–63; and from 1954–76 he held a research position at The Wellcome Institute for the History of Medicine. He is presently in retirement. Among his many publications are *Medicine in Medieval England* (London, 1967), and (with E. A. Hammond) *The Medical Practitioners in Medieval England* (London, 1965).

William A. Wallace, O.P.
Holds an M.S. in physics from Catholic University of America and doctorates in philosophy (1959) and theology (1961) from the University of Fribourg, Switzerland. He is professor of philosophy and history at Catholic University of America. His books include *The Scientific Methodology of Theodoric of Freiberg* (Fribourg, 1959); *The Role of Demonstration in Moral Theology* (Washington, 1962); *Causality and Scientific Explanation,* 2 vols. (Ann Arbor, 1972–74); *The Elements of Philosophy* (New York, 1977); and *Galileo's Early Notebooks: The Physical Questions* (Notre Dame, Ind., 1977). He was an associate editor of the *New Catholic Encyclopedia* and is presently Director General of the Leonine Commission charged with editing the *Omnia Opera* of Thomas Aquinas. His current research focuses on the transition from late medieval to early modern science.

James A. Weisheipl, O.P.
Received the Ph.D. in philosophy from the Pontifical University of St. Thomas Aquinas (Rome) in 1953 and the D. Phil. in history from Oxford University in 1957. He is professor of the history of medieval science at the Pontifical Institute of Mediaeval Studies in Toronto. He is past president of The American Catholic Philosophical Association and served as asso-

ciate philosophy editor of the *New Catholic Encyclo-pedia*; he was also founder and first director of the American Section of the Leonine Commission for the critical edition of the works of St. Thomas Aquinas. Among his many publications are *Nature and Gravitation* (River Forest, Ill., 1955); *The Development of Physical Theory in the Middle Ages* (New York, 1959); and *Friar Thomas d'Aquino: His Life, Thought and Works* (Garden City, N.Y., 1974). He is presently editing a volume dealing with the scientific work of Albertus Magnus.

Index